Music and Science from Leonardo to Galileo

Music, Science & Technology

General editor
Massimiliano Sala

Volume 5

Publications of the Centro Studi Opera Omnia Luigi Boccherini
Pubblicazioni del Centro Studi Opera Omnia Luigi Boccherini
Publications du Centro Studi Opera Omnia Luigi Boccherini
Veröffentlichungen des Centro Studi Opera Omnia Luigi Boccherini
Publicaciones del Centro Studi Opera Omnia Luigi Boccherini
Lucca

Music and Science from Leonardo to Galileo

edited by
Rudolf Rasch

BREPOLS
TURNHOUT
MMXXII

D/2022/0095/126

ISBN 978-2-503-60080-2

Printed in Italy

Contents

Library Sigla

A-Wn	Vienna, Österreichische Nationalbibliothek
B-Gu	Gent, Universiteitsbibliotheek
CDN-Tu	Toronto, University of Toronto, Library
CH-Bu	Basel, Universitätsbibliothek
D-Hs	Hamburg, Staats- und Universitätsbibliothek
D-Hvonbusch	Hamburg, Bibliothek Hans von Busch
D-Mbs	Munich, Bayerische Staatsbibliothek
D-Mdm	Munich, Deutsches Museum
D-W	Wolfenbüttel, Herzog-August-Bibliothek
E-Mn	Madrid, Biblioteca Nacional
E-Mp	Madrid, Biblioteca del Palacio
F-LYm	Lyon, Bibliothèque Municipale
F-Pa	Paris, Bibliothèque de l'Arsénal
F-Pi	Paris, Institut de France
F-Pm	Paris, Bibliothèque Mazarine
F-Pn	Paris, Bibliothèque nationale de France
GB-BSw	Berkshire, Windsor Castle, Royal Collection
GB-Lbl	London, British Library
GB-Lrsl	London, Royal Society of London, Library
GB-Lv	London, Victoria & Albert Museum
I-ARc	Arezzo, Biblioteca Città di Arezzo
I-Fn	Florence, Biblioteca Nazionale Centrale
I-Fr	Florence, Biblioteca Riccardiana
I-Ma	Milan, Biblioteca Ambrosiana
I-MOe	Modena, Biblioteca Estense

Library Sigla

I-Nn	Naples, Biblioteca Nazionale Vittorio Emanuele III
I-Pasu	Padua, Archivio Storico dell'Università
I-Pu	Padua, Biblioteca Universitaria
I-Rn	Rome, Biblioteca Nazionale Centrale
I-Ru	Rome, Biblioteca Universitaria Alessandrina
I-Rvat	See V-CVbav
I-Tr	Turin, Biblioteca Reale
I-TVco	Treviso, Biblioteca Comunale
I-Vas	Venice, Archivio di Stato
I-Vlevi	Venice, Biblioteca della Fondazione Ugo e Olga Levi
I-Vmc	Venice, Biblioteca Correr
I-Vmn	Venice, Biblioteca Nazionale Marciana
NL-Ar	Amsterdam, Rijksmuseum
NL-Au	Amsterdam, Universiteitsbibliotheek
NL-DHk	The Hague, Koninklijke Bibliotheek
NL-Gu	Groningen, Universiteitsbibliotheek
NL-Lu	Leiden, Universiteitsbibliotheek
NL-MIzb	Middelburg, Zeeuwse Bibliotheek
P-BRs	Braga, Arquivo da Sé
US-AA	Ann Arbor, MI, University of Michigan, Library
US-BES	Bethesda, MD, National Library of Medicine
US-CAh	Cambridge, MA, Houghton Library
US-CHH	Chapel Hill, NC, University of North Carolina, Music Library
US-MAL	Malibu, CA, J. Paul Getty Museum
US-NH	New Haven, CT, Yale University, Library
V-CVasv	Vatican City, Archivio di Stato Vaticano (formerly I-Rasv)
V-CVbav	Vatican City, Biblioteca Apostolica Vaticana (formerly I-Rvat)

Greek Writings

In the first column, the name of the author is given in English (if different from the Latin form), Latin (in italics), and Greek, with transliteration.

Author	English Title	Latin Title	Greek Title
Alypius Ἀλύπιος [Alypios]	*Introduction to Music*	*Introductio musica*	Εἰσαγωγὴ μουσική [Eisagōgē mousikē]
Aristides Quintilianus Ἀριστείδης Κοϊντιλιανός [Aristeidēs Koïntilianos]	*On Music*	*De musica*	Περὶ μουσικῆς [Peri mousikēs]
Aristotle | *Aristoteles* Ἀριστοτέλης [Aristotelēs]	*Metaphysics*	*Metaphysica*	Τά μετὰ τά φυσικά [Ta meta ta physica]
Aristotle	*On the Soul*	*De anima*	Περὶ ψυχῆς [Peri psykhēs]
Aristotle	*Organon*	*Organum*	Ὄργανον [Organon]
Aristotle	*Physics*	*Physica*	Φυσικὴ ἀκρόασις [Physikē akroasis]
Aristotle	*Posterior Analytics*	*Analytica posteriora*	Ἀναλυτικὰ ὕστερα [Analytica hystera]
Aristotle (Pseudo-)	*On Things Heard*	*De audibilibus*	Περὶ ἀκουστῶν [Peri akoustōn]
Aristotle (Pseudo-)	*Problems*	*Problemata*	Προβλήματα [Problēmata]
Aristoxenus Ἀριστόξενος [Aristoxenos]	*Harmonic Elements*	*Elementa harmonica*	Ἁρμονικά στοιχεία [Harmonika stoikheia]
Bacchius Βακχεῖος [Bakkheios]	*Introduction to the Art of Music*	*Isagoge musicae artis*	Εἰσαγωγὴ τέχνης μουσικῆς [Eisagōgē tekhnēs mousikēs]
(*Manuel*) *Bryennius* Μανουήλ Βρυέννιος [Manuel Bryennios]	*Harmonics*	*Harmonica*	Ἁρμονικά [Harmonika]
Cleonides Κλεονείδης [Kleoneides]	*Harmonic Introduction*	*Introductio harmonica*	Εἰσαγωγὴ ἁρμονική [Eisagōgē harmonikē]
Euclid | *Euclides* Εὐκλείδης [Eukleidēs]	*Elements*	*Elementa mathematica*	Στοιχεῖον [Stoikheion]
Euclid	*Division of the Canon*	*Sectio canonis*	Κατατομὴ κανόνος [Katatomē kanonos]

Greeks Writings

Author	English Title	Latin Title	Greek Title
Gaudentius Γαυδέντιος [Gaudentios]	*Harmonic Introduction*	*Introductio harmonica*	Ἁρμονικὴ εἰσαγωγή [Harmonikē eisagōgē]
Gregory of Nyssa *Gregorius Nyssenus* Γρηγόριος Νύσσης [Grēgorios Nyssēs]	*Dialogue on the Soul and Resurrection*	*Dialogus de anima et resurrectione*	Διάλογος περὶ ψυχῆς καὶ ἀναστάσεως [Dialogos peri psykhēs kai anastaseōs]
Longinus (Pseudo-) Λογγῖνος [Longinos]	*On the Sublime*	*De sublimitate*	Περὶ ὕψους [Peri hypsous]
Nicomachus Νικόμαχος [Nikomakhos]	*Harmonic Manual*	*Manuale harmonicum*	Ἐγχειρίδιον ἁρμονικόν [Enkheiridion harmonikon]
Plato Πλάτων [Platōn]	*Republic*	*De republica*	Πολιτεία [Politeia]
Porphyry | *Porphyrius* Πορφύριος [Porphyrios]	*Commentary on Ptolemy's «Harmonics»*	*Commentarius in Ptolemaei harmonica*	Εἰς τὰ ἁρμονικὰ Πτολεμαίου ὑπόμνημα [Eis ta harmonika Ptolemaiou hypomnēma]
Plutarch (Pseudo-) *Plutarchus* Πλούταρχος [Ploutarkhos]	*On Music*	*De musica*	Περὶ μουσικῆς [Peri musikēs]
(Claudius) Ptolemy (*Claudius*) *Ptolemaeus* Κλαύδιος Πτολεμαῖος [Klaudios Ptolemaios]	*Almagest*	*Syntaxis mathematica*	Μαθηματικὴ σύνταξις [Mathēmatikē syntaxis]
Ptolemy	*Harmonics*	*Harmonica*	Ἁρμονικά [Harmonika]
Themistius Θεμίστιος [Themistios]	*Paraphrase of Aristotle's «On the Soul»*	*In libros Aristotelis de anima paraphrasis*	Παράφρασις τῶν περὶ ψυχῆς Ἀριστοτέλους [Paraphrasis tōn peri psykhēs Aristotelous]

Abbreviations of Reference Works

Grove2	*The New Grove Dictionary of Music and Musicians, Second Edition*, edited by Stanley Sadie, London, Macmillan, 2001.
RISM	Repertoire International des Sources Musicales.
RISM A/I	*Einzeldrucke vor 1800. Band 1-9,* edited by Karlheinz Schlager, Kassel, Bärenreiter, 1971-1981.
RISM A/II	*Music Manuscripts after 1600,* Online database, <http://opac.RISM.info or www.RISM.info>.
RISM B/I	*Recueils imprimés, XVIe–XVIIe siècle,* edited by François Lesure, Munich, Henle, 1960.
RISM B/II	*Recueils imprimés, XVIIIe siècle,* edited by François Lesure, Munich, Henle, 1964.
RISM B/III/6	*The Theory of Music. Manuscripts from the Carolingian Era up to c. 1500. Addenda, Corrigenda,* edited by Christian Meyer, Munich, Henle, 2003.
RISM B/VII	*Handschriftlich überlieferte Lauten- und Gitarrentabulaturen des 15. bis 18. Jahrhunderts,* edited by Wolfgang Boetticher, Munich, Henle, 1978.
RISM B/XI	*Ancient Greek Music Theory: A Catalogue Raisonné of Manuscripts,* edited by Thomas J. Mathiesen, Munich, Henle, 1988.

Introduction

TODAY THE SCIENCES AND THE ARTS are usually regarded as belonging to different domains of human endeavour. In the sciences, in simplistic words, one proceeds from clearly defined concepts and fixed principles through undeniable reasonings to conclusions that are — if possible — generally valid. In the arts, equally simplistically said, concepts are often not as stringently defined as they ideally are in the sciences, and the outcome of artistic activities may include or even preferably includes fantasy and imagination, while the judgments pertaining to art are a matter of taste. Science is objective, arts are subjective. In terms of the Structure-of-Intellect model as developed by the American psychologist J. P. Guilford (1967), science is «convergent production», art is «divergent production»[1].

In the past, however, what we now call arts and sciences were not so strictly separated from one another. Before the Scientific Revolution of the sixteenth and seventeenth centuries, the two activities were closely related: art was considered to have its base in science, while science include quite a lot of speculation, if not fantasy. The words 'art' and 'science' and their cognates in the various applicable languages defined in fact a different dichotomy, one that can be succinctly described by the words 'theory' and 'practice', for science and arts respectively. Both science and art in the modern sense have their theory and their practice, which leads to four categories: the theory (science) and the practice (arts) of science and the theory (science) and the practice (art) of art. These categories are not as confusing as they might seem at first sight. The 'science of science' can be equated with natural and other forms of philosophy, and the 'art of science' with engineering. If we limit art to music, the 'science of music' is music theory, while the 'art of music' is composition and performance.

The period under consideration in the present book is described in the title as 'from Leonardo to Galileo', that is, from the time of Leonardo da Vinci (1452-1519; ILL. 0.1AB) to that of Galileo Galilei (1564-1642; ILL. 6.1), an era that can be objectively described as the 'long sixteenth century', and within the history of music as the 'Renaissance and early Baroque', and with regard to the history of science as the 'Scientific Revolution'.

---

1. GUILFORD 1967.

ILL. 0.1AB: Leonardo da Vinci, *Self Portrait* (sanguine on paper, 33.5 × 21.5 cm), and Francesco Melzi, *Leonardo da Vinci* (sanguine on paper, 27.5 × 29 cm). I-Tr and GB-BSw, MS 912726, respectively.

The two figures whose names demarcate this era share the distinction of usually being referred to by their first names only: Leonardo and Galileo. It is a rare honour, shared by the composer Josquin (Josquin Desprez, ca. 1460-1521) and the artists Michelangelo (Michelangelo Buonarroti; 1475-1564), Raphael (Raffaello Sanzio, 1483-1520), Titian (Tiziano Vecellio, c. 1490-1576) and Rembrandt (Rembrandt Harmenszoon van Rijn, 1606-1669), all living roughly in the time span covered by this book. We will refrain from drawing any conclusions from the fact that five of these seven laureates were artists and that five of the seven laureates were Italians[2].

[2]. 'Josquin' is a diminutive form of 'Josse'. It is interesting to note that there was a certain tradition of calling composers just by there first names in diminutive forms. For the sixteenth century 'Pierchon' (Pierre de La Rue) is another case in point, for the decades around 1700 one can think of Gasparini (Gasparo Visconti) and Nicolini (Nicolò Cosimi).

Introduction

Leonardo da Vinci

What place did music hold within Leonardo da Vinci's wide range of interests? First of all, he is considered to have been musician himself, and perhaps even a fine performer on the lira da braccio. An early anonymous biographer described Leonardo's musical capabilities as follows:

> He was an outstanding performer on the lira [da braccio], and he was the teacher of Atalante Migliorotti, whom he instructed on the instrument. From Lorenzo il Magnifico [de' Medici], he was sent to the Duke of Milan [Lodovico Sforza], together with Atalante Migliorotti, to present to him a lira, for he was unique in playing this instrument[3].

Giorgio Vasari — probably inspired by the anonymous biography just cited — further propagated Leonardo's fame as a musician in his *Le vite de' più eccellenti pittori, scultori, e architettori* (The Lives of the Most Excellent Painters, Sculptors, and Architects; Florence, 1550, 1568). Later biographies invariably mention Leonardo's musicianship, following these early leads.

Leonardo is not known to have been a composer, nor did he leave any structured text on music. Among his paintings, however, there is a portrait of a musician, a young man holding a sheet of paper on which some music notation is visible (Ill. 0.2ab). The identity of this person is unknown. The sitter has been variably identified as Franchinus Gaffurius (1451-1522), with whom Leonardo was probably acquainted during his Milan period[4], the above-mentioned Atalante Migliorotti (1466-1532), Josquin Desprez (c. 1455-1521)[5], or non-musicians such as Galeazzo di Sanseverino (1458-1525) or an anynomous Milanese nobleman[6].

In his book *Leonardo da Vinci as a Musician* (1982), Emanuel Winternitz has summarized what can be found about music in Leonardo's notebooks[7]. Notes and sketches with musical relevance are scattered through many of the notebooks (of which there are several dozen, not even counting the lost ones)[8], without any apparent system or order. The notes are undated and do not form any kind of consistent body of knowledge. Winternitz divides the musical contents

---

3. F-Pn, MS Magliabecchiano Classe xvii, 17 (Codice del Anonimo Gaddiano; Winternitz 1982, pp. xxii-xxiii): «Fu [...] raro sonatore di lira et fu maestro di quella d'Atalante Migliorotti. Dal detto Magnifico Lorenzo fu mandato al duca di Milano a presentarli insieme con Atalante Migliorotti una lira, che unico era in sonare tale extrumento». Translation from *Grove2*, article 'Leonardo da Vinci' by Emanuel Winternitz and Laurence Libin.

4. Beltrami 1905-1906.

5. Clercx-Lejeune 1972, Testolin 2007.

6. Fugelso 2010.

7. Winternitz 1982.

8. Today, the notebooks are spread over a fairly large number of libraries, including E-Mn, F-Pi, GB-BSw, GB-Lbl, GB-Lv, and I-Ma.

Ill. 0.2ab: Leonardo da Vinci, *Portrait of a Musician* (c. 1485?; oil on wood, 44.5 × 32 cm) and detail. Milan, Pinacoteca Ambrosiana.

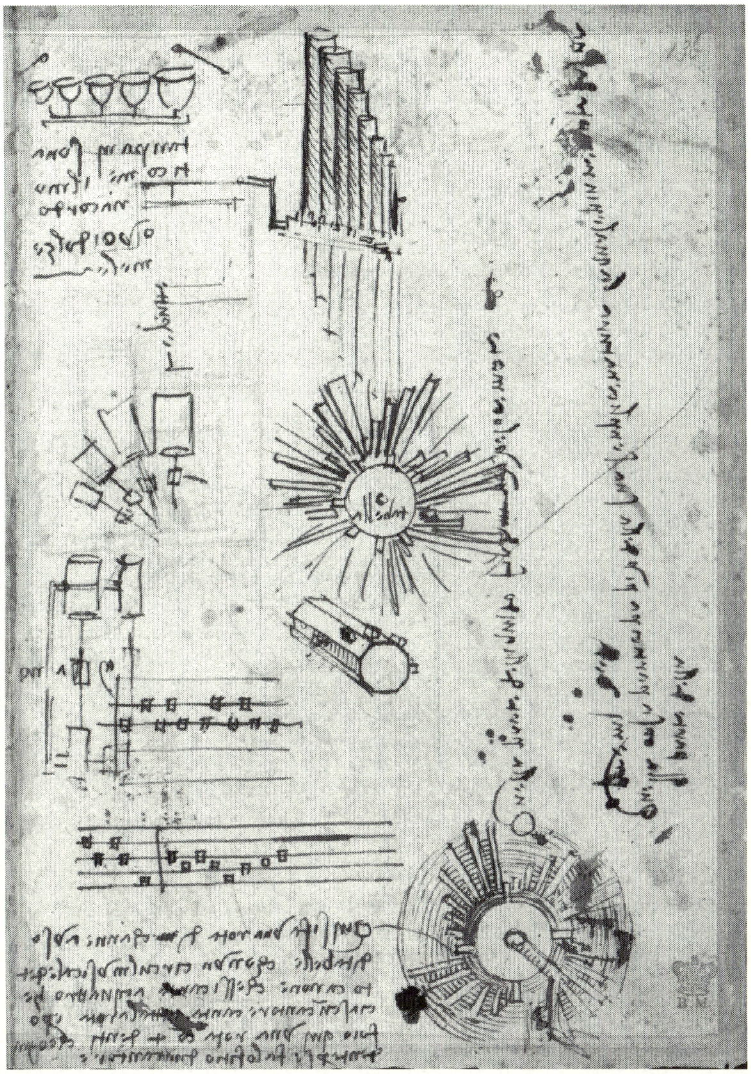

Ill. 0.3: Page from a notebook of Leonardo da Vinci, with sketches for a mechanical organ. Note Leonardo's handwriting, which runs from right to left. GB-Lbl, Arundel MS 263, f. 136r.

of the notebooks into several categories, including 'acoustics', 'viola organista', idiophones (struck instruments), and aerophones (blown instruments). The notes on acoustics deal with the origin of sound, its properties such as pitch and loudness, and its propagation, including observations concerning echoes and the speed of sound[9]. Reading through them, one is struck by their Aristotelian character: they depend on observation, introspection and intuition, and are completely qualitative. No attempt is made to arrive at any quantitative relation. Often sound is compared with the other categories of the '[sensible] species' (emanations from objects

[9]. Winternitz 1982, pp. 97-136.

that make them perceivable by the human senses), such as light, odour and even magnetism. Actually, the relation to music in these notes is nearly completely absent. A number of sketches in various notebooks are related to the 'viola organista', which must have been some kind of *Geigenwerk*, a cembalo-like instrument of which the strings were bowed by way of a sophisticated mechanism, in order to produce continuous tones[10]. The instrument was probably never built. Another group of sketches show contraptions connected to drums to make it possible to play them continuously[11]. A page of particular interest contains sketches of what must be some kind of mechanical organ (ILL. 0.3). The instrument apparently has four times seven (= 28) organ pipes, placed around a central circular frame. Each of the four groups comprises a seven-note (presumably diatonic) scale. In the middle of the frame there is a drum with octagonal cross-section fitted with pins that can be placed such that they activate the organ pipes when rotated. Cleverly programmed, the organ is capable of playing four-part circular canons, as described by Leonardo in the accompanying note:

> Here is a wheel with pipes for the use of a drum with a musical circle called a canon that is sung in four parts, such that every singer sings the entire wheel, And yet I make here a wheel with four pins such that every pin does the task of one singer[12].

The page is also interesting because of the music notation written on it. It is, however, 'fantasy music', not real music. There are merely series of ten, five and seven breves, without any recognizable pattern and without clefs.

Musical notation is also part of a fair number of the rebuses found in several sheets with notes and sketches now in the Royal Collection in Windsor[13]. The notation always has the form of a short piece of staff with a C-clef on the third line followed by one to six breves. When the breves are read as solmization syllables in the hard hexachord, they become part of the sentence that is represented by the rebus. As an example, one of the rebuses on the sheet Windsor 12692 can serve. It consists of a picture of a box (*cassa*), two pears (*pere*) and the three notes *la*, *fa* and *re* (*la fare*). Together this yields '*C'a sapere la fare*' (One must know how to do it). The syllables are used for complete words (possible for *mi*, *fa*, *sol*, and *la*) or for parts of words. An example with more notes is provided on Windsor 12697. First there is the series *re la sol mi fa re mi*, as

---

10. *Ibidem*, pp. 137-157. See also TIELLA 1989.

11. WINTERNITZ 1982, pp. 168-186.

12. GB-Lbl, Arundel MS 263, f. 136r: «Qui si fa una rota di canne a uso di tabella con un circul musicale detto canone, che si canta a quattro, e ciascun cantore canta tutta la rota. E però fo io qui una rota con 4 denti, che ogni dente per sé fa l'ofizio d'uno cantor». Translation from WINTERNITZ 1982, pp. 188-189.

13. About these rebuses, see MARINONI 1954 and WINTERNITZ 1982, pp. 90-93.

part of *Amore **la sol mi fa re**mirare* (Love makes me look up at her only), then *sol la mi fa sol*, as part of ***sol la mi fa sol**lecita* (Only she urges me; ILL. 0.4). The solmization syllable *fa* is use most frequently, often as part of a form of the verb *fare*, including *fa* itself. The syllable *ut* is never used. On the various sheets — Windsor 912691, 912693, 912695, 912697, and 912699 — there are eighteen such rebuses altogether (TABLE 0.1).

TABLE 1

Overview of the rebuses that include solmization syllables as found in the Windsor manuscripts (GB-BSw) with Leonardo's sketches. The first column refers to the numbering given to them in MARINONI 1954, with the pages where they are reproduced, transcribed, and interpreted. The second column gives the textual interpretation of the rebus, following *ibidem*, with the solmization syllables in bold face. The third column repeats the solmization syllables. When they are part of a word or words, the whole word or words are given between parentheses.

MARINONI 1954 NO. (PP.)	TEXT	SOLMIZATION SYLLABLES
	GB-BSw, MS 912692 recto	
29-33 (pp. 166-168)	Da' di preteriti insino a questo tempo non o **fa**tto mai alcuna opera ma io so che le presenti **mi fa**ccino triunfare (From the past days until now I have never made any work, but I know that the present ones will make me triumphant)	**fa** (fatto), **mi**, **fa** (faccino)
34 (p. 168)	Or ci cala **la** fortuna le vele (Fortune will now reef the sails)	**La**
55 (p. 178)	C'a sapere **la fare** (It is to know how to do it)	**la**, **fa re** (fare)
	GB-BSw, MS 912692 verso	
82 (p. 192)	Or, **fa mi** ... [damaged] (Well, make me ...)	**fa** (fa), **mi** (amico)
84 (p. 193)	**Sol la** fe' **mi fa** [incomplete?] (Only faith makes me [...])	**sol** (sol'), **la**, **mi**, **fa** (fa)
86 (p. 194)	Che posso **fare** se la femina mi trae'l core (What can I do if the woman draws my heart from me)	**fa re** (fare)
88 (p. 195)	L'amo**re mi fa sollazzare** (Love give me joy)	**re** (amore), **mi**, **fa**, **sol la re** (sollazzare)

Marinoni 1954 no. (pp.)	Text	Solmization syllables
94 (p. 199)	Pero, se la fortuna **mi fa** felice, tal viso asponerò [= esporrò] (Yet, if fate makes me happy, I'll show that face)	**mi fa**
95 (p. 200)	Cresca la **fama** (Let fame grow)	**fa** (fama)
117 (p. 211)	Or chi campa nella fiamme **del' a**more? (Well, who lives in the flame of love?)	**l'a** (l'amore)
127 (p. 215)	Or è chi vi vuole sol **fa** ... (Here is he who only wants you to do ...)	**fa** [**re**] (fare)
148 (p. 225)	**Fammi** libero (Make me free)	**fa mi** (fammi)
152 (p. 227)	Ch'io dicesse l'amo**re mi fa solla**zzo (That I would say, love makes me happy)	**re** (amore), **mi, fa, sol la** (sollazzo)
	GB-BSw, MS 912693	
(p. 228)	Chi attende a **fare** (Who is doing [this]?)	**fa re** (fare)
	GB-BSw, MS 912695 recto	
(pp. 229-230)	Dove è'l mi amare, **là** per nov'amore (Where is my loving, there, by a new love)	**La**
(pp. 229-230)	Dove **l'a**mare (Where is the loving)	**l'a** (l'amare)
	GB-BSw, MS 912697	
(p. 234)	Amo**re là sol mi fa remi**rare, **sol la mi fa sol**lecita (Love makes me look up at here only, only she urges me)	**re** (amore), **la, sol, mi, fa, re mi** (remirare), **sol** (sollecita)
	GB-BSw, MS 912699	
(pp. 235-236)	Amo**re la sol mi fa remi**rare (Love makes me look at here alone)	**re** (amore), **la, sol, mi fa, re mi** (remirare)

The notes written down in the rebuses have sometimes been read as musical notation and their sequence therefore as a primitive musical composition, but this is an incorrect interpretation. The use of the notes is the other way around: the musical notation is used to write down text, the resulting text is not to be read as music[14].

[14]. Leonardo's use of notes to represent syllables that are part of text was not unique. See Schiltz 2015, pp. 342-346.

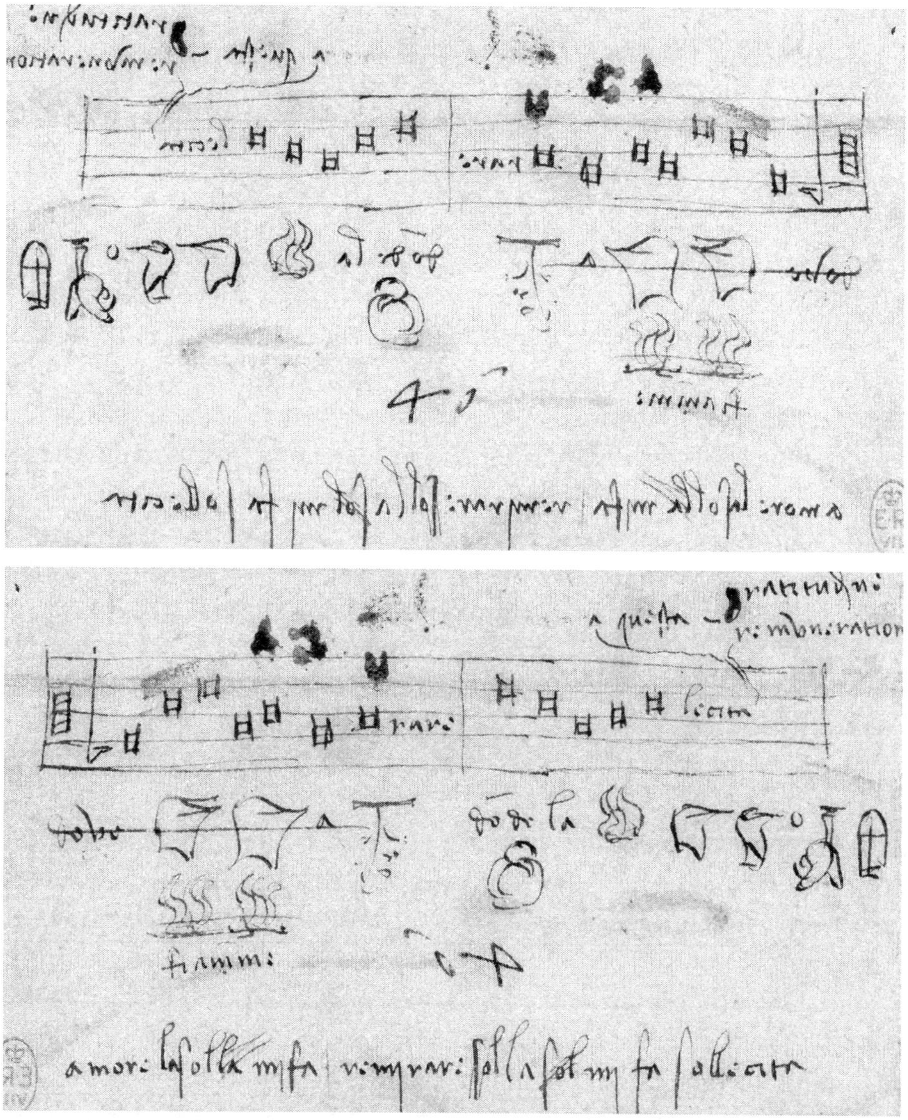

ILL. 0.4: Musical rebus by Leonardo da Vinci. GB-BSw, MS 912697, original orientation and mirror image.

There are also some rebuses that have a musical instrument in the series of pictures, such as bells (*campane*, in '*campane pochi*'), shawms (*pifferi*, in '*colpi feri*'), the viola (in '*violata*'), and the monochord (*menacordo*, in '*i' me n'accordo*')[15].

Music plays a distinctive role in a famous text entitled *Paragone di pittura, poesia, musica e scultura* (Comparison of Painting, Poetry, Music, and Sculpture). This is the first part of the so-called 'Trattato della pittura', a comprehensive treatment of the science and practice of

[15]. MARINONI 1954, nos. 107, 7, 106, and 123, respectively, pp. 156, 206, and 214.

painting composed from Leonardo's notes by his pupil Francesco Melzi, probably completed around 1540 and first published in incomplete form in Paris in 1651, in Italian (*Trattato della pittura*) and French (*Traitté de la peinture*, translated by Roland Fréart), and then, complete from Melzi's manuscript in Rome in 1817[16]. The basic aim of the *Paragone* is to demonstrate the superiority of painting over the other arts[17]. The main arguments for this position are the veridical representation of nature that painting provides, the simultaneity of the elements of such a representation, the permanence of visual works of art (as opposed, especially, to the ephemerality of music), and the harmonic proportions that could be incorporated within such works. Music is superior to poetry because it is built on simultaneous harmonic proportions (*un'armonico concento*), while poetry is not. But compared to painting, music fails on all accounts: musical sounds immediately decay after their appearance (while painting is permanent), music is boring because it consists of repetition, that is, it must be repeated in order to remain, and music deals with sound only, not with the world, as painting does. Of course, this is also part of a dialectical discourse, an exercise in providing arguments pro and contra certain statements. Nevertheless, there also resonates in it a remark jotted down in one of the notebooks:

> Music has two diseases, of which one is that of mortality, and the other that of decrepitude. The former is always connected with the moments following that of its creation; the latter one makes it odious and vile in its repetition[18].

After having worked through the many pages of sketches, drafts, notes and text, as compiled by Winternitz, for example, one begins to wonder if Leonardo was really interested in music. Direct references to music are practically absent, and musical terms are used only exceptionally; in the 'Paragone' music is merely part of an argument meant to prove the superiority of painting to all the other fine arts. The many sketches concerning musical instruments nearly always show contraptions that could be added to existing instruments, often in order to play them mechanically. In this light, the phrase «an intensive occupation with music», which has been used to describe Leonardo's musical interests, seems to be — regrettably, I must confess — an overstatement[19].

16. There are numerous modern editions. See, for example, LEONARDO 1992 and LEONARDO 1995.

17. WINTERNITZ 1982, pp. 204-223. See also FARAGO 1991.

18. I-Ma, Codex Atlanticus, f. 332 va (WINTERNITZ 1982, p. 220): «La musica ha due malattie, della quale l'una è mortale, l'altra è decrepitudinale. La mortale è sempre congiunta alle instante sequente a quel della sua creazione; le decrepitudinale la fa odiosa e vile nella sua replicazione».

19. *GROVE2*, article 'Leonardo da Vinci' by Emanuel Winternitz and Laurence Libin.

Galileo Galilei

What about Galileo's contribution to the field of science and music? Galileo was the son of a musician, Vincenzo Galileo (c. 1530-1591), lutenist, composer and music theorist. His older brother Michelagnolo (1575-1631) was a lutenist/composer, as were the latter's two sons Alberto Cesare (born 1617) and Cosimo (born 1621). Galileo played the lute himself.

In Galileo's works, however, discussions of musical matters appear to be practically limited to the last few pages of the 'Dialogo Primo' in the *Discorsi e dimostrazioni matematiche intorno a due nuove scienze* (1638), his last substantial publication[20]. Music, and more particularly the sympathetic vibrations of a string in response to another string in vibration, were called upon to help explain the resonance frequencies of pendulums. Salviati (Galileo's spokesman in the dialogue) then continues by explaining that to any given musical interval there corresponds, not a single proportion, but several. For example, if pitch is treated as a function of string length, then the octave has the proportion 2:1; but if the relevant variable is tension, then the proportion is instead 4:1; if linear density (*grossezza*), 1:4. There follows a brief exposition of the 'coincidence theory', the theory that explains consonance in terms of the coincidence of the 'vibrations and percussions' (*vibrazioni e percosse*; sound pulses) produced by the musical pitches comprising the consonant interval. Galileo's exposition of this includes descriptions of three experiments to support his explanations, none of which can have been in fact conducted because the outcome would have been different from what the text reports: (1) three pendulums with lengths in the proportion of 2:3:4 to illustrate the coincidence of the periods of intervals with string-length proportions of 2:1 and 3:2 (in reality the periods would have been in the proportion of ½:⅓:¼ or 6:4:3); (2) a goblet filled with water which, when rubbed harder on the edge, would form waves that would double in frequency (impossible to realize); and (3) a chisel scraping a copper plate with speeds in proportion of 3:2 and producing scratches distant from each other in the same proportion[21]. (Were this last experiment conducted accurately — which seems nearly impossible — the scratches would turn out instead to be equidistant.)

Several commentators have related Galileo's statements of 1638, in particular the quadratic relation between string tension and pitch, with the experiments (or perhaps rather: observations) of his father Vincenzo in the 1580s[22] as described in the *Discorso intorno all'opere di Messer Gioseffo Zarlino da Chioggia* (Florence, 1689, pp. 103-104) and in the *Discorso particolare intorno alla diversità delle forme del diapason* (MS, c. 1590)[23]. For sound-producing volumes such as organ pipes, Vincenzo postulated a cubic relation with pitch, which is incorrect. I see

---

20. Galileo Galilei, *Discorsi e dimostrazioni matematiche intorno a due nuove scienza* (Leiden, 1638), pp. 98-107.
21. Walker 1978, pp. 28-31, Cohen 1984, pp. 88-92.
22. Beginning with Palisca 1961, pp. 132-135.
23. I-Fn, MS Gal. 3, ff. 44r-54v, edited with translation in Palisca 1989, pp. 180-197.

little reason to relate Vincenzo's statements from the 1580s with those by Galileo half a century later. First, there is no reference to Vincenzo's work in any of Galileo's writings, and second, not everything said in in the *Discorsi* of 1638 can be found in Vincenzo's writings, whereas Galileo's statements can be easily traced back to Mersenne's *Harmonicorum libri XII* (Paris, 1636), the Latin version of the *Harmonie universelle*. In the *Harmonicorum libri XII*, the laws governing the pitch of strings — now commonly called 'Mersenne's laws' — are described in the 'Liber secundus de causis sonorum' (Propositions 6-8, pp. 11-12); the coincidence theory, in the 'Liber quartus de sonis consonis, seu consonantiis' (Proposition 25, p. 64)[24]. Sympathetic vibrations were a widely known phenomenon in the early seventeenth century. The little diagrams that Galileo included in the *Discorso* of 1638 to illustrate his theory (pp. 104-105) bear a striking resemblance to diagrams in the *Harmonicorum libri XII* (pp. 58 and 63)[25]. Mersenne had sent a copy of the *Harmonicorum libri XII* to Galileo by the end of 1637, as we learn from a letter of 23 December 1637 to Galileo from his brother Roberto[26], so that we may assume his familiarity with Mersenne's work.

Mersenne also wrote to Galileo directly[27], but the latter never answered, apparently also because he had trouble with Mersenne's handwriting[28]. (Everyone who has ever encountered Mersenne's handwriting with the task of reading or transcribing it will side with Galileo.)

Our conclusion must be, therefore, that Galileo did not develop any new ideas in the field of 'musical science'. Nevertheless, due to its publication in the *Discorsi* of 1638, the coincidence theory was often credited to him in later seventeenth-century literature on the subject.

Therefore, we can only conclude that the contributions of Leonardo and Galileo to the art and science of music were modest[29]. But if we accept a little expansion towards the end of the seventeenth century, their names serve well to demarcate the period considered in this book. Leonardo, at the beginning of the period in question, adhered still completely to Aristotelian views, whereas Galileo missed no occasion to reject them.

---

24. The coincidence theory of musical consonance was first proposed by Giovanni Battista Benedetti in his *Diversarum speculationum mathematicarum et physicarum liber* (Turin, 1585), pp. 277-283, but Mersenne probably got to know the theory by his correspondence with Isaac Beeckman and René Descartes. See COHEN 1984 and PALISCA 1985, pp. 258-261.

25. For these diagrams, Mersenne followed in his turn examples given by René Descartes, for example in the latter's letter to Mersenne of 18 December 1629 (DESCARTES 1897, p. 100; MERSENNE 1945, p. 349).

26. Roberto Galilei to his brother Galileo, 23 December 1637 (GALILEI 1906, p. 241; MERSENNE 1960, p. 343). The book went via Lyon to the French consul in Leghorn and then to Rome.

27. Marin Mersenne to Galileo Galilei, 27 November 1637 (GALILEI 1906, p. 226; MERSENNE 1960, pp. 339-340).

28. Galileo's letter to Elia Diodati, August 1638 (GALILEI 1906, p. 370; MERSENNE 1964, pp. 32-33).

29. The contents of the various chapters of this book would rather suggest a title somewhat like 'Music and Science from Gaffurius to Christiaan Huygens'.

Science in the Sixteenth and Seventeenth Centuries

The difference in the state of science around 1500, when compared with that of the early seventeenth century, is enormous. Around 1500 science was still completely dominated by Aristotelian thought, usually wrapped in a scholastic coat. The earth was the centre of the universe, with the sun, the moon and the planets moving around it in circular orbits. These heavenly bodies were subject to laws different from those that held on earth. On earth, different laws governed living and non-living bodies. Speed was supposed to be proportional to force, and inanimate bodies would come into motion only by an external force and would come to rest by themselves. No phenomenon was considered relevant that could not be observed by the naked eye. And natural philosophy — the historical term for what we now call physics — hardly made use of mathematics to describe relations between physical phenomena. The Bible as well as ancient authors served as infallible authorities.

In the sixteenth and seventeenth centuries many of these statements were seriously and successfully challenged by scientists all over Europe. Nicolaus Copernicus showed that the orbits of the planets could be described more efficiently when the Sun was supposed to be the centre of the universe[30], a theory that was further refined by Johannes Kepler, who concluded that the orbits were ellipses rather than circles[31]. Galileo proposed the principle of inertia, according to which a body in motion continues so unless acted on by an external force; he also developed formulas for the speed of falling bodies. A quantitative description of what happened in mathematical terms became more important than a qualitative description that provided words to name phenomena but no explanations. René Descartes would propose that animate and inanimate bodies would obey the same laws. Christiaan Huygens invented the pendulum clock, discovered the rings of Saturn, and contributed to the development of optics and acoustics. The transition from Aristotelian to classical physics found its culmination in the three laws of nature formulated by Isaac Newton (1642-1727) in his *Philosophiae naturalis principia mathematica* (London, 1687).

The new scientific insights were certainly not accepted without resistance. Jesuit teaching, so important in the Catholic countries of Europe, would continue to propagate Aristotelian science until well into the seventeenth century. Aristotelian concepts abound in sixteenth-century writings of every kind and are still found in a good many seventeenth-century texts, including those on music.

---

30. Nicolaus Copernicus, *De revolutionibus orbium caelestium* (Nuremberg, 1543).

31. Johannes Kepler, *Astronomia nova* (Heidelberg, 1609), *Epitome astronomiae Copernicanae* (Linz, 1619).

Rudolf Rasch

Music in the Sixteenth and Seventeenth Centuries

Like science, music also underwent significant changes from the late fifteenth to the early seventeenth century. In the decades around 1500, polyphony reigned supreme, as far as serious music was concerned. Four-part writing was most common, with the voices designated Cantus (or Superius), Altus (or Contratenor), Tenor and Bassus, but writing for fewer or more voices, roughly from three to six, was far from rare. Texture varied from homophonic, with all the voices in the same or almost the same rhythm, to polyphonic, featuring largely independent voices, each with its own contour and rhythm, held together by the rules of counterpoint. Countless liturgical and para-liturgical compositions — Masses and motets — set Latin texts drawn from the rites of the Catholic Church. Major composers in these genres include Pierre de La Rue (c. 1452-1518), Josquin Desprez (c. 1455-1521), and Jacob Obrecht (1457/1458-1505), all natives of the Southern Netherlands or the North of France. They and many others also composed numerous secular songs in the vernaculars of Italian, Spanish, French, English, Dutch, and German, less ambitious in scope and often simpler in texture than the sacred genres. Instrumental music was still largely an unwritten practice, dependent on memory and improvisation. Music theory continued its later-medieval bifurcation into speculative and practical disciplines, the former (*musica theorica* or *speculative*) concerned chiefly with mathematical modelling of pitch relations, the latter (*musica practica* or *activa*) with plainchant, notation, and counterpoint. Important theorists include Johannes Tinctoris (c. 1435-1511; *Liber de arte contrapuncti*, 1477) and Franchinus Gaffurius (1451-1522; *Theorica musice*, 1592, *Practica musice*, 1496). The year 1500 marks a new era in the dissemination of music with the publication in Venice of the first collection of printed music. Manuscript transmission remained important for several centuries after the invention of music printing, but through the printing press music soon began to find new and much larger audiences.

The 'musical landscape' of the early seventeenth century differed considerably from that of a century earlier. Although still important, the setting of sacred Latin texts is less predominant than it had been. Enormous secular repertoires in Italian and French (madrigals and chansons, respectively) had come into existence. The number of compositions on German, English and Spanish texts had also grown (that of Dutch text less so). Most of these repertoires employed more homophony than was customary at earlier times. The single most pregnant development was the advent, in the years just before the turn of the sixteenth to the seventeenth century, of a new technique called monody, consisting of a solo voice accompanied by a basso continuo, a bass line reinforced by a chordal instrument, such as organ, harpsichord or lute. The new monodic style enabled a much more flexible presentation of the text in the music so that it was particularly suitable for use in dramatic composition, what was later to be called opera. Opera — and later its sacred counterpart (though not scenic) oratorio — began in Italy (Florence, Rome, Manua, Venice) but spread over all of Europe during the seventeenth century.

During the sixteenth century, instrumental music had acquired considerable prestige, and from about 1600 composers such as William Byrd, Jan Pieterszoon Sweelinck, Giovanni Gabrieli and Girolamo Frescobaldi also became renowned for their instrumental music. In the meantime, Heinrich Glarean in his *Dodecachordon* (Basel, 1547) had revised the theory of the church modes to adapt it better to polyphony, while Gioseffo Zarlino in his *Institutioni musiche* (Venice, 1558) greatly expanded and deepened the theory of polyphonic composition while establishing the system we now call just intonation as the new theoretical norm for tuning. Zarlino's tremendously influential precepts retained much of their pedagogical authority until well into the seventeenth century (and on many points even later), even as the new monodic style and other developments gave rise to fresh theoretical ideas and pedagogical approaches, such as figured bass and, in place of the old modes, new systems of what we now call major and minor keys. The early seventeenth century also saw new syntheses of the theoretical aspect of music (the 'science of music'), deeply indebted to Zarlino but incorporating new emphases and ideas, such Michael Praetorius's *Syntagma musicum* (Wolfenbüttel, 1618-1619) and especially Marin Mersenne's vast *Harmonie universelle* (Paris, 1636).

The sixteenth century also saw important developments in fields other than just science and music. Perspective was now analysed in geometrical terms (see the cover illustration), and anatomy received new impetus from Andreas Vesalius's *De humani corporis fabrica* (Basel, 1543). We should also mention events of broader historical significance, such as the first European encounters with the Americas in the West and with South and East Asia and Australia in the East. The Reformation, begun by Martin Luther in Wittenberg in 1517, divided the religious map of Europe in two halves: the Southern countries largely continued their adherence to the Catholic Church, while in the North new religious offshoots became dominant, such as Lutheranism (mainly in Germany and Scandinavia), Calvinism (Switzerland, the Dutch Republic, Scotland) and Anglicanism (England). These new religious strands generated a new theology, new ethics and also new kinds of artworks and music.

Music and Science in the Sixteenth and Seventeenth Centuries

The relation between music and science in the sixteenth and seventeenth centuries can be understood in two ways. The first is by turning to the 'world of music' in order to see which musical phenomena were understood or explained by the use of scientific concepts. The second way is to observe the 'world of science' and to look for instances where science was inspired by music.

It is almost a tautology to speak of a scientific aspect of music before the seventeenth century. Music was one of the four sciences comprised in the quadrivium, the 'scientific' section of the seven liberal arts, which dominated the organization of knowledge since late Antiquity,

Ill. 0.5: The relation between pitch and the size of hammers, bells and glasses, the tension of strings and the lengths of pipes, according to Pythagoras and Philolaus. Woodcut in Franchinus Gaffurius, *Theorica musice* (Milan, 1492).

the other ones being geometry, arithmetic, and astronomy. The trivium, the other section of the liberal arts, contained grammar, logic and rhetoric, disciplines that we would today rather allocate to the humanities. (Note that in the context here described, these disciplines were called 'arts'.) And indeed, a reader opening an arbitrary page in Boethius's *De musica institutione*, a text written in the early sixth century, and the foundational text about music during the Middle Ages, will likely see more figures than text. Boethius summarized Greek music theory, and within that theory the proportions that describe the sizes of musical intervals played a major role. Pythagoras had, according to legend, first proposed the simple proportions of string lengths that would produce consonant intervals, such as the octave (2:1), the fifth (3:2) and the fourth (4:3; Ill. 0.5). The fact that musical intervals could be described by such simple numbers was, in addition, taken as evidence that many more aspects of the world would have such a background logic, be it hidden in most cases. Developments in sixteenth-century music theory brought about the addition of the major third (5:4) to the basic set, thereby creating the *senario* — the set of integer numbers from one to six — as the basis of a scientific, quantitative theory of music.

During the sixteenth and seventeenth centuries, music theory based on numbers was gradually supplemented by music theory based on notes, including notation, scales and modes, counterpoint, and (later) figured bass, as well as the consideration of the rhetoric and the emotional aspects of music.

An interesting by-product of the scientific interests of music theorists in the sixteenth and seventeenth centuries is the organization of texts on music in the form of mathematical treatises, following the model first provided by Euclid's *Elements*, with definitions, axioms and theorems. The earliest example in the field of music theory is Gioseffo Zarlino's *Dimostrazioni musicali* (Venice, 1571), which consists of four sections called reasonings (*ragionamenti*), each of which presents series of definitions (*definitioni*), axioms (*dignità*), and theorems (*proposte*). Here Zarlino may echo *Musica libris demonstrata quattuor* by Jacobus Faber Stapulensis (Jacques Lefèvre d'Étaples; Paris, 1496, 1552), which begins with ten *dignitates* (axioms) and eight *petitiones* (questions), which are then followed by the four books themselves, each one consisting of a long series of numbered demonstrations. Seventeenth-century examples of books on music in the form of mathematical treatises are Salomon de Caus's *Institutions harmoniques* (Frankfurt, 1615) and Marin Mersenne's *Harmonie universelle* (Paris, 1636; Chapter Twelve); a late, eighteenth-century example is Robert Smith's *Harmonics, or the Philosophy of Musical Sounds* (London, 1748, 1759). The seventeenth-century German polymath Joachim Jungius approached the theory of music along the lines set out in Aristotle's *Posterior Analytics*, with definitions, theorems, and demonstrations (Chapter Thirteen).

The reception of Greek musical writings within musical thought in the sixteenth and seventeenth centuries can also be considered to belong to the topic of 'music under the influence of science'[32]. Until the end of the fifteenth century Greek music was primarily known through Boethius's *De institutione musica*. The information contained in this work was supplemented by the passages about music in Macrobius's *Commentarius in Somnium Scipionis* (c. 400 AD), Martianus Capella's *De nuptiis Philologiae et Mercurii* (fifth century), and Cassiodorus's *Institutiones divinarum et humanarum litterarum* (fifth century). Boethius's *De institutione musica* was available in print from 1492 onward when it was included in the *Opera Boetii* (Venice); later editions followed. Before this time, there was a considerable dissemination of the work in manuscript, from the Carolingian era onward.

During the two centuries from the late fifteenth to the late seventeenth century there was a gradual but steady grow of the availability of the Greek texts either by publication in the original language or in Latin translation. The first round of Latin translations took place in the 1490s, when Francesco Burana translated Aristides Quintilianus's *On Music* and Bacchius's *Musical Introduction*, and Claudius Ptolemy's *Harmonics*, but these writings remained in manuscript and were therefore available to few (among them Franchinus

[32]. PALISCA 1985.

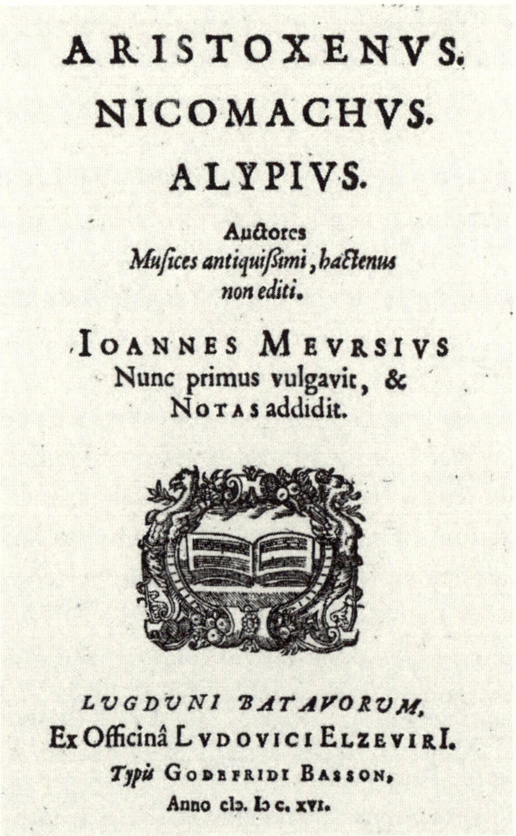

ARISTOXENVS.
NICOMACHVS.
ALYPIVS.

Auctores
Musices antiquissimi, hactenus
non editi.

IOANNES MEVRSIVS
Nunc primus vulgavit, &
NOTAS addidit.

LVGDVNI BATAVORVM,
Ex Officinâ LVDOVICI ELZEVIRI.
Typis GODEFRIDI BASSON,
Anno cIↄ. Iↄ c. XVI.

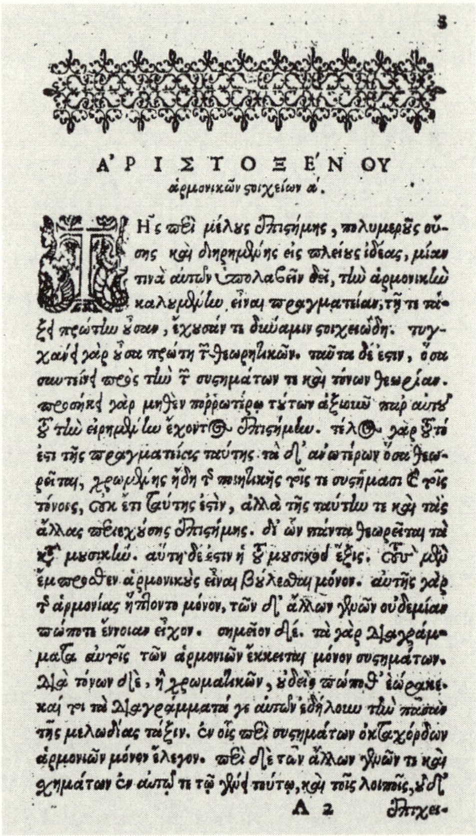

3

ἈΡΙΣΤΟΞΕΝΟΥ
ἁρμονικῶν στοιχείων α΄.

Α 2

Ill. 0.6ab: Johannes Meursius, *Aristoxenus, Nicomachus, Alypius, auctores musices antiquissimi* (Leiden, 1616), title page, and page 3, with the first page of Aristoxenus's *Harmonic Elements*. Copy B-Gu.

Gaffurius). Giorgio Valla's translation of Cleonides's *Musical Introduction*, on the other hand, was published, with other texts, as *Cleonidae Harmonicum introductorium, interprete Georgio Valla,* [...] (Venice, 1497).

Perhaps the two most important Greek writings on music, Aristoxenus's *Harmonic Elements* and Ptolemy's *Harmonics* became available in a Latin translation by Antonio Gogava published as *Aristoxeni* [...] *Harmonicorum elementorum libri III, Cl. Ptolemaei Harmonicorum, seu De musica Libri III* (Venice, 1562). But Girolamo Mei and Giovanni Battista Doni still made use primarily of manuscripts when they studied Greek music theory in the late sixteenth and early seventeenth century, respectively. It was only in the seventeenth century that Greek texts on music became relatively completely and easily available. Johannes Meursius edited, in Greek, the works of Aristoxenus, Nicomachus, and Alypius in his *Auctores musices antiquissimi, hactenus non editi* (Leiden, 1616; Ill. 0.6ab), Marcus Meibom the works, in Greek and Latin translation, of Aristoxenus, Euclid, Nicomachus, Alypius, Gaudentius, Bacchius, and Aristides Quintilianus, supplemented by Martianus Capella's *De nuptiis Philologiae et Mercurii*, in

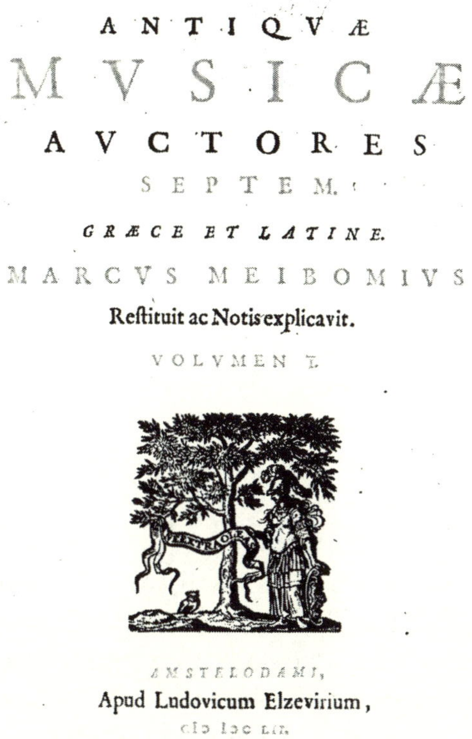

ANTIQVÆ
MVSICÆ
AVCTORES
SEPTEM.
GRÆCE ET LATINE.
MARCVS MEIBOMIVS
Reſtituit ac Notis explicavit.
VOLVMEN I.

AMSTELODAMI,
Apud Ludovicum Elzevirium,
CIϽ IϽC LII.

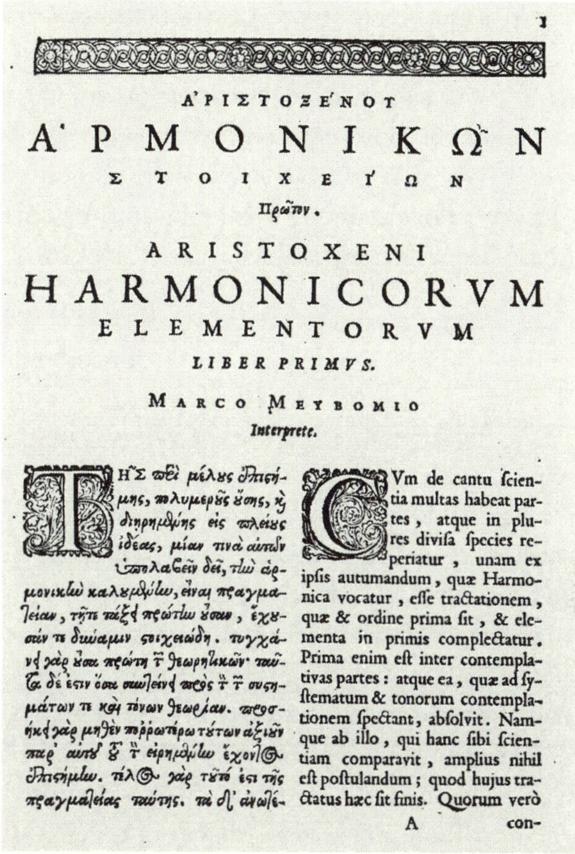

1

ἈΡΙΣΤΟΞΕΝΟΥ
ἉΡΜΟΝΙΚΩΝ
ΣΤΟΙΧΕΙΩΝ
Πρῶτον.

ARISTOXENI
HARMONICORVM
ELEMENTORVM
LIBER PRIMVS.
MARCO MEYBOMIO
Interprete.

Vm de cantu ſcientia multas habeat partes, atque in plures diviſa ſpecies reperiatur, unam ex ipſis autumandum, quæ Harmonica vocatur, eſſe tractationem, quæ & ordine prima ſit, & elementa in primis complectatur. Prima enim eſt inter contemplativas partes: atque ea, quæ ad ſyſtematum & tonorum contemplationem ſpectant, abſolvit. Namque ab illo, qui hanc ſibi ſcientiam comparavit, amplius nihil eſt poſtulandum; quod hujus tractatus hæc ſit finis. Quorum verò

A con-

Ill. 0.7ab: Marcus Meibom, *Antiquae musicae auctores septem, Graece et Latine* (Amsterdam, 1652), title page and page 1, with the first page of Aristoxenus's *Harmonic Elements*, Greek en Latin. Copy D-Mbs.

Antiquae musicae auctores septem, Graece et Latine (Amsterdam, 1652; Ill. 0.7ab). John Wallis provided a bilingual edition of Ptolemy's *Harmonics* in *ΚΛΑΥΔΙΟΥ ΠΤΟΛΕΜΑΙΟΥ ΑΡΜΝΙΚΩΝ ΒΙΒΛΙΑ Γ′ Claudii Ptolemaei harmonicorum libri tres* (Oxford, 1682). This edition was supplemented by editions of Porphyrius's *Commentary on Ptolemy's Harmonics* and Manuel Bryennius's *Harmonics*, when reprinted in Wallis's *Operum mathematicorum volumen tertium* (Oxford, 1699). Paradoxically, and perhaps also sadly for the editors and translators, the more and the better translations of these Greeks writings became available, the less was their relevance for contemporary music theory and musical practice[33].

During the period under consideration here, science could also derive its inspiration from music. First of all, there was the aforementioned scientific theory of music, based on proportions, also called 'harmonics'. Besides this, it was especially the physical counterpart of consonance that intrigued natural philosophers. Antiquity had already described the sound of

[33]. Rasch 2007.

a strings as a vibration or as a series of pulses, assumedly on the basis of the visual observation of strings producing low or very low tones. But what was the reason that otherwise identical strings with lengths related to one another by the proportion 2:1 would produce the interval of an octave, the most important consonance (not counting the unison)? The 'coincidence theory' that was formulated as an answer to this question occupied several sixteenth- and seventeenth-century scholars, among them Giovanni Battista Benedetti, Isaac Beeckman, René Descartes, Marin Mersenne, and Galileo.

Earlier Studies

The present book is not the first volume to deal with the relation between music and science in the Early Modern Period. In fact, it directly continues a tradition more than half a century old. A good starting point is Claude Palisca's essay on 'Scientific Empiricism in Musical Thought', published in 1961 in a collective volume on *Seventeenth Century Science and the Arts*[34]. Despite the title of the volume in which it appeared, Palisca's essay deals more with the sixteenth and eighteenth centuries, represented by Giovanni Battista Benedetti and Vincenzo Galilei (sixteenth century) and Jean-Philippe Rameau (eighteenth century), than with the seventeenth century. As examples of musical empiricism, Palisca describes how Benedetti, a natural scientist, related consonance to pulse patterns of sounds, how Vincenzo Galilei questioned the supremacy of just-intonation intervals by referring to musical practice, and how Rameau based his theory of tonality on the overtone series of musical sounds.

Palisca's approach was followed up by Stillman Drake in an article entitled 'Renaissance Music and Experimental Science', published in 1970[35]. Drake saw Galilei's observations that the relation between the tension of strings and the pitches produced by them should be described as squared proportions, not by simple linear proportions, as was done in all music theory up to that time, as the beginning of the experimental tradition in the natural sciences: never before had controlled experiments led to a 'law' that was not evident from known knowledge.

An often-quoted text on the interfaces between music and science is Daniel Walker's *Studies in Musical Science in the Late Renaissance* (London, 1978)[36], a collection of essays published somewhere else before (or after) the publication of the anthology, usually in a somewhat different form, and under a slightly different title. It contains chapters focusing on Vincenzo and Galileo Galilei, Gioseffo Zarlino, Johannes Kepler, Marin Mersenne, Joan Albert Ban and Giuseppe Tartini, as well as chapters devoted to certain topics, such as the harmony of spheres, the expressive values of intervals, and intonation systems. Many chapters in the present

[34]. Palisca 1961.

[35]. Drake 1970.

[36]. Walker 1978.

book refer to Walker's *Studies*. Walker introduced or at least contributed to the continued use of the phrase 'musical science' as a common denominator for research on the interaction between music and science.

It was only with the appearance of Floris Cohen's *Quantifying Music: The Science of Music at the First Stage of the Scientific Revolution, 1580-1650* in 1984 that the field saw perhaps the first systematic detailed treatment of the relations between music and science in the period under consideration here[37]. Cohen provides a concentration on theories of musical consonance, in particular on the 'coincidence theory', which describes the sensation of consonance as the result of coinciding sound pulses. The book contains in-depth discussions of 'musical theories' formulated by Kepler, Simon Stevin, Giovanni Battista Benedetti, Vincenzo and Galileo Galilei, Marin Mersenne, Descartes, Isaac Beeckman and Christiaan Huygens. Cohen's book has received both praise and criticism. Not surprisingly, many chapters in the present book refer to it.

In 1992 appeared *Music and Science at the Time of Galileo*, edited by Victor Coelho, with contributions derived from a conference held in 1989 in Calgary, Alberta[38]. Understandably, much attention is paid to father and son Vincenzo and Galileo Galileo, but some of the contributions also discuss other early scientists, among them Kepler, Descartes, Stevin, Beeckman and Christiaan Huygens.

Paolo Gozza edited two interesting collective volumes, each comprising a generous series of essays selected from various publications from the preceding decades. In 1989 appeared, in Italian, *La musica nella rivoluzione scientifica del Seicento*, in 2000, in English, *From Number to Sound: The Musical Way to the Scientific Revolution*[39]. There is some overlap between the two volumes, and there is overlap with the volumes mentioned above, but because of the sheer number of contributions — sixteen and twelve, duplicates included — they present a fine overview of the field of musical science from, say, 1500 to, say, 1650, as discussed in contributions from, say, 1970 to 2000.

Among the more recent studies we can mention the articles brought together in the collective volume *Music and Mathematics in Late Medieval and Early Modern Europe*, edited by Philippe Vendrix (Turnhout, 2008) and the section 'Music' in *Proportions: Science – Musique – Peinture & Architecture*, edited by Sabine Rommevaux, Philippe Vendrix and Vasco Zara (Turnhout, 2011)[40]. The studies in these two volumes less often concentrate on the works of single scientist or music theorist (or musician, for that matter) than the previously listed ones, but this does not exclude specific studies on, for example, Vincenzo Galilei, Nicola Vicentino, Simon Stevin, and Mersenne.

37. COHEN 1984.

38. COELHO 1992.

39. GOZZA 1989, GOZZA 2000.

40. VENDRIX 2008, ROMMEVAUX – VENDRIX – ZARA 2011.

Certainly, there are more studies that should have been mentioned here. But a selection should have to be made anyhow.

This Book

The fifteen chapters of this book are divided into four parts, which are defined by two dichotomies. The first is identified by the simple opposition of the key words 'music' and 'science'. 'Music' means here that the focal point of the chapter is in the realm of music: some musical aspect or phenomenon or theory is related to some kind of scientific theory. The chapters under the heading 'Science' work the other way around: here a scientific concept plays the central role, and the focus is on how music is called on or required to explain or at least to illustrate it.

The second dichotomy in the organization of this book is the separation of chapters that deal with the (long) sixteenth century from those related to the (long) seventeenth century. The distinction may seem purely chronological at first sight, but it is not. In the sixteenth century the science that is involved is, generally, Aristotelian science or, more general, scientific thought with a relation to Antiquity. In the chapters that deal with the seventeenth century, it is the 'new science' that is being referred to, theories, hypotheses, and ideas proposed by Kepler, Galileo, Descartes, and Christiaan Huygens. Chapter 8, however, on Giovanni Battista Doni, shows that the inspiration from Antiquity was not yet completely over in the seventeenth century.

The two distinctions create four groups of chapters. The first group contains a single chapter that discusses an example from the 'long sixteenth century', an unusual application of natural science to practical music theory found in the *Practica musice* by Franchinus Gaffurius (1496). The second group describes musical inspiration in the scientific domain in the sixteenth century. The third part of the book contains chapters that deal with 'scientific elements in music' in the seventeenth century, the fourth with 'musical aspects found in scientific contexts' in the same period of time.

'Scientific elements' can be recognized in practical music or in the theory of music in a variety of ways. Chapter One of this book discusses how a seemingly paradoxical element in the theory of music could be explained by theories about the perception of sounds. Chapter Six identifies a series of elements in early seventeenth-century Italian operas that can be understood as referring to Galileo's scientific work in the field of astronomy. Chapter Seven connects the composition of polymorphous canons in early seventeenth-century Rome with the contemporaneous interest in combinatorics visible in scientific writing. Chapter Eight proposes a method for analysing early seventeenth-century chromatic compositions making use of Greek modal theory. Chapter Nine demonstrates ways in which vocal composition could (or could not) follow extra-musical rhetoric principles. Chapter Ten, finally, discusses references to Galileo's discovery in a later seventeenth-century Mass composition.

The chapters that set out from the side of science can be briefly characterized as follows. Chapter Two describes the role of music theory in intellectual circles in Venice during the sixteenth century. Chapters Three and Four analyse references to music and musicians in the philosophical works of Philip Melanchthon and Giordano Bruno respectively, Chapter Five examines the role of music in medical studies. Passing to the seventeenth century, Chapter Eleven discusses music in the output of two scientist-philosophers, René Descartes and Isaac Beeckman; Chapter Twelve studies the role of music in establishing which musical intervals could be called natural; Chapter Thirteen describes how tuning theory could be formulated in Aristotelian terms, while chapters Fourteen and Fifteen demonstrate how by the end of the seventeenth century, music theory became — because of its necessarily aesthetic aspect — separated from physical theories concerning sound in general, a field now termed 'acoustics'.

The present book follows certain conventions in the presentation of the texts it contains. Names of musicians, composers and music theorists follow the spellings in the *New Grove Encyclopedia of Music and Musicians, Second Edition* (2001 or online). Names of other persons take the form given in the *New Oxford Style Manual* or the *Oxford Dictionary of English* if they are listed there. Otherwise, we follow Wikipedia or predominant usage on the Internet. If anglicized or italicized forms are in common use, they are used here too. If there are common variant forms of names, such as Latin forms of vernacular names (or the other way around), these may be added between parentheses the first time the person is mentioned. Years of birth and death are added to the first mention of a person only when that person is the focus of the discussion or when the dates are important for understanding the chronology of the discussion.

Titles of books are wherever possible quoted in the original language and provided with city and year of publication. A translation into the language of the article is provided when the meaning is not clear from the corresponding cognate words in the language of the article. An exception is made for Greek texts: these are cited with the usual English (or Italian) translations of their titles. A table listing the Greek texts cited and providing their titles in their standard English and Latin translations is included in the preliminary section of this book.

Quotations in the main text are always given in English (or Italian), with the original text (if in a different language) in a footnote. With very short quotations, the translation may follow in the main text. Translations may follow those available already in published form or be provided by the author of the chapter. Translations of Greek texts on music follow those in Andrew Barker's *Greek Musical Writings, Volume 2* (Cambridge, 1989)[41], unless the author of the chapter states otherwise.

In the original text of the quotation, the spelling used in the source is maintained as much as possible, if only because standardization — for instance with respect to u/v and i/j, to the addition or changing of diacritics, and so on — would imply very different sets of interventions

[41]. Barker 1989.

for the various texts. Abbreviations, however, have in most cases been resolved. We recognize that the lack of standardization may cause difficulties of comprehension. The Latin word '*musicē*', for example, may stand for '*musicae*' (genitive or dative singular or nominative plural of the noun '*musica*' or of the adjective '*musicus*' in the feminine gender) or for the adverbial form '*musicē*'; or it may be a transcription of the Greek word *μουσική* [*mousikē*][42]. Speaking of this Greek word: in early Latin writings, one also finds '*musices*' as a transcription of the genitive *μουσικῆς* [*mousikēs*] and '*musicen*' as a transcription of the accusative *μουσικήν* [*mousikēn*]. Similar forms may occur derived from the word *ἁρμονική* [*harmonikē*], as in the title of Johannes Kepler's *Harmonices mundi libri v* (Five Books about the Harmony of the World; Linz, 1619). One should also be aware of the meaning of the plural English form 'harmonics' in the context of Greek and Greek-related music theory. There, it generally means 'matters of harmony', that is, musical matters, or more specifically music theory concerned with pitch (as opposed to rhythm).

Most often, musical terminology will follow British usage; thus 'bar' (not 'measure'), 'quaver' (not 'eighth note'), and so on.

Since the word 'tone' has many meanings, it will always be specified as 'whole tone' when the reference is to that interval. The two varieties of the whole tone that are distinguished in proportional theories as larger versus smaller, major versus minor, or greater versus lesser whole tone, that is 9:8 versus 10:9, will be designated here by the first pair of opposites. Similarly, we will speak of larger semitones (16:15) and smaller semitones (25:24). A special case of the whole tone is called the 'mean tone', a term to be taken literally as denoting the (geometric) mean between the larger and the smaller whole tone, or (√(5):2 (or √(5:4)). This interval is characteristic for ¼-comma mean-tone temperament, in which the perfect fifth is tempered by a quarter of the syntonic comma (81:80)[43]. Other tunings and temperament systems referred to occasionally in this books are just intonation — with intervals based on just intervals, notably the octave (2:1), the fifth (3:2) and the major third (5:4) — and equal temperament — based on the division of the octave into twelve equal semitones ($2^{1/12}$:1). From the examples given it will be clear that proportions (that is, ratios) are nearly always written with the larger number first.

Each chapter in this book has its own list of references in the 'Bibliography' at the end of the chapter. I have tried to minimize references to internet sources. Often, these are not so permanent as one would wish; others are only accessible by paid subscription, and long URLs are impractical. In addition, the status of internet sources is not always guaranteed. Nevertheless, references to the Internet cannot be avoided completely. In those cases, addresses as compact as possible are provided, and it is expected that readers will find their further way by means of search windows and so on.

---

42. Or even a vocative of the masculine noun '*musicus*'.

43. The usage of the term mean-tone tuning (or mean-tone temperament) for temperaments with other fractions of the comma, common since Barbour 1951, is simply wrong: these temperaments do not have the mean-tone.

This book contains a fair number of illustrations, meant not only to support the text, but also and perhaps even particularly, to illustrate the nature of the source materials — primarily manuscripts, early printed books, letters, and so on — for the studies presented here.

Most chapters of this book are based on lectures delivered during the conference 'Music and Science from Leonardo to Galileo' held online from 13-15 November 2020 and organized by the Centro Studi Opera Omnia Luigi Boccherini in Lucca, Italy. Chapters 9 and 15 have a different background but were added because they provide useful supplementary points of view.

The editor wishes to thank the authors of the various chapters for their kind and fruitful collaboration and in particular the anglophone authors for their willingness to correct and improve the English wording of the contributions by non-anglophone authors, including the editor. Thanks go also to the Centro Studi Opera Omnia Luigi Boccherini in Lucca, Italy, and Brepols Publishers in Turnhout, Belgium, for making possible the publication of this book.

Rudolf Rasch
January 2022

Bibliography

Barbour 1951
Barbour, James Murray. *Tuning and Temperament: A Historical Survey*, East Lansing (MI), Michigan State University Press, 1951.

Barker 1989
Greek Musical Writings. 2: Harmonic and Acoustic Theory, edited by Andrew Barker, Cambridge, Cambridge University Press, 1989.

Beltrami 1905-1906
Beltrami, Luca. 'Il 'Musicista' di Leonardo da Vinci', in: *Raccolta Vinciana*, ii (1905-1906), pp. 75-80.

Clercx-Lejeune 1972
Clercx-Lejeune, Suzanne. 'Fortuna Josquini. A proposito di un ritratto di Josquin Desprez', in: *Nuova Rivista Musicale Italiana*, vi/3 (1972), pp. 315-341.

Coelho 1992
Music and Science in the Age of Galileo, edited by Victor A. Coelho, Dordrecht, Kluwer, 1992,

Cohen 1984
Cohen, H. Floris. *Quantifying Music: The Science of Music at the First Stage of the Scientific Revolution, 1580-1650*, Dordrecht, Reidel, 1984.

DESCARTES 1897
DESCARTES, René. *Correspondance I: avril 1622-février 1638*, edited by Charles Adam et Paul Tannery, Paris, Léopold Cerf, 1897.

DRAKE 1970
DRAKE, Stillman. 'Renaissance Music and Experimental Science', in: *Journal of the History of Ideas*, XXXI/4 (1970), pp. 483-500.

FARAGO 1991
FARAGO, Claire J. 'On Leonardo da Vinci's Defense of Painting against Poetry and Music and the Grounding of Aesthetic Experience', in: *Italian Culture*, IX (1991), pp. 153-170.

FUGELSO 2010
FUGELSO, Karl. 'Music as (Im)Mortality in Leonardo's «Portrait of a Man»', in: *Source: Notes in the History of Art*, XXX/1 (2010), pp. 24-28.

GALILEI 1906
GALILEI, Galileo. *Opere, Volume XVII: Carteggio 1637-1638*, edited by Antonio Favaro, Florence, Barbèra, 1906.

GOZZA 1989
La musica nella rivoluzione scientifica del Seicento, edited by Paolo Gozza, Bologna, Il Mulino, 1989.

GOZZA 2000
Number to Sound: The Musical Way to the Scientific Revolution, edited by Paolo Gozza, Dordrecht, Kluwer, 2000.

GUILFORD 1967
GUILFORD, Joy Paul. *The Nature of Human Intelligence*, New York, McGraw-Hill, 1967.

LEONARDO 1992
Leonardo da Vinci's Paragone: A Critical Interpretation with a New Edition of the Text in the Codex Urbinas, edited by Claire Farago, Leiden, Brill, 1992.

LEONARDO 1995
Leonardo da Vinci, Libro di Pittura. Codice Urbinate lat. 1270 nella Biblioteca Apostolica Vaticana, transcribed by Carlo Vecce, edited by Carlo Pedretti, Florence, Giunti, 1995.

MARINONI 1954
MARINONI, Augusto. *I rebus di Leonardo da Vinci raccolti e interpretati, con un saggio su «Una virtù spirituale»*, Florence, Olschki, 1954.

MERSENNE 1945
MERSENNE, Marin. *Correspondance du P. Marin Mersenne, réligieux minime. 2: 1628-1630*, edited by Paul Tannery, Cornelis de Waard and René Pintard, Paris, Presses Universitaires de France, 1945.

Mersenne 1960

Mersenne, Marin. *Correspondance du P. Marin Mersenne, réligieux minime. 6: 1636-1637*, edited by Cornelis de Waard, Paris, Éditions du Centre National de la Recherche Scientifique, 1960.

Mersenne 1964

Id. *Correspondance du P. Marin Mersenne, réligieux minime. 8: août 1638 - décembre 1639*, edited by Cornelis de Waard, Paris, Éditions du Centre National de la Recherche Scientifique, 1964.

Palisca 1961

Palisca, Claude V. 'Scientific Empiricism in Musical Thought', in: *Seventeenth Century Science and the Arts*, edited by Hedley Howell Rhys, Princeton (NJ), Princeton University Press, 1961, pp. 91-137.

Palisca 1985

Id. *Humanism in Italian Musical Thought*, New Haven (CT), Yale University Press, 1985.

Palisca 1989

Id. *The Florentine Camerata. Documentary Studies and Translations*, New Haven (CT), Yale University Press, 1989.

Rasch 2007

Rasch, Rudolf. '«Ongeschickte geslachten des gesanx». Simon Stevin en Joan Albert Ban over de muziektheorie van de oude Grieken', in: *Bronnen van inspiratie. Recepties van de klassieken in de vroegmoderne Nederlanden in muziek, literatuur en beeldende kunst*, edited by Jan Bloemendal and A. Agnes Sneller, Hilversum, Verloren, 2007 (= *De Zeventiende Eeuw*, xxiii [2007]), pp. 25-44.

Rommeveaux – Vendrix – Zara 2011

Proportions. Science – Musique – Peinture & Architecture, edited by Sabine Rommevaux, Philippe Vendrix, and Vasco Zara, Turnhout, Brepols, 2011.

Schiltz 2015

Schiltz, Katelijne. *Music and Riddle Culture in the Renaissance*, Cambridge, Cambridge University Press, 2015.

Testolin 2007

Testolin, Walter. 'Leonardo ritrae Josquin. Nuove conferme sull'identità del Musico dell'Ambrosiana', in: *Rivista Italiana di Musicologia*, lxii/2 (2007), pp. 309-321.

Tiella 1989

Tiella, Marco. 'Leonardo da Vinci's *viola organista*', in: *Report from the 9th International Meeting of the International Council for Traditional Music's Study Group on Folk Musical Instruments in Orta San Giulio, Italy 1986*, Stockholm, Musikmuseet, 1989, pp. 151-160.

Vendrix 2008

Music and Mathematiques in Late Medieval and Early Modern Europe, edited by Philippe Vendrix, Turnhout, Brepols, 2008.

WALKER 1978
WALKER, Daniel Pickering. *Studies in Musical Science in the Late Renaissance*, London, The Warburg Institute, 1978.

WINTERNITZ 1982
WINTERNITZ, Emanuel. *Leonardo da Vinci as a Musician*, New Haven (CT), Yale University Press, 1982.

The (Long) Sixteenth Century

Music

Chapter One

«Latet discordantia quartae»: An Early Natural-Scientific Explanation of Upper-Voice Fourths by Franchinus Gaffurius

David E. Cohen
(Max Planck Institute for Empirical Aesthetics, Frankfurt am Main)

The focus of this chapter is a passage in which Franchinus Gaffurius (1451-1522), the preeminent Italian music theorist of the decades around 1500, offers a physical, natural-scientific explanation for a particular aspect of musical practice; that is, an explanation employing a theory that we would now consider as belonging to the field of acoustics. The mere fact that he did so is of some interest, since to offer such an explanation for a specific feature of musical practice was quite an unusual thing for a theorist to do at that time — a point to which I shall return in my conclusion. The passage occurs in Gaffurius's *Practica musice* (The Practice of Music), published in Milan in 1496[1]. As its title indicates, this is a treatise concerned chiefly with techniques of practical music-making (solmization, the modes, mensural notation, and counterpoint) as opposed to the more speculative matters falling under the rubric of *musica theorica* or *speculativa*, which are largely mathematical but also to a lesser extent comprise other topics including what we now call acoustics. Gaffurius, as was usual at the

[1]. Franchinus Gaffurius, *Practica musice* (Milan, 1496). About the composition history of this work, see Miller 1968. English translation by Clement Miller in Gaffurius 1969. All translations in this chapter, unless otherwise identified, are mine.

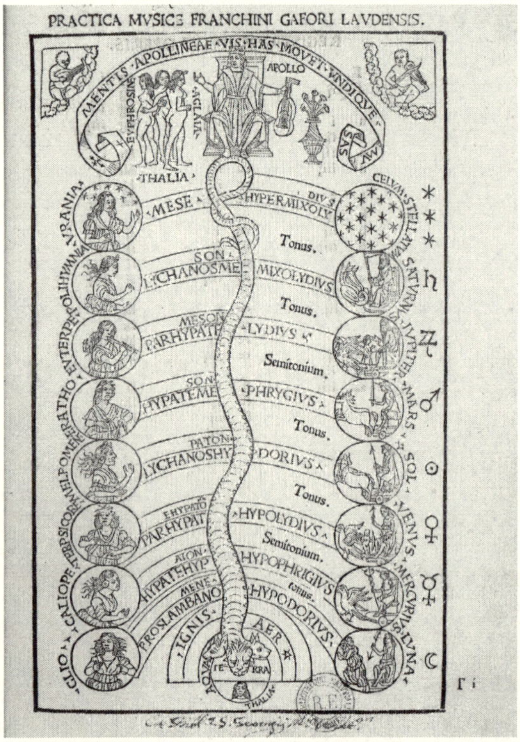

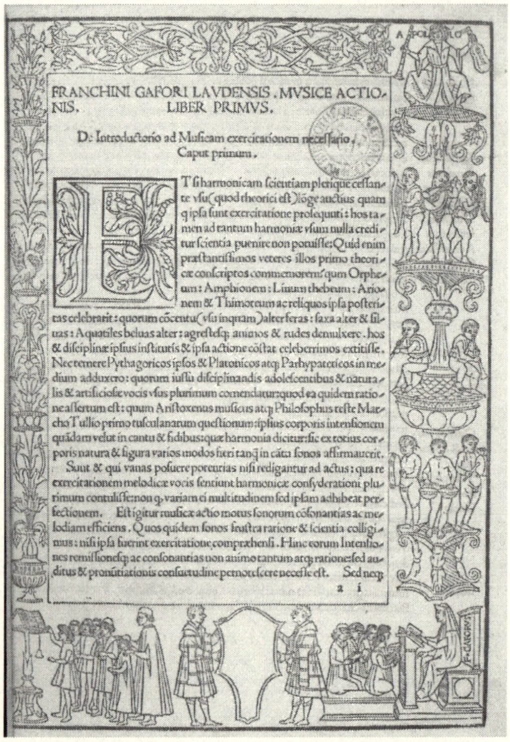

FRANCHINI GAFORI LAVDENSIS . MVSICE ACTIONIS. LIBER PRIMVS.

De Introductorio ad Musicam exercitationem necessario. Caput primum.

Etsi harmonicam scientiam plerique cessante usu (quod theorici est) longe auctius quam q ipsa sunt exercitatione prosequuti: hos tamen ad tantum harmoniae usum nulla creditur scientia puenire non potuisse: Quid enim praestantissimos veteres illos primo theoricae conscriptos commemorem: quum Orpheum: Amphionem: Linum thebeum: Arionem & Thimoteum ac reliquos ipsa posteritas celebrarit: quorum cōcentu (usu inquam) alter feras: saxa alter & silvas: Aquatiles beluas alter: agrestesq; animos & rudes demulxere. hos & disciplinae ipsius institutis & ipsa actione cōstat celeberrimos extitisse. Nec temere Pythagoricos ipsos & Platonicos atq; Parhypateticos in medium adduxero: quorum iussu disciplinandis adolescentibus & naturalis & artificiosae vocis usus plurimum comendatur: quod ea quidem ratione assertum est: quum Aristoxenus musicus atq; Philosophus teste Marcho Tullio primo tusculanarum questionum: ipsius corporis intensionem quādam velut in cantu & fidibus: quae harmonia dicitur: sic ex totius corporis natura & figura varios modos fieri tanq in cātu sonos affirmaverit.

Sunt & qui vanas posuere potentias nisi redigantur ad actus: quare exercitationem melodicae vocis sentiunt harmonicae consyderationi plurimum contulisse: non q. variam ei multitudinem sed ipsam adhibeat perfectionem. Est igitur musicae actio motus sonorum cōsonantias ac melodiam efficiens. Quos quidem sonos frustra ratione & scientia colligimus: nisi ipsa fuerint exercitatione comprehensi. Hinc eorum Intensiones remissionesq; ac consonantias non animo tantum atq; ratione: sed auditus & pronūtiationis consuetudine pernotescere necesse est. Sed neq;

a i

ILL. 1.1: Franchinus Gaffurius, *Practica musice* (Milan, 1496), title page and f. a i-recto. Copy F-Pn.

time, saw these as two quite distinct subdisciplines of the *ars musica*, having published a separate *Theorica musice* (The Theory of Music) four years prior to the appearance of the *Practica musice*[2]. In the dedication of the *Practica musice*, Gaffurius briefly characterizes that earlier treatise as one in which he «had followed the method of the ancients in the explanatory principles of this art, which the Greeks call 'theoretical'»[3]. Nearly all its matter clearly belongs to *musica theorica* (or *speculativa*), as that tradition had developed down to Gaffurius's time, and the emphasis therefore was on intervals and their correlative numerical proportions, the central topic of this style of theorizing.

There is, consequently, little overlap between the topics covered in the two books. My subject here is a rare exception to this rule. It concerns a theoretical problem posed in the *Practica musice* and affecting a particular aspect of counterpoint in multiple voices, a problem Gaffurius attempts to solve by invoking supposed physical facts pertaining to the mechanics of sound production. The problem comes down to an inconsistency in the rules governing the use

[2]. Franchinus Gaffurius, *Theorica musice* (Milan, 1492). English translation by Walter Kreyszig in GAFFURIUS 1993.

[3]. Gaffurius, *Practica musice* (1496), Dedication, ff. Γ iii-verso–iiij-recto: «librum etiam edidi antiquissimorum rationem secutus de huius artis speculatione: quam graeci Theoricam vocant».

of the interval of the perfect fourth (Ex. 1.1(a)). By Gaffurius's day it was firmly and universally accepted that, in the context of simple two-part counterpoint (*contrapunctus simplex*), the perfect fourth is a dissonance (or 'discord'), not to be used as a harmonic dyad (interval) between the two contrapuntal voice parts. When more than two voice parts are present, that rule becomes a prohibition against placing a note in any upper voice at the fourth (as well as its compounds) above a given note in the lowest voice (whether Tenor or Bassus; Ex. 1.1(b)). On the other hand, a fourth between the upper voices is permissible, and frequently occurs, in combinations in which the lowest voice is a third or a fifth below the fourth's lower note (or both; Ex. 1.1(c)).

Ex. 1.1: A perfect fourth and some of its possible three- and four-voice contexts.

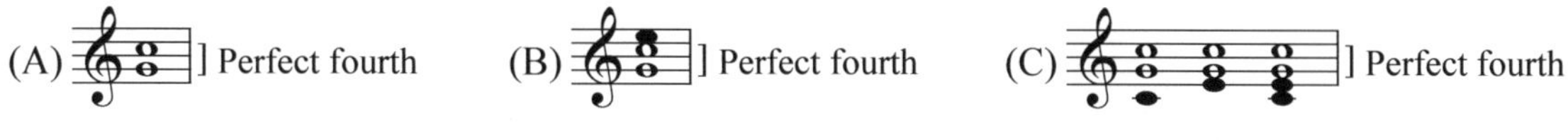

As Gaffurius puts it when he introduces this subject, «But that discord which arises from four mutually conjoined pitches — that is, the perfect fourth sounding [as] a *consonantia* — is admitted in counterpoint in two places»[4]. The first of these two exceptions to the general rule prohibiting fourths as dissonances is a combination with an octave between its outer voices, between which the middle voice lies a fifth above the lowest voice; there will then necessarily be a fourth between the middle and highest voices[5]. Using the anachronistic labeling system of figured bass, in which all intervals are specified with respect to the lowest voice, such a combination can be labeled an '8/5 chord' (Ex. 1.2a). It is the second exception, however, that is of interest to us here. In it, the outer voices are a sixth apart. In this case there are two possibilities, which we can call, using the same nomenclature, the '6/3 chord' and the '6/4 chord' (Exs. 1.2b and 1.2c). The 6/3 chord divides its outer-voice sixth with a middle voice at the third above the lowest voice: here, a fourth lies between the highest two voices (as in the 8/5 chord). In the 6/4 chord, conversely, the sixth is divided so that the fourth is between the lower two voices.

Ex. 1.2: (a) 8/5, (b) 6/3, and (c) 6/4 chords.

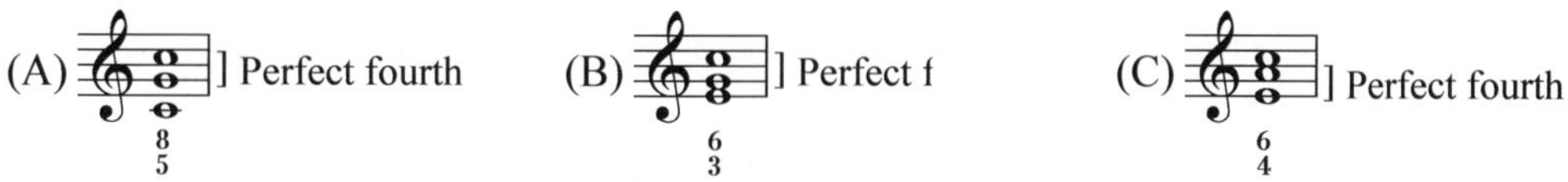

4. See *ibidem*, Liber iii, Cap. 5, f. dd iij-verso: «Sed quae ex quattuor coniunctis inuicem sonis discordantia consurgit: Diatessaron scilicet consonantiam sonans: duobus locis admittitur in contrapuncto».

5. Gaffurius discusses this situation — providing an explanation for the consonant effect of the upper-voice fourth that differs sharply from the one under consideration here — in *ibidem*, immediately after the sentence quoted in the preceding note.

Compositional practice and counterpoint pedagogy in Gaffurius's time treat the fourth in the 6/4 chord as a dissonance, while the fourth in the 6/3 chord is treated as a consonance. It is this inconsistency that Gaffurius seeks to explain in Chapter 6 of Book III of his *Practica musice*, having already provided an example of *fauxbourdon*, a contrapuntal technique that exploits 6/3 and 8/5 chords (Ex. 1.3)[6].

Ex. 1.3: Gaffurius's example of *fauxbourdon* (*Practica musice*, Liber III, Cap. 6, f. dd iiij-recto; Ill. 1.2a). Note values halved, barlines and figuring added. Clefs modernized (originally mezzosoprano, tenor, and tenor).

Gaffurius has previously made it clear that he considers the perfect fourth by itself a discord, that is, a dissonance[7]. He begins his explanation of the 6/3 chord by suggesting that the fourth is acceptable between its top two voices because its ordinarily discordant sound is «overshadowed» (*obumbrata*) by the chord's other two intervals, the sixth and the third, which are concords, «as something smaller [is overshadowed] by things that are greater»[8]. (An account in terms such as these is advanced, to my knowledge, in only one other source, a brief anonymous and unpublished text on the contratenor[9].) This hypothesis, however, fails

[6]. See *ibidem*, Liber III, Cap. 6, f. dd iiij-verso: «Quare Quarta inter medium sonum et acutiorem concordat: discordatque inter medium et grauiorem». (Why the fourth is concordant [when it lies] between the middle voice and the higher, but discordant [when it lies] between the middle and lower [voices].)

[7]. Gaffurius omits the fourth from his catalogue of concords, that is, consonant harmonic intervals or dyads, in *ibidem*, Liber III, Cap. 2, and includes it with the second and the seventh among the discords in Liber III, Cap. 4, even saying there that a fourth occurring as a relatively quick passing tone is «most notably discordant» (*notissime discordans*; f. dd iij-verso).

[8]. *Ibidem*, Liber III, Cap. 6, f. dd iiij-verso: «Et si sexta per tertiam supra tenorem mediata quartam inter medium terminum concorditer sustinet et acutum: quoniam [quarta] inter duas ipsas concordantias tertiam scilicet et sextam (quamvis imperfectas) disposita est ab ipsis tanquam minor a maioribus obumbrata». (Although the sixth mediated by a third above the tenor concordantly supports a perfect fourth between the middle and high[est] term, [this perfect fourth,] since it is arranged between those two concords — the third, that is, and the sixth — is overshadowed by them (even though [they are] imperfect), as something smaller by things that are greater.)

[9]. Incipit: «Cum arbitratus non nihil de contratenoris compositione», preserved in I-ARc, MS 216 (dated 1482; RISM B/III/6, pp. 429-432), ff. 14v-15r. The passage is transcribed in SACHS 1974, pp. 125-126. The

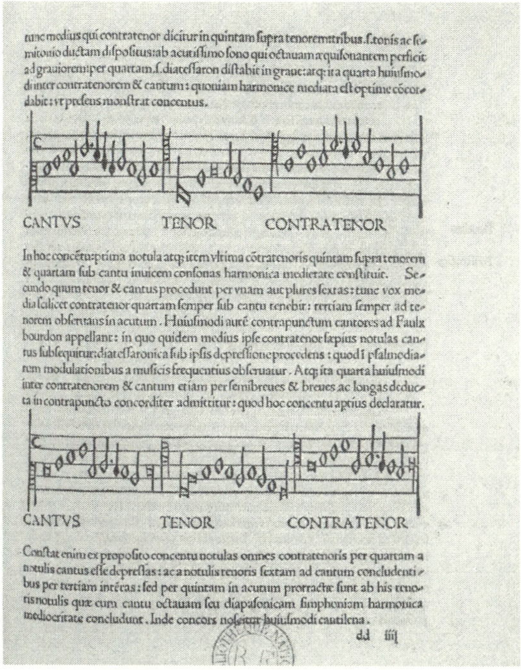

tunc medius qui contratenor dicitur in quintam ſupra tenorem tribus .ſ. tonis ac ſemitonio ductam diſpoſitus: ab acutiſſimo ſono qui octauam æquiſonantem perficit ad grauiorem: per quartam .ſ. diateſſaron diſtabit in graue: atq; ita quarta huiuſmodi inter contratenorem & cantum: quoniam harmonice mediata eſt optime cōcordabit: vt preſens monſtrat concentus.

In hoc concētu: prima notula atq; item vltima cōtratenoris quintam ſupra tenorem & quartam ſub cantu inuicem conſonas harmonica medietate conſtituit. Secundo quum tenor & cantus procedunt per vnam aut plures ſextas: tunc vox media ſcilicet contratenor quartam ſemper ſub cantu tenebit: tertiam ſemper ad tenorem obſeruans in acutum. Huiuſmodi autē contrapunctum cantores ad Faulx bourdon appellant: in quo quidem medius ipſe contratenor ſæpius notulas cantus ſubſequitur: diateſſaronica ſub ipſis depreſſione procedens: quod ī pſalmodiarum modulationibus a muſicis frequentius obſeruatur. Atq; ita quarta huiuſmodi inter contratenorem & cantum etiam per ſemibreues & breues ac longas deducta in contrapuncto concorditer admittitur: quod hoc concentu aptius declaratur.

Conſtat enim ex propoſito concentu notulas omnes contratenoris per quartam a notulis cantus eſſe depreſſas: ac a notulis tenoris ſextam ad cantum concludentibus per tertiam intēcas: ſed per quintam in acutum protractæ ſunt ab his tenoris notulis quæ cum cantu octauam ſeu diapaſonicam ſimphoniam harmonica mediocritate concludunt. Inde concors noſcitur huiuſmodi cantilena.

dd iiii

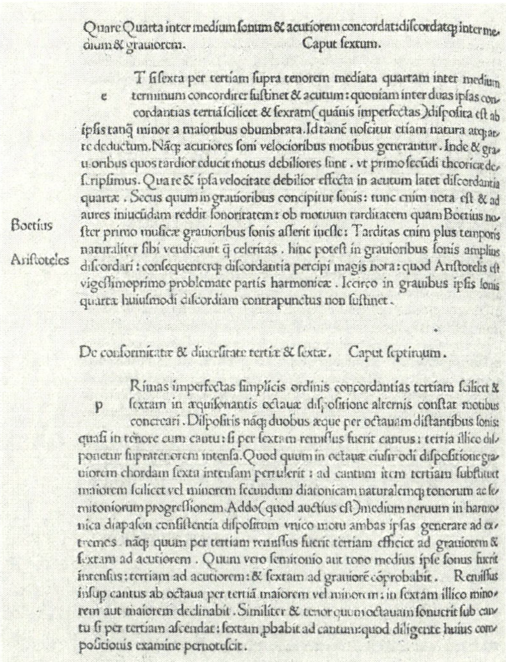

Quare Quarta inter medium ſonum & acutiorem concordat: diſcordatq; inter medium & grauiorem. Caput ſextum.

e T ſi ſexta per tertiam ſupra tenorem mediata quartam inter medium terminum concorditer ſuſtinet & acutum: quoniam inter duas ipſas concordantias tertiā ſcilicet & ſextam (quāuis imperfectas) diſpoſita eſt ab ipſis tanq̄ minor a maioribus obumbrata. Id tamē noſcitur etiam natura atq; arte deductum. Nāq; acutiores ſoni velocioribus motibus generantur. Inde & grauioribus quos tardior educit motus debiliores ſunt. vt primo ſecūdi theoricæ deſcripſimus. Quare & ipſa velocitate debilior effecta in acutum latet diſcordantia quartæ. Secus quum in grauioribus concipitur ſonis: tunc enim nota eſt & ad aures iniucūdam reddit ſonoritatem: ob motuum tarditatem quam Boetius noſter primo muſicæ grauioribus ſonis aſſerit ineſſe: Tarditas enim plus temporis naturaliter ſibi vendicauit q̄ celeritas. hinc poteſt in grauioribus ſonis amplius diſcordari: conſequenterq; diſcordantia percipi magis nota: quod Ariſtotelis eſt vigeſſimoprimo problemate partis harmonicæ. Iccirco in grauibus ipſis ſonis quartæ huiuſmodi diſcordiam contrapunctus non ſuſtinet.

Boetius

Ariſtoteles

De conformitate & diuerſitate tertiæ & ſextæ. Caput ſeptimum.

p Rimas imperfectas ſimplicis ordinis concordantias tertiam ſcilicet & ſextam in æquiſonantis octauæ diſpoſitione alternis conſtat motibus concreari. Diſpoſitis nāq; duobus æque per octauam diſtantibus ſonis: quaſi in tenore cum cantu: ſi per ſextam remiſſus fuerit cantus: tertia illico diſponetur ſupra tenorem intenſa. Quod quum in octauæ eiuſmodi diſpoſitione grauiorem chordam ſexta intenſam pertulerit: ad cantum item tertiam ſubſtituet maiorem ſcilicet vel minorem ſecundum diatonicam naturalemq; tonorum ac ſemitoniorum progreſſionem. Addo (quod auctius eſt) medium neruum in harmonica diapaſon conſiſtentia diſpoſitum vnico motu ambas ipſas generare ad extremes. nāq; quum per tertiam remiſſus fuerit tertiam efficiet ad grauiorem & ſextam ad acutiorem. Quum vero ſemitonio aut tono medius ipſe ſonus fuerit intenſus: tertiam ad acutiorem: & ſextam ad grauiorē cōprobabit. Remiſſus inſup cantus ab octaua per tertiā maiorem vel minorem: in ſextam illico minorem aut maiorem declinabit. Similiter & tenor quum octauam ſonuerit ſub cantu ſi per tertiam aſcendat: ſextam ꝓbabit ad cantum: quod diligente huius compoſitionis examine pernoteſcit.

ILL. 1.2AB: Franchinus Gaffurius, *Practica musice* (Milan, 1496), f. d iiii-recto and verso. The quotation in the title of this chapter is found on f. d iiii-verso, in lines 7-8 of Liber III, Cap. 6. Copy F-Pn.

to explain why the fourth can only lie between the upper two voices and not above the bass: if it were simply a matter of the fourth being outweighed by the two consonant intervals in the chord, then why does that not also occur in the 6/4 configuration?

Gaffurius therefore immediately goes on to ameliorate his unsatisfactory explanation by adding further detail of a more scientific nature[10]:

> For higher pitches are generated by faster motions. Hence, they are also weaker than lower pitches, which slower motion brings forth, as we have explained in the *Theorica* [*musice*], Book II, Chapter 1. Therefore, the discord of the fourth, made weaker in high [pitch] owing to this very speed, is concealed (*«latet discordantia quartae»*). Not so when it is brought forth in lower pitches, for then it is noticed and renders an unpleasant sound to the ears, owing to the slowness of [its] motions, which our [master] Boethius in *De institutione musica*, Book I, asserts is present in

problem addressed there — the fourth in the 8/5 chord — is solved in a different way by Gaffurius, as noted above (note 5).

[10]. Gaffurius introduces this more detailed explanation with the sentence «Id tamen noscitur etiam natura atque arte deductum» (This, however, is both known from nature and deduced by *ars*). In this context, the word *ars* is used with reference to acoustical theories that Gaffurius found stated in writings on the *ars musica*, a phrase which, in this context at least, clearly implies something that we could plausibly call 'the science of music'.

> lower pitches. For slowness naturally occupies more time than does swiftness. Hence [the fourth] can be more fully discordant in lower pitches, and consequently the discord is known to be more [clearly] perceived, which is the view of Aristotle in Problem 21 of the section on harmonic [science]. Therefore, counterpoint cannot allow the discord of this kind of perfect fourth in lower pitches[11].

Lower pitches, which are caused by slower motions, occupy 'more time' than the fast motions that cause high pitches, which are 'weaker'. Therefore, lower pitches are 'more perceived' (more audibly salient) than higher ones, and hence — a crucial corollary — the concordant or discordant qualities of harmonic intervals are more audible between lower pitches than between higher. The result is that a perfect fourth heard in lower pitches is more audibly discordant than one heard in higher pitches. More briefly, the essence of this explanation is that (other things being equal) lower pitches, caused by slower motions, are more audibly salient than higher pitches, caused by faster motions. For the sake of brevity, I shall call this the 'lower-slower-louder theory'.

Gaffurius, not surprisingly, adduces no experimental evidence in support of this lower-slower-louder theory. Rather, he invokes two authorities, Boethius and Aristotle, as well as that aforementioned work of his own, the *Theorica music*e. In fact, the work by 'Aristotle' that Gaffurius cites, the *Problems*, is now agreed to be a product not of the philosopher himself, but of his school. Gaffurius read this work in the thirteenth-century Latin translation of Bartolomeo da Messina, which, equipped with the *Expositio* (commentary) composed by Pietro d'Abano (Petrus Abanus) about 1300, had recently received its editio princeps[12]. In that edition, the passage to which he refers, from Book XIX, Problem 21, runs as follows:

> Why is it that, of those singing, those who sing lower notes are more noticeable if they are discordant than those who are singing high? And similarly [with respect to] rhythm: [that] those who go wrong at lower pitch are more noticeable? Is this because the low-pitched one takes more time? The latter, however, is more evident to

[11]. Gaffurius, *Practica musice* (1496), Liber III, Cap. 6, f. dd iiij-verso: «Namque acutiores soni velocioribus motibus generantur. Inde et grauioribus quos tardior educit motus debiliores sunt. vt primo secundi theoricae descripsimus. Qua re et ipsa velocitate debilior effecta in acutum latet discordantia quartae. Secus quum in grauioribus concipitur sonis: tunc enim nota est et ad aures iniucundam reddit sonoritatem: ob motuum tarditatem quam Boetius noster primo musicae grauioribus sonis asserit inesse: Tarditas enim plus temporis naturaliter sibi vindicat [*pro* vendicauit] quam celeritas. hinc potest in grauioribus sonis amplius discordari: consequenterque discordantia percipi magis nota: quod Aristotelis est vigessimoprimo problemate partis harmonicae. Iccirco in grauibus ipsis sonis quartae huiusmodi discordiam contrapunctus non sustinet».

[12]. Pseudo-Aristotle (translated by Bartolomeo da Messina, commentary by Pietro d'Abano), generally catalogued and cited under the title *Expositio problematum Aristotelis cum textu* (Mantua, 1475; Venice, 1482). See SIRAISI 1970 and PALISCA 1985, pp. 51-66.

> sense. Or is it because it offers, over a long span of time, much to be sensed? But that which is swift and high pitched [on the other hand] is hidden because of its speed[13].

One sees how this could have suggested to Gaffurius the explanation for 6/3 chords that I am calling the lower-slower-louder theory. The passage does not, however, contribute much to our understanding of that theory; nor, unfortunately, does Pietro d'Abano's commentary, which does little more than paraphrase the words of Problem 21. Let us turn then to Gaffurius's other citations, that is, to Boethius's *De institutione musica* and to his own earlier *Theorica musice*, which in fact incorporates the germane material from Boethius.

The relevant chapter in Gaffurius's *Theorica musice*, Book II, Chapter 1, is the first of several that deal at length with topics in what we would now call acoustics, including the nature, causes, and conditions of sound in general and of pitch and changes of pitch. These are among the traditional concerns of speculative music theory in the style we may broadly call Pythagorean — the *musica theorica* to which the book's title alludes. Gaffurius and his contemporaries, one assumes, would have understood these topics as belonging to the physical side of *musica*, seen as a science intermediate between physics and arithmetic[14]. One of the two principal sources in Chapter 1 of Book II of the *Theorica musice* is, as already noted, Boethius's *De institutione musica* — specifically, Book I, Chapter 3, most of which is paraphrased in Gaffurius's chapter[15]. Interspersed with this are sections of the *Paraphrase of Aristotle's «On the Soul»* by Themistius, an influential Byzantine author of the fourth century[16], which had recently been published in a Latin translation by the Venetian humanist Ermolao Barbaro[17].

---

13. Pseudo-Aristotle (tr. Bartolomeo da Messina), *Problemata* (1475), Liber XIX, Vigesimumprimum problema, f. 189-recto: «Propter quid cantantium grauius cantantes acute cantantibus si dissonant magis manifesti fiunt: similiter autem et rithimo: qui in grauiori peccantes manifesti magis: utrum quod plus tempus graui. hic autem magis sensibilis aut qui[a] in pluri tempore multum sensum exibet: uelox autem et acutum latet propter uelocitatem».

14. On the so-called 'mixed', 'intermediate', 'subalternate' or 'subordinate sciences' — mathematical astronomy, optics, mechanics, and mathematical harmonics — in Aristotle and in the Western medieval scholastics, especially Thomas Aquinas, see McKirahan, Jr. 1978, Livesey 1982, especially pp. 235-248, and Distelzweig 2013.

15. This includes, in Gaffurius 1993, the sentences numbered as 9-13 of Book II, Chap. 1 (p. 49), which are a close paraphrase of Boethius's *De institutione musica*, Liber I, Cap. 3 (Boethius 1867, p. 189:15-22).

16. Themistius (fourth century AD), *Paraphrase of Aristotle's «On the Soul»*. For the Greek text I used Themistius 1899. English translation by Robert Todd in Themistius 1996.

17. In: Themistius (translated by Ermolao Barbaro), *Paraphrasis in Aristotelis Posteriora, & physica, in libros item De anima, memoria et reminiscentia, somno et vigilia, insomniis, & divinatione per somnum* (Treviso, 1481, Venice, 1499, 1502). See Branca 1973, p. 226, Palisca 1985, pp. 199-200, 231, and 448-449, Bietenholz – Deutscher 1985, pp. 91-92, and Gaffurius 1993, pp. xxii and 49-53. For the Barbaro translation I cite here the edition of 1502.

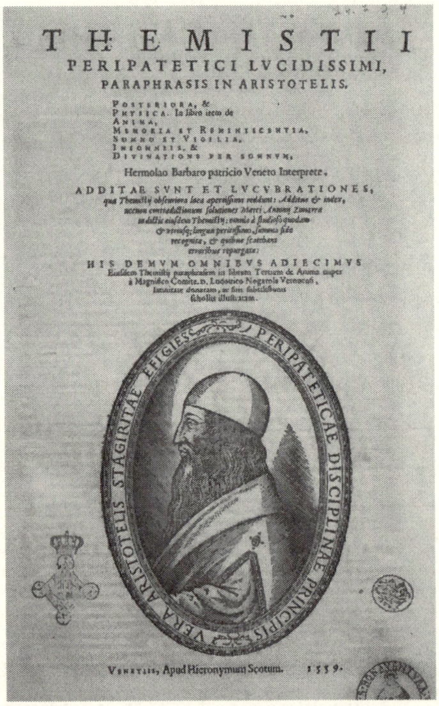
THEMISTII
PERIPATETICI LVCIDISSIMI,
PARAPHRASIS IN ARISTOTELIS.
POSTERIORA, &
PHYSICA. In libro item de
ANIMA,
MEMORIA ET REMINISCENTIA,
SOMNO ET VIGILIA,
INSOMNIIS, &
DIVINATIONE PER SOMNVM,
Hermolao Barbaro patricio Veneto Interprete.
ADDITAE SVNT ET LVCVBRATIONES,
HIS DEMVM OMNIBVS ADIECIMVS

ILL. 1.3: Themistius, *Paraphrasis in Aristotelis Posteriora*, [...] (Venice, 1559), title page, with a portrait of Aristotle: «Vera Aristotelis Stagiritae effigies, peripateticae disciplinae principis» (A true portrait of Aristotle the Stagirite [= from Stagira (near Thessaloniki)], the prince of peripatetic knowledge). Copy I-Rn.

The account of the nature and generation of sound and pitch presented by Gaffurius in the *Theorica musice* closely follows that of Boethius in his *De institutione musica*, Book I, Chapter 3, itself modeled on the prologue to the Euclidean *Division of the Canon*[18]. The essential points of this theory — already attested in Fragment 1 of the Pythagorean Archytas — are, first, that sound is generated by an impact of some kind, which requires a prior motion; second, that it is the speed of the motion that determines the sound's pitch; and third, that those speeds of the sound-generating motions that determine pitch are quantifiable in numbers that can be compared with one another in ratios. This, of course, is the line of thought that underwrites the entire project of Pythagorean or mathematical harmonics, and which structures the prologue to the Euclidean *Division of the Canon*. It is followed by most of the later writers who address this issue, including, most pertinently for us, Boethius, the source for much of Chapter 1 of Book II of Gaffurius's *Theorica musice*. But in this theoretical tradition, there was a long-standing tendency to associate, or even to conflate, higher pitch with greater force and auditory salience, owing to the greater speed of motion by which the higher pitch was produced. Indeed, this correlation is already present in Fragment 1 of Archytas, who writes:

[18]. Modern edition in JAN 1895, pp. 148-149; English translation in BARKER 1989, pp. 191-193.

> Thus, of things impinging on our sense of hearing, those that arrive swiftly and [strongly] from the impacts appear high-pitched, while those that come slowly and weakly appear to be low-pitched[19].

Similarly, Gaffurius in the *Theorica musice*, Book II, Chapter 1, writes:

> And so, every more infrequent and slower motion makes low-pitched sounds; for as slowness is close to standing still, so is lowness of pitch the nearest thing to silence. And this happens precisely because of the heaviness and slowness of the striking [of the air]. But a rapid, swift, and frequent motion produces high-pitched sounds by virtue of that very swiftness and rapidity of striking. [...] But a slacker string brings forth sparse, slow, and infrequent impacts owing to the very weakness of its striking [of the air]; nor does it vibrate very long[20].

This passage is borrowed, with minor alterations, from Boethius's *De institutione musica*, Book I, Chapter 3[21], except for the sentence «for as slowness is close to standing still, so is lowness of pitch the nearest thing to silence», which is from Book IV, Chapter 1, of the same work[22]. The conception it commends is one in which the lowest pitches, being produced by slow and feeble motions verging on lack of motion, themselves verge on complete absence of sound, that is, on silence. This certainly seems incompatible with the lower-slower-louder theory that Gaffurius propounds for the 6/3 chord four years later in the *Practica musice*, Book III, Chapter 6.

Gaffurius, in the *Theorica musice*, Book II, Chapter 1, also explicitly makes the complementary claim that higher pitches are more salient, in words taken from Themistius's paraphrase of Aristotle's *On the Soul* in the Barbaro translation[23]. Speaking of the Greek terms for high and low pitch, ὀξύς [oxys] and βαρύς [barys], which literally mean 'sharp' and 'heavy', Themistius (by way of Barbaro) first explains that these terms (in Latin *acutus* and *gravis*) are

---

19. Archytas, Fragment 1 (Diels – Kranz 1960, B47 B1, vol. I, p. 433:13-16), edited in Düring 1932, pp. 56-57: «τὰ μὲν οὖν ποτιπίπτοντα ποτὶ τὰν αἴσθασιν, ἃ μὲν ἀπὸ τᾶν πλαγᾶν ταχὺ παραγίνεται καὶ [ἰσχυρῶς], ὀξέα φαίνεται·τὰ δὲ βραδέως καὶ ἀσθενῶς, βαρέα δοκοῦντι ἦμεν». The word ἰσχυρῶς (strongly) is supplied as being parallel to ἀσθενῶς (weakly) in the next clause as well as clearly implied by the sequel. See Barker 1989, p. 41, and Huffman 2005, pp. 104:18-20 and 106.

20. Gaffurius, *Theorica musice* (1492), Liber II, Cap. 1, f. c i-recto: «Omnis itaque rarior et tardus motus sonos efficit graues: nam ut tarditas stationi proxima est ita grauitas contigua taciturnitati: fitque id ipsa grauitate et tarditate pellendi. Velox uero ac celer et spissus motus acutos reddit sonos ipsa celeritate et uelocitate pellendi. [...] laxior uero solutos ac tardos pulsus efferat atque raros ipsa imbecilitate feriendi: nec diutius tremat».

21. Boethius 1867, pp. 189:23-190:8.

22. *Ibidem*, p. 301:19-20. Gaffurius quotes this sentence verbatim.

23. Themistius 1899, p. 66:4-11.

words proper to the sense of touch, applied metaphorically to pitch[24]. The passage, as quoted by Gaffurius, continues:

> For just as that which has a form sharpened to a point stimulates [the touch] violently in a short time, but that which has a blunt shape does so slowly and sluggishly, so too a high and rapidly vibrating sound strikes and almost wounds the ear, as though stabbing it, [while] a low pitch strikes it bluntly and in a dispersed way as though shoving against it; and for this reason the former moves [the ear] rapidly and much, while the latter does so little and slowly[25].

And Gaffurius, following up (with Themistius and Barbaro) on this analogy between pitch and the shape of the body whose motion causes it, goes on to claim that

> [...] the motion of a sharp and pointed body is frequent and very swift, while that of a blunt and obtuse one is dilatory and slow. This is also why higher pitch is produced by smoother and wider bodies, because the air is moved by them more rapidly and disperses more swiftly; for which reason, too, the sense of hearing is [more] strongly and speedily stimulated[26].

---

[24]. Themistius (tr. Barbaro), *Paraphrasis* (1502), *De anima*, Liber II, Cap. 30, f. 84v: «haec nomina a tactilibus flexa per translationem sonis adiunximus»; quoted in Gaffurius, *Theorica musice* (1492), Liber II, Cap. 1, f. [b vi]-verso. Themistius's source for this passage is Aristotle, *On the Soul*, Book II, Chap. 8, 420a29-420b4 (BARKER 1989, pp. 78-79), where Aristotle writes that τό ὀξύ (the sharp) and τό βαρύ (the heavy), as used to denote (what we call) high and low pitch, «are said by way of a μεταφορά (transference) from tangible things, for the sharp moves the sense-perception a large amount in a short time, while the heavy moves it a little in a long time. The high pitch, to be sure, is not [itself] swift, nor the low pitch slow; but the motion of the former is of the kind it is owing to its speed, and that of the latter owing to its slowness. And there appears to be an analogy with what is sharp or blunt to the touch (καὶ ἔοικεν ἀνάλογον ἔχειν τῷ περὶ τὴν ἁφὴν ὀξεῖ καὶ ἀμβλεῖ). For the sharp pierces, as it were, while the dull pushes, since the former makes [something] move in a short time, the latter in a long time, so that it happens that the one is fast, the other slow». Compare BARKER 2015, pp. 228-229.

[25]. Gaffurius, *Theorica musice* (1492), Liber II, Cap. 1, ff. [b vi]-verso-c i-recto: «ut enim quod forma[m] in mucronem exacutam obtinet pauxillo tempore uehementer tentat: quod uero retusam habet sero & segniter: Ita et auditum exuibrissans & acuta uox icit: & prope modum uulnerat quasi pungens: grauis obtundit offunditque tanquam trudens: atque hoc ratione illa contingit quod altera multum celeriter mouet: altera parum praetarde». Compare the Aristotelian passage quoted above (n. 24) and Themistius (tr. Barbaro), *Paraphrasis* (1502), *De anima*, Liber II, Cap. 30, f. 84v.

[26]. Gaffurius, *Theorica musice* (1492), Liber II, Cap. 1, f. c i-recto: «ita euenit ut motus fastigati [*pro* fastigiati] turbinatique corporis frequens et praeceler sit: retusi uero atque hebetis cunctabundus et tardus: hinc etiam fit ut a corporibus polytioribus latioribusque acutior sonus edatur: quod aer ab iis uelocius mouetur et celerius dissilit: unde et auditus ualde et cito lacessitur». Compare Themistius (tr. Barbaro), *Paraphrasis* (1502), *De anima*, Liber II, Cap. 30, f. 84v. The source of this Greek language-based analogy between physical shapes and musical pitch is the passage from Aristotle's *On the Soul* quoted above. Themistius, and hence Gaffurius, here seem to me to be taking it

This association of faster motion and higher pitch with greater strength or salience of sound is so much the majority view that the only proponents I have found of the opposed 'slower-lower-louder' theory, which Gaffurius uses to explain the 6/3 chord in the *Practica musice*, Book III, Chapter 6, are Gaffurius himself in that very chapter, and the Pseudo-Aristotelian *Problems* passage that he cites there as his authority for it. But even more striking than his almost unique adoption of that theory is his apparent lack of concern over the inconsistency with his earlier account in the *Theorica musice* — an inconsistency heightened by the fact that a distinct trace of the majority view is present even in the *Practica musice* itself. For there, in the very first chapter, Gaffurius remarks that the lowest eight notes of the Guidonian Gamut are «as though next to silence and taciturnity»[27] — an unmistakable echo of the passage from Boethius's Book IV that he quotes in the *Theorica musice*. Yet, as we have seen, in Book III of the *Practica musice* he proposes the contrary theory.

Perhaps the explanation for the disparity in Gaffurius's views on this issue is that, confronted with the specific problem of the upper-voice fourth in 6/3 chords, he made a tactical decision to solve it by adopting, for that occasion, the slower-lower-louder theory set forth in Problem 21 of Book XIX of the Pseudo-Aristotelian *Problems*. That Gaffurius found this a sufficiently satisfactory theory is shown by his use of it again, a few pages later in the *Practica musice*, to explain the discordant effect of the interval of the thirteenth (and by clear implication, the sixth) when it occurs between the lower voices of a chord:

> And with respect to the thirteenth a similar discord [to that of the sixth] is sometimes perceived in the same lower pitches. Of this we believe the cause to be the [greater] quantity of time occurring in lower pitches, by which the asperity and dissonance of the imperfection more perceptibly reach and injure the ears[28].

Gaffurius, of course, cannot possibly be saying that lower pitches always have longer durations than higher ones; that would be absurd. Rather, he is conflating two correlative factors — speed of motion and the pitch caused thereby — by saying that a greater quantity of time occurs in lower pitches and omitting to say that what really takes more time is the correlative slower motion.

---

more literally than Aristotle had intended (see, however, BARKER 2015, pp. 228-229). This is of no consequence for the present argument, however, since my point here — the association of greater speed of motion, higher musical pitch, and greater aural salience — is either unaffected or, if affected, reinforced thereby.

27. Gaffurius, *Practica musice* (1496), Liber I, Cap. 1, f. a ij-recto: «octo priores quasi silentio et taciturnitati proximas: appellant graues».

28. *Ibidem*, Liber III, Cap. 11, f. [ee i]-recto: «Ac tertiaedecimae in grauioribus ipsis sonis consimilis quandoque percipitur discordia. Huiuscae rei causam esse credimus multitudinem temporis in grauioribus sonis consistentem: qua asperitas et dissonantia imperfectionis apertius auribus accedit afferens lesionem».

His doing so points to the fact that the slower-lower-louder theory has obvious problems. We can assume, I think, that the claim that a slower motion takes more time is to be understood in the obvious sense that a slower motion takes longer to traverse a given distance than a faster one. But just what motion is this? *What* is moving? And *how*? Here again there are two principal possibilities, both connected with the fundamental idea that pitch varies with speed of motion, which was axiomatic both for the *Practica music*e's lower-slower-louder theory *and* for the opposed, majority view that instead attributes greater auditory salience to faster motions and higher pitches.

The two theories regarding the nature of the motion whose speed determines a sound's pitch are these. First, there is the view that it is the speed of the sound's travel, the rapidity of its propagation through the air from its source to the listener's ear, that determines its pitch. This 'speed-of-travel theory' is already plainly evident in Archytas's Fragment 1 (quoted above), and it recurs in the great majority of the relevant texts, occasionally varied in ways which I must pass over here[29]. Much less common, indeed exceptional, is the second view, according to which what is actually subject to the variable of speed is the frequency of the impacts involved in the generation of the sound; for example, the rapidity with which a vibrating string strikes the air. The principal statement in Antiquity of this 'impact-frequency theory' is found near the start of the prologue to the Euclidean *Division of the Canon*[30]. Gaffurius's presentation of it in the *Theorica musice*, Book II, Chapter 1 (also borrowed from Boethius), includes the important proviso that the discrete impacts or percussions are separated by time intervals too short to be audible, so that our hearing perceives them as merging into a seemingly single and continuous sound:

> Neither should we think that only a single sound is produced whenever a string is struck; or that there is in these [sounds only] one impact, since the air is struck just as many times as the vibrating string strikes it. But because the speeds [of the motions] of the sounds are joined together, no intervening gap is perceived by the ears. For [only] one sound, high or low, strikes the sense [of hearing], although each [pitched sound] consists of many [individual impacts][31].

---

29. Andrew Barker notes that «it is [this] theory of Archytas, modified by Plato (*Timaeus*, 67b) and again by Aristotle (*On the Soul*, II.8, 420a26-b4), that was adopted by the great majority of writers through the rest of antiquity» (Barker 2015, pp. 231-232).

30. Euclid (?), *Division of the Canon*, Prologue: «Since all pitches are generated when an impact has occurred, but an impact is impossible unless there has first been a motion — but some motions are more frequent (πυκνότεραι), others more sparse (ἀραιότεραι), and the more frequent ones make the pitches higher (ὀξυτέρους), while the sparser make them lower (βαρυτέρους) — it is necessary that the pitches be higher when they are composed of more frequent and numerous motions, and lower when they are composed from less frequent and less numerous motions». Greek text edited in Jan 1895, pp. 148:6-149:3. On the impact-frequency theory, its textual witnesses, and possible date of currency, see Barker 2015, pp. 232-238.

31. Gaffurius, *Theorica musice* (1492), Liber II, Cap. 1, f. c i-recto: «Sed neque quoties corda pellitur unum tantum edi putemus sonum: aut unam in his esse percussionem. cum toties aer feriatur quoties eum corda percusserit

On this view, it seems to me, there would simply be nothing for the ear to register as taking up more time in the way required by the lower-slower-louder theory, since even with the lowest pitches we still hear an apparently single and continuous sound; the ear, therefore, cannot be registering the individual constituent motions, even if they are considerably slower than those of higher pitches.

The speed-of-travel theory, on the other hand, is perhaps compatible with the lower-slower-louder theory. The latter, after all, assumes that slower speed of propagation does register on the ear, precisely as lower pitch. This makes it possible to suppose that the interaction of the ear with the traveling sound might be such that the latter's speed will also affect its perceptibility because what is slower takes longer to be perceived. I admit, though, that I am unable to imagine just how that might work, and unfortunately, neither the Pseudo-Aristotelian *Problems* nor Gaffurius, our only two sources for this idea, gives us any help at all on that score.

That leaves us with the conflict — for such it seems to be — between the majority view that Gaffurius follows in the *Theorica musice* and in the first chapter of the *Practica musice*, and the lower-slower-louder theory that he finds in the Pseudo-Aristotelian *Problems* and uses in the sixth chapter of the first book of the *Practica musice* to explain the consonance of the upper-voice fourths in 6/3 chords, which asserts the greater salience of lower pitches with the scarcely intelligible explanation that slower motions require more time to be heard and therefore afford a greater object of perception to the hearing.

The difficulties with that account, and its contradiction of the majority view that Gaffurius himself endorses elsewhere, indicate that he adopted it as a purely *ad hoc* way of explaining the musical fact at issue — the problem of the upper-voice fourth in the 6/3 chord — and that he either failed to see the inherent problems of that theory and its inconsistency with the majority view, or was simply unconcerned by them. And the reason for that may simply have been that, for Gaffurius, it was sufficient that the lower-slower-louder theory was, as he believed, approved by the same philosopher, Aristotle (as expounded by Themistius) who maintained the standard theory. It was not his, Gaffurius's, job to solve the attendant problems. It was simply a matter of being able to invoke the name of an appropriate, and suitably weighty, authority. In this, Gaffurius appears under the aspect of a disappointingly typical second-rate medieval thinker.

But that is not the whole story, for the difficulty he is trying to resolve — the apparent contradiction between the classification of the perfect fourth as a dissonance and the undeniable fact that in the 6/3 chord it is treated, and heard, as a consonance — is not one that theorists prior

tremebunda: Verum quoniam iunctae sunt sonorum uelocitates: nulla intracapedo auribus sentitur: unus namque sonus sensum pellit uel grauis uel acutus quanquam uterque ex pluribus constat». Boethius, *De institutione musica*, Liber I, Cap. 3 (Boethius 1867, p. 190:8-14): «Neque enim quotiens chorda pellitur, unus edi tantum putandus est sonus aut unam in his esse percussionem, sed totiens aer feritur, quotiens eum chorda tremebunda percusserit. Sed quoniam iunctae sunt velocitates sonorum, nulla intercapedo sentitur auribus et unus sonus sensum pellit vel gravis vel acutus, quamvis uterque ex pluribus constet».

to him had raised[32]. On the contrary, that there was a problem, an inconsistency between theory and practice, in need of resolution, was, to all appearances, an original observation of Gaffurius's own. In seeking a way to resolve it, he evidently recalled and made use of something he had read in Book XIX of the *Problems* of (as he thought) Aristotle. But the fact remains that in applying the conjecture he found there — the lower-slower-louder theory — to the conundrum of the perfect fourth in the 6/3 chord, Gaffurius was anticipating by centuries something that was later to become a characteristic feature, first of the branch of music theory concerned specifically with harmony, and then of research in music psychology and cognition: the attempt to provide a natural-scientific explanation of some specific aspect of musical practice.

This aim, as I mentioned at the outset, was quite unusual at this time, though Gaffurius was not the first to essay something of the kind. His precursor in this was another important theorist (and one he knew personally), Johannes Tinctoris[33]. It is a corollary of Aristotle's concept of enforced (that is, externally motivated) motion that, contrary to natural motions, enforced motions lose intensity as they approach their end[34]. Tinctoris, in his *Liber de arte contrapuncti* (Book on the Art of Counterpoint)[35], attributes the requirement that dissonances be short in duration and placed on (what we would call) weaker metrical positions to the supposed fact that such short dissonances, «when they are located on the last parts of notes, do not present themselves to the hearing as vehemently as [they would] if they were placed on [their] first parts», that is, on the metrically stronger parts of the pulse. This is because

> [...] musical sounds are produced by enforced motions. If, therefore, it is the nature of enforced motion that it lessens in force as it comes to its end, it follows that the second parts of notes are of a less vehement sound than the first parts[36].

---

32. As noted above, one anonymous theorist did observe what is essentially the same problem with respect to the 8/5 chord; see above, note 9.

33. See MILLER 1970, pp. 372, n. 33, and 375-378, and ATLAS 1985, pp. 76 and 80.

34. See, for example, Thomas Aquinas, *Commentaria in octo libros physicorum Aristotelis* (Commentaries on Aristotle's Eight Books on Physics), Liber V, Lectio 10, No. 7 (THOMAS AQUINAS 1884, p. 265): «Sed in his quae moventur per violentiam, accidit contrarium [naturalibus motionibus]: quia semper [motus violentus] invenitur remissior, quanto magis appropinquat ad quietem». (But in those things which are moved by compulsion, the opposite [to what happens in natural motions] occurs, for [a motion under compulsion] is always found to be less intense the closer it comes to its rest.)

35. Johannes Tinctoris, *Liber de arte contrapuncti* (1477): modern edition by Albert Seay in TINCTORIS 1978, English translation by Albert Seay in TINCTORIS 1961.

36. Johannes Tinctoris, *Liber de arte contrapuncti*, Liber II, Cap. 31 (TINCTORIS 1978, p. 140): «Quaequidem discordantiae parvae ita vehementer se non repraesentant auditui, quom supra ultimas partes notarum collocantur, ut si supra primas assumantur. Soni nempe musici violento motu fiunt. Unde si motus violentus eius naturae sit, ut circa finem remittatur, consequens est secundas partes notarum non tam vehementis soni esse quam primas».

There are evident difficulties with this explanation, too. It is interesting that Gaffurius did not adopt it to account for syncopated dissonances in suspensions, contenting himself with the assertion that such a dissonance is «hidden, inflicting no harm on the ears»[37], while an explanation closely related to Tinctoris's of the legitimacy of such syncopated dissonances is given by Gioseffo Zarlino in his *Istitutioni harmoniche* (Venice, 1558)[38].

It is, I would argue, no mere coincidence that it was a problem concerning the use of dissonance specifically that motivated each of these two highly unusual attempts to press natural science into service to explain — and thereby justify — a facet of musical practice. For dissonance itself, as I have argued elsewhere, was always a fundamental problem for Western polyphonic theory, from the time of its inception in the *Enchiriadis* treatises in the late ninth century up to the advent of post-tonal theory in the twentieth[39]. In the fifteenth century, it seems, only an apparent inconsistency in dissonance treatment was able to push theorists to seek such a natural-scientific explanation.

Yet ever since Jean-Philippe Rameau invoked the harmonic series of what he called the *corps sonore* to explain numerous aspects of tonal harmony, from the triad on up, such acoustical explanations have been commonplace in that area of music theory. And natural-scientific explanation — as we now understand it — of human musical behavior is quite simply and literally the standard research program in the fields of music psychology and music cognition today. Of all this, Gaffurius's attempt, feeble though it may be, to enlist Aristotle to explain an aberrant but accepted use of dissonance, is a distant precursor.

BIBLIOGRAPHY

ATLAS 1985

ATLAS, Allan W. *Music at the Aragonese Court of Naples*, Cambridge, Cambridge University Press, 1985.

BARKER 1989

BARKER, Andrew. *Greek Musical Writings. 2: Harmonic and Acoustic Theory*, Cambridge, Cambridge University Press, 1989.

BARKER 2015

ID. 'Musical Pitch and the Enigmatic Octave in Problemata 19', in: *The Aristotelian Problemata Physica: Philosophical and Scientific Investigations*, edited by Robert Mayhew, Leiden, Brill, 2015, pp. 226-254.

37. Gaffurius, *Practica musice* (1496), Liber III, Cap. 4, f. dd iii-verso: «Atque iccirco discordantia huiusmodi sincopata latet nullam auribus afferens lesionem».

38. Gioseffo Zarlino, *Le istitutioni harmoniche* (Venice, 1558), Libro III, Cap. 42, p. 197.

39. For this argument, see COHEN 1993.

Bietenholz – Deutscher 1985

Bietenholz, Peter G. – Deutscher, Thomas B. *Contemporaries of Erasmus: A Biographical Register of the Renaissance and Reformation. 1: A-E*, Toronto, University of Toronto Press, 1985.

Boethius 1867

Boethius, Anicius Manlius Torquatus Severinus. *De insititutione arithmetica libri duo, De institutione musica libri quinque*, edited by Gottfried Friedlein, Leipzig, Teubner, 1867.

Branca 1973

Branca, Vittore. 'Ermolao Barbaro and Late Quattrocento Venetian Humanism', in: *Renaissance Venice*, edited by John Rigby Hale, London, Faber and Faber, 1973, pp. 218-243.

Cohen 1993

Cohen, David E. 'Metaphysics, Ideology, Discipline: Consonance, Dissonance, and the Foundations of Western Polyphony', in: *Theoria*, vii (1997), pp. 1-85.

Diels – Kranz 1960

Diels, Hermann – Kranz, Walter. *Fragmente der Vorsokratiker*, Berlin, Weidmann, 91960.

Distelzweig 2013

Distelzweig, Peter M. 'The Intersection of the Mathematical and Natural Sciences. The Subordinate Sciences in Aristotle', in: *Apeiron*, xlvi/2 (2013), pp. 85-105.

Düring 1932

Düring, Ingmar. *Die Harmonielehre des Klaudios Ptolemaios; Porphyrios Kommentar zur Harmonielehre des Ptolemaios*, Göteborg, Elanders, 1932.

Gaffurius 1969

Gaffurius, Franchinus. *Musica practica*, translated with introduction and notes by Clement A. Miller, Rome, American Institute of Musicology, 1969.

Gaffurius 1993

Id. *The Theory of Music*, translated with introduction and notes by Walter Kreyszig, New Haven (CT), Yale University Press, 1993.

Huffman 2005

Huffman, Carl A. *Archytas of Tarentum, Pythagorean, Philosopher and Mathematician King*, Cambridge, Cambridge University Press, 2005.

Jan 1895

Musici scriptores graeci, edited by Karl von Jan, Leipzig, Teubner, 1895.

Livesey 1982

Livesey, Steven John. *Metabasis: The Interrelationship of the Sciences in Antiquity and the Middle Ages*, Ph.D. Diss., Los Angeles (CA), University of California, 1982.

McKirahan, Jr. 1978
McKirahan, Jr., Richard D. 'Aristotle's Subordinate Sciences', in: *The British Journal for the History of Science*, xi/3 (1978), pp. 197-220.

Miller 1968
Miller, Clement A. 'Gaffurius's *Practica musicae*. Origin and Contents', in: *Musica Disciplina*, xx (1968), pp. 105-128.

Miller 1970
Id. 'Early Gaffuriana. New Answers to Old Questions', in: *Musical Quarterly*, lvi/3 (1970), pp. 367-388.

Palisca 1985
Palisca, Claude V. *Humanism in Italian Renaissance Musical Thought*, New Haven (CT), Yale University Press, 1985.

Sachs 1974
Sachs, Klaus-Jürgen. *Der Contrapunctus im 14. und 15. Jahrhundert. Untersuchungen zum Terminus, zur Lehre und zu den Quellen*, Wiesbaden, Franz Steiner, 1974.

Siraisi 1970
Siraisi, Nancy G. 'The *Expositio Problematum Aristotelis* of Peter of Abano', in: *Isis*, lxi/3 (1970), pp. 321-339.

Themistius 1899
Themistius. *Themistii in libros Aristotelis de anima paraphrasis*, edited by Richard Heinze, Berlin, Preussische Akademie der Wissenschaften, G. Reimer, 1899.

Themistius 1996
Id. *On Aristotle on the Soul*, translated by Robert B. Todd, London, Duckworth, 1996.

Thomas Aquinas 1884
Thomas Aquinas, Saint. *Commentaria in octo libros physicorum Aristotelis*, Rome, Ex Typographia Polyglotta, 1884.

Tinctoris 1961
Tinctoris, Johannes. *The Art of Counterpoint (Liber de arte contrapuncti)*, translated by Albert Seay, Rome, American Institute of Musicology, 1961.

Tinctoris 1978
Id. *De arte contrapuncti*, edited by Albert Seay, Rome, American Institute of Musicology, 1978.

The (Long) Sixteenth Century

Science

Chapter Two

Teoria musicale e *patronage* nobiliare a Venezia nel Cinquecento

Paolo Alberto Rismondo[*]
(Venezia)

La prestigiosa posizione che la teoria musicale occupava all'inizio del Cinquecento, tra le discipline del quadrivio, nella considerazione della nobiltà veneziana più colta, era senza dubbio il riflesso di un peculiare clima politico e culturale. Ne *La Republica, e i magistrati di Vinegia* (Venezia, 1542) il nobiluomo veneziano Gasparo Contarini accostava l'ordinamento sociale della Serenissima — lì ordinato (con evidente semplificazione) in doge, nobiltà e popolazione comune — alle diverse funzioni attribuite agli intervalli musicali, sicché «nella consonantia del Diapason le voci estreme con quelle di mezzo del Diatessaron, & Diapente insieme s'accordano»[1].

Tracce ancor più evidenti di quella temperie si possono rilevare in alcune opere d'arte figurativa. Alcuni soggetti di dipinti di artisti veneziani di spicco operanti nei decenni a cavallo tra Quattro e Cinquecento sviluppano una complessa tematica, ricca di allusioni neoplatonico-ermetiche, la cui interpretazione è a tutt'oggi materia di dibattito tra gli studiosi. Ne sono esempi

*. Mi è doveroso ringraziare Bonnie Blackburn (Oxford), Linda Borean (Udine), Enrico Valseriati (Trento) e Agostino Ziino (Roma) per l'assistenza fornita. Per la stesura del presente articolo è stata necessario l'esame di materiale librario e archivistico di archivi e biblioteche veneziane e non; si ringrazia particolarmente il personale delle biblioteche Marciana (=I-Vnm) e Correr (=I-Vmc), e dell'archivio di Stato di Venezia (=I-Vas).

1. Gasparo Contarini, *La Republica, e i magistrati di Vinegia* (Venezia, 1544), f. xxxiii-recto.

Ill. 2.1ab: Tiziano, *Il baccanale degli Andrii* (olio su tela, 175 × 193 cm; 1523-1526), e dettaglio con il canone di Willaert. Madrid, Museo del Prado.

il dipinto di Giovanni Bellini, raffigurante *San Francesco in un paesaggio*, commissionato da un componente della nobile famiglia Michiel, «*Messer* Zuan Michiel»[2], oggi nella collezione Frick di New York; o la cosiddetta *Allegoria sacra* dello stesso autore, oggi agli Uffizi di Firenze[3]. In qualche caso sono inserite raffigurazioni di codici musicali — come nel telero di Vittore

[2]. Lavin 2013, p. 16.

[3]. Verdier 1952-1953.

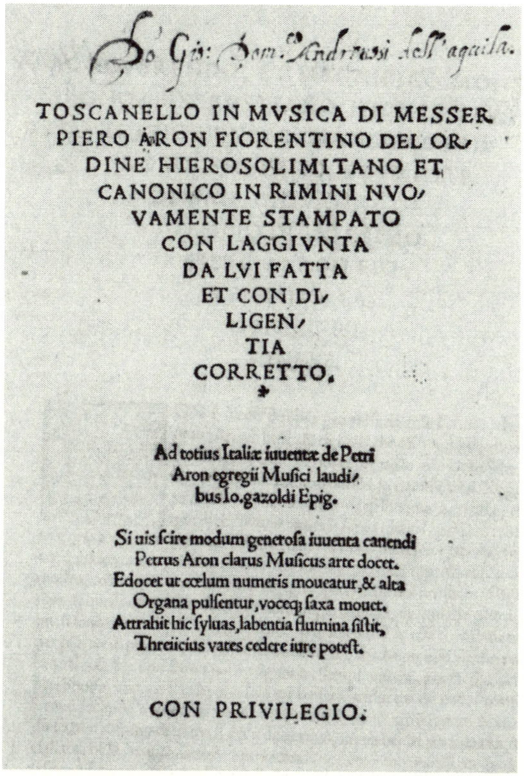
TOSCANELLO IN MVSICA DI MESSER PIERO ARON FIORENTINO DEL OR/DINE HIEROSOLIMITANO ET CANONICO IN RIMINI NVO/VAMENTE STAMPATO CON LAGGIVNTA DA LVI FATTA ET CON DI/LIGEN/TIA CORRETTO.

Ad totius Italiæ iuuentæ de Petri Aron egregii Muſici laudi/bus Io. gazoldi Epig.

Si uis ſcire modum generoſa iuuenta canendi
Petrus Aron clarus Muſicus arte docet.
Edocet ut cœlum numeris moueatur, & alta
Organa pulſentur, voceq; ſaxa mouet.
Attrahit hic ſyluas, labentia flumina ſiſtit,
Threiicius vates cedere iure poteſt.

CON PRIVILEGIO.

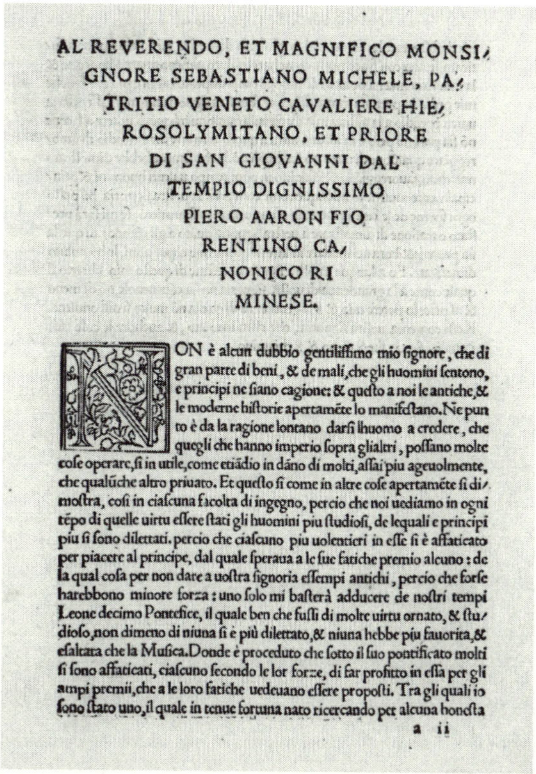
AL REVERENDO, ET MAGNIFICO MONSI/GNORE SEBASTIANO MICHELE, PA/TRITIO VENETO CAVALIERE HIE/ROSOLYMITANO, ET PRIORE DI SAN GIOVANNI DAL TEMPIO DIGNISSIMO PIERO AARON FIO/RENTINO CA/NONICO RI/MINESE.

NON è alcun dubbio gentiliſſimo mio ſignore, che di gran parte di beni, & de mali, che gli huomini ſentono, e principi ne ſiano cagione: & queſto a noi le antiche, & le moderne hiſtorie apertamēte lo manifeſtano. Ne punto è da la ragione lontano darſi lhuomo a credere, che quegli che hanno imperio ſopra glialtri, poſſano molte coſe operare, ſi in utile, come etiādio in dāno di molti, aſſai piu ageuolmente, che qualūche altro priuato. Et queſto ſi come in altre coſe apertamēte ſi di/moſtra, coſi in ciaſcuna facolta di ingegno, percio che noi uediamo in ogni tēpo di quelle uirtu eſſere ſtati gli huomini piu ſtudioſi, de lequali e principi piu ſi ſono dilettati. percio che ciaſcuno piu uolentieri in eſſe ſi è affaticato per piacere al principe, dal quale ſperaua a le ſue fatiche premio alcuno: de la qual coſa per non dare a uoſtra ſignoria eſſempi antichi, percio che forſe harebbono minore forza: uno ſolo mi baſterà addurre de noſtri tempi Leone decimo Pontefice, il quale ben che fuſſi di molte uirtu ornato, & ſtu/dioſo, non dimeno di niuna ſi è più dilettato, & niuna hebbe piu fauorita, & eſaltata che la Muſica. Donde è proceduto che ſotto il ſuo pontificato molti ſi ſono affaticati, ciaſcuno ſecondo le lor forze, di far profitto in eſſa per gli ampi premii, che a le loro fatiche uedeuano eſſere propoſti. Tra gli quali io ſono ſtato uno, il quale in tenue fortuna nato ricercando per alcuna honeſta

a ii

Ill. 2.2ab: Pietro Aaron, *Toscanello in musica* (Venezia, 1529), titolo e la dedica a Sebastiano Michiel. Copia US-Wc.

Carpaccio citato qui oltre — o brevi composizioni. Nel *Baccanale degli Andrii* di Tiziano, commissionato espressamente dal duca mantovano Alfonso i d'Este per la propria collezione personale, è inserito un canone *per tonos*, probabilmente opera di Adrian Willaert, la cui decifrazione richiede però considerevoli conoscenze teorico-musicali (Ill. 2.1)[4].

Evidentemente influenzati da questo clima culturale, alcuni nobili veneziani di spicco fornirono con continuità supporto di *patronage* ai teorici musicali operanti in città; e in particolare, nei primi decenni del Cinquecento, al gruppo di teorici e musicisti ben documentato nella nota corrispondenza Spataro - Aaron - Del Lago[5].

Nelle sue opere a stampa, Pietro Aaron (ca. 1480 - ca. 1550) si definiva «maestro di casa del Reverendo et magnifico cavaliere hierosolimitano messer Sebastiano Michele priore di Vinetia»; si dichiarava membro «del ordine hierosolimitano» nel frontespizio del *Toscanello in musica* nell'edizione 1529 (Ill. 2.2ab), ma non nel *Trattato della natura et cognitione di tutti gli tuoni di canto figurato* del 1525 (dedicato a Pietro Gritti, figlio naturale del doge Andrea

4. Lowinsky 1982.

5. Blackburn – Lowinsky – Miller 1991.

Gritti)[6]. Si può ragionevolmente presumere che l'ingresso del musicista in quell'ordine sia stato agevolato dal suo mecenate Sebastiano Michiel.

Il coevo cronista veneziano Marin Sanudo registrò con la sua consueta attenzione le presenze di questo nobiluomo durante numerose occasioni pubbliche dello stato veneziano[7], evidenziandone il ruolo di collegamento nella lega stretta tra Venezia, il Papato e Spagna, volta a contrastare l'avanzata della Francia nella val Padana. A suggellare quel suo ruolo, il 10 ottobre 1511 egli occupava una posizione di spicco nella solenne processione in piazza San Marco, ordinata dalla repubblica appunto per la «publicatione di la liga in questa terra»[8].

Sorprendentemente, data la sua importanza pubblica, Michiel non è identificabile in alcuno degli individui omonimi citati nelle genealogie nobiliari più accreditate; non è perciò possibile verificare il suo eventuale grado di parentela con il componente della stessa famiglia nobile, che aveva commissionato il citato dipinto di Bellini[9].

Ugualmente ben documentato è il mecenatismo artistico di Michiel. Ben riconoscibile nella sua veste caratteristica di priore dell'ordine gerosolimitano, è raffigurato in alcuni importanti dipinti della scuola veneta tra Quattro e Cinquecento: in quello di Giovanni Bellini, il cui soggetto è correntemente identificato come *Battesimo di Cristo con Sebastiano Michiel*, esposto tuttora nella chiesa del priorato dell'ordine gerosolimitano, San Giovanni del Tempio[10]; e, a pochi passi di distanza, in uno dei teleri dell'unico ciclo pittorico creato da Carpaccio rimasto *in situ* sino ad oggi, quello per la Scuola di San Giorgio degli Schiavoni[11]. Il ciclo è noto anche in ambito musicologico, perché in un altro telero, quello raffigurante *La visione di Sant'Agostino*, sono raffigurati alcuni codici musicali, dai quali è stato possibile trarre una trascrizione delle composizioni[12]. La datazione tradizionale del ciclo agli anni 1502-1507 è stata confermata dagli studiosi successivi, e collimerebbe perfettamente con il periodo di maggior auge e influenza di Michiel, del quale è stata suggerita l'azione mecenatesca relativamente a tutti i dipinti che lo compongono[13]. Considerando l'interesse del nobiluomo per la teoria musicale e la musica in generale, evidenziato poi dal mecenatismo nei confronti di Aaron, egli potrebbe essere stato

---

6. Due esemplari dell'edizione del *Toscanello in musica* comprensiva dell'aggiunta di 5 carte, datata 1529, sono in I-Vnm, con collocazioni D 227D 071.003-004 (legato con il *Trattato della natura et cognitione di tutti gli tuoni di canto figurato* di Aaron (Venezia, 1525) ed il *Tractato di musica* di Giovanni Spataro (Venezia, 1531)) e MUSICA 00097. Si veda inoltre Blackburn 2013.

7. Sanudo 1879-1903, vol. iii, coll. 404-405, 647; vol. iv, col. 682; vol. vi, col. 89; vol. xv, coll. 519-520; vol. xvi, col. 400, e qui oltre.

8. Sanudo 1879-1903, vol. xiii, col. 143.

9. I-Vas, Miscellanea codici, Storia veneta, ms. 21 («Genealogie Barbaro-Tasca», vol. v), pp. 79-132.

10. Modesti [2000], pp. 88-90.

11. Blackburn – Lowinsky – Miller 1991, pp. 1001-1002, e la bibliografia ivi citata.

12. Ludwig – Molmenti 1906, pp. 155-189 (a pp. 175-178 trascrizioni delle composizioni).

13. Perocco 1964, pp. 33-42, e Robuschi 2012, pp. 149-150, 379-380, 567.

l'ispiratore dell'inclusione della raffigurazione dei codici musicali nel telero raffigurante *La visione di Sant'Agostino*.

Le conoscenze teorico-musicali di Pietro Aaron erano notevoli per la sua epoca; originario di Firenze, ottenne un canonicato nel Duomo di Rimini, mentre non sono note le circostanze che lo spinsero a trasferirsi a Venezia[14]. Il suo sodale Giovanni Del Lago sembra fosse originario di Venezia; le testimonianze lo rivelano ben attestato nell'ambiente sociale ed ecclesiastico cittadino. Egli compilò per il nobiluomo veneziano Girolamo Molin un sunto della teoria musicale greco-antica[15]; Molin ebbe forse parte negli abboccamenti che, tra 1529 e 1531, il nobiluomo inglese Richard Croke intavolò a Venezia per conto di Enrico VIII d'Inghilterra per trovare giustificazione giuridica alla risoluzione del matrimonio del re con Caterina d'Aragona[16].

Del Lago dedicò ad un altro nobiluomo veneziano, Lorenzo Morosini, il suo unico, breve e poco significativo trattato a stampa, *Breve introduttione di musica misurata* (1540)[17]. La citata corrispondenza, e particolarmente il sunto da lui preparato per Molin, testimonia anche delle sue conoscenze teorico-musicali, discretamente ampie e variate sia nel campo della teoria musicale greco-antica, sia per quanto riguarda le conoscenze di teorici musicali a lui immediatamente precedenti[18].

Alla luce di queste testimonianze, si può anche ragionevolmente supporre che la disciplina teorico-musicale godesse in questi decenni di una piena compatibilità nel consueto corso di studi dei giovani nobili veneziani. In chiusura delle sue lettere ad Aaron, Giovanni Spataro scriveva spesso di comunicare le sue raccomandazioni ai «lucidi de virtù et nobilissimi figlioli» di Michiel, dei quali perciò si può supporre Aaron fosse precettore[19].

Tanto questi quanto Del Lago, comunque, poterono accedere soltanto a cariche e benefici ecclesiastici relativamente poco importanti. Aaron, alla morte del suo mecenate, si trasferì in un convento bergamasco; Del Lago aveva ottenuto una prelatura in una chiesa veneziana minore ma centrale, quella di Santa Sofia, e nelle sue immediate vicinanze risiedette sino alla morte[20].

14. Blackburn – Lowinsky – Miller 1991, pp. 74-100.

15. *Ibidem*, pp. 897-907 e 1002-1003, e Tomasi 2011.

16. Brewer 1876, numeri 6491, 6501, 6540, e 6786. Il personaggio vi è citato come «Hierome Molyns, a Venetian», e perciò l'identificazione non è assolutamente certa.

17. Giovanni Del Lago, *Breve introduttione di musica misurata* (Venezia, 1540; esemplare I-Vlevi, LEVI CF.E.24). Per Lorenzo Morosini, si veda I-Vas, Miscellanea codici, Storia veneta, ms. 21 («Genealogie Barbaro-Tasca», vol. v), p. 295.

18. Blackburn – Lowinsky – Miller 1991, pp. 143-182.

19. Giovanni Spataro a Pietro Aaron, 19 giugno, 24 ottobre (due lettere), 27 novembre 1531, 30 gennaio, 13 marzo, 5 aprile 1632 (*ibidem*, lettere 33-39, pp. 433-469).

20. *Ibidem*, pp. 127-129. Si veda anche Francesco di Pizoni a Giovanni del Lago, 2 giugno 1537, indirizzata «Al molto mio reverendo Messer Pre Zuane da Lago Diacono de Santta Sophia, [...] In Venetia in la contra[da] de Sancta Sophia in cale da le velle» (*ibidem*, lettera 91, pp. 872-873). La calle delle Vele, tuttora così denominata, è posta a fianco della chiesa di Santa Sofia.

Francesco Zorzi e l'ambiente politico e culturale veneziano del suo tempo

Tra i personaggi consultati da Croke durante le trattative citate poco sopra, uno dei più notevoli, per la notorietà e l'influenza che esercitò sul panorama culturale e sociale veneziano tra Quattro e Cinquecento, era senza dubbio Francesco Zorzi (Giorgi; 1466-1540), frate minore francescano, e discendente dall'antica omonima nobile famiglia veneziana[21]. Il suo nome fu, a più riprese, proposto per il patriarcato veneziano; l'abbondante documentazione coeva che di lui si è conservata testimonia della sua ampia cultura (fu incaricato dell'insegnamento della sacra scrittura e della lingua ebraica nel suo convento), della sua vicinanza con i più elevati livelli governativi veneziani, e delle sue missioni e dei segreti interventi diplomatici affidatigli dalla Serenissima nel primo decennio del Cinquecento. A Venezia, dall'inizio del secolo sino al 1530 circa, continuò a godere di indiscutibile fama e autorità grazie al prestigio conferitogli dalle sue opere, pubblicate in ricca veste tipografica, e dalle sue appassionate prediche nella chiesa statale di San Marco[22].

Seguendo i mutevoli andamenti della politica veneziana del tempo, Zorzi ebbe frequenti rapporti con gli ambasciatori francesi impegnati, tra terzo e quarto decennio del Cinquecento, a contrastare i disegni imperiali di dominio dell'Italia. Tramite loro egli poté venire a contatto con ambienti dell'evangelismo italiano (particolarmente del dissenso religioso veneto) e transalpino. Tra i più importanti esponenti di questi ultimi vi era Jacobus Faber Stapulensis (Jacques Lefèvre d'Étaples), della cui opera come teorico musicale si tratterà qui oltre; questi, peraltro, risiedette per breve tempo a Venezia e Padova, negli anni a cavallo tra Cinque e Seicento[23].

L'ambasciatore inglese a Venezia Giovanni Battista Casali agì da mediatore tra il citato agente inglese Richard Croke, e Zorzi. Quest'ultimo consultò a sua volta le sue conoscenze tra i dotti ebrei veneziani; ma incontrò ben presto forti resistenze, si scontrò duramente con il nunzio papale, e il suo tentativo non ebbe successo[24].

Se è probabile che tra gli eruditi contattati da Zorzi nella citata missione diplomatica vi fosse anche Aaron, è certo che Casali fu in contatto con il predetto circolo di musicisti e teorici, dove è spesso citato[25].

Già in anni precedenti la Repubblica aveva incentivato le sue relazioni con la corte inglese e particolarmente con il re Enrico VIII, facendo abilmente leva sulla sua passione musicale. Vi aveva inviato infatti l'organista marciano Dionisio Memmo (alla corte inglese dal 1516 al 1519

---

21. Per notizie biografiche su Zorzi, si veda Zorzi 2010, pp. [xi]-xxiv, e letteratura precedente ivi citata.

22. Vasoli 1974, pp. 168-173, 180, e 186-188.

23. Alonge 2019, *ad indicem*, e Gallo 1981, pp. 306-307.

24. Vasoli 1974, pp. 180-209; anche Brewer 1876, numeri 6372, 6412, 6413, 6425, e 6472.

25. Blackburn – Lowinsky – Miller 1991, pp. 984-985.

FRANCISCI GEORGII
VENETI
MINORITANAE FAMILIAE
DE HARMONIA
MVNDI TOTIVS
CANTICA
T R I A

Talia probarint, spiritus quibus spirat.

Prohibitus.

Donum S. D. H. Alexandri VII.

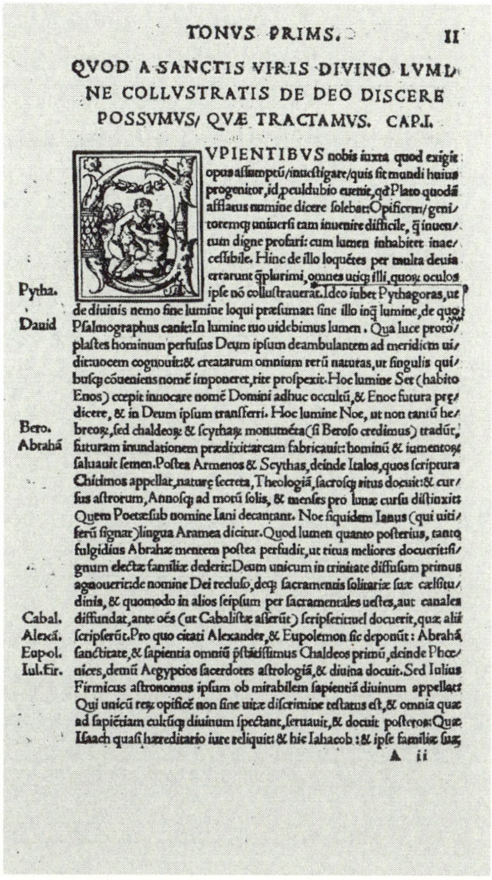

TONVS PRIMS. II

QVOD A SANCTIS VIRIS DIVINO LVMINE COLLVSTRATIS DE DEO DISCERE POSSVMVS, QVÆ TRACTAMVS. CAP. I.

ILL. 2.3AB: Franesco Zorzi, *De harmonia mundi totius cantica tria* (Venezia, 1525), titolo e f. II, con «Tonus prim[u]s». Nota l'iscrizioni manoscritte «Prohibitus» e «Donum S. D. H. Alexandri VII». Dalla collezione del Papa Alessandro VII (papa dal 1655 al 1667). Copia I-Ru.

almeno), che riscosse notevole successo presso il re, ma venne anche adoperato come informatore dall'ambasciatore veneziano alla corte inglese Sebastiano Giustinian. La sfortunata avventura del musicista (clavicembalista) Giovanni da Lezze (Da Legge) — anch'egli tra i corrispondenti del citato carteggio Spataro – Aaron – Del Lago — alla corte inglese, prima del 1525, è forse da considerarsi nel medesimo contesto[26].

Considerato tutto ciò, è evidente che l'opera letteraria di Zorzi dovette esercitare profonda influenza sul clima culturale veneziano a lui contemporaneo.

La sua opera più importante, *De harmonia mundi totius cantica tria* (Tre cantici sull'armonia dell'universo), venne edita a Venezia nel 1525 (ILL. 2.3A). Suscitò grande interesse anche in Francia, dove apparve nell'originale latino già nel 1545, e poi fu tradotta da Guy Lefèvre de la Boderie e pubblicata a Parigi nel 1578 sotto il titolo *Harmonie du monde divisée*

[26]. BOLCATO 2009; BLACKBURN – LOWINSKY – MILLER 1991, pp. 988-990.

en trois cantiques (L'armonia del mondo divisa in tre cantici). Le sezioni delle tre cantiche nelle quali è ordinata sono numerate secondo i vari toni del canto liturgico, «TONVS PRIMVS» «TONVS SECVNDVS», e così via (Ill. 2.3b). Vari passi testimoniano la peculiare visione della teoria musicale del suo autore, derivante dalla tradizione pitagorico-platonica; essa associava ad alcune consonanze perfette ottenute con procedimento razionale (cioè da rapporti tra numeri interi) un particolare significato mistico ed esoterico[27].

Ciò trova piena conferma nel progetto di Zorzi per la chiesa veneziana di San Francesco della Vigna, annessa al convento nel quale egli stesso risiedeva, da lui stesso nel 1535 su richiesta del doge Andrea Gritti; esso è basato su poche e semplicissime proporzioni, per le quali Zorzi indica il corrispettivo intervallo musicale, esaltandone profusamente le occulte proprietà[28]:

larghezza	9 (= 3^2)	riducibile a 3
Lunghezza	27 (= 9 x 3 = 3^3)	riducibile a 9
lunghezza presbiterio	9	riducibile a 3
larghezza presbiterio	6 (= 3 x 2)	riducibile a 2
larghezza cappelle laterali	3	riducibile a 1
altezza	12 (= 3 x 4)	riducibile a 4

La vicendevole combinazione di questi semplici rapporti può dar luogo a proporzioni di una certa complessità, ma nell'insieme la proposta di Zorzi sembra essere poco più che una meccanica riflessione della suggestiva idea di 'armonia del mondo', proclamata sin dal titolo della sua già citata opera a stampa. D'altro canto può invece apparire strano che Zorzi non abbia voluto trarre profitto dalle proporzioni ben più raffinate, e ben altrimenti armoniose, che originavano dall'uso di semplici costruzioni geometriche, note ed applicate già da secoli dagli architetti delle cattedrali della Francia del Nord (Laon, Reims) e, successivamente, dell'area germanofona dell'Europa (Strasburgo, Colonia, Friburgo), testimoniate dai disegni originali conservatisi sino ad oggi, e recentemente valorizzati[29]. È evidente il grande vantaggio pratico apportato dall'utilizzo di simili operazioni in architettura, in quanto risolveva contemporaneamente due problemi: ottenere le desiderate proporzioni (razionali o irrazionali che fossero) tra le parti, e riportarle con facilità 'in cantiere', alla scala voluta, ripetendo le stesse operazioni mediante l'uso di funi, pesi a piombo e paletti[30].

---

27. Edizione moderna in Zorzi 2010.

28. Foscari – Tafuri 1983; qualche interessante aggiornamento in Onda 2008, pp. 65-78.

29. Bork 2016.

30. *Ibidem*, p. 9, nota 13.

Ma è altrettanto evidente che la soluzione proposta da Zorzi si accordava meglio con la concezione della natura dei rapporti proporzionali da lui propugnata. Si può attribuire anche a questa influenza, se i teorici musicali veneziani successivi opposero una fortissima resistenza all'uso di costruzioni geometriche per la rilevazione e divisione degli intervalli. Inoltre, l'uso di strumentazione meccanica (ivi compresi, naturalmente, squadra e compasso) ripugnava alla nobiltà, veneziana e no.

I mercanti operanti da un certo tempo in città, ma provenienti da famiglie non autoctone, erano certo abilitati ad accedere alla nobiltà per il loro censo e ricchezza; ma la loro attività mercantile prevalente li portava a dover talvolta manipolare e misurare le loro merci, cosicché le loro richieste di aggregazione alla nobiltà correvano il rischio d'esser invalidate dal «pregiudizio meccanico» che colpiva persino quelle semplici operazioni. Ciò è attestato dalla documentazione originale, costituita dai 'processi di nobiltà', istituiti a Venezia, e nelle varie città sottoposte al suo dominio dalle amministrazioni cittadine[31]. Su territorio veneziano, una trattazione in merito fu stesa da un umanista bresciano ben noto per i suoi interessi teorico-musicali, Carlo Valgulio[32].

L'operazione geometrica relativa alla divisione di un qualsiasi intervallo — e quindi anche del tono intero maggiore di proporzione 9:8 — in due parti uguali era già stata esposta dal poligrafo francese Jacobus Faber Stapulensis nel suo trattato *Musica libris demonstrata quattuor* (La musica esposta in quattro libri), a stampa edito nel 1496, seguito poco dopo da altri teorici[33]. D'altra parte, le operazioni geometriche necessarie per eseguire l'operazione erano indicate nelle opere, ben note sin dall'antichità e poi per tutto il Medioevo sino ad Aaron e ai suo sodali, di Aristide Quintiliano e Tolomeo[34]. Il mesolabio, lo strumento più tardi proposto da Gioseffo Zarlino per ottenere il medio geometrico degli intervalli, era basato su un'operazione già esposta nell'antologia matematica di Pappo d'Alessandria[35].

Evidentemente, perciò, il rifiuto dell'uso di operazioni geometriche e della strumentazione relativa era una precisa scelta dei teorici musicali che la operavano; la continuazione delle loro attività dipendeva, in ultima analisi, dalla considerazione nella quale erano tenuti dal ceto che forniva loro adeguato sostegno di *patronage* — la nobiltà veneziana —, e perciò dovevano tener

---

[31]. Per Venezia si rimanda alla ricchissima, ma poco studiata documentazione nel fondo I-Vas, Avogaria di Comun, le serie Prove di Nobiltà e Processi per Nobiltà, (esempi significativi in b. 283, fasc. «Arrigoni/1625»). Per Padova e altre città Ventura 1997, pp. 205-224; per Brescia Valseriati 2017.

[32]. Valseriati 2017, pp. 197-207. Sull'importanza di Carlo Valgulio nell'ambito della teoria musicale, si veda Palisca 1985, pp. 88-110, e Gallo 1981, p. 313.

[33]. Jacobus Faber Stapulensis, *Musica libris demonstrata quattuor* (Parigi, 1496), p. g4[5]. Si veda Lindley 1987, pp. 140-143.

[34]. Hentschel 1998, Burnett 1986, e Barker 1989, pp. 319-320 e 498-499.

[35]. Garcia Pérez 2002, Pappus Alexandrinus 1876, e Barbieri 1996.

conto della mentalità connaturata a quel ceto.

Il purismo architettonico e teorico-musicale che informa la soluzione proposta da Zorzi aveva perciò due motivazioni, sul piano teorico-musicale e sul piano sociale, strettamente collegate tra loro.

L'Accademia Venetiana
e i suoi progetti editoriali in campo teorico-musicale

L''Accademia Venetiana o della Fama' venne fondata attorno al 1558. Il pieno appoggio concessole inizialmente dalle più importanti istituzioni consiliari dello stato veneziano — affidate sempre, com'è noto, a nobiluomini veneziani di spicco —, è documentato da una nutrita serie di atti ufficiali: la supplica presentata dal suo promotore, il nobiluomo Federico Badoer (1519-1593), alla magistratura statale dei Procuratori «de Supra» (i nobiluomini che sovrintendevano alla gestione della cappella musicale marciana, ma anche alla gestione di tutti gli edifici pubblici della piazza di S. Marco); la relativa concessione dell'uso del «vestibulo [della biblioteca Marciana] nel qual possa farsi le congregationi di tanti litterati huomeni con quei virtuosi modi che sonno sin hora stati fatti»[36]; gli statuti, corredati di altra documentazione atta a fornire giustificazione alla richiesta di appoggio da parte statale[37]; e il privilegio di stampa concesso dal doge Lorenzo Priuli, della durata di quindici anni[38].

Il sodalizio fu una delle pochissime accademie veneziane a prevedere sin dalla sua fondazione il pubblico insegnamento, e durante il breve arco della sua attività poté anche produrre una notevole quantità di documenti manoscritti ed a stampa, qualche edizione di opere dei suoi membri più notevoli, nonché gli statuti e un elenco dei membri e degli insegnanti[39]. Alcune di queste pubblicazioni contengono a loro volta elenchi di opere che l'Accademia Venetiana progettava di pubblicare, tra le quali sono presenti opere teorico-musicali greco-antiche e moderne (Ill. 2.4-5). Per le prime è stata giustamente notata l'evidente connessione con manoscritti presenti nella veneziana biblioteca Marciana, e ancor oggi lì conservati; tra le seconde, spicca significativamente quella di Faber Stapulensis. Una plausibile identificazione di

[36]. I-Vas, Procuratori di S. Marco, Procuratia «de supra», Chiesa, Registri, 129, s.d., ff. 34r-37r, e relativa delibera dei procuratori del 12 luglio [1560]. Si veda Fenlon 2002, nota 25.

[37]. I-Vnm, Ms. vii, 2535 (=12407), pp. 303-334. I fogli sono stati probabilmente estratti da qualche filza del senato o di altra magistratura dello stato veneziano.

[38]. I-Vnm, Ms. vii, 2535 (=12407), pp. 314-315 (4 ottobre 1558).

[39]. Un'edizione moderna complessivamente accurata di alcuni testi dell'accademia, concernenti la sua stessa attività, è in Pellegrini 1808. Proposte per l'identificazione di alcune opere effettivamente edite a cura dell'accademia sono in Guarna 2018, pp. 207-268 (nelle note al testo) e 269-273.

SOMMA DELLE OPERE
CHE IN TVTTE LE SCIENZE
ET ARTI PIV NOBILI, ET IN VARIE LINGVE
HA DA MANDARE IN LVCE
l'Academia Venetiana,
PARTE NVOVE, ET NON PIV STAMPATE,
parte con fedelissime tradottioni, giudiciose correttioni,
& utilissime annotationi riformate.

NELL'ACADEMIA VENETIANA,
M. D. LVIII.

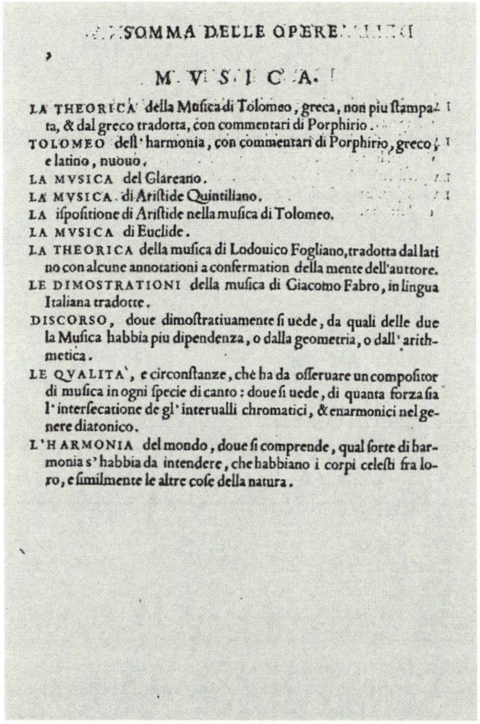

SOMMA DELLE OPERE

MVSICA.

LA THEORICA della Musica di Tolomeo, greca, non piu stampata, & dal greco tradotta, con commentari di Porphirio.

TOLOMEO dell'harmonia, con commentari di Porphirio, greco, e latino, nuouo.

LA MVSICA del Glareano.

LA MVSICA di Aristide Quintiliano.

LA ispositione di Aristide nella musica di Tolomeo.

LA MVSICA di Euclide.

LA THEORICA della musica di Lodouico Fogliano, tradotta dal latino con alcune annotationi a confermation della mente dell'auttore.

LE DIMOSTRATIONI della musica di Giacomo Fabro, in lingua Italiana tradotte.

DISCORSO, doue dimostratiuamente si uede, da quali delle due la Musica habbia piu dipendenza, o dalla geometria, o dall'arithmetica.

LE QVALITA', e circonstanze, chè ha da osseruare un compositor di musica in ogni specie di canto: doue si uede, di quanta forza sia l'intersecatione de gl'interualli chromatici, & enarmonici nel genere diatonico.

L'HARMONIA del mondo, doue si comprende, qual sorte di harmonia s'habbia da intendere, che habbiano i corpi celesti fra loro, e similmente le altre cose della natura.

Ill. 2.4ab: *Somma delle opere che in tutte le scienze ed arti più nobili, et in varie lingue, ha da mandare in luce l'Academia Venetiana*, Venezia, «nell'Academia Venetiana», 1558. Titolo e f. G[ii]-verso. Copia A-Wn.

molte delle opere elencate è stata proposta altrove[40]; le due seguenti però restano a tutt'oggi non identificate, probabilmente perché erano ancora manoscritte, e tali evidentemente restarono[41]:

> DISCORSO, dove dimostrativamente si vede, da quali delle due la Musica abbia piu dipendenza, o dalla geometria, o dall'arithmetica.
>
> *Disputatio in qua per demonstrationem ostenditur, an Musice à Geometria, uel Arithmetica pendeat.*

[40]. Si veda l'elenco in Fenlon 2002, pp. 128-131, numeri [1], [2]-[6]; le opere teoriche greco-antiche sono identificate a p. 131. La biblioteca Marciana possiede la copia manoscritta più antica pervenuta sino ad oggi dell'opera di Aristide Quintiliano, *Sulla musica*: I-Vnm, MS gr. app. vi 10 (RISM B/xi, n. 273), databile alla fine del dodicesimo secolo; si veda Aristide Quintiliano 1963, pp. viii-xi. Il numero [2], Porfirio, *Commentario su Tolomeo*, è I-Vnm, MS gr. 321 (coll. 894), ff. 99r-131r (RISM B/xi, n. 263).

[41]. Fenlon 2002, p. 129, ai punti [9] e [10]. Si è però preferito citare dalla *Somma delle opere che in tutte le scienze et arti più nobili, et in varie lingue ha da mandare in luce l'Academia Venetiana* ([Venezia], 1558; esemplare I-Vnm, Misc. 1232/2, edizione in Pellegrini 1808/xxiii, pp. 61-64; Ill. 4ab) e dalla *Summa librorum, quos in omnibus scientiis, ac nobilioribus artibus,* [...] *in lucem emittet Academia Veneta* ([Venezia], 1559; esemplare I-Vnm, Misc. 1224/2; Pellegrini 1808/xxii; Ill. 5ab).

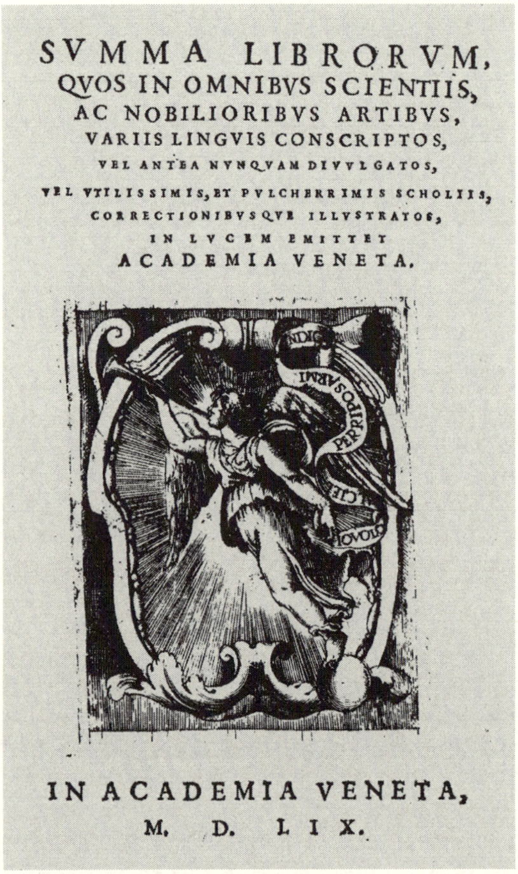

SVMMA LIBRORVM,
QVOS IN OMNIBVS SCIENTIIS,
AC NOBILIORIBVS ARTIBVS,
VARIIS LINGVIS CONSCRIPTOS,
VEL ANTEA NVNQVAM DIVVLGATOS,
VEL VTILISSIMIS, ET PVLCHERRIMIS SCHOLIIS,
CORRECTIONIBVSQVE ILLVSTRATOS,
IN LVCEM EMITTET
ACADEMIA VENETA.

IN ACADEMIA VENETA,
M. D. LIX.

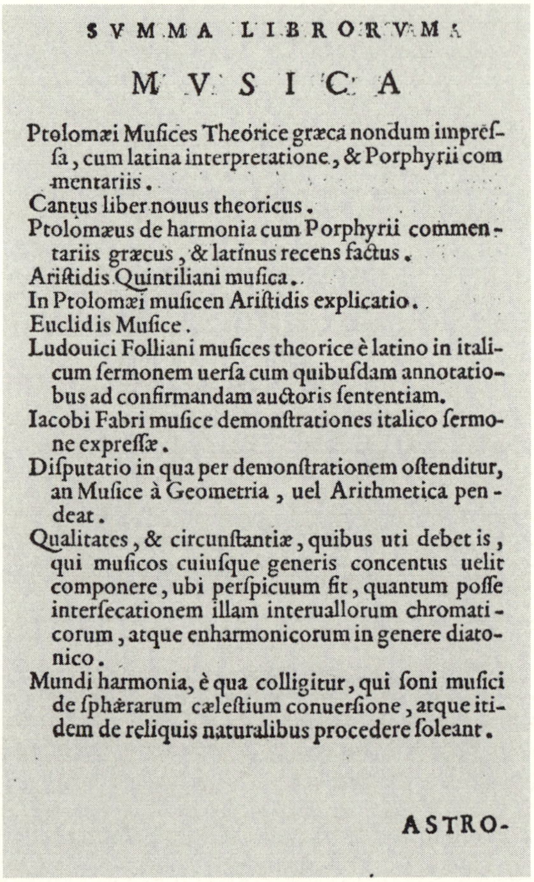

SVMMA LIBRORVM
MVSICA

Ptolomæi Musices Theorice græca nondum impressa, cum latina interpretatione, & Porphyrii commentariis.
Cantus liber nouus theoricus.
Ptolomæus de harmonia cum Porphyrii commentariis græcus, & latinus recens factus.
Aristidis Quintiliani musica.
In Ptolomæi musicen Aristidis explicatio.
Euclidis Musice.
Ludouici Folliani musices theorice è latino in italicum sermonem uersa cum quibusdam annotationibus ad confirmandam auctoris sententiam.
Iacobi Fabri musice demonstrationes italico sermone expressæ.
Disputatio in qua per demonstrationem ostenditur, an Musice à Geometria, uel Arithmetica pendeat.
Qualitates, & circunstantiæ, quibus uti debet is, qui musicos cuiusque generis concentus uelit componere, ubi perspicuum sit, quantum posse intersecationem illam interuallorum chromaticorum, atque enharmonicorum in genere diatonico.
Mundi harmonia, è qua colligitur, qui soni musici de sphærarum cælestium conuersione, atque itidem de reliquis naturalibus procedere soleant.

ASTRO-

Ill. 2.5ab: *Summa librorum quos in omnibus scientiis, ac nobilioribus artibus variis linguis conscriptos* [...] *in lucem emittit Academia Veneta*, Venezia, «in Academia Veneta», 1559. Titolo e f. D[i]-verso. Copia D-Mbs.

> LE QVALITÀ, e circostanze, che ha da osseruare un compositor di musica in ogni specie di canto: doue si uede, di quanta forza sia l'intersecatione de gl'intervalli chromatici, & enarmonici nel genere diatonico.
>
> *Qualitates, & circonstantiae, quibus uti debet is, qui musicos cuiusque generis concentus uelit componere, ubi perspicuum sit, quantum posse intersecationem illam interuallorum chromaticorum, atque enharmonicorum in genere diatonico.*

La prima è particolarmente interessante per la sua evidente connessione con un problema che preoccupava da secoli i teorici musicali, cioè la divisione degli intervalli (e particolarmente del tono intero maggiore di proporzione 9:8) in due parti proporzionalmente uguali, operazione detestata dalla corrente teorico-musicale 'neopitagorica'. La seconda è invece un evidente tentativo di sistemazione teorica della tendenza da tempo già operante — la si può constatare già nelle composizioni di Cipriano de Rore, Orlando di Lasso, e altri contemporanei —, ad interpretare il genere cromatico nell'ambito del temperamento mesotonico allora comunemente praticato. Si può soltanto supporre che avesse qualche relazione con l'opera teorica di Nicola Vicentino, e alla commistione dei *genera* (così com'erano intesi dalla teoria musicale coeva), suggerita da teorici successivi[42].

TESTIMONIANZE IN INVENTARI DI BIBLIOTECHE NOBILIARI VENEZIANE

Nel citato elenco dei trattati teorici moderni che l'Accademia progettava di pubblicare, è particolarmente interessante l'inclusione di quello di Faber Stapulensis[43]. Esso era peraltro già noto ad Aaron ed al suo circolo, come testimoniano tre lettere del citato carteggio, intercorse con l'erudito e musicista padovano Lorenzo Gazio[44].

La notevole diffusione dell'opera di Faber Stapulensis nell'ambiente veneziano è attestata dalla sua presenza nei fondi librari antichi di biblioteche cittadine; la biblioteca Marciana conserva tuttora varie copie dell'opera, sia nella più rara edizione originale, sia in altre edizioni più tarde[45].

Si trova citata nell'inventario della biblioteca del botanico tedesco Melchior Wieland (Melchiorre Guilandini), prefetto dell'Orto botanico di Padova e, dal 1567, professore di botanica nella locale università[46]. Questi era in stretto contatto con istituzioni e rappresentanti di spicco della Serenissima, tanto che per volontà testamentaria lasciò l'intera sua biblioteca allo stato veneziano. Alla sua morte fu perciò necessario redigerne l'inventario[47]. In esso è presente la citata opera di Faber Stapulensis («Fabri musica»), assieme a «Euclides musica» (una delle due opere già ascritte ad Euclide riportate nell'elenco di volumi da pubblicarsi dall'Accademia Venetiana), e all'«Institutio del Zarlino» (il noto trattato zarliniano[48]).

---

42. LEVENBERG 2016.

43. OOSTERHOFF 2018, pp. 152-161.

44. Giovanni Del Lago a Girolamo Molino, senza data (BLACKBURN – LOWINSKY – MILLER 1991, lettera 96, pp. 897-907). La tabella inserita è evidentemente copiata dal lavoro di Faber Stapulensis, compresi alcuni errori; si veda p. 906, nota 36. Pietro Aaron a [Lorenzo Gazio], 14 febbraio 1534 (*ibidem*, lettera 101, pp. 943-944); Gazio a [Pietro Aaron], 27 febbraio 1534 (lettera 102, pp. 945-948), dove sono notati errori nel trattato di Faber Stapulensis.

45. Due copie della prima edizione sono I-Vnm, INC. 0515, e I-Vmc, INC. F 029. Un esemplare della successiva edizione stampata a Parigi, «Henricus Stephanus suo grauissimo labore et sumptu, 1514 [...] die septima Septembris» è I-Vnm, RARI 0381.

46. HERRMANN 2015.

47. I-Vas, Miscellanea di carte non appartenenti ad alcun archivio, busta 15, fascicolo «Riformatori dello Studio di Padova, Inventario dei libri lasciati alla Repubblica da Melchiore Guilendini», datato 2 gennaio 1590.

48. FENLON 2002, p. 131, numeri [4]-[5], e Gioseffo Zarlino, *Le istitutioni harmoniche* (Venezia, 1558).

Alla fine del Seicento il nobiluomo veneziano, e cardinale Gregorio Barbarigo (1625-1697) possedeva ancora nella sua ricchissima biblioteca numerose opere teorico-musicali; si trattava certamente di materiali ereditati da componenti precedenti della stessa famiglia, o da altre famiglie nobili alle quali si era legata. L'inventario stilato alla sua morte elenca appunto il volume di Faber Stapulensis nell'edizione parigina del 1552 (mutilo del quarto libro), «antiquae musices auctores septem» (ovvero la ben nota edizione a cura di Marcus Meibom del 1652), «L'arte del contrapunto» (forse da identificarsi con l'omonima opera di Artusi), il citato volume di Zorzi, *De harmonia mundi totius*, e alcune opere a stampa, difficilmente identificabili, di Giovanni Battista Doni[49].

È stata suggerita la forte influenza del breve trattato di Faber Stapulensis sulla teoria musicale successiva in generale[50]. Le citazioni nell'indice dell'Accademia Venetiana e nella documentazione veneziana, e la presenza di alcuni esemplari nelle biblioteche pubbliche attuali (almeno in parte certamente provenienti da fondi librari nobiliari veneziani) testimoniano l'interesse che suscitò nell'ambiente culturale cittadino.

Successivamente, le opere teorico-musicali citate continuarono ad attrarre attenzione quasi soltanto nelle cerchie dei teorici musicali stessi, e per il loro proprio interesse personale. Assai significativa la nota di possesso, probabilmente autografa, dell'esemplare della prima edizione dell'opera di Lefévre conservato oggi presso la biblioteca civica di Treviso, secondo la quale il volume era posseduto «Ad solacium Baccalaureatis Zaccariae Taevi musices magistri», ossia Zaccaria Tevo, il noto teorico musicale attivo a Treviso tra Sei e Settecento[51].

Nicola Vicentino a Venezia

La presenza continuativa di Nicola Vicentino (1511-1576) a Venezia, tra la fine del 1560 ed inizio dell'anno successivo, — pochi mesi prima del fallimento dell'impresa dell'Accademia Venetiana — è testimoniata dalle sue quattro lettere inviate al vescovo di Arras, poi cardinale, Antoine Perrenot de Granvelle[52]. Il contatto tra Vicentino e l'ecclesiastico poté forse attivarsi grazie alle relazioni ben note che questi intratteneva con Willaert, anche mediante il nobiluomo veneziano Giovanni Francesco Dolfin[53].

---

49. Lovato 1999, pp. 557-558. Le edizioni sono identificabili in Marcus Meibom, *Antiquae musicae auctores septem, Graece et Latine* (Amsterdam, 1652; due esemplari in I-Vbnm, MUSICA 00383, e D 092D 136) e Giovanni Maria Artusi, *L'arte del contraponto* (Venezia, 1586, o una delle successive edizioni).

50. Barbour 1951, p. 139.

51. Collocazione I-TVco, ii 098.00G.13837.

52. Le quattro lettere, datate 14 settembre e 27 dicembre 1560, e 4 gennaio e 8 marzo 1561, sono in E-Mp, MS ii/2274, ff. 161r (1) e 207r (1), MS ii/2275, ff. 13r (1) e 212r (1). Si veda anche Bossuyt 1998, p. 442.

53. Bossuyt 1998, p. 442.

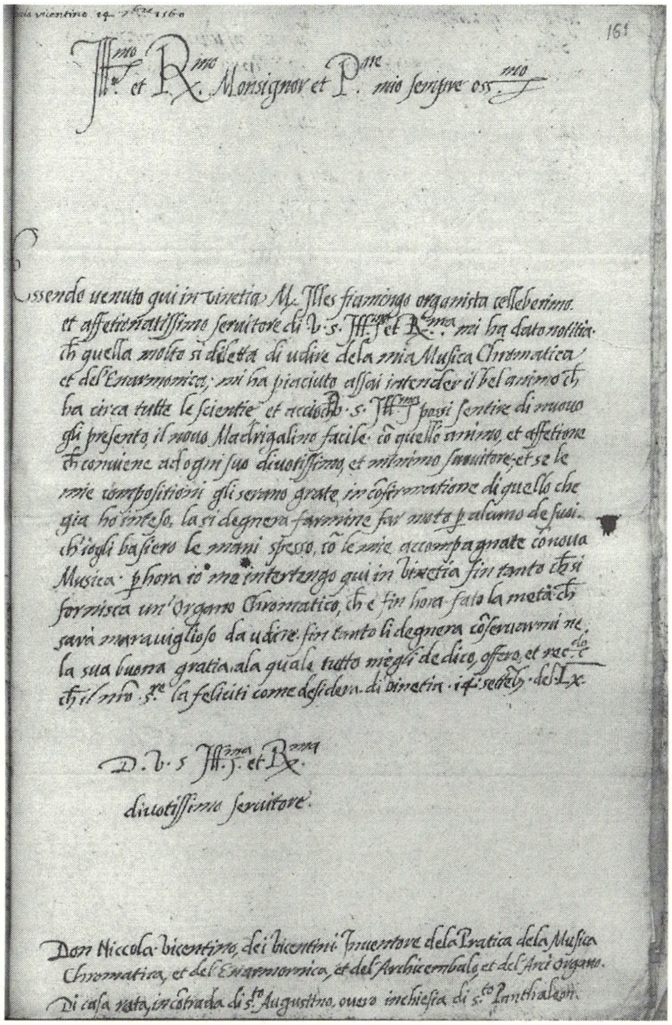

161

Ill.mo et R.mo Monsignor et P.ne mio sempre oss.mo

Essendo venuto qui in Vinetia M. Illes fiamingo organista celleberimo et affetionatissimo servitore di V. S. Ill.ma et R.ma mi ha dato notitia che quella molto si diletta di udire dela mia Musica Chromatica et del'Enarmonica, mi ha piaciuto assai intender il bel animo che ha circa tutte le scientie et acciochè V. S. Ill.ma possi sentire di nuovo gli presento il novo Madrigalino facile con quello animo, et affetione che conviene ad ogni suo divotissimo, et minimo servitore; et se le mie compositioni gli seranno grate, in confermatione di quello che gia ho inteso, la si degnerà farmine far moto per alcuno de suoi, che io gli basciero le mani spesso, con le mie accompagnate con nova Musica. per hora io me intertengo qui in Vinetia fin tanto che si fornisca un'Organo Chromatico, che è fin hora fato la metà che sarà maraviglioso da udire. fin tanto li degnerà conservarmi ne la sua buona gratia, ala quale tutto me gli dedico, offero, et racc.do et che il nro S.re la feliciti come desidera. di Vinetia. 14 sett.e del LX.

D. V. S. Ill.ma et R.ma

divotissimo servitore

Don Niccola Vicentino, dei Vicentini. Inventore dela Pratica dela Musica Chromatica, et del'Enarmonica, et del'Archicembalo, et del'Arci organo. Di casa [illegible], in contrada di S.to Augustino, overo in chiesia di S.to Panthaleon.

Ill. 2.6: Nicola Vicentino a Antoine Perrenot de Granvelle, 14 settembre 1561. E-Mp, MS ii/2274, f. 161r.

Il teorico e musicista, avendo inteso da tal «M. Illes fiamingo organista celeberrimo» il diletto che Granvelle prendeva nell'udire la sua «Musica Chromatica et del'Enarmonica», inviava al prelato alcune sue brevi composizioni; affermava inoltre di trattenersi «per hora [...] qui in Vinetia fin tanto che si fornisca un'Organo Chromatico», non ancora completato, «che sarà maraviglioso da udire».

Con caratteristica immodestia, si sottoscriveva «Don Niccola Vicentino, dei Vicentini. Inventore della Pratica della Musica Chromatica, et del'Enarmonica, et del'Archicembalo, et del'ArciOrgano», e dichiarava di risiedere «in contrada di santo Augustino, overo in chiesia di santo Panthaleon»; in altra lettera successiva di essere anzi «Maestro di Cappella di Santo Panthaleon» (Ill. 2.6).

Nella sua ultima lettera Vicentino non appone invece alcuna qualifica alla sua sottoscrizione. Le sue precedenti dichiarazioni sono di dubbia credibilità; non si sono conservate testimonianze, né archivistiche né memorialistiche, dell'esistenza di una sia pur ridotta cappella musicale in quell'antica chiesa veneziana.

È possibile che Vicentino si riferisse invece al piccolo gruppo di allievi che egli aveva raccolto attorno a sé; quasi una piccola cappella musicale il cui scopo precipuo era l'esecuzione e promozione delle composizioni 'enarmoniche' del fondatore, descritta da Vincenzo Galilei in un suo scritto più tardo. Il complesso portava «per tutte le principali città d'Italia» le composizioni enarmoniche di Vicentino, e Galilei ricordava di averlo udito «in diversi tempi et luoghi, più volte». Galilei però — conformemente al suo indirizzo pragmatico — continuava esprimendo forti dubbi sul duraturo successo di simili esperienze; e citava particolarmente l'esempio di un componente di quel gruppo, «detto per nome Jacomo Finetti, che oltre al sonare di tasti più che ordinariamente cantava di contralto molto bene». Questi, incontrato Galilei qualche anno più tardi, gli confidò che, per trovare un'occupazione nell'ambito musicale, aveva dovuto «lasciare da parte l'enharmonio del suo maestro»[54].

All'epoca di quelle esibizioni i componenti del complesso erano assai giovani; sempre secondo la testimonianza di Galilei, oltre alle composizioni enarmoniche, «quelli suoi giovani et fanciulli» cantavano «molte [altre] belle cantilene», sempre di Vicentino[55]. Tra essi vi era lo stesso Finetti e pertanto, almeno a titolo d'ipotesi, lo si può identificare nell'omonimo notevole compositore che, qualche decennio più tardi (almeno dal 1617 sino al 1629), fu maestro della cappella musicale della chiesa annessa al monastero dei Frari («Ca' Granda») a Venezia, posta a pochi passi dall'anzidetta chiesa di San Pantaleone (il che può fornire sostanza all'identificazione). Egli era semplice frate di quel convento; approfittando della sua presenza continuativa, era sovente chiamato a collaborare anche presso la vicina Scuola di San Rocco in occasione delle musiche festive annuali date il giorno del santo titolare, quando i più importanti compositori attivi a Venezia in quel periodo (Monteverdi, Alessandro Grandi ed altri) erano chiamati a dirigere il complesso musicale assoldato per l'occasione. Morì probabilmente durante la nota pestilenza del 1630. Le caratteristiche stilistiche delle sue numerose composizioni superstiti non fanno certo sospettare l'antico apprendistato presso l''arcimusico' Vicentino, il che però potrebbe trovare conferma nel caustico commento finale di Galilei. La testimonianza è interessante anche perché, se si accetta l'identificazione del giovane cantore citato da Galilei con il futuro frate e musicista, suggerisce la conoscenza, ma anche il mancato radicamento delle teorie di Vicentino a Venezia; e testimonia le evidenti difficoltà che i teorici musicali trovarono nei decenni tra Cinque e Seicento nell'ottenere adeguato *patronage*, particolarmente — sfumate le opportunità offerte dalle attività editoriali e didattiche dell'Accademia Venetiana — a Venezia.

[54]. I-Fn, MS Gal. 3, f. 9rv.

[55]. I-Fn, Ms. Gal. 3, f. 10r.

Continuità e discontinuità del *patronage*

Collegamenti tra la fase testimoniata dalla corrispondenza Spataro – Aaron – Del Lago e quella più tarda caratterizzata dalle pur brevi attività ispirate dall'Accademia Venetiana, si possono cogliere in numerosi aspetti di entrambe: Adrian Willaert, maestro della cappella ducale marciana ma anche insegnante di Gioseffo Zarlino — titolare della cattedra musicale dell'Accademia —, aleggia spesso nelle lettere dei tre teorici, e compare anche come destinatario di una lettera[56]. Paolo Manuzio, colui al quale si deve la conservazione della corrispondenza Spataro, fu lo stampatore ufficiale dell'Accademia, e alla sua officina si deve infatti la stampa degli atti ufficiali, dei cataloghi librari e dei pochi volumi promossi dall'accolita[57]. Girolamo Molin, il già citato mecenate di Giovanni Del Lago, fu anche tra i principali ispiratori dell'Accademia[58].

Non è possibile prendere qui in considerazione altri accadimenti e personaggi del periodo qui considerato, che costituiscono altrettanti notevoli trait d'union tra le due fasi: la traduzione vitruviana di Daniele Barbaro (tra gli ispiratori, se non un membro, dell'Accademia Venetiana)[59], il breve trattato teorico a lui attribuito[60], e la prolusione *Oratio de origine et dignitate musices* (Orazione sull'origine e sulla dignità della musica) del cantore marciano Pietro Gaetano (anch'essa verosimilmente concepita nell'ambito dell'accolita veneziana)[61].

Comunque, rimarrebbero confermati e semmai ribaditi i tratti generali che si sono fin qui esposti: lo stretto legame tra la fase documentata dalle attività della cerchia Spataro – Aaron – Del Lago, e quella successiva caratterizzata dalle attività dell'Accademia in campo teorico-musicale; nonché il comune obbiettivo — già evidente nella prima fase, ribadito nelle opere letterarie ed architettoniche di Zorzi e accanitamente poi difeso della stessa accademia — di mantenere la materia su un piano rigorosamente astratto, certamente in conformità con le aspettative della nobiltà veneziana più elevata, la sola che poteva fornire alle attività teorico-musicali adeguato *patronage*.

Tanto i teorici musicali che gravitavano attorno al nucleo testimoniato dalla corrispondenza Spataro, quanto quelli operanti nell'ambito dell'Accademia Venetiana e poco più tardi attorno a Zarlino (tra gli altri Nicola Vicentino e Vincenzo Galilei), avevano beneficiato di un peculiare clima culturale, attivato da un *patronage* nobiliare che consentiva la necessaria continuità nella ricerca e nello studio, e che informava di sé anche ambiti della cultura apparentemente distanti dalla teoria musicale, come l'architettura e l'arte figurativa. Non si trattava, in fondo, che di

[56]. Giovanni Maria Lanfranco ad Adrian Willaert, 20 ottobre 1531 (Blackburn – Lowinsky – Miller 1991, lettera 106, pp. 957-969). Se veda anche *ibidem*, pp. 1017-1019.

[57]. *Ibidem*, pp. 28-32.

[58]. Tomasi 2011, p. 361.

[59]. Vitruvio (traduzione e commento Daniele Barbaro), I dieci libri dell'architettura (Venezia, 1556).

[60]. Daniele Barbaro, 'Della musica' (Ms. I-Bc, B.26; I-Fl, Ms. Ashburnham 978, ff. 52r-79v).

[61]. I-Vmc, Ms. Cicogna 906.

uno dei portati dell'unione delle arti vagheggiata in epoca rinascimentale; un clima culturale che però, anche a Venezia, non durò a lungo[62]. Infatti, durante la seconda metà del Cinquecento questo sostrato culturale e politico andò dissolvendosi.

La carica di priore dell'ordine gerosolimitano che aveva ricoperto Sebastiano Michiel fu oggetto di contendere già in vita, sicché alla sua morte il papa Paolo III preferì conferirla non più ad un nobile veneziano, ma al figlio Ranuccio Farnese[63].

Le teorie di Francesco Zorzi erano ormai considerate apertamente eterodosse, e i suoi volumi furono messi all'indice; né lo poneva certo in miglior luce, agli occhi delle autorità ecclesiastiche, la citazione assai onorevole della sua figura nell'edizione veneziana del 1534 della commedia *Cortegiana* di Pietro Aretino, astutamente adattata dal'autore al clima veneziano del tempo[64]. Si può supporre che il riserbo attorno alla sua figura abbia coinvolto in qualche misura la sua visione della teoria musicale: l''armonia del mondo' da lui vagheggiata aveva sospette assonanze con una 'monarchia universale' incarnata di volta in volta nella monarchia francese o nell'impero asburgico-spagnolo. (Nelle menti dei nobili veneziani più accorti, era anche troppo evidente che ambedue avrebbero inevitabilmente sminuito, se non fagocitato, la Serenissima[65].) Le correzioni che, quasi un secolo dopo, tentò di apportare all'opera di Zorzi l'erudito francese — ben noto anche come teorico musicale —, Marin Mersenne, erano un tentativo di aggiornare, in pieno Seicento, alcuni aspetti di un panorama politico-culturale ormai definitivamente tramontato[66].

L'Accademia Venetiana, recuperando le opere di Zorzi, non si poneva certo in luce migliore agli occhi della nobiltà veneziana più accorta, che aveva accesso alle massime cariche statali. L'interesse dimostrato dall'accolita nei confronti delle opere teorico-musicali di Faber Stapulensis diventava sospetto, alla luce delle opinioni espresse nelle sue opere teologiche, accomunate a quelle di Zorzi nella censura ecclesiastica[67].

I soggetti d'ispirazione sacra delle opere d'arte furono sottoposti, dopo il concilio tridentino, alla stretto controllo del clero. Comunque, i soggetti vagamente esoterici delle citate opere di Bellini e Tiziano non furono più in uso, a Venezia, neppure per la devozione privata.

Lo stato veneziano non poteva vedere certo di buon occhio i legami di Faber Stapulensis con circoli ereticali presenti sul suo territorio[68], così come quelli testimoniati dalla citata corrispondenza intrattenuta da Dolfin e Vicentino con Granvelle (stabilmente a servizio della

62. Kristeller 1947; si veda in particolare, per Aaron e Spataro, ivi, pp. 263, 265.

63. Robuschi 2012, p. 150.

64. Vasoli 1974, p. 223, nota 239, e 2001.

65. Vasoli 1998, pp. 209-210.

66. Marin Mersenne, *Observationes et emendationes ad Francisci Georgii Veneti problemata* (Parigi, 1623).

67. Vasoli 2001, p. 61.

68. Evidenziati in Alonge 2019, *ad indicem*.

corona spagnola)[69], mentre urgeva invece, da una parte, mantenere buoni rapporti con la Santa Sede, d'altra parte mantenere l'indipendenza della Serenissima da Spagna ed Impero.

L'Accademia finì ingloriosamente la propria esistenza pochi anni dopo la fondazione, nel 1561; lo stesso Badoer andò incontro ad un vero e proprio fallimento, che ne mise in pericolo, secondo le parole d'un contemporaneo (già membro dell'accademia), «l'honore et forse la vita»[70].

Negli ambienti nobiliari italiani, la teoria musicale — così come l'avevano intesa i teorici attivi tra Quattro e Cinquecento — perse progressivamente la posizione di prestigio che era riuscita ad acquisire con fatica, tra Quattro e Cinquecento, come scienza del quadrivio. Già precedentemente un erudito in contatto con gli ambienti dell'Accademia, Giovanni Battista Benedetti, aveva mostrato — forse sollecitato dal lavoro teorico di Faber Stapulensis — di essere pienamente consapevole dell'impossibilità di applicazione coerente delle teorie intervallari di ispirazione pitagorico-platonica nella pratica musicale del tempo[71]. In pieno Seicento il processo giunse a maturazione, e quella stessa nobiltà colta che aveva tanto appassionatamente sostenuto la reviviscenza delle teorie e pratiche musicali antiche suggeriva ora, tutt'al più, d'impiegare i *genera* antichi soltanto in brevi episodi d'una composizione — quasi fossero *curiosa* di una Wunderkammer —, per il resto concepita nel più comune «accordo participato» (cioè intonazione mesotonica)[72].

Le suggestioni mistico-pitagoriche (forse ereditate dall'opera di Zorzi) che Zarlino vedeva nelle proporzioni numeriche espresse in alcuni intervalli musicali, erano considerate con impaziente sufficienza da Vincenzo Galilei[73]. Com'è noto, la nobiltà veneziana riconobbe per tempo l'importanza del lavoro scientifico del figlio Galileo: Giovan Francesco Sagredo era uno dei protagonisti del *Dialogo sopra i due massimi sistemi del mondo* (Firenze, 1632), e copie del *Dialogo* stesso furono inviate a vari nobili veneziani, primo fra tutti Zaccaria Sagredo, fratello di Giovan Francesco, nonché Angelo Contarini, Francesco Duodo e Domenico Molin[74]; e certamente le opinioni e obiezioni teorico-musicali del padre non erano ignote, né era possibile ignorarle.

Per finire, la disciplina si orientò sempre più verso l'elaborazione di una manualistica di pronto impiego per il cantore e lo strumentista, disdegnata dalla nobiltà più colta. Significativo

69. Bossuyt 1998, p. 433.

70. Lettera di Luca Contile al [Filippo] Binaschi, 4 febbraio 1560, in *Delle lettere di Luca Contile primo volume* (Pavia, 1564), f. 200r. Si veda inoltre ivi, Contile al Signor Tomaso Macchiavelli, 21 agosto 1560 (ff. 265r-266r: f. 266r).

71. Cohen 1987.

72. «NOTE | DI PIETRO DELLA VALLE | NEL DISCORSO | Sopra la Musica Antica e Moderna | Indrizzato al Signor Nicolò Farfaro | Che và a nome di | Giorgio Mazzaferro», I-Vnm, Ms. it. VI, 181 (= 5841), ff. 99r-117v: f. 115rv. Lo scritto è oggetto di Ziino 1969.

73. Vincenzo Galilei, *Discorso intorno all'opere di Messer Gioseffo Zarlino da Chioggia* (Firenze, 1589).

74. Pagano 2009, pp. cxxvii-cxxviii, p. cxlvi, nota 406.

quanto accadde, secondo il suo stesso autore, alle copie del primo libro del trattato di Lodovico Zacconi, *Prattica di musica* (Venezia, 1596). Le cinquecento copie già stampate a Venezia presso Girolamo Polo furono tenute dall'editore «per quattro anni in magazzino, senza che la lasciasse mai veder da nessuno»; furono poi rilevate «di grazia» dal «signor Sebastiano Combi [...] poco meno che per cartaccia», e quegli le distribuì con la dicitura «Venezia, appresso Girolamo Carampelli» (ma molte copie furono donate dall'autore «a questo e quello»)[75].

Alcuni particolari della narrazione di Zacconi, peraltro, lasciano appena trasparire il suo sospetto che qualcuno abbia brigato per impedire la diffusione della sua opera; accusa già lanciata, com'è noto, e ben più apertamente, da Vincenzo Galilei nei confronti di Zarlino[76].

La ricerca storica ha da tempo evidenziato la tendenza del pensiero politico veneziano, sviluppatasi in seguito al secolare ruolo svolto dalla Serenissima nel quadro degli stati italiani ed europei, al mantenimento di un equilibrio politico basato su compromessi; e conseguentemente, la sua tenace volontà di evitare accuratamente qualsiasi polemica che potesse mettere in discussione quell'equilibrio.

Non potrà perciò stupire se, dopo gli aspri scontri polemici che caratterizzarono la fase tardo-cinquecentesca della teoria musicale a Venezia, nel secolo successivo il *patronage* nobiliare veneziano si applicasse soprattutto alla nascente 'opera in musica'. Questa fu plasmata a perfetta immagine dell'edonismo e pragmatismo tipico della mentalità nobiliare veneziana, sempre meno convinta della possibilità (e dell'opportunità) di una sua adesione ad un 'impero universale', e quindi di un suo legame duraturo con l'una o l'altra delle maggiori potenze europee; e sempre più orgogliosamente (e insieme prudentemente) cosciente delle proprie specificità.

Bibliografia

Alonge 2019

Alonge, Guillaume. *Ambasciatori. Diplomazia e politica nella Venezia del Rinascimento*, Roma, Donzelli, 2019.

Aristide Quintiliano 1963

Aristidis Quintiliani De Musica libri tres, a cura di Reginald P. Winnington-Ingram, Lipsia, Teubner, 1963.

Barbieri 1996

Barbieri, Patrizio. 'Il mesolabio e il compasso di proporzione. Le applicazioni musicali di due strumenti matematici (1558-1675)', in: *Musica, scienza e idee nella Serenissima durante il Seicento*, a cura di Francesco Passadore e Franco Rossi, Venezia, Fondazione Levi, 1996, pp. 201-220.

75. Zacconi 2005, pp. 97-98 (le citazioni sono state leggermente modernizzate). Si veda anche Blackburn 2013.

76. Vincenzo Galilei, *Discorso* (1589), p. 12; 'Al [...] M. Gioseffo Zarlino da Chioggia' (dedica), ff. A2[v]-A3[r].

BARBOUR 1951
BARBOUR, James Murray. *Tuning and Temperament: A Historical Survey*, East Lansing (MI), Michigan State College Press, 1951.

BARKER 1989
Greek Musical Writings. 2: Harmonic and Acoustic Theory, a cura di Andrew Barker, Cambridge, Cambridge University Press, 1989.

BLACKBURN 2013
BLACKBURN, Bonnie B. 'Publishing Music Theory in Early Cinquecento Venice and Bologna. Friends and Foes', in: *Music in Print and Beyond: Hildegard von Bingen to The Beatles*, a cura di Craig A. Monson e Roberta Montemorra Marvin, Rochester (NY), University of Rochester Press, 2013, pp. 36-61.

BLACKBURN – LOWINSKY – MILLER 1991
EAD. – LOWINSKY, Edward E. – MILLER, Clement A. *A Correspondence of Renaissance Musicians*, Oxford, Clarendon Press, 1991.

BOLCATO 2009
BOLCATO, Vittorio. 'Memmo, Dionisio', in: *Dizionario Biografico degli Italiani. 73*, Roma, Istituto per l'Enciclopedia Italiana, 2009, pp. 418-420 (<https://www.treccani.it/enciclopedia/dionisio-memmo_(Dizionario-Biografico)/>, visitato a maggio 2022>).

BORK 2016
BORK, Robert. *The Geometry of Creation: Architectural Drawing and the Dynamics of Gothic Design*, Abingdon-New York, Routledge, ^{2}2016.

BOSSUYT 1998
BOSSUYT, Ignace. '«O socii durate». A Musical Correspondence from the Time of Philip II', in: *Early Music*, XXVI/3 (1998), pp. 433-444.

BREWER 1876
Letters and Papers Foreign and Domestic, of the Reign of Henry VIII, Preserved in the Public Record Office, the British Museum, and Elsewhere in England. 4/III (1529-1530), a cura di John Sherren Brewer, Londra, Longman & Trübner, 1876.

BURNETT 1986
BURNETT, Charles. 'The Use of Geometrical Terms in Medieval Music. Elmuahim and elmuarifa and the Anonymous IV', in: *Sudhoffs Archiv, Zeitschrift für Wissenschaftsgeschichte*, LXX/2 (1986), pp. 198-205.

COHEN 1987
COHEN, H. Floris. 'Benedetti's Views on Musical Science and Their Background in Contemporary Venetian Culture', in: *Cultura, scienze e tecniche nella Venezia del Cinquecento*, a cura di Antonio Manno, Venezia, Istituto veneto di scienze lettere ed arti, 1987, pp. 301-310.

FENLON 2002

FENLON, Iain. 'Gioseffo Zarlino and the Accademia Venetiana della Fama', in: ID. *Music and Culture in Late Renaissance Italy*, Oxford, Oxford University Press, 2002, pp. 118-138.

FOSCARI – TAFURI 1983

FOSCARI, Antonio – TAFURI Manfredo. *L'armonia e i conflitti. La chiesa di San Francesco della Vigna nella Venezia del '500*, Torino, Einaudi, [1983].

GALLO 1981

GALLO, Francesco A. 'La trattatistica musicale', in: *Storia della cultura veneta, dal primo Quattrocento al concilio di Trento. 3/III*, [Vicenza], Neri Pozza, 1981, pp. 297-314.

GARCIA PÉREZ 2002

GARCIA PÉREZ, Amaya Sara. 'El Temperamento en las Teorias Musicales de Salinas y Zarlino. Uso y aplicaciòn del Mesolabio', in: *Revista de Musicologìa*, XXV/2 (2002), pp. 347-361.

GUARNA 2018

GUARNA, Valeria. *L'Accademia Veneziana della Fama (1557-1561). Storia, cultura e editoria*, Manziana, Vecchiarelli, 2018.

HENTSCHEL 1998

HENTSCHEL, Frank. 'Die Unmöglichkeit der Teilung des Ganztones in zwei gleiche Teile und der Gegenstand der *musica sonora* um 1300', in: ID. *Musik und die Geschichte der Philosophie und Naturwissenschaften im Mittelalter. Fragen zur Wechselwirkung von musica und philosophia im Mittelalter*, Leida, Brill, 1998, pp. 39-60.

HERRMANN 2015

HERRMANN, Sabine. 'Ein Preuße in Venedig: Der Botaniker Melchior Wieland (ca. 1520-1589), Pionier der botanischen Feldforschung in der Levante', in: *Sudhoffs Archiv, Zeitschrift für Wissenschaftsgeschichte*, XCIX/1 (2015), pp. 1-14.

KRISTELLER 1947

KRISTELLER, Oskar Paul. 'Music and Learning in the Early Italian Renaissance', in: *Journal of Renaissance and Baroque Music*, I/4 (1947), pp. 255-274.

LAVIN 2013

LAVIN, Marilyn Aronberg. 'The Joy of St. Francis. Bellini's Panel in the Frick Collection', in: *Beyond the Text: Franciscan Art and the Construction of Religion*, a cura di Xavier Seubert e Oleg Bychkov, Saint Bonaventure (NY), Franciscan Institute Publications, 2013, pp. 14-39.

LEVENBERG 2016

LEVENBERG, Jeffrey. 'Worth the Price of the *Musurgia universalis*. Athanasius Kircher on the Secret of the «Metabolic Style»', in: *Recercare*, XXVIII/1-2 (2016), pp. 43-88.

Lindley 1987

Lindley, Mark. 'Stimmung und Temperatur', in: *Geschichte der Musiktheorie*, a cura di Friederich Zaminer, Darmstadt, Wissenschaftliche Buchgesellschaft, 1987, pp. 130-331.

Lovato 1999

Lovato, Antonio. 'Il Barbarigo e l'insegnamento della musica sacra a Padova', in: *Gregorio Barbarigo, patrizio veneto, vescovo e cardinale nella tarda Controriforma (1625-1697). 1*, a cura di Liliana Billanovich e Pierantonio Gios, Padova, Istituto per la storia ecclesiastica padovana, 1999, pp. 553-613.

Lowinsky 1982

Lowinsky, Edward E. 'Music in Titian's Bacchanal of the Andrians. Origin and History of the *Canon per tonos*', in: *Titian: His World and His Legacy*, a cura di David Rosand, New York, Columbia University Press, 1982, pp. 191-282.

Ludwig – Molmenti 1906

Ludwig, Gustav – Molmenti, Pompeo. *Vittore Carpaccio. La vita e le opere*, Milano, Hoepli, 1906.

Modesti 2000

Modesti, Paola. [scheda in] *Lungo il tragitto crociato della vita*, a cura di Laura Corti, Sovrano Militare Ordine di Malta, Gran Priorato di Lombardia e Venezia, Venezia, Marsilio-Editoriale Generali, [2000], pp. 88-90.

Onda 2008

Onda, Silvano. *La Chiesa di San Francesco della Vigna e il Convento dei frati minori. Storia, arte, architettura*, Venezia, Parrocchia di San Francesco della Vigna, [2008].

Oosterhoff 2018

Oosterhoff, Richard J. *Making Mathematical Culture: University and Print in the Circle of Lefevre d'Etaples*, Oxford, Oxford University Press, 2018.

Pagano 2009

Pagano, Sergio. *I documenti vaticani del processo di Galileo Galilei (1611-1741)*, Città del Vaticano, Archivio Segreto Vaticano, 2009.

Palisca 1985

Palisca, Claude V. *Humanism in Italian Renaissance Musical Though*, New Haven (CT), Yale University Press, 1985.

Pappus Alexandrinus 1876

Pappi Alexandrini collectionis quae supersunt, a cura di Fridericus Hultsch, Berlino, Weidmann, 1876.

Pellegrini 1808

Pellegrini, Domenico Maria. 'Breve dissertazione previa al sommario dell'Accademia Veneta della Fama', in: *Giornale dell'Italiana Letteratura*, xxii (luglio 1808), pp. 3-32, 113-128, 193-212; xxiii (ottobre 1808), pp. 49-68.

Perocco 1964

Perocco, Guido. *Carpaccio nella Scuola di S. Giorgio degli Schiavoni*, Venezia, Ferdinando Ongania, 1964.

Robuschi 2012

Robuschi, Luigi. *Il rapporto tra Repubblica di Venezia e Ordine di Malta e le relazioni economiche tra Sei e Settecento*, tesi di dottorato, Padova, Università di Padova, 2012.

Sanudo 1879-1903

Sanudo, Marino. *I diarii di Marino Sanuto*, 58 voll., Venezia, A spese degli editori, 1879-1903 (vol. iii, 1880; vol. iv, 1880; vol. vi, 1881; vol. xiii, 1886; vol. xv, 1886; vol. xvi, 1886).

Tomasi 2011

Tomasi, Franco. 'Molin, Girolamo', in: *Dizionario Biografico degli Italiani. 75*, Roma, Istituto per l'Enciclopedia Italiana, 2011, pp. 359-362 (<https://www.treccani.it/enciclopedia/girolamo-molin_(Dizionario-Biografico)/>, visitato a maggio 2022).

Valseriati 2017

Valseriati, Enrico. 'Il superamento del pregiudizio meccanico. Mobilità sociale e geografica a Brescia tra prima e seconda dominazione veneziana', in: *La mobilità sociale nel Medioevo italiano: 2. Stato e istituzioni*, a cura di Andrea Gamberini, [Roma], Viella, 2017, pp. 189-214.

Vasoli 1974

Vasoli, Cesare. 'Intorno a Francesco Giorgio Veneto e all' «armonia del mondo» ', in: *Profezia e ragione: Studi sulla cultura del Cinquecento e del Seicento*, Napoli, Morano, 1974, pp. 129-403.

Vasoli 1998

Id. 'Il tema musicale e architettonico della *Harmonia Mundi* da Francesco Giorgio veneto all'accademia degli Uranici e a Gioseffo Zarlino', in: *Musica e Storia*, vi/1 (1998), pp. 193-210.

Vasoli 2001

Id. 'Nuovi documenti sulla condanna all'indice e la censura delle opere di Francesco Giorgio veneto', in: *Censura ecclesiastica e cultura politica in Italia tra Cinquecento e Seicento*, a cura di Cristina Stango, Firenze, Olschki, 2001, pp. 55-78.

Ventura 1997

Ventura, Angelo. *Nobiltà e popolo nella società veneta del Quattrocento e Cinquecento*, Milano, Unicopli, 1997.

Verdier 1952-1953

Verdier, Philippe. 'L'allegoria della Misericordia e della Giustizia di Giambellino agli Uffizi', in: *Atti dell'Instituto Veneto di Scienze, Lettere e Arti*, cxi (1952-1953), pp. 97-116.

Zacconi 2005

Zacconi, Ludovico. *Vita con le cose avvenute al padre baccelliere fra Ludovico Zacconi da Pesaro dell'Ordine Eremitano di S. Agostino. Fatta così da lui come si vede e scritta di proprio pugno*, a cura di Fernando Sulpizi e Cinzia Lanti, Perugia, Hyperprism Edizioni, 2005.

Ziino 1969

Ziino, Agostino. '«Contese letterarie» tra Pietro della Valle e Nicolò Farfaro sulla musica antica e moderna', in: *Nuova Rivista Musicale Italiana*, iii/1 (1969), pp. 101-120.

Zorzi 2010

Zorzi, Francesco. *L'armonia del mondo*, a cura di Saverio Campanini, Milano, Bompiani, 2010.

Chapter Three

Nascent Mechanism in Philip Melanchthon's References to Josquin des Prez

Alexander Jakobidze-Gitman[*]
(Witten/Herdecke University)

Philip (Philipp) Melanchthon (1497-1560) was a prominent German humanist and the intellectual leader of the Lutheran Reformation (Ill. 3.1)[1]. For his invaluable contributions to the Lutheran educational system, he has been called «the teacher of Germany» (*praeceptor Germaniae*). Melanchthon was a staggeringly productive and astonishingly versatile author. Despite his contributions to a great many academic disciplines, however, he wrote no major work on music. Moreover, during his long tenure as president of the University of Wittenberg, he paid little attention to it. Music was not explicitly listed in the statutes that he revised, no professorship for theoretical music was established, and the university was more concerned with musicians for practical instruction[2]. As Sachiko Kusukawa (1995) has noted, of the quadrivial arts, music had the least significance in Melanchthon's educational program. Although Luther, who assigned to music the highest place after theology,

*. I would like to thank Alexius Belov (Moscow) for checking my translations of Melanchthon's Latin texts and Rhian Jones (London) for her help in reading and checking the text of this article. I naturally remain responsible for any remaining errors.

1. Melanchthon's original name was Philipp Schwarzert. In English-language writings his first name is usually spelled with a single p at the end.

2. Groote 2013, p. 238.

Ill. 3.1: Monogrammist 1B (after Lucas Cranach), *Philip Melanchthon* (engraving, 8.5 × 7 cm; 1530). NL-Ar, RP-P-OB-3082.

entrusted Melanchthon with the mission of educational reform, all the circumstantial evidence testifies to the fact that the latter did not share the former's high estimation of music[3].

And yet, as an ambitious polymath and the intellectual leader of the German Reformation, Melanchthon was eager to leave his trace in practically every field. Since music played an important role in the Lutheran Reformation, he could scarcely overlook the opportunities provided by the spread not only of book printing but also of music printing. Accordingly, his thoughts on music are best recognized in his prefaces to various editions of music, that is, in occasional writings. In these brief texts, Melanchthon argued that the power of music was provided not by its numerical proportions, as the canonist (Pythagorean) tradition had claimed, but by the support that music lent to the Divine Word. Correspondingly, Melanchthon is considered one of the key figures who contributed decisively to the paradigm shift from a speculative-mathematical, Boethian-influenced theory to a «rhetorical», text- and expression-oriented conception of music[4]. The hypothesis that Melanchthon might have been the originator of the term *musica poetica* (long before Joachim Burmeister) was proposed by Hans Heinrich Eggebrecht in 1959[5].

3. Kusukawa 1995, p. 192.
4. Groote 2013, p. 227.
5. Eggebrecht 1959, p. 33. There is, however, no documentary evidence for this suggestion.

Melanchthon mentioned Josquin des Prez (c. 1450/1455-1521) on several occasions in works published in the 1540s. This seems perfectly understandable since at that time the composer was widely considered the greatest musician in the history of Christian Europe. Jessie Ann Owens (1997) has held that the reason why Josquin eventually acquired the special status he enjoyed at that time was that «the invention of printing as a commercially viable way of preserving and transmitting music [...] coincided with Josquin's moment in time». That is why

> Josquin, and not Machaut or Vitry or Landini or Dufay or Ockeghem became the first 'classical' composer mentioned by the Swiss humanist Heinrich Glarean. Glarean, whose *Dodecachordon* was published in 1547 but completed in about 1530, [...] had contempt for contemporary music, whose only virtue was its newness, and argued that music had reached perfection twenty-five years earlier — that is, with Josquin and his generation. Glarean considered Josquin the Virgil of music and referred to Josquin and his contemporaries as 'classics'[6].

Glarean's views, however, conformed to the position of the Catholic Church. As for the Protestant world, Josquin was mostly praised specifically for the way his musical settings preserved the comprehensibility of the words while enhancing its emotional impact, features which retrospectively were numbered among the virtues of Lutheran music. The prominent music publisher Hans Ott extolled Josquin in his foreword to the *Secundus tomus novi operis musici* (Nuremberg, 1538) for the motet *Miserere mei, Deus* because, there, music had submitted itself particularly successfully to the words[7]. Owens has suggested that the 'German Renaissance' of Josquin was due to some aspects of his style that «must have lent themselves to the taste that prevailed in the 1530s and 40s, thus remaining fashionable longer than usual. The close relationship between music and text, the profound expression of the rhetorical structures on every level in the music, must have resonated strongly in a culture where 'the Word' was all-important»[8].

In this context, Melanchthon's position appears to be quite unusual. Firstly, he made a comparison akin to the Catholic Glarean: his repeated mentioning of Josquin along with Ovid and Homer as examples of musical and poetic talent implied that the modern musician Josquin was as much a classical author as the foremost ancient poets[9]. Secondly, his references to Josquin contained nothing about music's buttressing of the power of the Divine Word and hence do not correspond to the scholarly picture of Josquin's reception in Reformation Germany.

---

6. OWENS 1997, p. 273.

7. *Secundus tomus novi operis musici* (Nuremberg, 1538; RISM B/I, 1538-3), 'Serenissimo Principi [...] Ferdinando Romanorum regi semper augusto' (dedication by Hans Ott, dated 15 October 1538). Ott referred to Josquin's *Miserere* as published in the preceding *Novum et insigne opus musicum* (Nuremberg, 1537; RISM B/I, 1537-1), no. 13. See SCHLAGEL 2006, p. 117.

8. OWENS 1997, p. 278.

9. GROOTE 2013, p. 242.

It is important to stress that, apart from the preface to Johannes Reusch's *Zehen deudscher Psalm Davids* (Ten German Psalms of David; 1552)[10], Melanchthon's references to Josquin occur in his major academic treatises, where music did not play a significant role. Josquin seems to have instantiated for Melanchthon something which did not directly relate to music. These references appear all the more enigmatic since they neither associated music with rhetoric nor concerned the emotional impact of music. I shall argue here that they must remain obscure as long as we interpret them within the common contexts of Lutheranism or Renaissance Humanism. Instead, they will make sense and even reveal new meanings if we contextualize them within the nascent Scientific Revolution.

Sachiko Kusukawa (1995) has claimed that in Melanchthon's legacy natural philosophy occupied the same place as music did in Martin Luther's thought. Indeed, Melanchthon was not only a theologian and philologist, but also a devoted advocate of natural sciences such as medicine and anatomy. My main claim here will be that, while Melanchthon's discussions of the role of music reflect his identity as a humanist, his references to Josquin express his interest in natural philosophy and even help to highlight the unusual character of his perspective in this area. For, as I will also argue, although Melanchthon generally remained within an Aristotelian framework, his references to Josquin testify to the advent of a mechanistic worldview. This suggestion is at odds with the accepted view that the mechanistic philosophy had its advent with the firm dismissal of Aristotelian physics by Francis Bacon and René Descartes.

To articulate this argument, I structure my essay as follows. First, I clarify the points of convergence and divergence between Melanchthon and Luther regarding the role of music. The next three sections are dedicated to the three major works by Melanchthon in which the name Josquin appears: the revised edition of the *Philosophiae moralis epitome* (Strasbourg, 1540), the *Erotemata dialectices* (Wittenberg, 1547), and the *Initia doctrinae physicae* (Wittenberg, 1549). I distinguish among those cases in which it is a matter of propagating Aristotelian philosophy, those in which Melanchthon takes up the ideas of his humanist predecessors, and those in which he reacts to contemporary breakthroughs in science, such as the seminal 1543 publications by Nikolaus Copernicus (*De revolutionibus orbium coelestium*; On the Revolutions of Heavenly Spheres; Nuremberg) and Andreas Vesalius (*De humani corporis fabrica*; On the Fabric of the Human Body; Basel). Furthermore, I discuss in what respects Melanchthon's preoccupations with astrology, astronomy and anatomy might be relevant for understanding his references to Josquin. Finally, I juxtapose the implications of Melanchthon's image of Josquin with the main tenets of the later fully-fledged mechanistic worldview, as well as with late eighteenth-century criticisms of mechanical philosophy.

10. Johannes Reusch, *Zehen deudscher Psalm Davids* (Wittenberg, 1552; RISM A/I, R 1208).

Music in the German Reformation: Melanchthon versus Luther

Melanchthon's references to Josquin are very different from his general statements on music. In the latter he seems to share Martin Luther's views. Firstly, Melanchthon was at one with Luther in considering music as a liberating «gift of God»[11]. In his preface to a collection of 52 motets published in 1538 by Georg Rhau, cantor at St. Thomas in Leipzig, which, apart from pieces by Josquin, included pieces by Ludwig Senfl, Heinrich Isaac, Pierre de la Rue, Philippe Verdelot and others, Luther wrote:

> Therefore, the gift of language combined with the gift of song was given, rather than to other creatures, to man alone, to let him know that he should praise God with both word and music, namely through sonorous preaching and by providing words with sweet melodies[12].

Melanchthon's foreword to Johannes Reusch's *Zehen deudscher Psalm Davids* also picked up and developed further an essentially Lutheran view, namely that since all words remained much more firmly in the memory when they were put into song; that it was God's will that music be preserved; and that young people should learn to practise and understand this beautiful joy ordered by God[13]. In his letter of 1 August 1537 to Eobanus Hessus on the latter's Latin translation of the Psalter[14], Melanchthon emphasized the affinity between verse and music and argued that verse — like music — had a stronger effect on the recipient than mere prose. For Inga Mai Groote (2013), Melanchthon's prefaces demonstrated that he elaborated much more systematically than Luther on the power of music over man and provided figures of thought that promoted the combination of words and music as a particularly effective means of religious instruction[15].

---

11. Groote 2013, p. 232.

12. *Symphoniae iucundae atque adeo breves quatuor vocum* (Wittenberg, 1538; RISM B/i, 1538-8), 'Martinus Luther musicae studiosis' (preface): «Denique homini soli prae caeteris, sermo voci copulatus donatus est, ut sciret, se Deum laudare oportere verbo & musica, scilicet sonora praedicatione, et mixtis verbis suavi melodiae». Translation by Rudolf Rasch.

13. Groote 2013, p. 234.

14. Eobanus Hessus, *Psalterium universum carmine elegiaco redditum atque explicatum* (Marburg, 1537). Melanchthon's letter to Hessus is included in later editions of Hessus's translation, entitled *Psalterium Davidis carmine redditum* (Leipzig, 1575, 1584), ff. B3-recto-B4-recto.

15. Groote 2013, p. 238: «Im Vergleich zu Luthers eher theologisch und eher verstreuten Argumenten bot Melanchthon eine viel systematischere Erklärung der Macht der Musik über den Menschen und lieferte Denkfiguren, die die Kombination von Text und Musik als besonders wirksames Mittel der religiösen Unterweisung propagierten und sich damit hervorragend in einen Bildungsplan integrieren ließen, der alle Disziplinen der Theologie dienstbar machen sollte».

However, Melanchthon's approach to the impact of music deviated from Luther's. In the above-quoted letter to Hessus, he explicitly stated that music was particularly effective in arousing or soothing affections[16]. Here Melanchthon relied heavily «on an Aristotelian-influenced philosophical system to grasp [music's] quality as a morally improving art»[17] and revived the Greek doctrine of musical ethos. This corresponded with Melanchthon's general attitude to Aristotle: despite Luther's hostility towards Aristotelian scholasticism, which had led to its brief expulsion from the curriculum, Melanchthon reintroduced the teaching of Aristotelian texts at the University of Wittenberg[18].

Groote has distinguished two trains of thought in Melanchthon's occasional writings on music:

> The first is theological, which addresses the functions of the arts in relation to religion. [...] The second strand is genuinely humanistic. On the one hand, it draws upon the Aristotelian notion that music entertains, instructs, and improves morals, and on the other hand, it ties in with the idea that there is a close connection of the soul with harmonic relationships and with music, through which it can approach the divine order, and, thirdly (this idea also comes from Plato's *Republic*), that word and music have a stronger effect in combination[19].

Unlike the early humanists, Melanchthon did not focus on the philological work of translating, commenting, or preparing for print ancient manuscripts, but on applying creatively some ideas from the ancient Greek heritage to contemporary material. Melanchthon resorted to the ancient doctrine of musical ethos to explain the psychological and even physiological impact of modern music.

---

16. Philip Melanchthon to Eobanus Hessus, 1 August 1537, in: Eobanus Hessus, *Psalterium Davidis carmine redditum* (Leizpig, 1575), f. B3-recto (Groote 2013, p. 236): «Magna enim vis est Musices ad affectus in animis excitandos, aut sedandos».

17. Groote 2013, p. 232.

18. Hamid 2018, pp. 29-30.

19. Groote 2013, p. 237: «Vereinfacht ausgedrückt, sind also zwei Gedankengänge in Melanchthons eigenen Schriften zu erkennen. Der eine ist der theologische, der sich auf die Funktionen der Künste in Bezug auf die Religion bezieht. [...] Der zweite Strang ist genuin humanistisch: Er rekurriert einerseits auf die aristotelische Vorstellung, dass die Musik unterhält, belehrt und die Sitten verbessert. [...] Zum anderen knüpft sie an die Vorstellung an, dass es eine enge Beziehung der Seele zu harmonischen Verhältnissen und zur Musik gibt, durch die sie sich der göttlichen Ordnung annähern kann, und dass drittens (auch dieser Gedanke stammt aus Platons Staat) Wort und Musik in Kombination eine stärkere Wirkung haben».

Philosophiae Moralis Epitome (1540)

Given how Melanchthon drew on the Aristotelian link between music and morals, it should not seem strange that he first mentioned Josquin in the context of moral philosophy. Curiously enough, in the first edition of the *Philosophiae moralis epitome* (Summary of Moral Philosophy) of 1538 Josquin's name is still absent. Melanchthon first cited him in the revised and enlarged second edition, which was published in 1540, the same year in which, according to Johannes Mathesius, Luther uttered his famous praise of Josquin (Ill. 3.2)[20]. The cursory appearance of Josquin in the *Philosophiae moralis epitome* may be regarded as indirect evidence of Melanchthon's then fresh acquaintance with the composer's oeuvre. Furthermore, Josquin is mentioned in a context that seems very far from music: in the elucidation of the ancient Greek notion of *prohairesis* (προαίρεσις), variously translated as 'moral character', 'will', 'intention', or 'moral choice'[21], which Melanchthon however left untranslated:

> [...] virtue is twofold: common and heroic. In the case of heroic virtue, great men, by dint of an extraordinary divine impetus, are roused to certain distinguished deeds, from which arises the excellent virtue that exceeds common human capabilities, such as the heroic courage of Achilles, Alexander [the Great], Caesar and Scipio, that is, [virtue] superior to any virtue that can exist in humans without [such] an extraordinary divine impulse. Thus, although *prohairesis* (moral choice) is responsible for common virtue, for heroic virtue an extraordinary divine impetus is required which supports the *prohairesis*, that is, it both renders the mind more perspicacious in judgment and rouses the will, just as in the arts the heroic artist is vastly superior to the mediocre artists, as in music **Josquin** is vastly superior to the common musicians[22].

---

20. Mathesius 1841, p. 266.

21. Melanchthon defined προαίρεσις in his *Philosophiae moralis epitome* (1540), p. 58, as follows: «Est igitur quartus gradus προαίρεσις, cum recto iudicio & sponte aliquid agitur, ut Fabius videt ubi sit cedendum, ubi sit pugnandum, et cum opus est, non detrectat periculum». (The fourth degree therefore is προαίρεσις, when something is done with the right judgement and willingly, as Fabius sees where he should yield, where he should fight, and, when it is necessary, does not shy away from danger.)

22. *Ibidem*, p. 59 (Melanchthon 1850, cc. 57-58; emphasis added): «[...] duplex est virtus, communis et heroica. Est autem heroica, cum singulari divino impetu magni viri ad certas honestas actiones extimulantur, unde fit excellens virtus, supra communem captum hominum, ut in Achille, Alexandro, C. Caesare, Scipione fuit heroica fortitudo, id est, excellentior, quam qualis existere potest in hominibus sine singulari divino motu. Itaque etsi προαίρεσις efficit communem virtutem, tamen ad illam excellentem virtutem, quae vocatur heroica, opus est singulari divino motu, qui adiuvat προαίρεσιν, hoc est, et mentem reddit perspicaciorem in iudicando, et voluntatem efficit excitatiorem, ut in artibus heroicus artifex longe praestat mediocribus artificibus, ut in Musica **Iosquinus** longe superat vulgares Musicos».

PHILIPPI
MELANCHTONIS PHILO
sophiæ moralis epitomæ libri duo
emendati & aucti.
NAM TRES VIRTVTES ADDI-
tæ, Quintus Aristotelis totus immutatus, ac-
cessit etiam tractatus De autoritate prin
cipum, & alter De arbore consang
uinitatis & affinitatis, &
index copiosus.
AN. M. D. XL.
Cum gratia, & Priuilegio Imp.

PHILOS. MORALIS 59
dati diuini, & nostræ uocationis. Cæterum du-
plex est uirtus, communis & Heroica. Est autem
Heroica, cum singulari diuino impetu magni uiri
ad certas honestas actiones extimulantur, unde fit
excellens uirtus, supra communem captum homi-
num, ut in Achille, Alexandro, C. Cæsare, Sci-
pione fuit Heroica fortitudo, id est, excellentior
quàm qualis existere potest in hominibus sine sin-
gulari diuino motu. Itaque & si προαίρεσις ef-
ficit communem uirtutem, tamen ad illam excel-
lentem uirtutem, quæ uocatur Heroica opus est
singulari diuino motu, qui adiuuat προαίρεσιν,
hoc est, & mentem reddit perspicatiorem in iu-
dicando, & uoluntatem efficit excitatiorem, ut
in artibus Heroicus artifex longe præstat medio-
cribus artificibus, ut in Musica Iosquinus longe
superat uulgares Musicos. Atque hi sunt fortu-
nati, etiamsi interdum extra regulam aliquid fa-
ciunt, sicut Græci dicunt. Herculem remigando in
Argo naui remos fregisse, eoq; deinceps non ad-
hibitum esse ad remigandum, & dulcissime in-
quit Pindarus de Heroica uirtute, gratum initi-
um & dulcis finis est impellente Deo. Tales sunt
& sanctorum uirtutes, quæ à spiritu sancto ex-
uscitantur, ut Abrahæ, Dauidis, Pauli & ali-
orum. Hos gradus actionum & uirtutum te-
nere ad

Sanctorum uirtutes su Heroicæ.

Ill. 3.2ab: Philip Melanchthon, *Philosophiae moralis epitomae libri duo* (Strasbourg, 1540), title page and page 59. Copy D-Mbs.

Today, Melanchthon's treatment of morals appears strikingly naturalistic for a Christian theologian, since he suggested at several places that moral conduct was heavily influenced by physiological processes. Actually, his comparison of creative artists with heroic warriors seems to be taken from the Pseudo-Aristotelian *Problems*, Book xxx, Problem 1[23]. Generally,

23. Pseudo-Aristotle, *Problems*, Book xxx, Problem 1: «[953a10-12] Why is it that all men who have become outstanding in philosophy, statesmanship, poetry or the arts are melancholic. [...] [26-28] [such as] Empedocles, Plato, Socrates, and many other well-known men. [...] [29-30] For many such men have suffered from diseases which arise from this mixture in the body. [...] [954a12-14] Such melancholic humour is already mixed in nature, for it is a mixture of hot and cold for nature consists of these two elements. [...] [21-25] Now black bile, which is naturally cold and does not reside on the surface when it is in the condition described, if it is in excessive quantity in the body, produces apoplexy or torpor, or despondency or fear; but if it becomes overheated, it produces cheerfulness with song, and madness, and the breaking out of sores and so forth». Translation from Aristotle 1957, pp. 155-163.

Melanchthon relied heavily on the humoural theory, which was elaborated most thoroughly not so much by Aristotle as by the physician and philosopher Galen (129-210 AD). In this well established and widely accepted view, both mental and physical proclivities of human beings could be traced back to the proportions and disposition of the four humours — blood, phlegm, yellow bile, and black bile — within the body. However, Melanchthon's image of Josquin was inconsistent with Pseudo-Aristotle's suggestion that all great men were prone to melancholy due to an excess of black bile (*melaina cholê*). This may be easily explained by the fact that Josquin was supposed to represent Christian virtues, while in the sixteenth century melancholy was widely associated with *acedia* (apathetic lethargy) — one of the seven capital sins.

EROTEMATA DIALECTICES (1547)

The second instant when Melanchthon cited Josquin was the first book of the *Erotemata dialectices* (Dialectical Questions; Frankfurt, 1547), an extremely popular book, which enjoyed 46 editions from 1547 to 1600 (ILL. 3.3)[24]. At that time the term 'dialectic' was understood either as synonymous with logic or as a method of discussing entitites or phenomena whose causes could not be demonstrated. In addition, dialectic was understood in the Aristotelian sense as an art of reasoning about a wide variety of subjects without being able to prove the truth or falsity of any thesis. Dialectic understood in this latter sense proceeded from 'received opinions' and was not aimed at acquiring certain knowledge[25].

Melanchthon himself defined the subject as follows:

> Dialectic is the art or method of instructing someone in a correct, clear, and orderly manner [about a subject-matter], which is done by means of correct definitions and distinctions as well as through the construction of valid arguments and the uncovering and refutation of unsound or false arguments[26].

Melanchthon was innovative compared to Aristotelian and scholastic traditions. He claimed that dialectic was an art of training, developing, and perfecting a *natural* faculty. Melanchthon adopted the term 'dialectica nativa' (natural or innate dialectic), implying that

---

24. MACK 1993, p. 320.

25. According to Aristotle, a deduction was dialectical if its premises were 'received opinions' (ἔνδοξα [endoxa]); Aristotle contrasted dialectic arguments, which were 'according to opinion' (κατὰ δόξαν [kata doxan]), with demonstrative ones, which are 'according to truth' (κατ'ἀλήθειαν [kat'alētheian]). See SMITH 1993, p. 336.

26. Philip Melanchthon, *Erotemata dialectices* (Frankfurt, 1547), f. 1r. (MELANCHTHON 1846, c. 513): «Dialectica est ars seu via, rectè, ordine, et perspicuè docendi, quod fit rectè definiendo, dividendo, argumenta vera connectendo, et malè cohaerentia seu falsa retexendo et refutando». Translation from WELS 2008, p. 142.

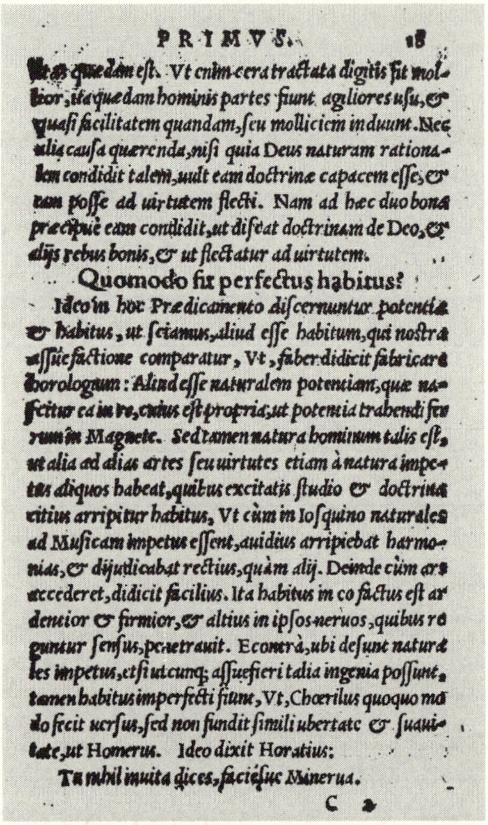

PRIMVS. 18

...quedam est. Vt enim cera tractata digitis fit mollior, ita quædam hominis partes fiunt agiliores usu, & quasi facilitatem quandam, seu mollicicm induunt. Nec alia causa quærenda, nisi quia Deus naturam rationalem condidit talem, uult eam doctrinæ capacem esse, & eam posse ad uirtutem flecti. Nam ad hæc duo bona præcipuè eam condidit, ut discat doctrinam de Deo, & alijs rebus bonis, & ut flectatur ad uirtutem.

Quomodo fit perfectus habitus?

Ideo in hoc Prædicamento discernuntur potentia & habitus, ut sciamus, aliud esse habitum, qui nostra assuefactione comparatur, Vt, faber didicit fabricare horologium: Aliud esse naturalem potentiam, quæ nascitur ea in re, cuius est propria, ut potentia trahendi ferrum in Magnete. Sed tamen natura hominum talis est, ut alia ad alias artes seu uirtutes etiam à natura impetus aliquos habeat, quibus excitatis studio & doctrina citius arripitur habitus, Vt cùm in Iosquino naturales ad Musicam impetus essent, auidius arripiebat harmonias, & dijudicabat rectius, quàm alij. Deinde cùm ars accederet, didicit facilius. Ita habitus in eo factus est ardentior & firmior, & altius in ipsos neruos, quibus reguntur sensus, penetrauit. Econtrà, ubi desunt naturales impetus, etsi utcunq; assuefieri talia ingenia possunt, tamen habitus imperfecti fiunt, Vt, Chœrilus quoquo modo fecit uersus, sed non fundit simili ubertate & suauitate, ut Homerus. Ideo dixit Horatius:

Tu nihil inuita dices, faciésue Minerua.

C 2

ILL. 3.3AB: Philip Melanchthon, *Erotemata dialectices* (Frankfurt am Main, 1547), title page and f. 18r. Copy D-Mbs.

dialectic was not a system of norms that people would have to acquire 'artificially' at school, but that everyone had from birth a certain ability to reason. He regretted that in the scholastic educational system «natural dialectic had degenerated into an artificial form of logic». Instead of relying on that natural capacity, medieval dialectic «had lost touch with any reference to reality and dealt with problems that no longer had any basis in everyday practices», as Volkhard Wels (2008) has summarized Melanchthon's viewpoint[27].

Melanchthon elaborated on the relations between natural faculties and acquired skills in the chapters devoted to the notion of *habitus*. Melanchthon used it (as it was often used) as a translation of the Greek term *hexis* (ἕξις), which designates an arrangement or disposition, as, for example, a person's health or character. *Habitus*, however, could signify knowledge as well, both of a propositional and a practical sort[28]. Melanchthon stressed that *habitus* was

27. WELS 2008, p. 142.

28. That would be, in the terminology of the British philosopher Gilbert Ryle, both «knowledge that» and «knowledge how».

a prerogative of rational beings, that is, consciously acquired[29]. It followed that he meant a habit or a skill, or even the process of their acquisition, for he distinguished *habitus* from *potentia* ('power'):

> In this category, powers and habits are distinguished so that we may know that a habit is one thing, which is acquired by training, for example in the way a craftsman has learned how to make clocks, while a natural power that is inherent in something and proper to it, like the force of a magnet to attract iron, is a different thing. Human nature, however, is such that certain people have a natural inclination to particular arts and virtues, and this, when it has been activated, [enables them] by diligent application and instruction to acquire the [relevant] skill more swiftly than others; as, for example, **Josquin**, since he had a natural inclination to music, grasped harmonies more passionately, and distinguished them more correctly than other people[30].

Later on, I will come back to this association with a clockmaker. For now, it seems more urgent to clarify how Melanchthon accounted for the emergence of «natural inclination». This inclination was not simply a spiritual substance with which God endowed the chosen ones. It had a corporeal substratum that resulted from physical causes. Here Melanchthon relied heavily on several notions from Greek Antiquity which he sought to reconcile with the Christian doctrine of Divine Providence.

In relation to Josquin, the most important notion was that of fate (*fatum*). In Antiquity, it was inextricably connected with the ancient pagan religions: the fate of humans was thought to be regulated by the power of the stellar constellations, which were seen as the activities of the pagan gods. Melanchthon sought to resurrect astrology but also to strip it of its pagan associations. Hence, he subjected the power of the stars to the will of the Christian God. The second instant in this treatise where he referred to Josquin aimed at elucidating this point:

> Fate means a cause, that is, either a divine command (assignment) or the power of the light of the stars. This is what Ptolemy calls the fate caused by physical reasons[31]. When we say: the Romans attained monarchy by fate, we are signifying a

[29]. Melanchthon, *Erotemata* (1547), f. 17v (Melanchthon 1846, c. 534).

[30]. Melanchthon, *Erotemata* (1547), f. 18r (Melanchthon 1846, c. 535; emphasis added): «Ideo in hoc praedicamento discernuntur potentiae et habitus ut sciamus, aliud esse habitum, qui nostra assuefactione comparatur, ut faber didicit fabricare horologium, aliud esse naturalem potentiam, quae nascitur ea in re, cuius est propria, ut potentia trahendi ferrum in magnete. Sed tamen natura hominum talis est, ut alia ad alias artes seu virtutes etiam a natura impetus aliquos habeat, quibus excitatis studio et doctrina citius arripitur habitus, ut, cum in **Iosquino** naturales ad Musicam impetus essent, avidius arripiebat harmonias, et diiudicabat rectius, quam alii».

[31]. As the historians of geography T. W. Freeman and Philippe Pinchemel (1979) have pointed out, «The *fatum physicum* was nothing more than a revelation of how the world existed, and therefore the first kind of fate,

> cause. The basis of this cause is God who ordained it and urged the Romans to move in this direction. Its end is the attainment of monarchy. But when we say: **Josquin** is destined by physical reasons to become a musician, the foundation of this fate is the position of the stars, which brings about in **Josquin** a temperament better suited to music than in other people. The end is to grasp harmonies rapidly, and to make sweeter music than other people[32].

Here, Melanchthon refers not to «natural inclinations» but to «temperaments», addressing nonetheless one and the same phenomenon. We see that God did not endow Josquin with the gift of divine musicality directly, but rather via the mediation of natural causes — the ones that accompany God's will in the view of late scholastics. The power of stellar light was a means to bring about the divine assignment. In Albert Blaise's (1975) dictionary of medieval Latin, the term *fatum* has been defined as «the order of secondary causes destined to accomplish the order of Providence»[33]. That means that God or Divine Providence occasioned the stars to adopt their unique position at the hour of Josquin's birth.

This citation testifies that Melanchthon remained within the framework of traditional humoural theory. In contrast to the *Epitome*, however, he did not confine himself to explaining the individual predispositions of men by the combination of their humours. Apparently, he drew inspiration here not from Greek authors but from another Renaissance philosopher, Marsilio Ficino, who undertook a systematic attempt to synthesize humoural theory with the astrological doctrines of Antiquity. In Ficino's *De vita libri tres* (Three Books on Life; 1480-1489) it was stated that the humours were heavily influenced by the stars, and that a musician who knew the proportions between the stars could heal bodies and souls by implementing the same proportions in music[34]. Grantley McDonald (2015) has noted that Melanchthon «imported an element of astrological determinism conspicuous in *De vita*. [...] Like Ficino, Melanchthon suggested that the natural temperament of a child could be influenced by the planetary conjunctions under which he or she was born»[35].

---

deliberate decision, was another name for the providence of God». FREEMAN – PINCHEMEL 1979, p. 95.

32. Melanchthon, *Erotemata* (1547), ff. 31v-32r (MELANCHTHON 1846, c. 553; emphasis added): «Fatum significat causam, id est, aut ordinationem divinam, aut vim luminis stellarum. Id vocat Ptolemaeus fatum physicum. Cum dicimus: Romani fato consecuti sunt monarchiam, hic fatum significat causam. Et fundamentum est Deus ordinans hanc rem, et movens Romanos. Terminus est ipse eventus. Sed cum dicimus: **Iosquinus** est fato physico Musicus, hic fati fundamentum est positus stellarum, tale temperamentum in **Iosquino** efficiens, quod convenientius est Musicae, quam temperamenta sunt in aliis. Terminus est, celeriter arripere harmonias, et suavius canere quam alii».

33. BLAISE 1975, p. 375: «Fatum (scol.), ordre des causes secondes destinées à accomplir les décrets de la Providence».

34. About Ficino and music, see TOMLINSON 1993 and BOCCADORO 2000.

35. MCDONALD 2015, pp. 118-119.

Notwithstanding his indebtedness to Ficino, Melanchthon did not say a word about the impact of music on the humours or spirits. Nor did he borrow anything from Ficino's celebrated music therapy, where music was capable of bringing humours into balance. Ten pages before the above quotation, Melanchthon had touched upon the theory of affections[36]. Although he treated the affections from a physiological standpoint, he did not establish any connection to music whatsoever. Only in his (much later) foreword to Johannes Reusch's *Zehen deudscher Psalm Davids* did he list Josquin's psalm settings among other musical works with the power to lift the heart of one in distress[37]. But in the *Erotemata dialectices* it was neither the beauty nor the impact of Josquin's music that Melanchthon was interested in, only the composer's extraordinary natural capabilities. Overall, it is striking how this treatise on the art of reasoning and discussion is so full of ideas pertaining to natural philosophy. Yet, although music and its impact by no means figure in the otherwise very broad scope of topics handled in the *Erotemata dialectices*, it is in that work that Melanchthon invoked Josquin to exemplify the natural ability or talent that he calls *habitus*, which he understood as the consequence of astrological influences on the humoural disposition of the body: a physical, natural-scientific explanation.

Initia doctrinae physicae (1549)

In his next major textbook, the *Initia doctrinae physicae* (Principles of Physical Knowledge; Wittenberg, 1549; Ill. 3.4), Melanchthon was no longer willing to refer to the impact of the stars on humours and in general distanced himself from the host of arguments that pertained to the paradigm of «Renaissance magic». Instead of astrology, Melanchthon now advocated astronomy. Moreover, the *Initia* is often cited as one of the first pieces of evidence of the upheaval caused by the publication of Nikolaus Copernicus's *De revolutionibus orbium coelestium* (1543). The assumption that acquaintance with Copernicus's work had caused a conversion in Melanchthon's outlook is, however, without any foundation. Melanchthon had propagated astronomy in Wittenberg at least since 1530, as he «supervised an editorial program that produced several innovative textbooks, notably in astronomy, written or edited by the professors and produced in the university print shop»[38]. His enthusiasm about the general revival of astronomy is documented in his prefaces to editions of older astronomical treatises, such as the one for the 1538 Wittenberg edition of *De sphaera mundi* (On the Earth Globe) by Johannes de Sacrobosco (c.1230; first published in Ferrara in 1472)[39]. In this light, it is not

36. Melanchthon, *Erotemata* (1547), f. 24r (Melanchthon 1846, c. 543).

37. Groote 2013, p. 235.

38. Pantin 2020, p. 288.

39. Hammer 1951, p. 318.

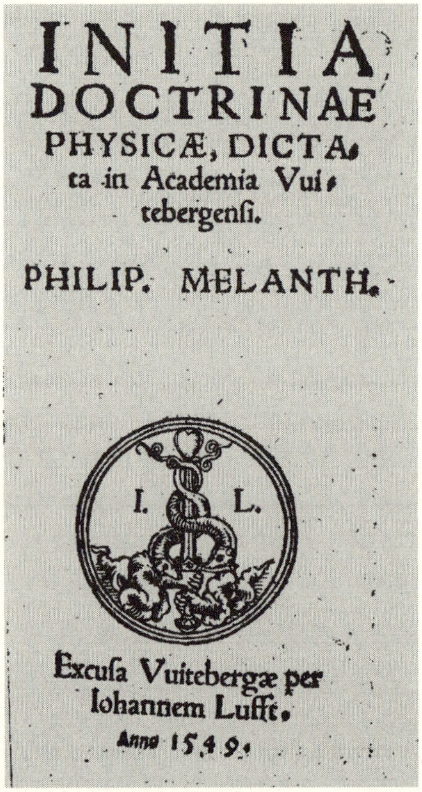

INITIA
DOCTRINAE
PHYSICÆ, DICTAta in Academia Vuitebergensi.

PHILIP. MELANTH.

Excusa Vuitebergæ per Iohannem Lufft.
Anno 1549.

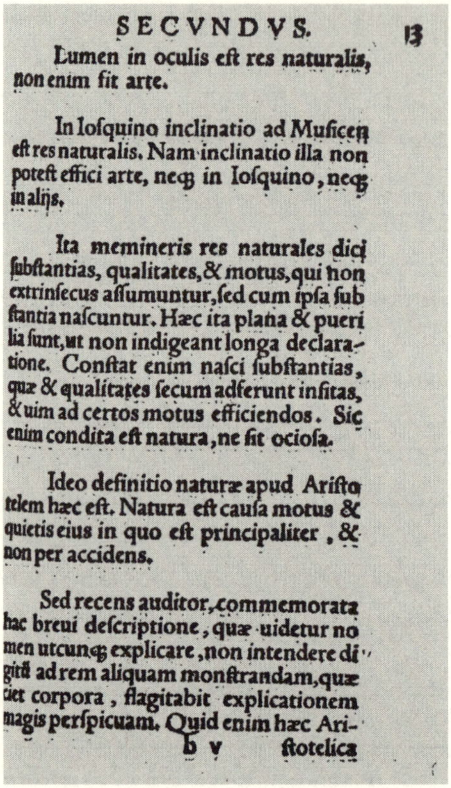

SECVNDVS. 13

Lumen in oculis est res naturalis, non enim fit arte.

In Iosquino inclinatio ad Musicen est res naturalis. Nam inclinatio illa non potest effici arte, neq; in Iosquino, neq; in aliis.

Ita memineris res naturales dici substantias, qualitates, & motus, qui non extrinsecus assumuntur, sed cum ipsa substantia nascuntur. Hæc ita plana & puerilia sunt, ut non indigeant longa declaratione. Constat enim nasci substantias, quæ & qualitates secum adferunt insitas, & uim ad certos motus efficiendos. Sic enim condita est natura, ne sit ociosa.

Ideo definitio naturæ apud Aristotelem hæc est. Natura est causa motus & quietis eius in quo est principaliter, & non per accidens.

Sed recens auditor, commemorata hac breui descriptione, quæ uidetur nomen utcunq; explicare, non intendere digitū ad rem aliquam monstrandam, quæ ciet corpora, flagitabit explicationem magis perspicuam. Quid enim hæc Aristotelica

b v

ILL. 3.4AB: Philip Melanchthon, *Initia physicae doctrinae* (Wittenberg, 1549), title page and Liber secundus, f. 13r. Copy A-Wn.

surprising that he familiarized himself with Copernican theory even before Copernicus himself made up his mind to publish *De revolutionibus*: the latter's student Georg Joachim Rheticus had sent his summary of his teacher's theory, published as *De libris revolutionum Copernici narratio prima* (Gdańsk, 1540), to Melanchthon already in the year of publication[40]. In the *Initia* Melanchthon rejected the heliocentric model, describing the new theory as «an old and absurd 'paradox', which Aristarchus had once defended, and which young students ought to stay away from since it conflicted with Scripture»[41]. And yet, in other chapters of the *Initia*, he praised Copernicus's lunar theory and cited his observations and calculations[42].

[40]. WESTMAN 1975, p. 172.

[41]. Philip Melanchthon, *Initia doctrinae physicae* (Wittenberg, 1549), Liber primus, f. 47v (MELANCHTHON 1846, c. 216): «Sed hic aliqui vel amore novitatis, vel ut ostentarent ingenia, disputarunt moveri terram, et contendunt nec octavam sphaeram, nec Solem moveri, cum quidem caeteris coelestibus orbibus motum tribuant, Terram etiam inter sidera collocant. Nec recens hi ludi conficti sunt. Extat adhuc liber Archimedis de numeratione arenae, in quo narrat, Aristarchum Samium hoc paradoxum tradidisse, Solem stare immotum, et terram circumferri circa Solem». Quoted from WESTMAN's (1975, p. 173) paraphrase.

[42]. Melanchthon, *Initia* (1549), Liber primus, f. 75r-78v, 97v (MELANCHTHON 1846, cc. 241-244, 262).

As for physics, Melanchthon was still very far from regarding it as an empirical science in the modern sense of the word. He treated many speculative issues still within the framework of ancient and medieval natural philosophy. He defined physics as a doctrine «that inquires into and reveals the succession, qualities, and motions of all bodies and species[43] in nature, as well as the causes of generation, corruption and others motions in the elements and in other bodies which arise from the mixture of elements»[44].

The two references to Josquin in the *Initia doctrinae physicae* are very different in character. The first one does not contain anything substantially new compared to the earlier citations:

> Temperament or 'mixture' is a natural thing, for it comes into being along with the substantial form of a living being.
>
> Light in the eyes is a natural thing, for it does not come to be by artifice.
>
> **Josquin**'s inclination to music is a natural thing. For that inclination cannot be caused by artifice, neither in **Josquin**, nor in anybody else[45].

Here Melanchthon seems merely to reiterate points that he had already made in the *Erotemata dialectices*, equating «natural inclination» with «temperament». The opposition between artifice and nature is akin to the opposition between habit and inclination. It is reasonable to think that «artifice» implied the result of a «mechanical art» and had nothing to do with any «liberal art».

Three pages further, Josquin emerges once again in a similar context. Now, however, his genius is used to illustrate the opposition between an automaton and a living entity:

> Natural motion in the body is much more powerful than artificial motion, as when an automaton is stirred into motion by artifice. In the case of **Josquin**, suited for music by nature, the ability to make music is much more outstanding than in a non-musical person, even one who has, by means of study (*arte*), tries as hard as he can to make music his own[46].

43. 'Species' is a scholastic term to describe the immaterial emanations of bodies (such as light and sound) that reach the sense organs so that the body can be perceived.

44. Melanchthon, *Initia* (1549), Liber primus, f. 9r (Melanchthon 1846, c. 179): «Quid est physica doctrina? Est quae seriem, qualitates, et motus omnium corporum et specierum in natura, et causas generationum, corruptionum et aliorum motuum in elementis et aliis corporihus, quae ex elementorum commixtione oriuntur, inquirit et patefacit».

45. Melanchthon, *Initia* (1549), Liber secundus, ff. 12v-13r (Melanchthon 1846, c. 303; emphasis added): «Temperamentum seu κρᾶσις est res naturalis. Nascitur enim cum forma substantiali animantis. Lumen in oculis est res naturalis, non enim fit arte. In **Iosquino** inclinatio ad Musicen est res naturalis. Nam inclinatio illa non potest effici arte, neque in **Iosquino**, neque in aliis».

46. Melanchthon, *Initia* (1549), Liber secundus, f. 16v (Melanchthon 1846, c. 306; emphasis added): «Naturalis motus in corpore longe potior est, quam artificialis, ut cum arte cietur aliquod automaton. In **Josquino**

The first and foremost instantiation of an automaton in Melanchthon's time was the clock. In the sixteenth and the seventeenth centuries, clocks featuring automata reached great popularity. Melanchthon's contemporary, the famous clockmaker Hans Bullmann of Nuremberg (died 1535), was believed to have constructed a completely mechanical figure that simulated a living human or an animal. In addition, he produced numerous successful figures of men and women that moved and played musical instruments[47]. Both imitated the actions of living beings, often accompanied by mechanical music, as Otto Mayr (1986) has pointed out[48]. Apart from the magical effects that these self-moving machines produced, they also had a symbolic dimension: automated machinery exemplified ruling power and authority. According to Mayr, the regular running of clocks made them «a symbol of any authority that brought order into human life»[49]. The tolling of the tower clock commanded one to submit to the day's schedule, even if it ran counter to a given individual's current inclination or emotional state.

In the *Erotemata dialectices*, Josquin had already been juxtaposed with a clockwork. The present context reveals that Melanchthon's worshipping of Josquin was at its core akin to Martin Luther's well-known exuberant praise of the composer, even though the two Reformers used very different rhetorics. Luther's praise showed his appreciation of Josquin as a free-creating artist not bound by the rules. For Luther, Josquin was an external example of the contrast between *lex* and *evangelium*, between old compulsion and new freedom, which was fundamental for his theology[50]. It is reasonable to surmise that in his second Josquin citation in the *Initia*, Melanchthon was referring to an automaton not only as an instantiation of an unnatural motion, but also as a symbol of compulsion[51]. In Melanchthon's thought, however, the metaphoric uses of «automata» and «machines» bore great significance, and their spectrum of meanings deserves a closer examination.

THE ARGUMENT FROM DESIGN

From the beginning of his career, Melanchthon was concerned about reconciling certain theories of pagan Antiquity with the dogmata of Christian faith. Towards the mid-sixteenth

natura ad musicam idoneo, multo praestantior facultas est canendi, quam in homine amuso, qui tamen arte, quamtum potest musicam sibi accersit».

47. BEDINI 1964, p. 26.

48. MAYR 1986, p. 21.

49. *Ibidem*, p. 43.

50. STAEHELIN 1986, p. 3; cited in GROOTE 2013, p. 233.

51. In this context, Josquin proves to have stood in opposition to his almost exact contemporary Leonardo da Vinci, whose life-time fame and exceptional status made the comparison between the two so tempting. In contrast to Leonardo, Josquin's genius was confined to music and had no connections to the *artes mechanicae*. Nor does the history of Josquin reception testify to any connection between the composer and the history of science.

century, his new concern became the reconciliation of both pagan philosophies and Christian religion with the recent ground-breaking scientific discoveries.

Melanchthon championed what in modern intellectual history has been called the 'argument from design'. It was an argument by analogy, «claiming that since the universe and (say) a clock resembled each other in some respects, they probably resembled each other in the further respect of being the product of design»[52]. Philosophers eager to reconcile science and Christian theology posited God as a supreme designer.

By the mid-sixteenth century, the likening of the Universe with automata or a clock already had a long tradition. Particularly famous became the comparison of celestial motion to a mechanical clock drawn by the fourteenth-century philosopher Nicole Oresme soon after this contrivance had been invented[53]. In the *Initia doctrinae physicae*, Melanchthon referred to the system of planets as a «machine in the sky», especially in the chapter devoted to the lunar theory of Copernicus[54]. Sachiko Kusukawa (1995) has noted that Melanchthon justified the propagation of astronomical knowledge because it «taught about Providence»[55].

In contrast to the late medieval philosophers, however, Melanchthon applied the design argument not only to the macrocosm (the Universe) but also to the microcosm (the human body). In his revised commentary on Aristotle's *On the Soul*, published in 1553 as *Liber de anima* (A Book on the Soul), he insisted on the theological importance of anatomical knowledge:

> Looking at this wonderful variety of work and these designs of God from without and through a thick darkness, we are struck dumb and lament that we cannot look into nature and discern causes. But then at last, when we discern the 'idea' of nature in the divine mind, we shall look into that whole machine as if from the inside, and we shall understand the designs of the Maker and the causes of all divine works. Now, through this incomplete consideration, we know that God is the Architect, and we should be inflamed by the desire for that perfected wisdom[56].

---

[52]. BLACKBURN 2008, p. 130.

[53]. BONER 2013, p. 17: «"If the heavens display such a high degree of rationality and regularity", Oresme wrote, "how much more ought we to think of that architect who, it is said, made all things in number, weight, and measure?"».

[54]. Melanchthon, *Initia* (1549), Liber primus, f. 78v.

[55]. KUSUKAWA 1995, pp. 12-13.

[56]. Philip Melanchthon, *Liber de anima* (Wittenberg, 1553), f. 51v: «Hanc mirificam operis varietatem, et haec Dei consilia, foris et per densam caliginem aspicientes, obstupescimus, ac dolemus, nos non penitus introspicere naturam, et causas videre posse. Sed tunc demum, cum idaeam naturae in mente divina cernemus, totam hanc machinam velut intus aspiciemus, et opificis consilia et causas omnium operum divinorum intelligemus. Nunc inchoata hac consideratione Deum architectum agnoscamus, et illius perfectae sapientiae desiderio accendamur». English translation (uncredited) from KORNU 2015, p. 106.

While Melanchthon's response to Copernicus was mixed, his instant reaction to Andreas Vesalius's anatomical treatise *De humani corporis fabrica* was very enthusiastic. In his *Liber de anima*, he incorporated many references to Vesalius's work, including anatomical illustrations to accompany the text. He even wrote a poem on Vesalius's book entitled 'De consideratione humani corporis' (On the Contemplation of the Human Body)[57]. With considerable effort, Melanchthon succeeded in establishing *De humani corporis fabrica* as the main textbook on anatomy at the University of Wittenberg[58].

Melanchthon's endeavour to employ anatomy for theological purposes led to a more radical rhetoric compared to Vesalius. Melanchthon emphatically called the human body a «machine»:

> Although we trace the beginnings of other physical parts from those that are more obvious to our senses, in this doctrine, before we talk about forces, it is useful to regard the human body and its different members as if they were machines. For just as, after having considered a mill or a cannon with its ball and powder, we can more easily deduce causes, so after pointing out the machines in our bodies we can better understand and judge their powers[59].

Kimbell Kornu (2015) has noted in this connection that «Melanchthon was not alone in calling the human body a machine but was the most theologically influential»[60]. In resorting to this metaphor, Melanchton was drawing on ancient sources; his chief source, however, was not a pagan author such as Galen, but one of the Fathers of the Eastern Church, Gregory of Nyssa (Gregorius Nyssenus; fourth century AD)[61]. In his *Dialogue on the Soul and Resurrection*, Gregory spoke of «the organic machine of our natural bodies»[62], justifying this metaphor by the observations that the components of a living body behaved in a manner contrary to their nature: air being harnessed to produce sound, water impelled to move upward, and so forth. Moreover, Gregory regarded the human body as a miniature, harmonious version of the cosmos as a whole[63].

---

[57]. Handwritten in the copy of Vesalius's book in US-BES.

[58]. Nutton 1993.

[59]. Melanchthon, *Liber de anima* (1553), f. 15r.: «Ut autem alias physicae partes ordimur ab iis, quae magis sensui obvia sunt, ita in hac doctrina, priusquam de potentiis dicemus, prodest aspicere corpus hominis, et diversa membra et organa, quasi machinas considerare. Ut enim conspecta mola seu bombarda et globo, ac pulvere tormentario, facilius raciocinari causas possumus, ita in corpore nostro monstratis machinis, potentias facilius intelligere et iudicare possumus».

[60]. Kornu 2015, p. 107.

[61]. On Melanchthon's study of Gregory of Nyssa among the other Cappadocian Fathers, see Hall 2014, pp. 241-243.

[62]. Gregory of Nyssa, *Dialogue on the Soul and Resurrection*, Chap. 13: «τὸ μηχανικὸν τοῦτο τῆς φύσεως ἡμῶν ὄργανον». Translation from Gregory of Nyssa 1893, p. 435.

[63]. Gregory of Nyssa 1995, pp. 441-444.

The Rise and Fall of the Mechanistic Worldview

The analogies with mechanism pertaining to the argument from design were largely aimed at sparking admiration for divine wisdom. Setting the organic motion of Josquin's music against the artificial motion of an automaton, by contrast, aimed at emphasizing the latter's lifelessness and inanimateness, features that were likely to reduce one's enthusiasm. We see that Melanchthon's thought in a nutshell foreshadowed the general course of development of the mechanistic worldview.

In the seventeenth century, the scope of phenomena compared to «artificial, man-made devices and the ways the various parts of these artefacts interacted and moved in relation to each other»[64] was expanding greatly. The seventeenth-century mechanists not only likened the Universe and the human body to machines, but also deliberately sought to abolish the Aristotelian framework with its notion of the tripartite soul (vegetative, sensual, and rational) and its doctrine of four causes (the formal, the material, the final and the efficient). By contrast, Melanchthon continued to adhere to the Aristotelian notion of natural things and maintained that human organs had the cause of their motion in themselves. His view of the human body as a machine did not appear uncanny to anybody because its conception was not radically reductionistic (as was later the case with René Descartes). But despite this conservative stance, Melanchthon managed to prefigure the crucial insight that was articulated only towards the end of the eighteenth century: the mechanical lifelessness of perfectly executed human actions.

Two hundred and fifty years after Melanchthon's death, Georg Wilhelm Friedrich Hegel wrote in his *Wissenschaft der Logik* (The Science of Logic; 1812-1816):

> This is what constitutes the character of mechanism, namely, that whatever the connection that obtains between the things combined, the connection remains one that is alien to them, that does not affect their nature, and even when a reflective semblance of unity is associated with it, the connection remains nothing more than juxtaposition, mixture, aggregate, etc. Spiritual mechanism, like its material counterpart, also consists in the things connected in the spirit remaining external to one another and to spirit. A mechanical mode of representation, a mechanical memory, a habit, a mechanical mode of acting, mean that the pervasive presence that is proper to spirit is missing in the person who perceives or executes this[65].

---

[64]. Gouk 2013, p. 36.

[65]. Georg Wilhelm Friedrich Hegel, *Wissenschaft der Logik* (Nuremberg, 1812-1816), vol. ii, pp. 202-203: «Dies macht den Charakter des Mechanismus aus, daß, welche Beziehung zwischen den Verbundenen stattfindet, diese Beziehung ihnen eine fremde ist, welche ihre Natur nichts angeht und, wenn sie auch mit dem Schein eines Eins verknüpft ist, nichts weiter als Zusammensetzung, Vermischung, Haufen usf. bleibt. Wie der materielle Mechanismus, so besteht auch der geistige darin, daß die im Geiste Bezogenen sich einander und ihm selbst äußerlich bleiben. Eine mechanische Vorstellungsweise, ein mechanisches Gedächtniß, die Gewohnheit, eine

Accordingly, Hegel approved the use of the term «mechanism» and its derivatives in everyday language and in the figurative sense: «It is quite right to speak of 'mechanical' memory, and of all manner of 'mechanical' activities, such as reading, writing, and playing music, for example»[66]. Hegel's passage testifies that around the turn of the nineteenth century «mechanism» had become a pejorative metaphor.

As the unique possessor of a divine gift, Josquin was contrasted in Melanchthon's writings not only to an automaton but also to a «normal» musician who merely employed acquired habits. In Melanchthon's view, no perfect skills could ever substitute for that mysterious quality that we call musical talent. His praise of artistic sensitivity in contrast to artificial motion and mechanical skills foreshadowed the stance of the advocates of the *Empfindsamkeit* movement. Before the writings on music by Carl Philipp Emanuel Bach and Jean-Jacques Rousseau, neither natural talent nor the inner world were seen as decisive factors in achieving a high degree of emotional impact in the arts.

Although the above-quoted passage by Hegel does not directly refer specifically to either music or the arts, it reflects the changing attitude to the acquisition of skills in rhetoric and music during the early modern period. Rhetoric, which enjoyed a great revival in the seventeenth century, was understood as an art of influencing the souls of the orator's audience, while the state of the soul of the orator himself was of little account. Towards the end of the eighteenth century, however, the persuasive power of speech was no longer attributed to a particular skillfully applied technique but instead to its source in the heart or the soul of the orator. Immanuel Kant confessed his aversion to rhetoric: «the reading of the best discourse [...] has always been mingled with an unpleasant feeling of disapprobation of a treacherous art, which means to move men in important matters *like machines* to a judgement that must lose all weight for them on quiet reflection»[67]. Likewise, the musical doctrine of affects, which had arisen in a tight coupling with rhetoric, lost its dominance: «the conventionalized representations of the affects [...] began to seem stereotyped and unnaturally static», as George Buelow (2001) has pointed out[68]. The masterful possession of artificial skills began to be compared to a mechanical action. It was supposed that profound souls would remain completely unaffected. If they were superficial, however, the case could be very different, as was shown by E. T. A. Hoffmann in a satirical scene in *Der Sandmann* (The Sandman; 1816), in which the public was enraptured by the perfect execution of melodies by the mechanical doll Olympia.

---

mechanische Handlungsweise bedeuten, daß die eigenthümliche Durchdringung und Gegenwart des Geistes bey demjenigen fehlt, was er auffaßt oder thut». Translation (modified) from Hegel 2010, p. 631.

66. Hegel 1986, pp. 355-356: «Man spricht mit Enzyklopädie der philosophischen Wissenschaften I 581 Recht vom mechanischen Gedächtnis und von den allerhand mechanischen Betätigungen, wie z. B. Lesen, Schreiben, Musizieren usw.». Translation from Hegel 1991, p. 275.

67. Kant 1914, p. 217 (emphasis added).

68. *Grove2* 2001, article 'Rhetoric and Music: ii. After the 1750' by George J. Buelow.

Concluding Remarks

The works by Melanchthon in which Josquin is mentioned pertain to very different fields of knowledge. The image of Josquin depicted by Melanchthon is nonetheless quite consistent: an extraordinary artist exhibiting a divine gift that required explanation. Unlike Luther, Melanchthon did not specify what exactly the composer's superiority was, and nowhere did he touch on Josquin's mastery in setting words to music.

Although Josquin embodied for him natural strength, natural movement, and natural temperament, Melanchthon nonetheless seems to have seen him merely as a divine tool and a mediator of the stars, with nothing genuinely human about him. Therefore, notwithstanding Melanchthon's prophetic insight into the common mechanistic nature of automata and artistic habits, it would be far-fetched to claim that in his image of Josquin he prefigured the romantic aesthetic of the artistic genius. Unlike typical images of geniuses by Rousseau and the Romantics, Melanchthon's Josquin suffered neither spiritual turmoil nor creative crises.

Actually, it is not Josquin but Melanchthon himself who appears inconsistent and even eclectic. In the *Philosophiae moralis epitome* he utilized the composer to promote the Aristotelian version of the ethos doctrine applied to music. In the *Erotemata dialectices* he appears as a passionate advocate of Renaissance magic. In the *Initia doctrinae physicae* he seems to be a proponent of empirical science. His four references to Josquin des Prez in three treatises, which were written within less than one decade, eloquently testify to the co-existence, in the mid-sixteenth century, of diverse mindsets that appear now as barely compatible.

Bibliography

Aristotle 1957

Aristotle. *Problems, Books xxii-xxx*, translated by W. S. Hett, London, William Heinemann, 1957.

Bedini 1964

Bedini, Silvio A. 'The Role of Automata in the History of Technology', in: *Technology and Culture*, v/1 (1964), pp. 24-42.

Blackburn 2008

Blackburn, Simon. *The Oxford Dictionary of Philosophy*, Oxford, Oxford University Press, 2008.

Blaise 1975

Blaise, Albert. *Dictionnaire latin-français des auteurs du Moyen-Age*, Turnhout, Brepols, 1975.

Boccadoro 2000

Boccadoro, Brenno. 'Marsilio Ficino. The Soul and the Body of Counterpoint', in: *Number to Sound: The Musical Way to the Scientific Revolution*, edited by Paolo Gozza, Dordrecht, Kluwer, 2000, pp. 99-134.

Boner 2013

Boner, Patrick J. *Kepler's Cosmological Synthesis: Astrology, Mechanism and the Soul*, Leiden, Brill, 2013.

Eggebrecht 1959

Eggebrecht, Hans Heinrich. *Heinrich Schütz. Musicus poeticus*, Gottingen, Vandenhoeck & Ruprecht, 1959.

Freeman – Pinchemel 1979

Geographers: Biobibliographical Studies. 3, edited by Thomas Walter Freeman and Philippe Pinchemel, Bloomsbury Academic, 1979.

Gouk 2013

Gouk, Penelope. 'Clockwork or Musical Instrument? Some English Theories of Mind-Body Interaction before and after Descartes', in: *Structures of Feeling in Seventeenth-Century Cultural Expression*, edited by Susan McClary, Toronto, University of Toronto Press, 2013, pp. 36-59.

Gregory of Nyssa 1893

Saint Gregory of Nyssa. 'On the Soul and Resurrection', translated by William Moore and Henry Austin Wilson, in: [*A Select Library of*] *Nicene and Post-Nicene Fathers of the Christian Church, Second Series, Vol. 5*, edited by Philip Schaff and Henry Wace, Buffalo (NY), Christian Literature Publishing, 1893, pp. 428-470.

Gregory of Nyssa 1995

Id. *On the Inscriptions of the Psalms*, translated by Ronald E. Heine, Oxford, Clarendon Press, 1995.

Groote 2013

Groote, Inga Mai. 'Musikalische Poetik nach Melanchthon und Glarean: Zur Genese eines Interpretationsmodells', in: *Archiv für Musikwissenschaft*, lxx/3 (2013), pp. 227-253.

Grove2 2001

New Grove Dictionary of Music and Musicians, Second edition, edited by Stanley Sadie, 29 vols., Oxford, Oxford University Press, 2001.

Hall 2014

Hall, H. Ashley. *Philip Melanchthon and the Cappadocians: A Reception of Greek Patristic Sources in the Sixteenth Century*, Göttingen, Vandenhoeck & Ruprecht, 2014.

Hamid 2018

Hamid, Nabeel. *Being and the Good. Natural Teleology in Early Modern German Philosophy*, Ph.D. Diss., Philadelphia (PA), University of Pennsylvania, 2018.

Hammer 1951

Hammer, William. 'Melanchthon, Inspirer of the Study of Astronomy, With a Translation of His Oration in Praise of Astronomy (*De Orione*, 1553)', in: *Popular Astronomy*, lix (1951), pp. 308-319.

Hegel 1986

Hegel, Georg Wilhelm Friedrich. *Enzyklopädie der philosophischen Wissenschaften. [1830], Erster Teil. Die Wissenschaft der Logik. Mit den mündlichen Zusätzen*, edited by Eva Moldenhauer and Karl Markus Michel, Frankfurt am Main, Suhrkamp, 1986.

Hegel 1991

Id. *The Encyclopaedia Logic (with the Zusatze)*, translated by Theodore F. Geraets, W. A. Suchting and H. S. Harris, Indianapolis (IN), Hackett Publishing Company, 1991.

Hegel 2010

Hegel, Georg Wilhelm Friedrich. *The Science of Logic*, translated and edited by George di Giovanni, Cambridge, Cambridge University Press, 2010.

Kant 1914

Kant, Immanuel. *Critique of Judgement*, translated by J. H. Bernard, London, Macmillan, 1914.

Kornu 2015

Kornu, Kimbell. '«Know thyself»: The Soul of Anatomical Dissection', in: *The Resounding Soul. Reflections on the Metaphysics and Vivacity of the Human Person*, edited by Samuel Kimbriel and Eric Austin Lee, Cambridge, The Lutterworth Press, James Clarke & Co. Ltd., 2015, pp. 93-113.

Kusukawa 1995

Kusukawa, Sachiko. *The Transformation of Natural Philosophy: The Case of Philip Melanchthon*, Cambridge, Cambridge University Press, 1995.

Mack 1993

Mack, Peter. *Renaissance Argument*, Leiden, Brill, 1993.

Mathesius 1841

Mathesius, Johannes. *Leben Dr. Martin Luthers, in siebzehn Predigten*, edited by A. J. D. Rust, Berlin, G. Crantz, 1841.

Mayr 1986

Mayr, Otto. *Authority, Liberty, & Automatic Machinery in Early Modern Europe*, Baltimore, Johns Hopkins University Press, 1986.

McDonald 2015

McDonald, Grantley. 'Melanchthon's Theory of Spirit as a Bridge between Galen, Ficino and Luther', in: *Aesthetics of the Spirits: Spirits in Early Modern Science, Religion, Literature and Music*, edited by Steffen Schneider, Göttingen, V&R Unipress, 2015, pp. 111-128.

Melanchthon 1846

Melanchthon, Philip. *Opera quae supersunt omnia, Volumen* XIII, edited by Karl Gottlieb Bretschneider, Halle, C. A. Schwetschke, 1846.

Melanchthon 1850

Id. *Opera quae supersunt omnia, Volumen xvi*, edited by Karl Gottlieb Bretschneider and Heinrich Ernst Bindseil, Halle, C. A. Schwetschke, 1850.

Nutton 1993

Nutton, Vivian. 'Wittenberg Anatomy', in: *Medicine and the Reformation*, edited by Ole Peter Grell and Andrew Cunningham, Abingdon-New York, Routledge, 1993, pp. 11-32.

Owens 1997

Owens, Jessie Ann. 'How Josquin Became Josquin. Reflections on Historiography and Reception', in: *Music in Renaissance Cities and Courts: Studies in Honor of Lewis Lockwood*, edited by Jessie Ann Owens and Anthony M. Cummings, Sterling Heights (MI), Harmony Park Press, pp. 271-280.

Pantin 2020

Pantin, Isabelle. 'Borrowers and Innovators in the History of Printing Sacrobosco. The Case of the In-Octavo Tradition', in: «*De sphaera*» *of Johannes de Sacrobosco in the Early Modern Period*, edited by Matteo Valleriani, Cham (Switzerland), Springer Open, 2020, pp. 265-312.

Schlagel 2006

Schlagel, Stephanie P. 'A Credible (Mis)Attribution to Josquin in Hans Ott's «Novum et insigne opus musicum». Contemporary Perceptions, Modern Conceptions, and the Case of «Veni sancte Spiritus»', in: *Tijdschrift van de Koninklijke Vereniging voor Nederlandse Muziekgeschiedenis*, lvi/2 (2006), pp. 97-126.

Smith 1993

Smith, Robin. 'Aristotle on the Uses of Dialectic', in: *Synthese*, xcvi/3 (1993), pp. 335-358.

Staehelin 1986

Staehelin, Martin. 'Luther über Josquin', in: *Festschrift Martin Ruhnke zum 65. Geburtstag von den Mitarbeitern des Instituts für Musikwissenschaft der Universität Erlangen-Nürnberg*, Stuttgart, Carus, 1986, pp. 326-338.

Tomlinson 1993

Tomlinson, Gary. *Music in Renaissance Magic: Toward a Historiography of Others*, Chicago, University of Chicago Press, 1993.

Wels 2008

Wels, Volkhad. 'Melanchthon's Textbooks on Dialectic and Rhetoric as Complementary Parts of a Theory of Argumentation', in: *Scholarly Knowledge: Textbooks in Early Modern Europe*, edited by Emidio Campi, Simone De Angelis, Anja-Silvia Goering and Anthony Grafton, Geneva, Droz, 2008, pp. 139-156.

Westman 1975

Westman, Robert S. 'The Melanchthon Circle, Rheticus, and the Wittenberg Interpretation of the Copernican Theory', in: *Isis*, lxvi/2 (1975), pp. 164-193.

Chapter Four

«Intra quaternarii limites musicam intelligunt omnem consonantiam»: La musica nel pensiero di Giordano Bruno

Carlo Bosi
(Paris-Lodron-Universität Salzburg)

Nell'atomismo vitalistico e, in quanto tale, 'animistico' della filosofia di Giordano Bruno 'da Nola' (1548-1600), l'intera materia non è permeata che da un'unica vita, o principio formale, grazie alla quale, nel continuo cangiare vicissitudinale delle forme e delle specie, essa acquisisce sempre nuove apparenze, pur restando nella sua totalità sempre identica a sé stessa, come il Nolano espone in maniera particolarmente pregnante nel dialogo *De la causa, principio et uno*, pubblicato a Londra nel 1584[1]. In questo contesto ogni singolo essere, pur mantenendo una propria distinta individualità, è e rimane in qualche modo connesso in una rete fittissima di corrispondenze, di 'vincoli', con tutti gli altri esseri. Inoltre per Bruno la Deità include la Materia o 'Natura', in quanto suo fondamento inesauribile, è quindi in qualche modo anche Materia/Natura, visto che quest'ultima ne è «il grande simulacro»[2] e, allo stesso tempo, è causa immanente della materia, pur senza coincidere con essa[3].

Questa concezione è radicalmente contraria alle dottrine aristoteliche, ancora in parte dominanti, seppure in retrocessione, alla fine del XVI secolo, dottrine che fra l'altro Bruno, sin

1. Giordano Bruno, *De la causa, principio et uno* (Londra, 1584) (Bruno 2002a, pp. 591-746).
2. Bruno, *De la causa* (1584), p. 83 (Bruno 2002a, p. 693).
3. Carannante 2019, voce 'Dio, dèi (Idio; *Deus, dii*)' (Salvatore Carannante), p. 345.

dai suoi anni di noviziato presso il convento di San Domenico Maggiore a Napoli, recepisce soprattutto con il filtro dei commenti, assai problematici dal punto di vista dell'ortodossia cristiana, di Averroè[4]. Inoltre per Bruno, in netta contrapposizione all'Umanesimo, l'uomo non occupa più un posto ontologicamente privilegiato rispetto agli altri esseri. Egli in effetti sarebbe, al pari di ogni altra entità, animato dallo stesso principio vitale universale, la differenza sostanziale fra l'uomo e gli altri esseri dipendendo esclusivamente dalla sua particolare conformazione fisica, e, in particolare, dal fatto di possedere un organo che, come la mano, è capace di modificare il suo ambiente ben al di là di ogni altro essere[5]. Ed è solo grazie ai propri sforzi e a una tensione costante verso una sempre maggiore 'perfezione', che l'essere umano può cercare di avvicinarsi sempre maggiormente alla comprensione del cosmo nella sua infinitezza, pur senza mai raggiungerla appieno, divenendo in tal modo, in una sorta di 'furia' gnoseologica, 'Eroe'[6], senza per questo usufruire di un'imponderabile 'Grazia' divina o di un'altrettanto insondabile Provvidenza.

Questa concezione essenzialmente 'egalitaria' di tutti gli esseri viventi, compreso l'uomo, va letta anche nell'ambito di una rivalutazione dei filosofi presocratici da parte del Nolano, in particolar modo Parmenide, per cui «ogni atto cognitivo, in qualunque tipo di animale o di intelligenza separata [esso si manifesti], sarebbe l'istanza di un singolo potere cognitivo derivante dall'Intelletto Universale e dunque in ultima analisi da Dio in quanto Mente»[7]. Oltre all'influsso presocratico, non va naturalmente dimenticato quello platonico e neoplatonico, sebbene, a differenza che in Marsilio Ficino o in Niccolò Cusano, altri punti di riferimento importanti per l'iter filosofico di Bruno[8], Platonismo e Neoplatonismo vengano recepiti in un'ottica del tutto estranea ad ogni tentativo di cristianizzazione e ferma restando la fondamentale critica bruniana al concetto di 'Idea' platonica separata dal mondo corruttibile. In effetti il Dio bruniano e i semi intellettivi da Lui promananti (le Idee platoniche) sono sì sostanze eterne, ma sono allo stesso tempo coesistenti con il cosmo materiale illimitato e non ontologicamente da esso separate[9]. In Bruno sembra inoltre essere implicito una sorta di politeismo enoteistico, dove l'unità del divino si manifesta nella molteplicità di Dei, a loro volta esplicantisi negli astri e riflettentisi nella ricchezza cangiante e molteplice di 'demoni' e delle relative specie naturali a loro sottoposte[10].

---

[4]. *Ibidem*, voce 'Aristotele (Aristotile)' (Ilenia Russo), p. 450.

[5]. Montano 2000 e Carannante 2019, voce 'Mano (*manus*)' (Simonetta Bassi), pp. 284-288.

[6]. Bruno 2002b (*Degli eroici furori*).

[7]. Knox 2019: «All cognitive acts, in whatever animal or indeed separate intelligence, were instantiations of a single cognitive power deriving from the Universal Intellect and ultimately, therefore, from God as Mind».

[8]. Su Ficino e Cusano in Bruno, si veda Carannante 2019, voce 'Ficino Marsilio' (Salvatore Carannante), pp. 520-528, e voce 'Cusano Niccolò' (Salvatore Carannante), pp. 501-514, rispettivamente.

[9]. *Ibidem*, voce 'Platone' (Salvatore Carannante), pp. 441-443.

[10]. *Ibidem*, voce 'Dio, dèi (Idio; *Deus, dii*)' (Salvatore Carannante), pp. 347-348.

«Intra quaternarii limites musicam intelligunt omnem consonantiam»

Gnoseologia, cabala/magia e immagini musicali

Che anche l'acquisizione della conoscenza dipenda da una molteplicità di oggetti e d'immagini delle categorie più disparate, tutte a loro modo provviste di una loro dignità, è chiaro non appena si prendano in esame gli scritti bruniani dedicati all'«ars reminiscendi», in particolare *De umbris idearum*, uscito a Parigi nel 1582, la prima pubblicazione bruniana giunta sino a noi (Ill. 4.1A)[11]. Come ha ampiamente sottolineato Frances A. Yates (1966), l''ars memoriæ' nella 'nova philosophia' del Nolano va ben oltre le applicazioni pratiche, che, per esempio, possono essere di beneficio per una convincente «ars perorandi»[12]. Piuttosto, le immagini cariche di simboli dell'arte della memoria, con le loro ruote combinatorie, dovrebbero fungere da potente impalcatura su cui erigere e gestire la costruzione delle scienze, intese nel senso più ampio del termine (Ill. 4.1B)[13]. Del resto in gran parte degli scritti bruniani salta subito all'occhio l'importante ruolo che le immagini e di conseguenza la vista giocano nel suo sistema gnoseologico. Naturalmente non c'è nulla di nuovo in questo, perché la vista e l'immaginazione — cioè la capacità di formare immagini — hanno avuto un ruolo privilegiato in campo epistemologico almeno sin da Aristotele, secondo il quale «l'anima non pensa mai senza immagini»[14], enunciato poi riformulato da Tommaso d'Aquino secondo l'adagio «niente è nell'intelletto che prima non sia stato nel senso»[15]. Anche in Ficino, in molti aspetti un modello per Bruno, l'acquisizione di nuove conoscenze si basa in gran parte su 'εἴδωλα' [eidōla], ovvero immagini, e persino la musica viene colta da principio come un tipo speciale di immagine[16]. C'è però una differenza fondamentale tra le gnoseologie dei due maghi-filosofi: mentre in Ficino la Divinità impianta i semi della conoscenza o delle idee direttamente nell'anima umana (Ill. 4.2A), secondo Bruno l'uomo non può che percepire le immagini in quanto tracce delle Idee presenti non in sé stesso, bensì nella Natura (Ill. 4.2B)[17]. In tal modo l'uomo perderebbe la sua posizione privilegiata nel cosmo, diventando un anello come tanti nella catena degli esseri. L'essere umano, cioè, di norma non può che conoscere le ombre delle Idee, non arrivando mai a confrontarsi con le Idee stesse; tutt'al più può tendervi al massimo in un atto di sforzo supremo (gli 'eroici furori', per l'appunto).

[11]. Bruno 2004, pp. 2-586 (*De umbris idearum / Le ombre delle idee*).

[12]. Yates 1966, Cap. xi-xiv.

[13]. Carannante 2019, voce 'Natura (*natura*)' (Maria Elena Severini), p. 30.

[14]. Aristotele, *Sull'anima*, iii, 7, 431a16-17 (Aristotele 1907, p. 140): «νοεῖ ἄνευ φαντάσματος ἡ ψυχή».

[15]. Tommaso d'Aquino, *Quaestiones disputatae de veritate*, Quaestio 2, Articulus 3, Argumentum 19: «Nihil est in intellectu quod non sit prius in sensu».

[16]. Tomlinson 1993, p. 121.

[17]. Carannante 2019, voce 'Ficino Marsilio' (Salvatore Carannante), p. 523.

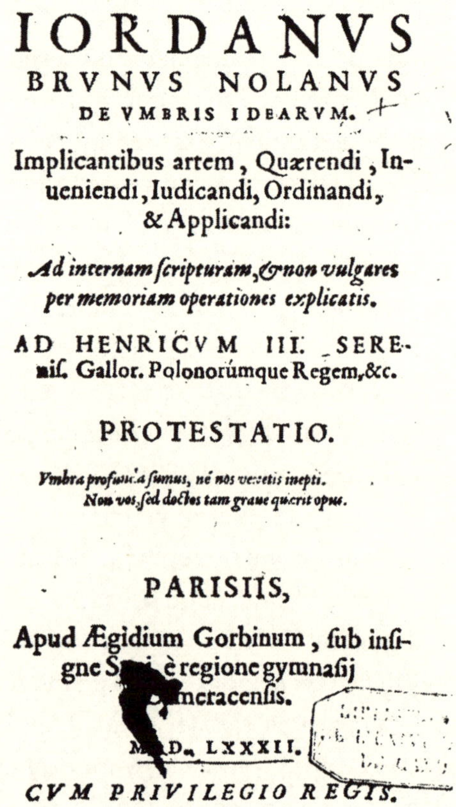

IORDANVS
BRVNVS NOLANVS
DE VMBRIS IDEARVM.

Implicantibus artem, Quærendi, Inueniendi, Iudicandi, Ordinandi, & Applicandi:

Ad internam scripturam, & non vulgares per memoriam operationes explicatis.

AD HENRICVM III. SERENISS. Gallor. Polonorúmque Regem, &c.

PROTESTATIO.

Vmbra profunda sumus, ne nos vexetis inepti. Non vos, sed doctos tam graue quærit opus.

PARISIIS,

Apud Ægidium Gorbinum, sub insigne S[illegible] è regione gymnasij [illegible]meracensis.

M.D. LXXXII.

CVM PRIVILEGIO REGIS.

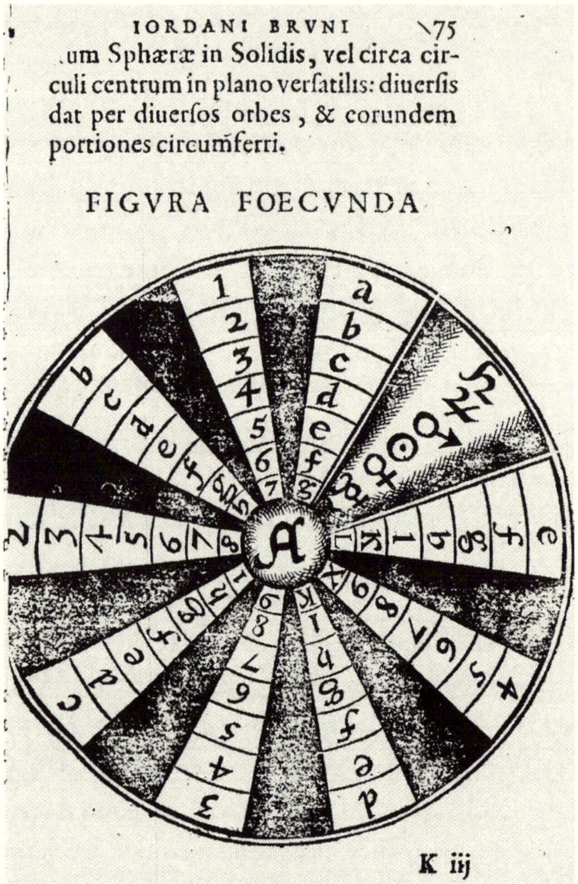

IORDANI BRVNI 75

.um Sphæræ in Solidis, vel circa circuli centrum in plano versatilis: diuersis dat per diuersos orbes, & eorundem portiones circumferri.

FIGVRA FOECVNDA

K iij

Ill. 4.1ab: Giordano Bruno, *De umbris idearum* (Parigi, 1582), frontespizio e p. 75 dell'*Ars memoriae*, con una ruota combinatoria. Copia D-Mbs.

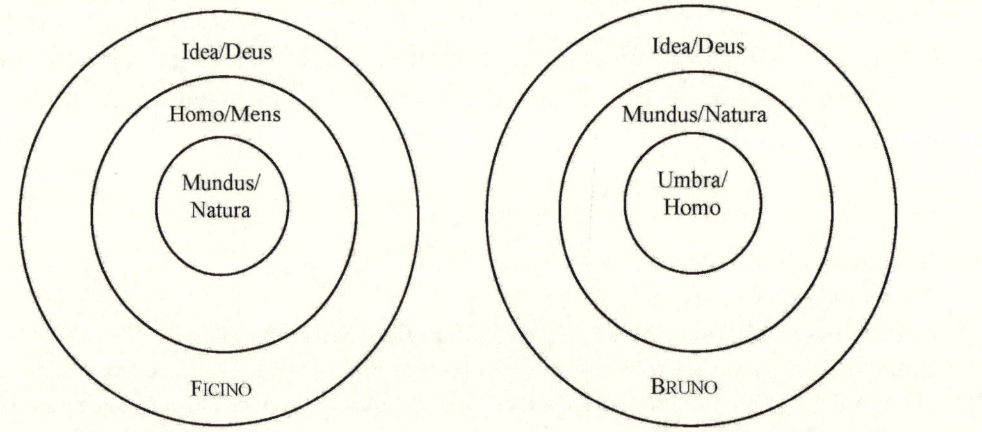

Ill. 4.2ab: gerarchie gnoseologiche in Ficino (con la Mente/Anima dell'Uomo mediante tra Divinità e Natura/Mondo) e Bruno (con il Mondo/Natura mediante tra Divinità/Idea e Ombra/Uomo).

Ecco perché per Bruno l'esistenza dell'uomo si caratterizza per una sua sostanziale 'umbratilità'. Tuttavia, questa condizione non è da giudicare negativamente, perché lo stato di 'umbratilità' acquista una dimensione tutta particolare in Bruno. L'ombra è infatti conseguenza della luce, «come traccia della luce, partecipe della luce, luce non piena»[18], ed essa stessa è una traccia delle Idee Divine impresse nel 'Libro della Natura'. È vero che l'uomo non potrà mai valicare l'incolmabile abisso tra il finito e l'infinito, tra la fisicità cangiante e l'eternamente divino; egli non può sostanzialmente superare la sua 'umbratilità'; eppure è proprio grazie alla sua esistenza umbratile, fatta di un miscuglio di luce e di tenebra, che può godere della libertà di esperienze infinite, alcune di esse potendo addirittura aiutarlo ad estendere al massimo i suoi limiti ontologici, acquisendo in tal modo lo status di 'eroe'[19]. Ma per raggiungere questo status, come già indicato, l'uomo deve guardare direttamente nel 'Libro della Natura', perché, contrariamente a quanto sostenuto da Ficino, le Idee posseggono un grado di verità più elevata nel mondo, nella Natura, e un grado minore nella mente umana, essendo la Natura più vicina alla monade divina, e di conseguenza, meno umbratile[20]. Da qui la grande importanza gnoseologica delle molteplici immagini tratte dalla Natura, che per Bruno comprendono anche, e questo è fondamentale, idee e concetti astratti, assimilando i quali l'uomo saggio può edificare, passo dopo passo, una struttura complessa e stratificata, manipolando la quale può persino sperare di agire direttamente sulla realtà. Questo è perché «tra *ordo idearum* ed *ordo rerum*, tra l'attività immaginativa umana e l'essere del mondo, vi è un forte legame, se è vero che la varietà infinita di enti che si producono nell'universo corrisponde alle infinite combinazioni di forme realizzate dall'attività fantastica»[21].

Un esempio palese sarebbe costituito da raggruppamenti o serie numeriche che, ricorrendo in Natura a un livello quasi radicale, sembrerebbero fornire una chiave di lettura analogica della Natura stessa[22]. E questo perché «[a]ttraverso i numeri [...] certo si dischiude per noi il cammino dalle immagini dei corpi e dalle ombre, che sono oscure realtà sensibili, alle idee»[23]. Si consideri, ad esempio, il passo seguente, tratto da uno degli ultimi trattati pubblicati da Bruno prima del suo lungo processo, e cioè il *De monade, numero et figura* uscito a Francoforte nel 1591 (Ill. 4.3ab):

18. Giordano Bruno, *De umbris idearum* (Parigi, 1582), f. e ii-verso (Bruno 2004, p. 46): «[...] ut lucis vestigium, lucis particeps, lux non plena».

19. Carannante 2019, voce 'Ombra (*umbra*)' (Nicoletta Tirinnanzi), p. 20; inoltre Ciliberto 1999.

20. Carannante 2019, voce 'Natura (*natura*)' (Maria Elena Severini), pp. 23-31; inoltre Védrine 1967, Brockmeier 1980, Carannante 2018 e Tirinnanzi 2000.

21. Carannante 2019, voce 'Natura (*natura*)' (Maria Elena Severini), p. 29.

22. *Ibidem*, voce 'Matematica (*mathematica*)' (Marco Matteoli), p. 377; inoltre De Bernart 2002 e Bönker-Vallon 2006.

23. Giordano Bruno, *Sigillus sigillorum* (Londra, 1583), p. 197 (Bruno 2009, p. 263): «Nobis sane a corporum imaginibus et umbris, quae sunt obscura sensibilia, per mathemata [...] ad ideas [...] datur accessus [...]».

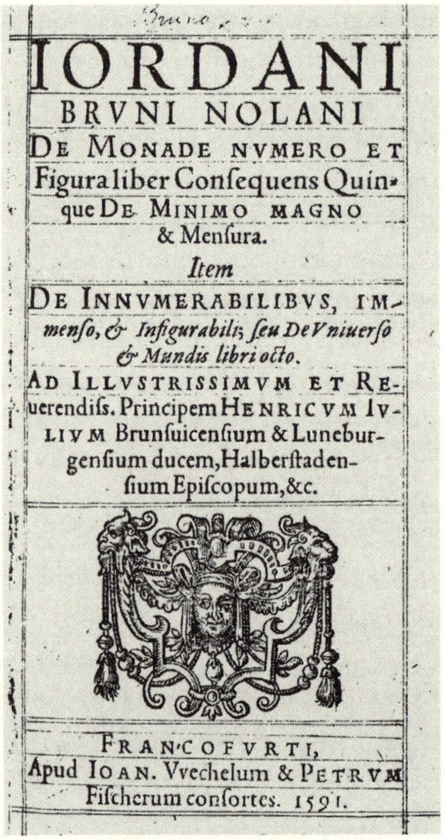

IORDANI
BRVNI NOLANI
De Monade nvmero et
Figura liber Consequens Quinque De Minimo magno
& Mensura.
Item
De Innvmerabilibvs, immenso, & Infigurabili; seu De Vniuerso & Mundis libri octo.
Ad Illvstrissimvm et Reuerendiss. Principem Henricvm Ivlivm Brunsuicensium & Luneburgensium ducem, Halberstadensium Episcopum, &c.

Francofvrti,
Apud Ioan. Vvechelum & Petrvm Fischerum consortes. 1591.

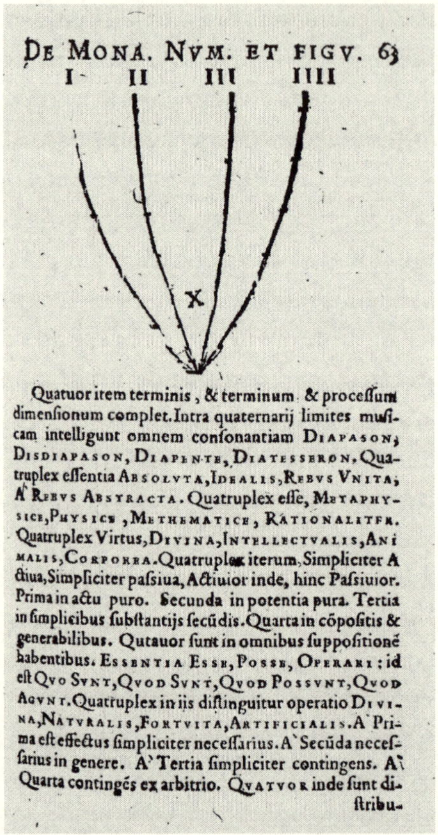

De Mona. Nvm. et figv. 63

Quatuor item terminis, & terminum & processum dimensionum complet. Intra quaternarij limites musicam intelligunt omnem consonantiam Diapason, Disdiapason, Diapente, Diatesseron, Quatruplex essentia Absolvta, Idealis, Rebvs Vnita, A Rebvs Abstracta. Quatruplex esse, Metaphysice, Physice, Mathematice, Rationaliter. Quatruplex Virtus, Divina, Intellectvalis, Animalis, Corporea. Quatruplex iterum, Simpliciter Actiua, Simpliciter passiua, Actiuior inde, hinc Passiuior. Prima in actu puro. Secunda in potentia pura. Tertia in simplicibus substantijs secũdis. Quarta in cõpositis & generabilibus. Qutauor sunt in omnibus suppositionẽ habentibus. Essentia Esse, Posse, Operari: id est Qvo Svnt, Qvod Svnt, Qvod Possvnt, Qvod Agvnt. Quatruplex in ijs distinguitur operatio Divina, Natvralis, Fortvita, Artificialis. A' Prima est effectus simpliciter necessarius. A' Secũda necessarius in genere. A' Tertia simpliciter contingens. A' Quarta contingẽs ex arbitrio. Qvatvor inde sunt distribu-

Ill. 4.3ab: Giordano Bruno, *De monade, numero et figura liber* (Francoforte, 1591), frontespizio e p. 63. Copia GB-Lbl.

Parimenti con quattro termini definisce il termine ed il processo delle dimensioni. Nell'ambito dei limiti del quaternario, intendono ogni consonanza musicale («*Intra quaternarii limites musicam intelligunt omnem consonantiam*»): il Diapason, il Disdiapason, il Diapente e il Diatessaron. Quadrupla è l'essenza: Assoluta, Ideale, Unita alle Cose, Separata dalle Cose. Quattro sono i modi dell'essere: Metafisico, Fisico, Matematico, Razionale. Quattro sono le potenze: Divina, Intellettuale, Animale, Corporea, vale a dire: Semplicemente Attiva, semplicemente Passiva, Maggiormente Attiva, Maggiormente Passiva. [...] Nelle cose si distinguono quattro aspetti: l'Essenza, l'Essere, il Potere e l'Operare, vale a dire: Ciò di cui Sono, Ciò che Sono, Ciò che Possono e Ciò che Fanno. Vi si distingue inoltre una quadruplice capacità operativa: Divina, Naturale, Casuale, Artificiale [...][24].

---

[24]. Giordano Bruno, *De monade, numero et figura* (Francoforte, 1582), p. 63 (Bruno 1884, p. 388): «Quatuor item terminis, et terminum et processum dimensionum complet. Intra quaternarii limites musicam intelligunt omnem consonantiam Diapason, Disdiapason, Diapente, Diatessaron. Quatruplex essentia Absoluta, Idealis, Rebus Unita, A Rebus Abstracta. Quatruplex esse, Metaphysice, Physice, Mathematice,

Questo passaggio è inserito in una discussione complessiva sul senso metafisico e metaforico del Quaternario o della Tetrade ed è incluso nel suo complesso in un discorso più ampio nel quale vengono messi in evidenza i numeri e le figure ad essi legati con cui la Natura imprime nel mondo le proprietà universali delle cose e «le leggi delle loro vicissitudini»[25]. Ciò è legato imprescindibilmente alla questione, di per sé di venerabile tradizione magico-cabalistica, dei segni occulti che la Divinità lascerebbe nelle cose, segni che sarebbero sintetizzabili nei numeri che manifestano le loro proprietà intrinseche negli esseri e negli oggetti naturali[26]. Così il quinario o pentade, manifestantesi ad esempio nella mano umana con le sue cinque dita (Ill. 4.3d), esplicherebbe «la Legge di tutta la vita», poiché «ogni movimento ed azione, sia interni che esterni, sia sensibili che insensibili, si compiono attraverso la mano e quindi attraverso il quinario, esterno o interno, visibile o invisibile»[27]. In tal senso la strutturazione dei corpi e degli oggetti di natura secondo determinate regole e numeri, non sarebbe casuale, ma sarebbe una spia, una traccia che reinvia in ultima analisi alla struttura stessa del cosmo. Questa associazione basata sul 'numero' sarebbe dunque in un certo senso rivelatrice di una loro comunanza d'intenti, di una loro comune 'intenzionalità'.

Per tornare alla tetrade, essa sarebbe la prima espressione sensibilmente concreta della volontà di manifestarsi della Natura/Divinità, ossia «la pienezza del corpo mondano» e «dell'animo»[28]; questo anche perché in essa «è rintracciata la natura del solido»[29]. Pertanto, ogni struttura tetradica simbolizzerebbe la volontà materiale di manifestazione. Il fatto che dunque le consonanze fondamentali di ottava, doppia ottava, quarta e quinta vengano elencate insieme ad altre quadruplicità, come le essenze, le operazioni, le virtù, ecc., le contrassegnerebbe

---

Rationaliter. Quatruplex Virtus, Divina, Intellectualis, Animalis, Corporea. Quatruplex iterum, Simpliciter Activa, Simpliciter passiva, Activior inde, hinc Passivior. [...] Quatuor sunt in omnibus suppositionem habentibus. Essentia, Esse, Posse, Operari: id est Quo Sunt, Quod Sunt, Quod Possunt, Quod Agunt. Quatruplex in iis distinguitur operatio, Divina, Naturalis, Fortuita, Artificialis [...]». Traduzione italiana da Bruno 1980, pp. 346-347.

[25]. Bruno 1980, p. 28.

[26]. Sul rapporto ambivalente di Bruno con la cabala e sulla presenza di concetti cabalistici negli scritti bruniani, in particolare proprio nel *De monade*, si veda Carannante 2019, voce 'Cabala, cabalista, cabalistico (caballistico; *cabalisticus*)' (Fabrizio Meroi), pp. 299-304; in aggiunta a Spang 1999, il quale, fra l'altro, mette in evidenza paralleli fra passaggi numerologici nel *De monade* e in *De verbo mirifico* (Basilea, 1494) del cabalista Johannes Reuchlin (1455-1522).

[27]. Bruno, *De monade* (1591), pp. 93-94 (Bruno 1884, pp. 418-419): «legem omnis vitae [...] omnis motio et actio, sive interna, sive externa, sive sensibilis, sive non sensibilis, per manum, et quinarium externum, vel internum, visibilem peragitur, vel invisibilem». Traduzione italiana da Bruno 1980, p. 372.

[28]. Bruno, *De monade* (1591), p. 61 (Bruno 1884, p. 386): «plenitudinem mundani tum corporis, tum animi». Traduzione italiana da Bruno 1980, p. 345.

[29]. Bruno, *De monade* (1591), p. 56 (Bruno 1884, p. 380): «Et tetrade est primum solidi natura reperta». Traduzione italiana da Bruno 1980, p. 341.

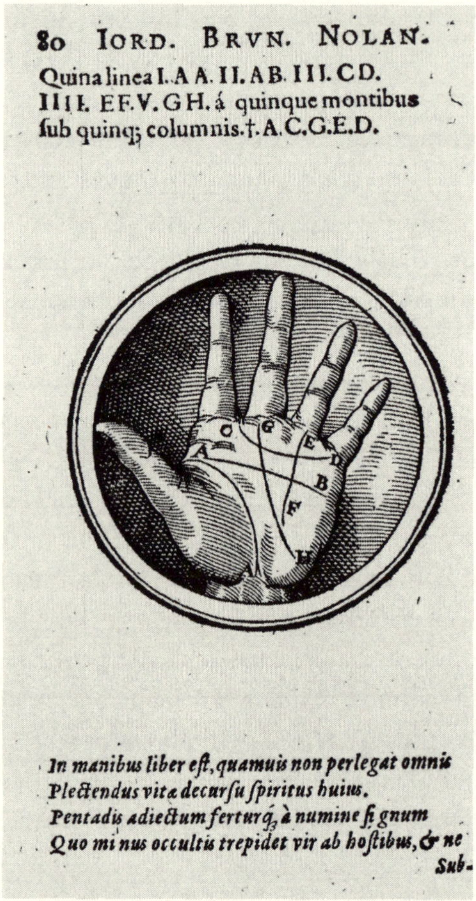

ILL. 4.3CD: Giordano Bruno, *De monade, numero et figura liber* (Francoforte, 1591), pp. 55 (con la tetrade, inscritta in un cerchio con quattro punti equidistanti dal centro) e 80 (con la pentade, manifestata attraverso la mano). Copia GB-Lbl.

come una delle manifestazioni della corporeizzazione ovvero estroflessione/esplicazione mondana del Divino grazie alla tetrade e quindi, in un certo senso e da un certo punto di vista, come fondanti per l'espressione e la manifestazione fisica e corporea del cosmo stesso. Se poi si considera che le figurazioni o immagini simboliche che Bruno fornisce come rappresentazione geometrica della tetrade hanno in comune di essere entrambe inscritte in un cerchio con quattro punti equidistanti dal centro (ILL. 4.3C), si deduce che per il filosofo le quadruplicità di volta in volta illustrate non siano gerarchicamente ordinate fra loro e che quindi anche le quattro consonanze scelte per rappresentare la tetrade in ambito musicale abbiano la stessa dignità, pur essendo ovviamente dissimili, come i quattro punti sulla circonferenza sono dissimili per posizione. Se dunque la tetrade simbolizza il primo fondamento della 'materialità', a maggior ragione l'ogdoade, come raddoppiamento della tetrade, ne sancirà in maniera ancora più solidamente fondante la sua affermazione. Così

120 IORD. BRVN. NOLAN.

VII. Phanes——Erus——Vmbra

VIII. Chaos——Nox——Tenebra.

Et in Musicis soliditatem quandam constituere comperitur.

SECVNDVS ORDO.

OCtonarius numerus in genere corporeorum atq; sensibilium;ut pote perfectè par,atq; quadratus est,ut in ędificijs nostris, quatuor latera octo terminantur lineis punctisq; primó, proindeq; á numeris qui ab octonarij radice prodeunt, ut sunt sexdecim, trigintaduo,64.& alij proinde in eodem pariter pares ordine.Sed pro loco alius numerus alijs debetur locatorum speciebus.Exagonas enim apes sibi domos constituunt. Aranearum quædam propter numerum tibiarũ exagonas, quædam tetragonas. Serpentum genus auium & ferarum plurium rotundas. Reis trilatera loca & tri-

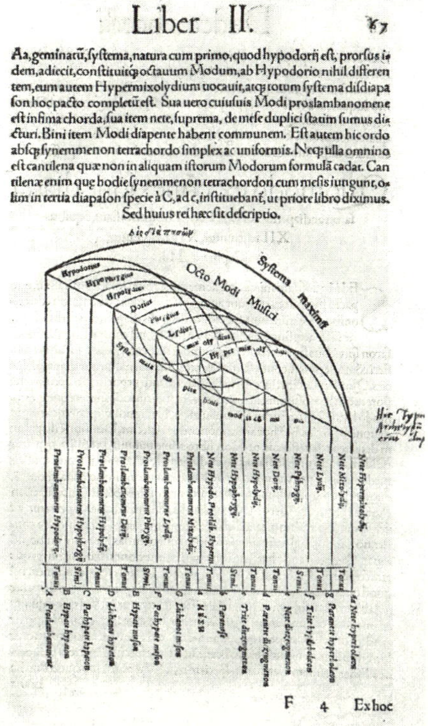

Liber II. 67

Aa,geminatũ,systema,natura cum primo,quod hypodorij est, prorsus idem,adiecit,constituitq; octauum Modum,ab Hypodorio nihil differentem,cum autem Hypermixolydium uocauit,atq; totum systema disdiapason hoc pacto completũ est. Sua uero cuiusuis Modi proslambanomene est infima chorda,sua item nete,suprema, de mese duplici statim sumus dicturi. Bini item Modi diapente habent communem. Est autem hic ordo absq; synemmenon tetrachordo simplex ac uniformis. Neq; ulla omnino est cantilena quæ non in aliquam istorum Modorum formulã cadat. Cantilenæ enim quę hodie synemmenon tetrachordon cum mesis iungunt, olim in tertia diapason specie à C,ad c,instituebant, ut priore libro diximus. Sed huius rei hæc sit descriptio.

F 4 Ex hoc

ILL. 4.4AB: Giordano Bruno, *De monade, numero et figura liber* (Francoforte, 1591), p. 120 (con gli «Octo modi musici»), e Enrico Glareano, *Dodecachordon* (Basilea, 1547), Liber secundus, p. 67. Copie GB-Lbl and I-Rn, rispettivamente.

> Il numero otto, nel genere delle cose sensibili, è un numero perfettamente pari e quadrato; anche nei nostri edifici, i quattro lati sono determinati da otto linee e punti o secondo numeri che procedono dalla radice dell'ottonario, come il sedici, il trentadue, il sessantaquattro, e così via, nello stesso ordine altri parimente pari. [...] Anche nella musica il numero otto costituisce il fondamento[30].

E questa solidità nelle cose musicali è guarda caso proprio testimoniata dagli otto modi musicali, i fondamenti ultimi, per così dire, della pratica musicale, così come le quattro consonanze primarie ne costituiscono una base più remota e più radicale (ILL. 4.4A). Va tra l'altro detto che, mentre le figure geometriche rappresentanti le altre categorie numeriche sono di pugno di Bruno, l'immagine degli otto modi è tratta invece dal *Dodecachordon* di Glareano,

30. Bruno, *De monade* (1591), p. 120 (BRUNO 1884, p. 447): «Octonarius numerus in genere corporeorum atque sensibilium, utpote perfecte par, atque quadratus est, ut in aedificiis nostris, quatuor latera octo terminantur lineis punctisque primo, proindeque a numeris qui ab octonarii radice prodeunt, ut sunt sexdecim, trigintaduo, 64; et alii proinde in eodem pariter pares ordine. [...] Et in Musicis soliditatem quondam constituere comperitur». Traduzione italiana da BRUNO 1980, pp. 392-393.

pubblicato per la prima volta nel 1547 (Ill. 4.4b). Ciò potrebbe sembrare paradossale, visto che l'opera di Glareano si propone di 'riformare' la modalità classica con l'aggiunta di quattro modi ulteriori, ma evidentemente a Bruno, per cui non è nota una specifica formazione o uno specifico interesse musicale, durante il suo breve soggiorno francofortese, importava trovare un'illustrazione atta a trasmettere al meglio il suo messaggio e il trattato di Glareano dovette essere ancora abbastanza diffuso in Germania alla fine del secolo[31].

Capita però anche molto spesso che le immagini di cui ci serviamo per rappresentare dati di natura siano fasulle, nel senso che non riflettano analogicamente caratteristiche veramente radicali, fondamentali degli oggetti che presumono di descrivere. Per Bruno uno dei casi più palesi di finzione o di illusione generata da immagini 'ingannevoli' sarebbe la proiezione arbitraria di figure terio- o antropomorfe nel cielo stellato onde facilitare l'orientamento nel cielo notturno, comprendendo con ciò lo zodiaco come pure quasi tutti gli altri catasterismi. Nello *Spaccio della bestia trionfante*, redatto e stampato a Londra nel 1584, l'autore mette in bocca a Giove l'intento di riformare questa follia astrale sostituendo i segni tradizionali con altri associati a rappresentazioni simboliche di vizî e virtù, più utili alla civiltà umana. E tuttavia il supremo Dio olimpico si rende conto che togliendo repentinamente segni di orientamento ormai tanto noti, senza cioè un passaggio intermedio, gli umani rimarrebbero sconcertati e senza punti d'appiglio. Dunque i tre dialoghi che costituiscono lo *Spaccio*

> [...] son stati messi e distesi sol per materia e suggetto d'un artificio futuro; perché, essendo io in intenzione di trattar la moral filosofia secondo il lume interno che in me ave irradiato ed irradia il divino sole intellettuale, mi par espediente prima di preponere certi preludii a similitudine de musici; imbozzar certi occolti e confusi delineamenti ed ombre, come gli pittori; ordire e distendere certa fila, come le tessetrici; e gittar certi bassi, profondi e ciechi fondamenti, come gli grandi edificatori [...][32]

In questo passaggio l'autore afferma in effetti di proporre nuove figure celesti partendo da e facendo uso di immagini già note, sebbene illusorie e false — i segni zodiacali e le altre costellazioni — ben sapendo che la memoria e la capacità immaginifica hanno un ruolo fondamentale nel trasmettere, anche per analogia, determinati contenuti gnoseologici. Fra le varie metafore artistiche adottate, Bruno paragona questo processo a quello di un musico che introduce un nuovo brano 'preludiando', partendo cioè da contenuti — forse scale e passaggi musicali facenti parte di una tradizione orale diffusa e quindi familiare — o schemi già in parte noti. La metafora musicale di Bruno sembra essere tratta da un passaggio del libro iii della *Retorica* di Aristotele in cui il filosofo paragona il preludiare di un flautista («proaulion»)

31. Su l'immagine di Glareano in Bruno, si veda Bosi 2015, pp. 85-86.

32. Giordano Bruno, *Spaccio della bestia trionfante* ('Parigi' [= Londra], 1584), Epistola esplicatoria, p. 8 (Bruno 2002b, pp. 178-179).

al proemio di un retore prima di attaccare con il discorso vero e proprio[33]. Come il flautista introduce l'uditorio all'argomento 'preludiando', cioè tastando il terreno con scale, passaggi figurati, etc., familiari a un pubblico provvisto anche di conoscenze basilari della musica della propria epoca, così il retore prepara la strada al tema vero e proprio del discorso che intende pronunciare con un preambolo o proemio farcito di figure e immagini familiari all'uditorio. Bruno effettua dunque una *traslatio* di questa metafora dall'ambito retorico del trattato aristotelico a quello 'scientifico' o comunque gnoseologicamente fondante dello *Spaccio*. Più tardi, quando arriva il momento di edificare la nuova costruzione 'scientifica', Bruno adotta la metafora musicale del 'concertare':

> Or, mentre ciò si mette in esecuzione [vale a dire il cominciare a far piazza pulita degli ingannevoli asterismi terio- e antropomorfi], se vedete vituperar cose che vi paiono indegne di vitupèro, spreggiate cose degne di stima, inalzate cose meritevoli di biasimo; e per il contrario; abbiate tutto per detto (anco da quei che possono nel suo grado dirlo) indefinitamente, come messo in difficultade, posto in campo, cacciato in teatro, che aspetta di essere essaminato, discusso e messo al paragone, quando si consertarà la musica, si figurarà la imagine, s'intesserà la tela, s'inalzarà il tetto[34].

Come dunque il 'preludiare' prepara l'uditorio e gli astanti gradualmente alla nuova sistemazione catasteristica, così il concertare ne costituisce la vera e propria messa in opera. È interessante comunque notare che una delle poche costellazioni tradizionali ad essere risparmiate è, guarda caso, la Lira in quanto:

> Dove si vede la Lira di nove corde, monta la madre Musa con le nove figlie, Aritmetrica, Geometria, Musica, Logica, Poesia, Astrologia, Fisica, Metafisica, Etica; onde, per conseguenza, casca l'Ignoranza, Inerzia e Bestialitade[35].

In una straordinaria 'mise en abyme' iconologica una costellazione squisitamente e palesemente musicale, come la Lira simbolizza, con le sue nove corde, nove discipline una delle quali è proprio la musica, che quindi gode, o continua a godere, di un posto privilegiato nel cielo riformato. Del resto già nel *De umbris idearum*, Bruno si era servito dell'immagine della Lira apollinea come di un simbolo supremamente importante nella sua cosmologia:

> Poiché in tutte le cose si dà ordine e connessione, così che gli enti inferiori fanno seguito a quelli intermedi e gli intermedi ai superiori, le realtà composte

[33]. Aristotele, *Retorica*, Libro III, Cap. 14, 1414b (ARISTOTELE 1999, p. 185).

[34]. Bruno, *Spaccio* (1584), Epistola esplicatoria, pp. 8-9 (BRUNO 2002B, pp. 179-180).

[35]. Bruno, *Spaccio* (1584), Epistola esplicatoria, p. 17 (*Ibidem*, p. 189).

> sono unite alle semplici, le semplici alle più semplici, le materiali si vincolano alle spirituali, le spirituali a quelle totalmente immateriali, di modo che uno sia il corpo dell'ente universale, uno l'ordine, uno il governo, uno il principio, uno il fine, uno il primo, uno l'estremo; e poiché (come non ignorarono i principi dei platonici) dalla luce alle tenebre si dà una trasmigrazione perenne (in quanto alcune delle menti con un movimento di attrazione verso la materia e allontanamento dall'atto vanno ad assoggettarsi alla natura e al fato), nulla impedisce che al suono della cetra suonata dall'Apollo universale le realtà inferiori siano gradatamente richiamate a quelle superiori, e le inferiori, passando per le intermedie, penetrino nella natura delle superiori, così come i sensi ci mostrano che la terra si attenua in acqua, l'acqua in aria, l'aria in fuoco e che allo stesso modo il fuoco si addensa in aria, l'aria in acqua, l'acqua in terra[36].

Anche se la menzione di Apollo e della sua lira («universalis cythara») è ripresa dalla *Theologia Platonica* di Ficino (IV, 1), l'immagine del suono della lira apollinea, che può attivamente invertire la catabasi generale da luce divina a materia 'oscura' e questa viceversa in un'ascesa (anabasi) verso la luce, manca nel corrispondente passaggio di Ficino — «Ed è necessario fondare una sola mente, essendo essa sia mente che unità, e che non sia o l'uno o l'altro, ovvero una mente o un'anima, ma l'Uno stesso che Pitagora chiama l'Apollo universale»[37] — dove l''Apollo universale' funziona solo come rappresentazione dell'unità universale. In Bruno la lira 'universale' rappresenta quindi da un lato l'unità fondamentale dell'insieme, dall'altro la forza magica e vitalistica che mette tutto in moto e mantiene tutto vivo.

Epilogo

In conclusione, sebbene l'opera di Giordano Bruno, a differenza di quelle di altri maghi-filosofi rinascimentali, non abbondi certo di riferimenti musicali, probabilmente anche a causa

36. Bruno, *De umbris* (1582), f. e iiii-verso («Intentio septima. G»): «Cum vero in rebus omnibus ordo sit atque connexio, ut inferiora mediis et media superioribus succedant corporibus, composita simplicibus, simplicia simplicioribus uniantur, materialia spiritualibus, spiritualia prorsus immaterialibus adhaereant, ut unum sit universi entis corpus, unus ordo, una gubernatio, unum principium, unus finis, unum primum, unum extraemum; cumque (ut non ignoraverunt Platonicorum principes) demigratio detur continua a luce ad tenebras (cum mentium quaedam per conversionem ad materiam et aversionem ab actu subeunt naturam atque fatum), nihil impedit quominus ad sonum cytharae universalis Apollinis ad superna gradatim revocentur inferna et inferiora per media superiorum subeant naturam, quemadmodum et sensu constat terram in aquam, aquam in aerem, aerem in ignem rarefieri, sicut ignis in aerem, aer in aquam, aqua in terram densabatur». Traduzione italiana da Bruno 2004, pp. 50-52.

37. Ficino, *Theologia Platonica*, Liber IV, Cap. 1, 16 (Ficino 2011, pp. 124-125): «Et una mens, quia & mens est & unum, simpliciter est erigenda, quod non sit unum hoc unum illud, ceu vel una mens vel una anima, sed unum ipsum, quod vocat Pythagoras universalem Apollinem».

di uno scarso interesse ovvero una scarsa propensione musicale del Nolano, oltre al ruolo largamente preponderante dell'immaginazione visiva nella sua «nova filosofia», è tuttavia significativo che immagini a carattere musicale servano ad esplicare e a comunicare in maniera pregnante snodi gnoseologici decisivi del sistema filosofico bruniano. E anche se spesso, come è il caso dell'appena discussa «universalis cythara» apollinea, ma anche per le altre, queste immagini vengano mutuate da un repertorio mitografico e iconologico ben noto, esse, come abbiamo visto, vengono quasi sempre sottoposte ad una torsione e trasformazione sufficiente da renderle veicoli efficaci alla comunicazione delle per molti versi rivoluzionarie nozioni del filosofo di Nola. In questo del resto, non si differenziano da immagini e concetti ripresi da altri campi del sapere e delle arti, immagini che, in una sorta di abilissima *ars combinatoria*, vengono riutilizzate e ricomposte in maniera tale da servire alla reinterpretazione anche profonda della realtà e della percezione proposte da Bruno.

Bibliografia

Aristotele 1907

Aristotele. *De anima*, traduzione, introduzione e note di Robert Drew Hicks, Cambridge, Cambridge University Press, 1907.

Aristotele 1999

Id. *Rhetorik*, a cura e traduzione di Gernot Krapinger, Stoccarda, Philipp Reclam, 1999, ^{2}2007.

Bönker-Vallon 2006

Bönker-Vallon, Angelika. 'Matematica', in: *Enciclopedia bruniana e campanelliana. 1*, a cura di Eugenio Canone e Germana Ernst, Pisa, Istituti Editoriali e Poligrafici Internazionali, 2006, coll. 105-119.

Bosi 2015

Bosi, Carlo. 'Musik in den Werken Giordano Brunos: Eine erste Erkundung', in: *Archiv für Musikwissenschaft*, lxxii/2 (2015), pp. 77-97.

Brockmeier 1980

Brockmeier, Jens. *Die Naturtheorie Giordano Brunos. Erkenntnistheoretische und naturphilosophische Voraussetzungen des frühbürgerlichen Materialismus*, Francoforte, Campus Verlag, 1980.

Bruno 1884

Bruno, Giordano. *Opera latine conscripta. 1/ii*, a cura di Francesco Fiorentino, Napoli, Morano, 1884 (pp. 319-484: *De monade, numero et figura*).

Bruno 1980

Id. *Opere latine*, a cura di Carlo Monti, Torino, UTET, 1980 (pp. 293-413: *La monade, il numero e la figura*).

Bruno 2002a

Id. *Opere italiane. 1*, a cura di Nuccio Ordine, Torino, UTET, 2002 (pp. 591-746: *De la causa, principio et uno*).

Bruno 2002b

Id. *Opere italiane. 2*, Torino, UTET, 2002 (pp. 169-404: *Spaccio de la bestia trionfante*, a cura di Maria Pia Ellero; pp. 485-753: *De gli eroici furori*, a cura di Miguel Ángel Granada).

Bruno 2004

Id. *Opere mnemotecniche, Tomo i*, a cura di Michele Ciliberto, Milano, Adelphi, 2004 (pp. 2-586: *De umbris idearum / Le ombre delle idee*).

Bruno 2009

Id. *Opere mnemotecniche, Tomo ii*, a cura di Michele Ciliberto, Milano, Adelphi, 2009 (pp. 186-303 [testo], pp. 431-481 [commento]: *Sigillus sigillorum*).

Carannante 2018

Carannante, Salvatore. *Unigenita natura. Dio e universo in Giordano Bruno*, Roma, Edizioni di Storia e Letteratura, 2018.

Carannante 2019

Giordano Bruno. Filosofia, magia, scienza, a cura di Salvatore Carannante, Giovanni Licata, e Pasquale Terracciano, Pisa, Edizioni della Normale, 2019.

Ciliberto 1999

Ciliberto, Michele. *Umbra profunda. Studi su Giordano Bruno*, Roma, Edizioni di Storia e Letteratura, 1999.

De Bernart 2002

De Bernart, Luciana. *«Numerus quodammodo infinitus». Per un approccio storico-teorico al «dilemma matematico» nella filosofia di Giordano Bruno*, Roma, Edizioni di Storia e Letteratura, 2002.

Ficino 2011

Ficino, Marsilio. *Teologia platonica*, a cura di Enrico Vitale, Milano, Bompiani, 2011.

Knox 2019

Knox, Dilwyn. 'Giordano Bruno', in: *The Stanford Encyclopedia of Philosophy*, Stanford University, Department of Philosophy, 2019, <https://plato.stanford.edu/entries/bruno/>.

Montano 2000

Montano, Aniello. *La mente e la mano. Aspetti della storicità del sapere e del primato del fare in Giordano Bruno*, a cura di Michele Ciliberto, Napoli, La Città del Sole, 2000.

SPANG 1999

SPANG, Michael. 'Brunos *De monade, numero et figura* und christliche Kabbala', in: *Bruniana & campanelliana*, v/1 (1999), pp. 67-94.

TIRINNANZI 2000

TIRINNANZI, Nicoletta. *'Umbra naturae'. L'immaginazione da Ficino a Bruno*, Roma, Edizioni di Storia e Letteratura, 2000.

TOMLINSON 1993

TOMLINSON, Gary. *Music in Renaissance Magic: Toward a Historiography of Others*, Chicago, University of Chicago Press, 1993.

VÉDRINE 1967

VÉDRINE, Hélène. *La conception de la nature chez Giordano Bruno*, Parigi, Vrin, 1967.

YATES 1966

YATES, Frances A. *The Art of Memory*, Londra, Routledge & Kegan Paul, 1966.

Chapter Five

Dissezioni anatomiche con musica a Padova nella prima età moderna

Gioia Filocamo[*]
(Swedish Collegium for Advanced Study, Uppsala /
Istituto superiore di Studi musicali 'G. Briccialdi', Terni)

In memoria di Albert Dunning
(1936-2005)

The *Taming of the Shrew* (La bisbetica domata) di Shakespeare è ambientata a Padova, città definita da Lucenzio «nursery of arts» (culla delle arti)[1]. L'incerta datazione della commedia inglese rimanda alla fine del Cinquecento, e il termine «arts» impiegato all'epoca va dunque riferito all'insieme del sapere ufficiale articolato in arti

[*]. Dedico questo saggio alla memoria di Albert Dunning (Arnhem, 5 agosto 1936 - Loenen aan de Vecht, 2 giugno 2005), il celebre musicologo olandese che ho avuto la fortuna di incontrare a Cremona oltre vent'anni fa. Il prof. Dunning fu vittima di una terribile malattia del cui decorso ebbe lucida consapevolezza. Con la generosità che lo contraddistingueva, stabilì che dopo la sua scomparsa vari studenti italiani e stranieri avrebbero beneficiato dei suoi libri, e che al momento della morte il suo corpo sarebbe stato donato alla scienza, per il progresso della medicina. Fui enormemente impressionata da quest'ultima decisione, che seppi direttamente dall'interessato. Mi colpì molto l'idea che la sua morte non sarebbe stata onorata con un funerale, e che nessuna tomba avrebbe custodito le sue spoglie. Al contrario, quello stesso corpo segnato dalla patologia sarebbe stato un ultimo regalo utile a tutti i vivi. Ammirai questo radicale proponimento che giudicai un raro esempio di onore scientifico di alta coerenza: la conoscenza aveva permesso allo studioso di vivere una vita ricca e piena, e donare i propri resti mortali agli anatomopatologi, affinché ne imparassero qualcosa di nuovo, richiudeva in un certo senso il cerchio epistemologico della sua vita. Questo è stato il più alto insegnamento che ho ricevuto da Albert Dunning.

[1]. William Shakespeare, *The Taming of the Shrew* (Londra, 1623), I, 1: «Lucentio: *Tranio*, since for the great desire I had | To see faire *Padua*, nurserie of Arts, | I am arriv'd for fruitfull *Lumbardie*, | The pleasant garden of great *Italy*, | [...]» («Lucenzio: Tranio, dacché per il gran desio che avevo di visitare la bella Padova, culla delle arti, son giunto nella fertile Lombardia, ridente giardino della grande Italia, | [...]»).

Ill. 5.1: immagine del costume di uno studente. Dall'*Album amicorum* di Bernardus Paludanus (Berend ten Broecke (1550-1633), medico e collezionista olandese che studiò medicina a Padova). NL-DHk, MS 133 M 63, c. 199r.

liberali, cioè quelle a prevalente attività intellettuale, e arti meccaniche, dove invece risaltava l'applicazione manuale, come la medicina[2]. A quel tempo Padova era un centro universitario di prima importanza che vantava già tre secoli di storia e prestigio. A séguito della guerra perduta nel 1404-1405 contro Venezia, la città era divenuta parte integrante della Repubblica Serenissima, e con la *Bolla d'oro* (30 gennaio 1406) Venezia le aveva completamente demandato la funzione di cuore culturale dello Stato. Padova poté dunque mantenere il ruolo egemonico già acquisito nel campo degli studi universitari europei[3].

Nato da una secessione dall'Università di Bologna avvenuta nel 1222, lo Studio patavino ne riprodusse l'organizzazione di base che divideva le *universitates* (associazioni, corporazioni) studentesche in 'legiste' (cioè costituite da studenti di diritto civile, diritto canonico e teologia) e 'artiste' (che raggruppavano chi studiava astronomia, dialettica, filosofia, grammatica,

[2]. Sulla storia delle 'arti' a partire dalla strutturazione del sapere enunciata da Marziano Capella all'inizio del secolo V nella favola mitologica *De nuptiis Philologiae et Mercurii* (Sulle nozze di Filologia e Mercurio) si veda Frugoni 1991 .

[3]. Sul rapporto tra le due città venete si veda Melchiorre 2012.

retorica e medicina) (Ill. 5.1). L'aggiunta dell'*universitas artistarum* avvenne dopo il 1399, sotto la signoria di Francesco Novello da Carrara (Francesco ii), che durò fino alla perdita dell'indipendenza di Padova del 1405. Ulteriori distinzioni esistevano tra le *universitates ultramontane* (d'oltralpe) e quelle *citramontane* (cioè italiane), all'interno delle quali le varie *nationes* riunivano in forma di corporazione gli studenti iscritti in base alla lata provenienza geografica, non secondo la loro reale nazionalità[4]. L'appartenenza a una *natio* era determinante per ottenere varie forme di assistenza o tutela.

Gli studenti stranieri sono sempre stati presenti nelle università italiane. Il fenomeno delle *peregrinationes academicae* era piuttosto diffuso nel Medioevo, e riguardava tanto gli studenti stranieri quanto quelli italiani[5], ma per arginare il pericolo della 'nefanda' influenza morale dei protestanti sui cattolici, dopo la Riforma comparve un potente deterrente *ad hoc*. Nella bolla controriformistica *In sacrosancta* del 13 novembre 1564 emessa da papa Pio iv — pubblicata a Padova il 4 marzo 1565 — si trova infatti la *forma professionis fidei* da pronunciare obbligatoriamente prima del conseguimento dei gradi accademici su suolo cattolico. La *professio fidei Tridentina* (o *Credo di Pio iv*) conteneva il Credo Niceno-Constantinopolitano e una serie di articoli definiti dal Concilio di Trento: la funzione primaria della professione di fede era evidentemente di riaffermare la dottrina cattolica come diversa dalla protestante. Eccezioni a tal riguardo esistevano già, come le dispense papali che consentivano di conferire lauree agli ebrei fin dall'inizio del Quattrocento. Gli ebrei non potevano né laurearsi regolarmente né ricoprire cariche pubbliche, in conformità al divieto emanato dal giurista Bartolo da Sassoferrato (1314-1357) nel suo commento al *Corpus iuris civilis*, una normativa più volte ribadita. A Padova, inoltre, anche l'imperatore e i conti palatini potevano attribuire lauree 'esterne' allo Studio, peculiarità di cui gli studenti germanici si giovarono molto e che perdurò fino al 1615[6]. Dal dovere di enunciare la *professio fidei Tridentina* furono poi esenti i Collegi dottorali veneti appositamente creati — nel 1616 quello 'artista', nel 1635 quello 'giurista' —, col potere di conferire gradi accademici *auctoritate Veneta* anche a studenti non cattolici, *in primis* protestanti. Della deroga si avvantaggiarono non solo i protestanti, ma anche ebrei, cristiani ortodossi e atei[7]. L'afferenza alla Repubblica di Venezia comportò quindi, per l'università patavina, l'eccezionale prerogativa di elargire agli studenti non cattolici titoli dottorali senza far uso di privilegi imperiali o palatini, anzi, col preciso intento di scoraggiarne l'utilizzo. Tale atteggiamento ufficialmente 'ecumenico' non inibiva però una certa insofferenza reciproca nella vita pratica fra studenti protestanti e cattolici[8].

4. Brizzi 1996, p. 179.

5. Rosso 2012. Sulla *peregrinatio medica* in Europa si veda in particolare p. 113, n. 6.

6. Su questa questione, si veda Martellozzo Forin 1997.

7. Brizzi 1996, p. 191.

8. Sulla loro non sempre facile convivenza a Padova si veda Brugi 1893-1894. Molte questioni legate alla vita pratica degli affiliati alla *natio Germanica* padovana si leggono in Costa 1913.

Gli intensi e fruttuosi rapporti di Venezia col mondo tedesco erano già attestati dall'attività del Fòndaco (*Fondego*) dei Tedeschi affacciato sul Canal Grande, un vero e proprio centro commerciale e finanziario eretto all'inizio del secolo XIII dove approdavano merci provenienti dal Nord Europa e risiedevano i mercanti 'alemanni' (cioè generalmente oltremontani). Il Fòndaco dei Tedeschi fu inoltre la prima *enclave* luterana in Italia, e una delle più antiche fuori dai territori tedeschi: vi era consentito praticare la fede protestante, purché ciò avvenisse con la dovuta discrezione e senza alcuna velleità di proselitismo nei confronti dei cattolici veneziani[9]. Questa eccezionale licenza era frutto della tradizionale indipendenza della Chiesa veneziana da quella di Roma, e della conseguente consuetudine che teneva convenientemente distinta la sfera politica da quella religiosa. Fin dall'epoca della sua erezione (828) la Basilica di San Marco fu sottoposta a giurisdizione dogale mediante l'istituto del primicerato: il doge vi assumeva le funzioni di vescovo, e il primicerio rappresentava il doge in quanto primo canonico del capitolo della basilica. Tutto ciò ebbe fine con la caduta della Repubblica di Venezia, nel 1797. Di tale tolleranza benevola si giovarono i numerosi studenti luterani oltremontani nell'Università di Padova, città di cui essi dimostrarono spesso di apprezzare anche la vivacità culturale e musicale, sacra e profana[10]: varie Accademie locali patrocinavano musica[11] e vi erano presenti diversi liutai, molti dei quali tedeschi, come i celebri Wendelin Tieffenbrucker (*ante* 1551 - *post* 1611) e Michael Hartung (*fl.* ca. 1591 - ca. 1627)[12].

L'insegnamento della medicina fu una delle punte di diamante dell'ateneo patavino, almeno fin dalla seconda metà del XIV secolo. L'eccellenza e la fama si dovettero anche alla specifica attenzione qui dedicata alla dissezione anatomica, che vi divenne parte integrante del *curriculum* studentesco, com'era già a Bologna dall'inizio del Trecento[13]. Alla pratica della dissezione umana si era opposta la bolla *Detestandae feritatis abusum* (Detestando gli abusi selvaggi) emanata nel 1299 da papa Bonifacio VIII, documento interpretato in modo estensivo, ma che in verità vietava solo la diffusa usanza di bollire pezzi di cadaveri al fine di ricavarne ossa trasferibili con pochi rischi igienici. Questa pratica serviva tanto a produrre reliquie dal corpo dei santi, quanto a consentire il trasporto dei resti di defunti nel luogo in cui dovevano essere inumati. La legittimazione definitiva della dissezione per fini medici fu sancita da una lettera del 2 aprile 1482 firmata da Giuliano della Rovere — Cardinale Penitenziario Maggiore della Penitenzieria Apostolica[14], nonché nipote del papa regnante Sisto IV (e futuro papa Giulio II) —,

9. Rieder 2010, pp. 99-102.

10. Comparin 2020, p. 85, n. 8.

11. Fenlon 2018, pp. 102-108.

12. Si vedano i saggi di Bonnie Blackburn, Jeffrey Kurtzman ed Eleanor Selfridge-Field contenuti in Schiltz 2018.

13. Qualche utile approfondimento bibliografico sulla preminenza acquisita da Padova, che surclassò Bologna in campo medico, si legge in Rosso 2012, p. 112, n. 3, e p. 178.

14. La Penitenzieria Apostolica è un dicastero tuttora esistente istituito nel secolo XII allo scopo di coadiuvare il papa a evadere le richieste di assoluzione da pene e censure provenienti da tutto il mondo concernenti il cosiddetto

che autorizzava le autopsie nell'Università di Tubinga. Il pronunciamento ufficiale fu sollecitato personalmente da Eberardo v detto 'il Barbuto' (1445-1496), conte di Württemberg-Urach fondatore proprio dell'Università di Tubinga nel 1477, e contribuì ad alleviare le comprensibili remore morali legate all'atto di incidere i cadaveri[15].

Tra i docenti che praticarono la dissezione anatomica nello Studio di Padova in età moderna troviamo alcuni fra i nomi più noti del panorama europeo: oltre al fiammingo Andreas Vesalius (1514-1564), vanno ricordati almeno il cremonese Realdo Colombo (1516-1559), il modenese Gabriele Falloppio (1523-1562), e Girolamo Fabrici d'Acquapendente (ca. 1533 - 1619), titolare della cattedra di anatomia e chirurgia dal 1565 al 1613. Fu proprio quest'ultimo a promuovere l'edificazione a Padova del primo teatro anatomico permanente al mondo, tra il 1583 e il 1584, tuttora esistente e visitabile nel Palazzo del Bo, cuore dell'università (Ill. 5.2). Il teatro anatomico patavino fu inaugurato il 16 gennaio 1595 e vi si praticarono autopsie fino al 1872. La struttura lignea, progettata dal servita veneziano Paolo Sarpi, è a forma di cono rovesciato privo di finestre verso l'esterno; può ospitare poco più di duecento persone (Ill. 5.5ab).

Ill. 5.2: Padova, Palazzo del Bo, che ospita il teatro anatomico. Da Giacomo Filippo Tomasini, *Gymnasium Patavinum* (Padova, 1654), fra le pagine 40 e 41. Copia I-Ru.

'foro interno', cioè i casi di coscienza. Un'utile sintesi della sua storia e funzioni si legge in Saraco 2012. Sulla struttura delle lettere della Penitenzieria d'antico regime si veda Salonen 2007.

15. Il documento è conservato nell'Universitätsarchiv Tübingen, UAT AS 20 / 7a: No. 1. Si veda Carlino 1994, p. 217.

Ill. 5.3: Andreas Vesalius, *De humani corporis fabrica* (Basilea, 1543), frontespizio (xilografia da Jan Stefan van Calcar). Copia collezione privata.

DE HVMANI CORPORIS FABRICA LIBER II. 237

DE INSTRVMENTIS, QVAE SECTIONIbus administrandis parari possunt. Caput VII.

ANATOMICORVM INSTRVMENTORVM DELINEATIO.

CHARACTERVM SEPTIMI CAPITIS FIGVRAE INDEX.

PRAESENTI *figura mensæ cuidam incumbentem finximus asserem, quo in uiuorum sectionibus opportunè utimur, dein huic asseri omnia propemodum accõmodauimus, quibus in dissectionum administrationibus, adeoq; tota Anatome quis posset uti. Quo autem singula leuiori opera assequaris, huic etiam figuræ characteres, ac demum eorum indicem adhibere non grauatus sum. Indicetur itaq;*

A, A *Mensa, cui reliqua omnia modò seriatim indicanda supersternuntur.*
B, B *Asser uiuis sectionibus administrandis idoneus.*
C, C *Varia foramina, quibus laqueos pro animalis mole adhibemus, quũ femora et brachia uincimus.*
D, D *Eiusmodi anuli, summis manibus pedibusq; ligandis adaptantur.*
E *Huic anulo maxilla superior, libera inferiori, catenula alligatur, ut caput immotum seruetur, ac interim neq; uox, neq; respiratio uinculorum occasione præpediantur.*
F, F *Diuersa nouacularum genera, quibus spongia accumbit.*
G *Cultelli ad earum speciem formati, quibus calami adaptantur.*
H *Vulgaris qui mensæ adhibetur culter.* I. *Grandis ac ualidus culter.*
K *Cultri è buxo parati.* L *Hamuli.* M *Varij styli unà cum siphone.*
N, N *Obliquatæ acus cum filo crasſiore, quibus literarum fasciculos colligamus.*
n *Minores acus, quas uulneribus suendis accommodamus.*
O *Serra* P *Forficula.* Q *Malleus ligneus.*
R *Arundines inflandis pulmonibus, & alijs quibusdam partibus idoneæ.*
S *Filum æneum, osſibus nectendis aptum.* T *Subula forandis osſibus parata.*
V *Varia subularum ferra.* X *Forpex intorquendis filorum extremis comparatus.*
Y *Forpex, quo intorti, & ossa iam committentis fili reliquias præscindimus.*

V 4 QVAN-

Ill. 5.4: Andreas Vesalius, *De humani corporis fabrica* (Basilea, 1543), p. '237' [=235]. Copia CH-Bu.

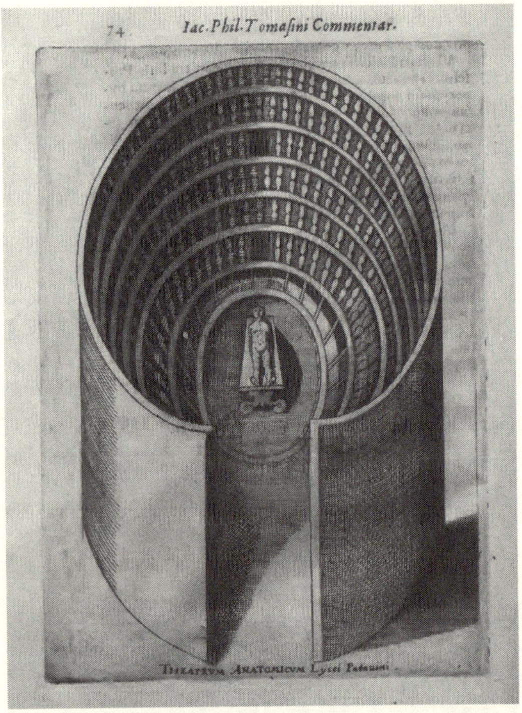

Ill. 5.5ab: il teatro anatomico di Padova: vuoto (da Giacomo Filippo Tomasini, *Gymnasium Patavinum*, Padova, 1654), p. 74, e pieno (frontespizio di Johannes Wesling, *Syntagma anatomicum*, Padova, 1647). Copie I-Ru e A-Wn, rispettivamente.

L'interesse degli studenti per le dissezioni anatomiche fu sempre forte, al punto che dal Cinquecento in poi la pratica divenne curricolare nelle università di tutt'Europa. Tale entusiasmo trasse linfa vitale dalla cosiddetta 'rivoluzione vesaliana', la nuova impostazione scientifica promossa da Vesalius che sovvertì il canone vigente fino a metà Cinquecento. Abitualmente il *lector*, cioè il docente ufficiale più alto in grado, sedeva lontano e in alto rispetto al cadavere, *ex cathedra*, e declamava la descrizione del corpo umano leggendola da un trattato classico d'impostazione galenica; l'*ostensor*, che era un docente di grado inferiore o uno studente esperto, indicava con una bacchetta le parti anatomiche menzionate dal *lector* sul corpo nudo; infine il *dissector* incideva materialmente il cadavere, iniziando con un taglio dallo sterno al pube: si trattava di un chirurgo-barbiere, figura professionale tradizionalmente dedita al contatto fisico coi pazienti, che praticava i salassi e curava le malattie veneree. La dimostrazione serviva a far sì che gli studenti presenti fissassero nella memoria quanto leggevano sui libri canonici grazie all'esperienza visiva. Fino al secolo xvi, dunque, la dissezione non serviva a indagare il corpo umano, bensì unicamente a confermare le descrizioni diffuse e accettate nei testi ufficiali. La cosiddetta 'rivoluzione vesaliana' rovesciò i termini di tale prospettiva. Nell'affollato frontespizio del trattato di Andreas Vesalius, *De humani corporis fabrica* (Sulla struttura del corpo umano; Basilea, 1543; Ill. 5.3 e Ill. 5.4), è raffigurato lo stesso Vesalius, cioè il docente, che si sporca le mani mentre incide il cadavere. Sparisce anche lo scranno del *lector*, poiché «è il corpo stesso a diventare il libro da consultare»[16]. Da metà Cinquecento, quindi, l'approccio accademico alla dissezione anatomica diventa esplorativo, investigativo. E va ricordato che tanto la medicina teorica quanto quella pratica erano percepite in totale complementarità alla 'filosofia naturale' aristotelica, soprattutto in certe università europee tra cui primeggiava Padova[17]. Ecco perché chi si laureava in medicina veniva appellato '*philosophus et medicus*'[18]. Nonostante la riscoperta cinquecentesca di platonismo, scetticismo, epicureismo e stoicismo, l'aristotelismo rimarrà comunque a base della formazione universitaria almeno fino a inizio Seicento, quando compariranno le prime cattedre d'ispirazione platonica[19].

La musica risuonava in alcuni teatri anatomici d'età moderna (a Bologna e Leida, per esempio; Ill. 5.6) durante ricorrenze celebrative come lezioni inaugurali dell'anno accademico o feste carnevalesche, eventi ai quali si accedeva spesso in maschera pagando un biglietto d'ingresso, e che comprendevano anche un'autopsia al fine di dimostrare l'eccellenza dei docenti universitari locali[20]: gli *studia universitatis* ricorrevano a ogni mezzo possibile per farsi concorrenza. Ma il caso di Padova documenta un contesto diverso: qui le anatomie si eseguivano in quaresima, non a carnevale come avveniva invece a Bologna o Pisa[21]. Gli studenti tedeschi iscritti a Padova richiesero e ottennero la presenza della musica all'interno di vere lezioni pubbliche di anatomia, presenza che si protrasse per vari giorni. I passi cronachistici che testimoniano tali eventi musicali si trovano negli *Atti della nazione Germanica artista*[22], vero e proprio diario della vita universitaria redatto dagli studenti, riferiti al 12 dicembre 1597 e all'8 gennaio 1600[23]. Vi si menzionano esecuzioni di *fidicines*, ma null'altro specificano su quanto noi gradiremmo sapere: quanti erano gli strumentisti? dove si posizionarono? che musica eseguirono? che compenso ricevettero? Di certo nessun astante pagò il biglietto d'ingresso al teatro anatomico patavino, poiché tale tassa era stata abolita da Fabrici nel 1596.

---

16. Kemp 2004, p. 84.

17. Sulla tradizione aristotelica dell'Università di Padova, si veda Rosso 2012, p. 120, n. 23.

18. Bylebyl 1979, pp. 337-338.

19. Bianchi 2015, p. 27.

20. Su Bologna, si veda Ferrari 1987. Quanto a Leida, si veda Rupp 1990.

21. Lazzerini 1994, p. 194, e Klestinec 2011, p. 93.

22. Dal 1553 la *natio Germanica* padovana, che era la più numerosa fra le corporazioni che tutelavano gli studenti, si scisse in séguito a dissidi interni in due anime: *natio Germanica iuristarum* e *natio Germanica artistarum*, ognuna delle quali con archivio e biblioteca propri. Quest'ultima radunava gli studenti di filosofia, medicina e teologia; si veda Rossetti 1979, p. 171.

23. Entrambe le citazioni provengono dal MS 471 (*Acta inclytae nationis Germanicae artistarum qui sunt Patavii*) dell'Archivio storico dell'Università di Padova (I-Pasu). Esse sono trascritte (con imprecisioni) in Favaro 1912, pp. 111 e 171. Sono riconoscente a Letizia Pellegrini e Leofranc Holford-Strevens per aver sciolto alcuni dubbi nella traduzione italiana letterale.

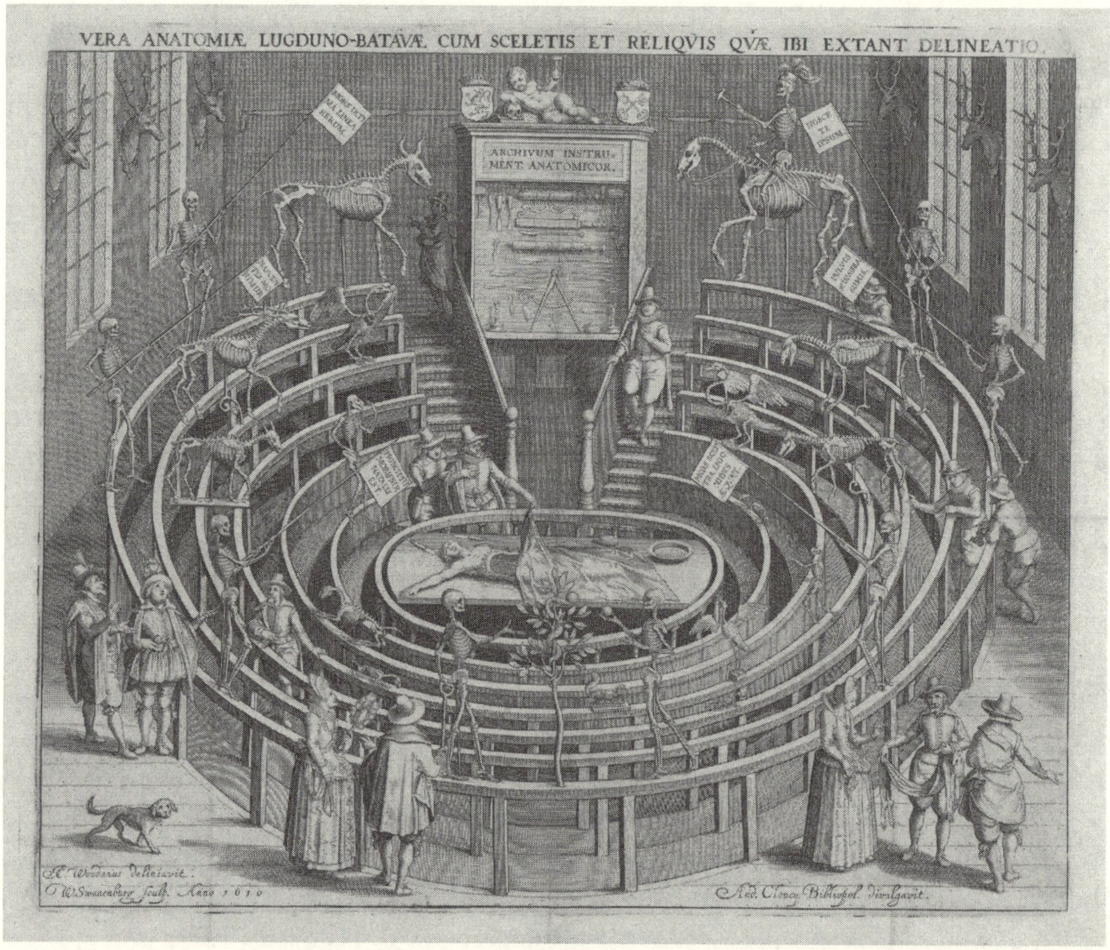

Ill. 5.6: Willem van Swanenburg (da Jan Corneliszoon van 't Woudt [Johannes Cornelius Woudanus]), *Il teatro anatomico di Leida* (incisione, 33 × 39.5 cm; 1610). NL-Ar, RP-P-1894-A-18588.

Qui di séguito le due annotazioni in questione. La prima, del 12 dicembre 1597:

> Il 12 dicembre, affinché gli spettatori dell'anatomia si rallegrassero e ricreassero gli animi dalla triste visione, come da antica usanza (che tuttavia era stata interrotta negli anni immediatamente precedenti), furono assoldati *fidicines* [cioè strumentisti a corda] dagli anatomisti e condotti nel teatro: cosa di cui si occupò con zelo D[avid] Placotomus[24]; e questi musicisti furono presenti anche nei molti giorni

[24]. È molto probabile che il personaggio qui menzionato sia David Placotomus (Brettschneider, 1565-1602) di Gdańsk (Danzica), che nel 1592 ottenne il titolo di dottore in medicina nell'Università di Padova ma vi figurava già iscritto nel 1581 (Jaśniewicz 2019, p. 71 e n. 13). Nella stessa università David Placotomus, *Licentiatus* in Medicina, fu nominato Prorettore degli Artisti *pro tempore* a partire dal 2 agosto 1598 («mdxcviii. postridie kal.

> seguenti, certamente le spese a conto loro non sono da rimpiangere, dal momento che mentre gli spettatori li osservano e li ascoltano, essi sogliono soprassedere da ogni tumulto e riluttanza, tranquillità in virtù della quale il teatro anatomico potrà resistere imperterrito per alcuni anni[25].

La seconda annotazione negli *Atti* è dell'8 gennaio 1600:

> Incombendo già i grandi freddi, e cadendo ormai la neve dal cielo, le brine e il ghiaccio, avendo l'Eccellentissimo Acquapendente compiuta, anzi compiutissima, nel pubblico uditorio la dimostrazione tanto su esseri umani quanto su altri animali, come pure su uccelli o su ossa di pennuti, ed essendo pronti tre corpi o soggetti, due maschili e uno femminile, si presentò alla dissezione in pompa solenne, con i *fidicines* condotti dagli anatomisti, e per alcuni giorni praticò la dissezione in pubblico sempre in presenza di numerosi ascoltatori, e infine, dopo una magistrale dissezione dell'occhio, pose fine all'operazione[26].

Sarebbe interessante riuscire a precisare cosa questi studenti tedeschi intendessero con *fidicines*. Certamente si riferivano a suonatori di strumenti a corda, ma non nell'accezione ereditata dal mondo classico presente anche negli scritti di Sant'Agostino[27]. Quali strumenti risuonarono allora nel teatro anatomico patavino durante le lezioni di anatomia? Anche il compositore e teorico tedesco Michael Praetorius (1571-1621) adopera il termine latino in senso piuttosto generico. Nel suo trattato di organologia *Syntagmatis musici tomus tertius* (1619) Praetorius menziona il *chorus vel capella fidicinum* (o *fidicinia*), descrivendolo come *ensemble* di

---

sextil. David Placotomus Dantiscanus, tametsi jam Licentiatus in Medicina esset, & de alieno viveret, attamen pro temporum conditione Prorector creatus est. Ex Actis Cod. III. p. 90». Si veda Jacopo Facciolati, *Fasti Gymnasii Patavini* (Padova, 1757), Tomus II, p. 222).

25. I-Pasu, MS 471, pp. 142-143: «a. d. 12 Xbris ad exhilarandos Anatomiae spectatores recreandosque ex tristi aspectu animos, ex vetusta consuetudine (quae tamen superioribus aliquot annis proximis interrupta[)], Fidicines ab Anatomisti[ci]s conducti et in Theatrum reducti fuere: Procurante hanc rem sedulo D. Placotomo; Aderantque Musici isti etiam sequentibus diebus quamplurimis, sumtus certe, qui illis irrogantur, minime poenitendi, siquidem dum ipsis attendunt et auscultant spectatores, ab omni tumultu et calcitratione supersedere solent, cuius tranquillitatis gratia Theatrum Anatomicum aliquot annos diutius inconcussum durare poterit».

26. I-Pasu, MS 471, p. 206: «Ingruentibus iam magnis frigoribus, coeloque iam nives, pruinas, glaciesque demittente, Excellentissimus Aquapendens absoluta prius absolutissima in publico auditorio tam humanorum, quam ceterorum animalium, ut et volucrium seu pennatorum ossium ostensione, posteaque tria extarent corpora seu subiecta, duo virilia, muliebre unum, ad sectionem solemni pompa cum fidicinibus ab Anatomisticis conductis accessit, eamque aliquot dierum spacio, frequenti semper auditorum corona admodum evidenter administravit, et tandem post luculentam oculi dissectionem sectioni finem imposuit».

27. Per Sant'Agostino *fidicen* era sinonimo di *citharista*; si veda Ferrari Barassi 1979, pp. 67 (con citazione dal *De musica*, Liber I, Art. 7) e 77, n. 55. Filtrato dal Medioevo, il termine risulta impiegato anche da Franchino Gaffurio (1451-1522) con riferimento generico a vihuele o viole; si veda Ferrari Barassi 2016, p. 409.

«Geigen oder Lauten / und alle besaeittete Instrument» (violini o liuti e tutti strumenti a corda)[28]. Nel terzo capitolo se ne legge una spiegazione più dettagliata qui letteralmente tradotta:

> L'ho chiamato capella fidicinia perché è meglio formare il gruppo con strumenti a corda come violini, liuti, arpe, e tutti gli altri, in particolare viole da gamba (se disponibilie, in mancanza delle quali, altrimenti violini), poiché il suono e la sonorità di viole e violini è continuo, particolarmente piacevole, e privo delle interruzioni dovute al respiro che non possono essere evitate da tromboni e altri strumenti a fiato. Tuttavia, per variare, si possono talvolta usare quattro tromboni [...]. O tre tromboni e un flauto tenore, o un cornetto al soprano, oppure un fagotto e tre flauti[29].

Praetorius riteneva che il gusto musicale tedesco, non avvezzo alla 'nudità' sonora prodotta in Italia dalle voci soliste che cantavano accompagnate dal solo strumento a tastiera, preferisse la sonorità più compatta dei gruppi di strumenti omogenei come quelli a corda[30].

Preciso appare invece il riferimento al termine latino nelle *Communiones totius anni* del polacco Mikołaj Zieleński, composizioni pubblicate a Venezia nel 1611 e realizzate su modello veneziano, il cui indice «tra le indicazioni esecutive riporta le seguenti parole: "Fidicen, id est Violino Italice vocatus"»[31].

Tornando a Padova, ciò che sinteticamente si apprende dalle annotazioni presenti negli *Atti* della *natio Germanica* è che le esecuzioni musicali sollecitate dagli studenti tedeschi durante le lezioni di anatomia avvennero lungo i vari giorni nei quali si protrassero le dissezioni, e servirono tanto a risollevare il morale degli astanti sensibili quanto a placare gli animi dei facinorosi. La musica fu molto gradita dai presenti; tale usanza esisteva già in passato; i musicisti assoldati erano solo *fidicines* (liutisti/violisti/violinisti), e la spesa affrontata per pagarli fu considerata un ottimo investimento.

Per cercare di capire cosa abbia portato gli studenti tedeschi a richiedere, e di conseguenza a sovvenzionare *performances* musicali in momenti così particolari come le dissezioni anatomiche,

---

28. Michael Praetorius, *Syntagmatis musici tomus tertius* (Wolfenbüttel, 1619), Dritter Theil, Cap. 1, p. '127' [=107]. Sul *Tomus tertius* si veda Toffetti 2019.

29. Praetorius, *Syntagmatis musici tomus tertius* (1619), Dritter Theil, Cap. 3, pp. '136-137' [=116-117]: «Daß ich diese Capellam darumb, Fidiciniam genennet, dieweil es besser ist, mit Besäiteten Instrumenten, als Geigen, Lauten, Harffen und allen andern, und sonderlich mit Violn de Gamba, wo man die haben kan, In mangelung aber derselben, mit Violen de Bracio, dieselbige Capellam zu bestellen. Denn der Sonus und Harmonia der Violen und Geigen continuiret sich immer nach einander mit sonderbahrer Liebligkeit, ohne einige respiration, deren man uff Posaunen und andern blasenden Instrumenten nicht entrathen kan. Jedoch pro variatione kan man bißweilen 4 Posaunen [...]. Oder 3 Posaunen, und eine Tenorfloeite, oder ein Cornett zum Discant, oder aber einen Fagott, und 3 Flöiten zur umbwechselung darzu gebrauchen». Il passaggio è tra quelli tradotti in inglese da Kite-Powell 2019, p. 31.

30. Praetorius, *Syntagmatis musici tomus tertius* (1619), Dritter Theil, Cap. 3, p. '136' [=116].

31. Zwolińska 2015, p. 70.

mi sono posta essenzialmente due domande: (1) che valenza aveva la musica per quegli studenti nello specifico contesto della dissezione anatomica? E poi: (2) il desiderio di musica in tale inquietante circostanza poteva essere specificamente collegato alla fede religiosa che connotava l'ambiente socioculturale di provenienza dei protestanti?

Musica e dissezione anatomica

Non c'è dubbio che, tra fine Cinque e inizio Seicento, la medicina in quanto scienza sia stata influenzata dalle filosofie non aristoteliche, *in primis* da quella neoplatonica. Il trattato *De vita cœlitus comparanda* (Sulla vita da vivere in conformità col cielo)[32] del filosofo neoplatonico toscano Marsilio Ficino (1433-1499), per esempio, esalta le qualità medicinali della musica sull'animo umano, interpretandole come dirette emanazioni energetiche dell'amore divino presente nel cosmo: «il peculiare potere della musica è dovuto alla affinità tra il mezzo materiale in cui è trasmessa, l'aria, e lo spirito dell'uomo», e il canto viene considerato il più potente imitatore di tutte le cose[33]. Il canto per Ficino è sostanzialmente contemplazione del divino. Ma in un contesto così peculiare come quello dei teatri anatomici è possibile che la musica, anziché agente riequilibratore dei disordini umani, fosse invece fruita alla stregua di un farmaco emetico risanatore.

Lo storico della cultura Piero Camporesi (1983) ci ha lasciato riflessioni illuminanti sull'interscambiabilità delle funzioni di medico ed esorcista fra Cinque e Seicento, poiché entrambi promuovevano tecniche di evacuazione dal corpo per purificarlo dalla «distemperanza degli umori» tipica dell'uomo galenico[34]. Medicamenti vomitorii e purganti, tra cui i frequenti salassi, attuavano esorcismi e promuovevano guarigioni per ristabilire la sanità umana e allungare la vita[35]. Diverse attestazioni confermano che, sebbene ritenessero la dissezione anatomica indispensabile alla loro formazione, gli studenti di medicina la reputavano una pratica non solo inquietante, ma anche vagamente immorale[36]. Tale disagio era chiaramente condiviso dalle autorità preposte al regolare svolgimento delle dissezioni, autorità che in tutt'Europa le consentivano per fini di studio quasi unicamente sui corpi dei condannati a morte, e a patto che questi non fossero originari della stessa città in cui si sarebbero praticate le autopsie. Allora: perché non attribuire alla musica richiesta dagli studenti tedeschi una valenza 'purgativa'

---

32. È il terzo libro del *De vita libri tres* o *De triplici vita* (scritto ca. 1480-1489, stampato a Firenze nel 1489), manuale medico e astrologico su come mantenere salute e vigore fisico. I primi due libri della trilogia si intitolano *De vita sana* e *De vita longa*. Sul potere medicinale della musica si veda Gozza 2004.

33. Walker 1989, pp. 90 e 94.

34. Camporesi 1994, p. 115.

35. *Ibidem*, pp. 138 e 156-157.

36. Lazzerini 1994.

nei confronti dell'inevitabile turbamento personale causato dalla pratica dissettoria? Per sopportarne il forte peso psicologico forse non era soltanto necessario alleggerire la tristezza prodotta negli animi degli astanti; serviva piuttosto mitigarne al massimo l'effetto traumatico, al fine di favorirne un atteggiamento 'scientifico' neutrale e lucido. Ed è proprio questo il senso della spersonalizzazione estrema proiettata sul corpo da dissezionare, chiamato con distacco *subjectum* (soggetto) nei manuali e nelle cronache come quella registrata dalla *natio Germanica artista* a Padova[37]. Credo quindi che alla musica risuonante si attribuisse proprio il potere di intervenire attivamente sul turbamento emotivo, al fine di ristabilire l'impassibilità personale necessaria a uno scienziato che osserva un corpo inerte.

Purtroppo, nulla sappiamo sulle effettive musiche che risuonarono nel teatro anatomico patavino. Ma esistono varie antologie musicali circolanti da fine Cinquecento che, attraverso le dedicatorie, attestano lo stretto contatto fra ambienti tedeschi e veneti, spesso motivato dalla presenza dell'Università a Padova. Qui di séguito alcuni esempi di libri del genere in cui mi sono imbattuta, nei quali si noterà la netta preponderanza di dediche ai nobili Khisl da Kaltenbrunn, in Stiria (Austria), famiglia assai legata all'ambiente universitario padovano[38]. Le raccolte musicali contengono un repertorio che, anche se non compare sotto forma di intavolatura liutistica, poteva però essere facilmente intavolato, e la presenza di molti compositori minori in questi libri attesta il radicamento nell'ambiente veneto-padovano. Forse ciò che gli studenti ascoltarono durante le dissezioni provenne proprio da qualcuno dei volumi qui elencati in ordine cronologico:

- Giacomo Gorzanis, *Intabolatura di liuto, Libro primo* (Venezia, 1561)[39]. Contiene 27 brani, con dedica (non datata) del cieco Gorzanis (pugliese, ma cittadino triestino) a «Giovanne Khisl detto da Kolten Pruen, della Cesarea M. Dignissimo Conseglier & Supremo Pagator alla Frontere, & confini de Croatia»;
- Giacomo Gorzanis, *Il primo libro di napolitane che si cantano et sonano in leuto* (Venezia, 1570)[40]. Contiene 25 'villanelle alla napolitana': 23 di Gorzanis, una di Pietro Santucci, e una d'autore incerto. Con dedica a «Giorgio Kis de Koltemprun, figliuolo del clariss. et generoso S. Gio. Kisl benemerito Cavalliero, aureato, Hereditario epifaro dell'Illustre Contado de Goritio, et Cesareo et Arciducale consegliero, et vicecapitanio della provincia del Cragno»;

---

37. Bohde 2003, p. 30.

38. Sulla famiglia Khisl (Khiesel) von Kaltenbrunn, si vedano *Allgemeine Deutsche Biographie*, xv, 1882, p. 708, Kokole 2007, Comparin 2020, e *Deutsche Biographie* (<https://www.deutsche-biographie.de>, voce 'Khiesel', visitato a maggio 2022). Nelle dediche stampate all'interno dei libri che seguono sono menzionati i Khisl seguenti: Johann/Hans (Giovanni; 1530-1591), *Hofkammerpräsident* alla corte asburgica dell'arciduca Carlo ii e consigliere dell'imperatore Rodolfo ii; e i suoi figli: Georg (Giorgio; 1558-1605); Veit (Guido; 1560-1609); Hans (Gian); Hans Jakob (Giovan Giacomo; 1565-1638), compositore dilettante che nel 1587-1588 era *consiliarius* della *natio Germanica* a Padova, dal 1613 al 1619 siniscalco alla corte di Stiria a Graz e poi *Kriegspräsident* dell'Austria Interiore (Federhofer 1955, p. 185, n. 880); e Karl (Carlo; 1570 - *post* 1648).

39. RISM A/i, GG 3031.

40. RISM A/i, G 3035.

• Claudio Merulo, *Primo libro de ricercari da cantare a quattro voci* (Venezia, 1574)[41]. Contiene 20 ricercari, con dedica (datata Venezia, 22 novembre 1574) a «Illustribus Wolfgango Baroni a Stubmberg et Georgio Khisl a Khaltemprun»;

• Filippo (Philippe) de Duc, *Il primo libro de madrigali a cinque et sei voci* (Venezia, 1586)[42]. Contiene 21 madrigali con dedica (datata «Di Padova il dì 15 di Genaro 1586») indirizzata a «Gio: Giacomo et Carlo Chiseli da Caltenprun & Gonobiz», cioè Hans Jakob e Karl Khisl da Kaltenbrunn e Gonobiz». Il primo brano, con incipit *Giovan Giacomo & Carlo*, mette addirittura in musica i nomi dei dedicatari (Ill. 5.7abc);

• *Canzonette a tre voci di diversi eccellentissimi musici, Libro primo* (Venezia, 1587)[43]. La raccolta (datata «Di Venetia il dì 10 di Giugno 1587») contiene 21 brani, dove figurano alcuni compositori ben integrati nel mondo accademico padovano[44]. Fu curata da Angelo Barbato, musicista dilettante attivo in quegli anni a Padova, e dedicata a «voi tre Illustrissimi Signori fratelli», cioè Hans Jakob, Karl e Veit Khisl;

• Lodovico Balbi, *Musicale essercitio* (Venezia, 1589)[45]. Contiene 27 madrigali a cinque voci che sono rielaborazioni musicali basate sulle voci superiori di noti compositori, con dedica (datata Venezia, 4 marzo 1589) a «Giovanni Chisel da Coltemprum & Gonoviz Consigliero di S.M. Cesarea, Presidente della Camera Aulica del Ser. Arciduca Carlo d'Austria [...] et alli [...] suoi figlioli Giorgio, Guido, Gian, Gioan Giacomo & Carlo Chiseli»[46];

• *Laudi d'amore, madrigali a cinque voci de diversi eccellenti musici di Padova* (Venezia, 1598)[47]. La raccolta di 16 madrigali fu curata dal padovano Girolamo Boni, come si legge nella dedicatoria (datata 15 febbraio 1598) indirizzata a «Guglielmo Adorne Borusso» (Guglielmo da Dorn Prussiano), studente tedesco di diritto nell'Università di Padova[48]. Alessandra Andreotti (1996) collega il raffinato simbolismo cabalistico e l'ermetismo riscontrabili nella raccolta all'intellettualismo in voga nei circoli accademici universitari patavini;

• *Madrigali de diversi a quattro voci raccolti da Gio. Maria Radino* (Venezia, 1598)[49]. Contiene 20 brani di diversi autori, con dedica (datata «Di Padoa il dì 14 di Marzo 1598») al «sig. Giorgio Guglielmo et sig. Carlo Iorgeri», studenti nobili dell'Austria Superiore: Georg Wilhelm (ca. 1580-1617) e Karl (1584-1623) Jörger von Tollet;

• Romano Micheli, *Musica vaga et artificiosa, continente motetti con oblighi & canoni diversi* (Venezia, 1615)[50]. Raccolta di 59 mottetti e canoni su soggetti di vari musicisti menzionati nelle intestazioni dei brani[51], con dedica (datata 18 giugno 1615) dell'allora maestro di canto nella Basilica di S. Marco a Venezia «al Sig. Gio. Giacomo Khisl».

---

41. RISM A/i, M 2374.

42. RISM A/i, D 3613.

43. RISM B/i, 1587[7].

44. Comparin 2020, p. 93.

45. RISM A/i, B 742; BB 742.

46. Vogel 1947, pp. 301-303, e Spanò 1993-1994.

47. RISM B/i, 1598[7].

48. Guglielmo da Dorn risulta iscritto fra gli studenti non nobili dell'Università di Padova il 28 novembre 1594; si veda Andreotti 1996, p. 219, e n. 47.

49. RISM B/i, 1598[9].

50. RISM A/i, M 2683; RISM B/i, 1615[3].

51. Di Pasquale 2017, p. 184.

CANTO

DI FILIPPO DE DVC FIAMENGO

IL PRIMO LIBRO DE MADRIGALI A CINQVE ET Sei Voci, Nouamente composti, & dati in luce.

IN VENETIA

Presso Giacomo Vincenzi, & Ricciardo Amadino, compagni.

M D LXXXVI.

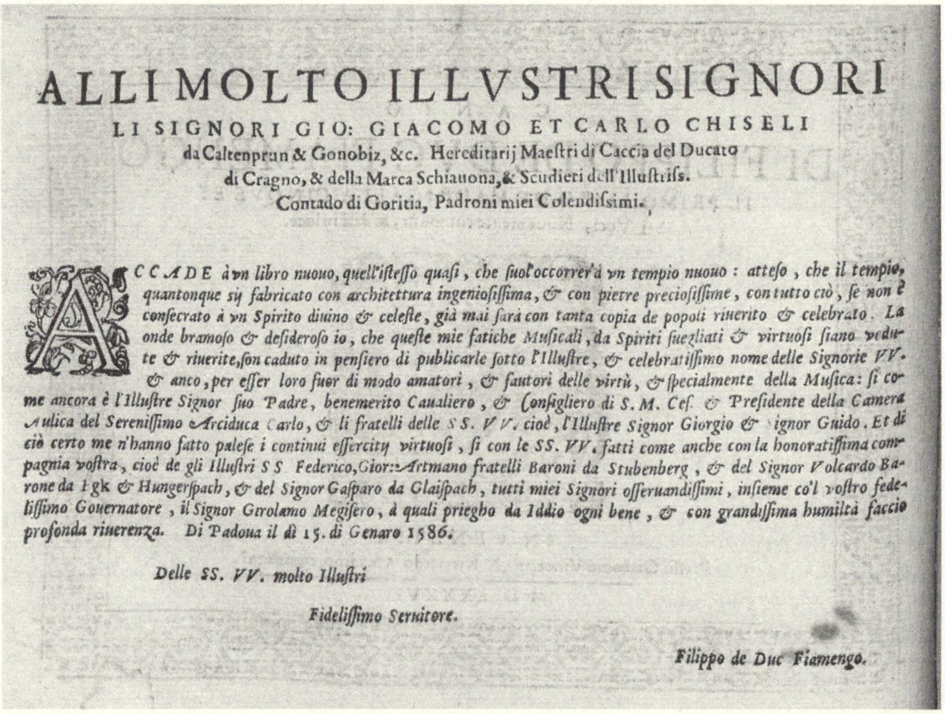

ALLI MOLTO ILLVSTRI SIGNORI

LI SIGNORI GIO: GIACOMO ET CARLO CHISELI

da Caltenprun & Gonobiz, &c. Hereditarij Maestri di Caccia del Ducato di Cragno, & della Marca Schiauona, & Scudieri dell'Illustriss. Contado di Goritia, Padroni miei Colendissimi.

ACCADE à vn libro nuouo, quell'istesso quasi, che suol'occorrer'à vn tempio nuouo: atteso, che il tempio, quantonque sij fabricato con architettura ingeniosissima, & con pietre preciosissime, con tutto ciò, se non è consecrato à vn Spirito diuino & celeste, già mai sarà con tanta copia de popoli riuerito & celebrato. La onde bramoso & desideroso io, che queste mie fatiche Musicali, da Spiriti suegliati & virtuosi siano vedute & riuerite, son caduto in pensiero di publicarle sotto l'Illustre, & celebratissimo nome delle Signorie VV. & anco, per esser loro fuor di modo amatori, & fautori delle virtù, & specialmente della Musica: si come ancora è l'Illustre Signor suo Padre, benemerito Caualiero, & Consigliero di S. M. Ces. & Presidente della Camera Aulica del Serenissimo Arciduca Carlo, & li fratelli delle SS. VV. cioè, l'Illustre Signor Giorgio & Signor Guido. Et di ciò certo me n'hanno fatto palese i continui essercitij virtuosi, si con le SS. VV. fatti come anche con la honoratissima compagnia vostra, cioè de gli Illustri SS Federico, Gior: Artmano fratelli Baroni da Stubenberg, & del Signor Volcardo Barone da Egk & Hungerspach, & del Signor Gasparo da Glaispach, tutti miei Signori osseruandissimi, insieme co'l vostro fedelissimo Gouernatore, il Signor Girolamo Megisero, à quali priegho da Iddio ogni bene, & con grandissima humiltà faccio profonda riuerenza. Di Padoua il dì 15. di Genaro 1586.

Delle SS. VV. molto Illustri

Fidelissimo Seruitore.

Filippo de Duc Fiamengo.

Ill. 5.7abc: Filippo (Philippe) de Duc, *Il primo libro de madrigali a cinque e sei voci* (Venezia, 1586), frontespizio, dedica a Hans Jacob (Giovan Giacomo) e Karl (Carlo) Khisl von Kaltenbrunn, e Canto del primo brano, *Giouan Giacomo e Carlo, Spirti felici & rari*. Copia D-Mbs.

Quasi vent'anni fa (2003), inoltre, il liutista americano Terrell Stone pubblicò un CD registrato all'interno del teatro anatomico di Palazzo del Bo, con musica per liuto tratta da tre raccolte certamente circolanti in Veneto fra Cinque e Seicento: un'ipotesi avventurosa su ciò che risuonò realmente nel contesto delle dissezioni anatomiche. Ma almeno una delle tre antologie da cui Stone ha estrapolato le sue incisioni rimanda al contatto stretto fra ambiente accademico patavino e ambiente musicale: il 'manoscritto Herold' (Padova, 1602), copia di un'intavolatura liutistica in notazione francese compilata dallo studente tedesco Christoph Herold (nato ca. 1576-1578), originario di Halle[52]. Studente di giurisprudenza a Leida dal 20 settembre 1598, passò poi all'Università di Padova, dove risulta iscritto dal 16 dicembre 1601. A Leida prese lezioni di liuto dal fiammingo Joachim van den Hove, che lì risiedette fino al 1614. Traccia visibile del rapporto fra i due musicisti è nel 'manoscritto Herold', che contiene alcune composizioni del maestro[53].

[52]. D-Hvonbusch; RISM B/vii, p. 374. Del manoscritto esiste facsimile a cura di Andreas Schlegel e François-Pierre Goy (Monaco di Baviera, 1991).

[53]. Burgers 2013, p. 110, e Burgers 2016, pp. 63-65, dove il 'manoscritto Herold' viene considerato un prodotto di Leida.

Il vivo interesse degli studenti verso la musica per liuto è attestato anche altrove. Ne è un esempio l'antologia manoscritta per liuto che contiene musica di provenienza internazionale assemblata da Emanuel Wurstisen (1572-1619), studente di medicina all'Università di Basilea tra fine Cinque e inizio Seicento[54].

C'è almeno un altro caso interessante di cui sono venuta a conoscenza grazie ad Agnieszka Leszczyńska: il liutista e compositore borgognone Jean-Baptiste Bésard (ca. 1576 - *post* 1617), dopo aver studiato legge nell'Università di Dôle, in Borgogna, studiò medicina a Roma a fine Cinquecento; lì diventò allievo del famoso liutista Lorenzino (Lorenzo Tracetti), a cui dedicò l'antologia di brani per liuto *Thesaurus harmonicus* (Colonia, 1603).

Nello stesso periodo, anche a Bologna esistevano fertili relazioni fra musicisti locali e studenti universitari tedeschi, veri appassionati di musica: ne è esempio il caso del compositore Ottavio Vernizzi (Invernizzi; 1569-1649), che dedicò ai consiglieri della *natio Germanica* locale (senza il consueto previo accordo reciproco) la sua raccolta *Armonia ecclesiasticorum concertuum qui duabus, tribus et quatuor concinuntur vocibus, Opus authoris secundum* (Venezia, 1604)[55].

Richiesta di musica e sensibilità protestante

Il fatto che la singolare esigenza di musica nel teatro anatomico di Padova tra fine Cinque e inizio Seicento sia scaturita da studenti tedeschi mi ha spinto a chiedermi se ciò avesse in qualche modo a che fare con la sensibilità religiosa protestante. Per Lutero, infatti, l'impatto della musica sull'animo umano andava definito in termini non cosmologici né pedagogici, ma piuttosto comunicativi, poiché questo importante dono divino ha la capacità di muovere i cuori[56]. La necessaria 'semplicità' invocata dal riformatore tedesco coincide probabilmente con l''estrema comunicabilità' e 'naturalezza' della musica stessa: al contrario di Sant'Agostino, molto preoccupato della distrazione che i suoni musicali potevano indurre nell'uomo, Lutero invitava tanto ad abbandonarsi senza scrupoli alla musica, quanto a diffidare di coloro che si dichiarano insensibili ad essa, poiché tale indifferenza non è umana. Addirittura, la libertà del cristiano verso Dio era, secondo Lutero, assimilabile all'atteggiamento verso l'ammiratissima musica di Josquin, recepita come scorrevole, ben comprensibile, non dura, eppure mai scontata. Da queste considerazioni veniva però esclusa la musica strumentale praticata in chiesa, contro cui i riformatori protestanti del Cinquecento polemizzarono ponendosi sulla scia umanistica che enfatizzava il ruolo dei padri della chiesa e il canto come forma di preghiera[57]. Per costoro era evidente il forte valore etico intrinseco al canto, un sacrificio spirituale da offrire a Dio.

---

[54]. CH-Bu, Ms. F.ix.70; si veda Leszczyńska 2020.

[55]. RISM A/i, V 1293; si veda Vernizzi 2016.

[56]. Anttila 2019.

[57]. Kim 2019.

Conclusione

In conclusione, ritengo quindi che la richiesta di musica strumentale documentata a fine Cinquecento dagli studenti protestanti a Padova durante le lezioni anatomiche non sia stata motivata dalla particolare sensibilità frutto della diversa fede. (E anche il pregiudizio circolante fra gli storici secondo cui i protestanti sarebbero stati molto più ostili alla pratica della dissezione rispetto ai cattolici riceve oggi qualche contestazione[58].)

Trovo comunque molto suggestiva la riflessione di Andrea Carlino (1994) sul contesto accademico dell'epoca: «Non è forse un caso che la rottura epistemologica operata da Vesalio, la discussione dell'autorità antica e la disputa tra galenisti e anti-galenisti nei termini in cui è stata condotta, abbiano coinciso temporalmente con i conflitti teologici tra cattolici e protestanti su temi strutturalmente analoghi e in toni non dissimili»[59]. Più probabilmente, invece, gli studenti tedeschi benestanti ben sapevano cosa la città di Padova offrisse anche in termini di piacevolezza, e ne pagavano volentieri il prezzo. La musica rientrava assolutamente tra i godimenti che gli studenti stranieri facoltosi non si lasciavano sfuggire, soprattutto in Veneto, ed essi sapevano procurarsela a complemento delle occupazioni accademiche. Lo testimonia anche la presenza fra i libri elencati nella biblioteca della *natio Germanica artista* di due raccolte di brani a tre voci, entrambe di compositori cinquecenteschi legati alla Baviera: Orlando di Lasso e Andrea Gabrieli[60]. Sfortunatamente, la genericità delle intitolazioni indicate rende pressoché impossibile l'identificazione certa di entrambi i libri musicali.

Bibliografia

Andreotti 1996

Andreotti, Alessandra. '*Laudi d'amore*. Cabala ed ermetismo in una miscellanea madrigalistica padovana', in: *Musica e Storia*, iv (1996), pp. 189-226.

Anttila 2019

Anttila, Miikka. '*Geistliche, liebliche Lieder*. In Search of Aesthetic Criteria for Music in Luther's Theology', in: *Music and Theology in the European Reformations*, a cura di David J. Burn, Grantley McDonald, Joseph Verheyden e Peter De Mey, Turnhout, Brepols, 2019, pp. 141-148.

---

[58]. Si veda, per esempio, Kinzelbach 2018.

[59]. Carlino 1994, pp. 252-253.

[60]. Favaro 1912, p. 186: «*Cantiones Orlandi di Lusso* [*sic*; errore di trascrizione di Favaro?], 3 vocum» e le «*Cantiones And. Gabrielis*, 3 vocum». Questi volumi furono donati alla biblioteca nel 1601 da Friedrich Seyller, *consiliarius* della nazione; si veda *ibidem*, pp. 185-186. La biblioteca della *natio Germanica artistarum* prese forma nel 1586. Ne esistono indici a stampa secenteschi: *Bibliotheca medico-philosophico-philologica inclytae Nationis Germanae Artistarum* (Padova, 1673, 1677) e *Bibliotheca medico-philologica inclytae Nationis Germanae Artistarum* (Padova, 1685). I libri superstiti sono oggi conservati in I-Pu.

Bianchi 2015

Bianchi, Luca. 'La tradizione aristotelica nell'Italia del Quattrocento', in: *La filosofia e le sue storie. L'età moderna*, a cura di Umberto Eco e Riccardo Fedriga, Roma, Laterza, 2015, pp. 27-32.

Bohde 2003

Bohde, Daniela. 'Skin and the Search for the Interior. The Representation of Flaying in the Art and Anatomy of the Cinquecento', in: *Bodily Extremities. Preoccupations with the Human Body in Early Modern European Culture*, a cura di Florike Egmond e Robert Zwijnenberg, Aldershot, Ashgate, 2003, pp. 10-47.

Brizzi 1996

Brizzi, Gian Paolo. 'Le università italiane tra Rinascimento ed età moderna', in: *Il pragmatismo degli intellettuali. Origini e primi sviluppi dell'istituzione universitaria*, a cura di Roberto Greci, Torino, Scriptorium, 1996, pp. 175-200 [saggio già apparso in *Dal Rinascimento alle riforme religiose*, a cura di Gian Paolo Brizzi e Jacques Verger, Cinisello Balsamo, Silvana, 1991, pp. 21-53].

Brugi 1893-1894

Brugi, Biagio. 'Gli studenti tedeschi e la S. Inquisizione a Padova nella seconda metà del secolo xvi', in: *Atti del Reale Istituto Veneto di Scienze, Lettere ed Arti*, tomo lii (serie vii, tomo v), 1893-1894, pp. 1015-1033.

Burgers 2013

Burgers, Jan W. J. *The Lute in the Dutch Golden Age: Musical Culture in the Netherlands ca. 1580-1670*, Amsterdam, Amsterdam University Press, 2013.

Burgers 2016

Id. 'Leiden, an International Lute Centre in the Golden Age', in: *The Lute in the Netherlands in the Seventeenth Century: Proceedings of the International Lute Symposium Utrecht, 30 August 2013*, a cura di Jan W. J. Burgers, Tim Crawford e Matthew Spring, Newcastle upon Tyne, Cambridge Scholars Publishing, 2016, pp. 53-72.

Bylebyl 1979

Bylebyl, Jerome J. 'The School of Padua. Humanistic Medicine in the Sixteenth Century', in: *Health, Medicine, and Mortality in the Sixteenth Century*, a cura di Charles Webster, Cambridge, Cambridge University Press, 1979, pp. 335-370.

Camporesi 1994

Camporesi, Piero. *La carne impassibile. Salvezza e salute fra Medioevo e Controriforma*, Milano, Garzanti, 1994 (Gli elefanti. Saggi) [prima edizione: Milano, il Saggiatore, 1983].

Carlino 1994

Carlino, Andrea. *La fabbrica del corpo. Libri e dissezione nel Rinascimento*, prefazione di Adriano Prosperi, Torino, Einaudi, 1994.

Comparin 2020

Comparin, Chiara. 'Music for the Khisls from the Paduan Milieu. The *Canzonette a tre voci* (1587) and Ludovico Balbi's *Musicale Essercitio* (1589)', in: *De Musica Disserenda*, xvi/2 (2020), pp. 83-99.

Costa 1913
Costa, Emilio. 'Gli atti delle due Nazioni germaniche nello Studio di Padova', in: *Archivio Storico Italiano*, s. 5, L/268 (1913), pp. 314-337.

Di Pasquale 2017
Di Pasquale, Marco. 'Romano Micheli e Venezia. Due episodi senza conseguenze', in: *Tra musica e storia. Saggi di varia umanità in ricordo di Saverio Franchi*, a cura di Giancarlo Rostirolla ed Elena Zomparelli, Roma, Ibimus, 2017, pp. 179-196.

Favaro 1912
Atti della nazione germanica artista nello Studio di Padova. 2, a cura di Antonio Favaro, Venezia, Reale Deputazione Veneta di Storia Patria, 1912.

Federhofer 1955
Federhofer, Hellmut. 'Graz Court Musicians and Their Contributions to the *Parnassus musicus Ferdinandaeus* (1615)', in: *Musica Disciplina*, IX (1955), pp. 167-244.

Fenlon 2018
Fenlon, Iain. 'Music and the Academies of Venice and the Veneto', in: *A Companion to Music in Sixteenth-Century Venice*, a cura di Katelijne Schiltz, Leida, Brill, 2018, pp. 99-121.

Ferrari 1987
Ferrari, Giovanna. 'Public Anatomy Lessons and the Carnival. The Anatomy Theatre of Bologna', in: *Past & Present*, CXVII (1987), pp. 50-106.

Ferrari Barassi 1979
Ferrari Barassi, Elena. *Strumenti musicali e testimonianze teoriche nel Medio Evo*, Cremona, Fondazione Claudio Monteverdi, 1979.

Ferrari Barassi 2016
Ead. 'Giovanni Maria Lanfranco teorico degli strumenti musicali e il suo tempo', in: *Philomusica Online*, XV/1 (2016), pp. 353-422.

Frugoni 1991
Frugoni, Chiara. 'Arti liberali e meccaniche', in: *Enciclopedia dell'arte medievale*, Roma, Istituto della Enciclopedia Italiana, 1991 (<https://www.treccani.it/enciclopedia/arti-liberali-e-meccaniche_(Enciclopedia-dell'-Arte-Medievale)/>, visitato a maggio 2022).

Gozza 2004
Gozza, Paolo. 'Platone e Aristotele nel Rinascimento. La psicologia della musica di Ficino e Giacomini', in: *Il saggiatore musicale*, XI/2 (2004), pp. 233-252.

Jaśniewicz 2019
Jaśniewicz, Aleksandra. 'Model do medalu Daniela Placotomusa w kontekście medali osobistych w XVI-wiecznym Gdańsku', in: *Porta Aurea*, n. 18 (dicembre 2019), pp. 68-80.

KEMP 2004

KEMP, Martin. '«Il mio bell'ingenio». L'anatomia visiva nel *Theatrum totius animalis fabricae* di Fabrici', in: *Il teatro dei corpi. Le pitture colorate d'anatomia di Girolamo Fabrici d'Acquapendente*, a cura di Maurizio Rippa Bonati e José Pardo-Tomás, Milano, Mediamed, [2004] (catalogo della mostra tenuta a Venezia nel 2004-2005), pp. 83-107.

KIM 2019

KIM, Hyun-Ah. 'Singing, Prayer, and Sacrifice. The Neo-Platonic Revival of 'Musica humana' in the Swiss Reformation', in: *Music and Theology in the European Reformations*, *op. cit.*, pp. 413-436.

KINZELBACH 2018

KINZELBACH, Annemarie. 'Dissecting Pain. Patients, Families and Medical Expertise in Early Modern Germany', in: *Pathology in Practice: Diseases and Dissections in Early Modern Europe*, a cura di Silvia De Renzi, Marco Bresadola e Maria Conforti, Abingdon-New York, Routledge, 2018, pp. 170-187.

KITE-POWELL 2019

KITE-POWELL, Jeffery. 'Michael Praetorius's Variable Opinions on Performance', in: *De Musica Disserenda*, XV/1-2 (2019), pp. 29-46.

KLESTINEC 2011

KLESTINEC, Cynthia. *Theaters of Anatomy: Students, Teachers, and Traditions of Dissection in Renaissance Venice*, Baltimora (MD), Johns Hopkins University Press, 2011.

KOKOLE 2007

KOKOLE, Metoda. '«Sequamini o socii» ali vesela glasbena druščina s Kranjske in Štajerske. Prva knjiga pet- in šestglasnih madrigalov Filippa de Duca (1586)', in: *Muzikološki Zbornik*, XLIII/1 (2007), pp. 67-90.

LAZZERINI 1994

LAZZERINI, Luigi. 'Le radici folkloriche dell'anatomia. Scienza e rituale all'inizio dell'età moderna', in: *Quaderni Storici*, Nuova Serie, LXXXV (1994), pp. 193-233.

LESZCZYŃSKA 2020

LESZCZYŃSKA, Agnieszka. 'Emanuel Wurstisen, His Tablature and Links to Poland Lute Music with *Medicine* in the Background', in: *Muzyka*, LXV/2 (2020), pp. 3-24.

MARTELLOZZO FORIN 1997

MARTELLOZZO FORIN, Elda. 'Conti palatini e lauree conferite per privilegio. L'esempio padovano del sec. XV', in: *Annali di Storia delle Università Italiane*, III (1997), pp. 79-119.

MELCHIORRE 2012

I patti con Padova (1405-1406). Dalla guerra alla Bolla d'oro, a cura di Matteo Melchiorre, Roma, Viella, 2012.

Rieder 2010

Rieder, Maximiliane. 'Cosmopoliti sull'Adriatico. Mercanti ed industriali tedeschi a Venezia e Trieste', in: *Qualestoria*, n. 1 (giugno 2010), pp. 99-134.

Rossetti 1979

Rossetti, Lucia. 'L'Archivio storico dell'Università di Padova', in: Semenzato, Camillo *L'Università di Padova. Il palazzo del Bo. Arte e storia*, con saggi di Lucia Rossetti e Gilda Mantovani, Trieste, LINT, 1979, pp. 153-174.

Rosso 2012

Rosso, Paolo. 'Studiare e insegnare «in Studiis alienis». La 'peregrinatio medica' toscana negli 'Studia generalia' dell'Italia settentrionale (Padova, Pavia, Torino, secoli xiv-xv)', in: *Umanesimo e Università in Toscana (1300-1600). Atti del Convegno Internazionale di Studi (Fiesole-Firenze: 25-26 maggio 2011)*, a cura di Stefano U. Baldassarri, Fabrizio Ricciardelli ed Enrico Spagnesi, Firenze, Le Lettere, 2012, pp. 111-182.

Rupp 1990

Rupp, Jan C. C. 'Matters of Life and Death. The Social and Cultural Conditions of the Rise of Anatomical Theatres, with Special Reference to Seventeenth-Century Holland', in: *History of Science*, xxxviii (1990), pp. 263-287.

Salonen 2007

Salonen, Kirsi. 'La Penitenzieria apostolica e le «partes»', in: *Offices, écrit et papauté (xiiie-xviie siècle)*, a cura di Armand Jamme e Olivier Poncet, Roma, École française de Rome, 2007, pp. 253-265.

Saraco 2012

Saraco, Alessandro. 'La Penitenzieria Apostolica e il suo archivio storico', in: *Anuario de Historia de la Iglesia*, xxi (2012), pp. 423-434.

Schiltz 2018

A Companion to Music in Sixteenth-Century Venice, a cura di Katelijne Schiltz, Leida, Brill, 2018.

Spanò 1993-1994

Spanò, Roberto. 'Il *Musicale esercitio* di Lodovico Balbi. Un singolare accostamento di scritture diverse nella polifonia profana della fine del secolo xvi', in: *Rassegna Veneta di Studi Musicali*, ix-x (1993-1994), pp. 107-129.

Toffetti 2019

Toffetti, Marina. 'Italian Musical Culture and Terminology in the Third Volume of Michael Praetorius's *Syntagma musicum* (1619)', in: *De Musica Disserenda*, xv/1-2 (2019), pp. 71-103.

Vernizzi 2016

Vernizzi, Ottavio. *Armonia ecclesiasticorum concertuum. Mottetti a 2, 3 e 4 voci con basso continuo, Op. 2, Venezia 1604*, a cura di Paola Dessì, Padova, CLEUP, [2]2016.

Vogel 1947

Vogel, Emil. 'Bibliography of Italian Secular Vocal Music Printed between the Years 1500-1700', *Notes*, 2da serie, IV/3 (giugno 1947), pp. 301-316.

Walker 1989

Walker, Daniel Pickering. 'La teoria dello spirito musicale di Ficino', in: *La musica nella rivoluzione scientifica del Seicento*, a cura di Paolo Gozza, Bologna, Il Mulino, 1989, pp. 89-95.

Zwolińska 2015

Zwolińska, Elżbieta. 'Zieleński e la ricezione dell'italianità nella cultura musicale polacca', in: *Italian Music in Central-Eastern Europe around Mikołaj Zieleński's «Offertoria» and «Communiones» (1611)*, a cura di Tomasz Jeż, Barbara Przybyszewska-Jarmińska e Marina Toffetti, Venezia, Edizioni Fondazione Levi, 2015, pp. 65-72.

The (Long) Seventeenth Century

Music

Chapter Six

Galileo and Opera: Music, Science, Religion, and Politics in the Seventeenth Century

Daniel Martín Sáez
(Universidad de Granada, Spain)

Like musical theatre in every age, themes of mythology, cosmology and theology permeate Florentine theatre. The *ingegni* of Filippo Brunelleschi, Bernardo Buontalenti and others for *sacre rappresentazioni* and *intermedi* sought to embody the great ideas of their time through theatrical architecture, machinery, costumes, music, and other arts. In metaphysical terms, theatre was an ideal place to convey the idea of totality. It is difficult to imagine a better artform to visualize Dante's universe than the *intermedi* of the sixteenth century[1]. It is also not surprising that the operas of the next century were full of infernal, heavenly and earthly scenes, which helped shaping a comprehensive *theatrum mundi*[2].

Although it is often taken for granted that opera is an artform in which music-theatrical content is central to any analysis, the metaphor of the theatre of the world also leads us to understand opera as a stage for the truths of philosophy, science, theology, cosmology, and politics, something that the myth of art tends to hide. The role of this metaphor is especially illuminating in the seventeenth century when the lines between the different arts, and those

---

1. Mamone 1981, Mamone 2003, Saslow 1996, and Fenlon 2002.
2. Martín Sáez 2020a.

Ill. 6.1: Ottavio Leoni, *Galileo Galilei* (engraving, 14 × 11 cm; 1624). Copy NL-Ar, RP-P-1950-421.

between art, science, and religion, became blurred. In Florence, there are many examples of this cross-fertilisation of the arts, especially when the Medici achieved their greatest power after Tuscany was elevated to a Grand Duchy. The Uffizi building, which was simultaneously a space for government offices, an art museum, a spectacular theatre (where several operas were performed), and a scientific cabinet, might serve as an example of this synthesis.

Seen in this light, it would be strange if opera and the new cosmology of the Early Modern Age, which were developed in the same city and even in the same court, did not end up coinciding. At least eleven operas performed in Florence and Rome from 1613 to 1638 can be directly or indirectly connected to the new discoveries and theories of Galileo Galilei (1564-1642; Ill. 6.1). This is particularly interesting in view of the interaction between Florence and Rome, which ultimately represented the most influential institutions of their time, namely the State and the Church. In fact, these 'Galilean operas' coincided with key episodes in Galileo's life such as the theological debates of 1613-1615, the condemnation of Copernicanism in 1616, the disputes with the Jesuits about the comets of 1618 which culminated in Galileo's *Il saggiatore* (The Assayer; 1623), his conviction by the Inquisition in 1633, and the publication of his *Discorsi* in 1638. These episodes should be placed against the background of the tension between Protestants and Catholics over their interpretation of the Bible and, from 1618 onward, the Thirty Years' War. Accordingly, historic research can serve to clarify the history of opera, and vice-versa: opera can help to illuminate some aspects of what has come to be called the 'Galileo affair'.

Table 6.1: Popes in Galileo's Lifetime

Pontificate	Pope	Personal name
1592-1695	Clement (Clemens) viii	Ippolito Aldobrandini (1536-1605)
1605	Leo (Leo) xi	Alessandro Ottaviano de' Medici (1535-1605)
1605-1621	Paul (Paulus) v	Camillo Borghese (1550-1621)
1621-1623	Gregory (Gregorius) xv	Alessandro Ludovisi (1554-1623)
1623-1644	Urban (Urbanus) viii	Maffeo Barberini (1588-1644)
1644-1655	Innocent (Innocens) x	Innocenzo Pamfilj (1574-1655)
1655-1667	Alexander (Alexander) vii	Fabio Chigi (1599-1667)

Galilean and Anti-Galilean Operas

References to Galileo and his controversies in opera prove to be particularly relevant considering that there seems to be no similar case in other artforms. No other art seems to have been as close to science as opera in this period. Yet, of the eleven Galilean or anti-Galilean operas discussed in this chapter, only three have been connected to Galileo to date. Frederick Hammond (1992) has interpreted *Sant'Alessio* (1634) as a smokescreen to mitigate the criticism caused by the condemnation of 1633[3]; Mario Biagioli (1993) has studied *Amor pudico* (1614) in his work about the astronomer as a courtier[4]; and Paolo Fabbri (2003) has suggested a connection between Endymion and Galileo in *Diana schernita* (1629)[5]. Other scholars studied the relationship between opera and science in an indirect way: some investigated Galileo's musical environments, his generic references to music or the relationship with his father, the lutenist, composer and music theorist Vincenzo Galilei[6]. Fred Kersten (1997) has compared the

[3]. Hammond 1992.

[4]. Biagioli 1993.

[5]. Fabbri 2003.

[6]. Coelho 1992, Fabbri 2008, and Cypess 2016.

birth of opera with Galileo's new science from a phenomenological perspective, but he did not refer to any of these operas[7]. The subject does not seem to have interested art historians either, possibly due to the lack of multidisciplinarity in their approaches[8]. Given this situation, it may be useful to offer an overview of the Galilean operas first. I propose that these can be organized into six groups.

Table 6.2

Overview of the Galilean and anti-Galilean operas discussed in this chapter. In the column 'Authors', names in roman font are librettists, names in italic font composers. In the column 'Libretto', place-plus-year markings refer to printed librettos; the figures between paratheses, to Sartori 1990-1992. In the column 'Music', place-plus-year markings refer to printed editions. In the columns 'Librettos' and 'Music', RISM sigla refer to manuscript librettos and scores.

Year(s)	Place	Title	Authors	Libretto	Music
Group (1)					
1613	Florence	Barriera	Jacomo Cicognini	Florence, 1613	
1614	Rome	*Amor pudico*	Jacomo Cicognini *Cesare Marotta* *Pellegrino Mutij* *Hipp. Macchiavelli*	Viterbo, 1614 (S 1455, 1456)	
Group (2)					
1622	Rome	*Apotheosis sive consecratio SS. Ignatii et Francisci Xaverii*	Orazio Grassi *Gio. Gir. Kapsberger*		A-Wn, MS Mus. 16013; D-Mbs, MS Coll. mus. Max. 98
Group (3)					
1623	Florence	*Il Medoro*	Andrea Salvadori *Jacopo Peri*	Florence, 1623 (S 15377)	
1624	Florence	*La regina Sant'Orsola*	Andrea Salvadori *Marco da Gagliano*	Florence, 1624 (S 19704, 19705)	
1625	Florence	*La liberazione di Ruggiero*	Ferd. Saracinelli *Francesca Caccini*	Florence, 1625 (S 14223)	
1628	Florence	*La Flora*	Andrea Salvadori *Marco da Gagliano*	Florence, 1628 (S 10734, 10735)	
Group (4)					
1629	Rome	*Diana schernita*	Francesco Parisani *Giacinto Cornacchioli*	None	Rome, 1629

7. Kersten 1997.

8. See, for example, Reeves 1996.

Group (5)					
1632-1634	Rome	*Sant'Alessio*	Giulio Rospigliosi *Stefano Landi*	I-Rvat, various	Rome, 1634
1633-1637	Rome	*Erminia sul Giordano*	Giulio Rospigliosi *Michelangelo Rossi*	I-Rvat, various	Rome, 1637
1637	Florence	*Le nozze degli dei*	Carlo Coppola *five composers*	Florence, 1637 (S 16712)	
Group (6)					
1638-1639	Rome	*San Bonifatio*	Giulio Rospigliosi *Virgilio Mazzocchi*	I-Rvat, Lat. 10192	I-Rvat, Ottob. 3394

(1) The earliest staged work is a Florentine *barriera* (a danced combat) of 1613 designed by Jacopo Cicognini. The relation to Galileo is clear since the moons of Jupiter, whose discovery he reported in 1610, appear on the stage as companions of Jupiter. Furthermore, the libretto refers to him explicitly by name. A similar element is found in Cicognini's opera *Amor pudico* (1614), performed in Rome in 1614. Both pieces fell in the period between the publication of Galileo's *Sidereus nuncius* (Venice, 1610) and the condemnation of Copernicanism in 1616.

(2) The creation of a Roman melodrama coincided with Galileo's dispute with the Jesuit Orazio Grassi over the nature of comets from 1618 to 1623. Grassi wrote a libretto to demonstrate the power of the Jesuit order in the *theatrum mundi* entitled *Apotheosis sive consecratio SS. Ignatii et Francisci Xaverii*, on the occasion of the canonization of Saint Ignatius of Loyola and Saint Francis Xavier in 1622. It was set to music by Giovanni Girolamo Kapsberger.

(3) In 1623, when Maffeo Barberini was elected as Pope Urban VIII and Galileo's *Saggiatore* was published, Galileo strengthened his relations with Rome. From 1623, the Florentine bookseller and printer Pietro Cecconcelli published several opera librettos with a printer's mark on the title page showing Jupiter and its four moons. As a deliberate reference to Galileo, Cecconcelli called his shop «alle stelle Medicee» (at the Medicean Stars) in the imprint of his editions. The first libretto to appear under Cecconcelli's imprint was Andrea Salvadori's *Medoro* (1623). Librettos of three more operas have the printer's mark with the Medicean stars: *La regina Sant'Orsola* by Andrea Salvadori (text) and Marco da Gagliano (music; 1624); *La liberazione di Ruggiero dall'isola d'Alcina* by Ferdinando Saracinelli (text) and Francesca Caccini (music; 1625); and *La Flora* by Andrea Salvadori (text) and Marco da Gagliano (music; 1628).

(4) In 1629 *Diana schernita* by Giacomo Francesco Parisani (text) and Giacinto Cornacchioli (music) was performed in Rome. The mythical love between Endymion and the

Moon, who gave her lover a telescope so that he could observe her every night, refers to Galileo and his observations of the Moon.

(5) The condemnation of Galileo in 1633 was a direct consequence of the publication of his *Dialogo sopra i due massimi sistemi del mondo, Tolemaico e Copernicano* the year before. According to Frederick Hammond (1992), the condemnation left its traces in two operas: *Sant'Alessio* (1634 version) by Cardinal Giulio Rospigliosi (future Pope Clement IX; text) and Stefano Landi (music), and *Erminia sul Giordano* (1637 version) by Rospigliosi (text) and Michelangelo Rossi (music)[9]. Here I offer further support for Hammond's thesis.

(6) The last 'anti-Galilean opera', *San Bonifatio*, by Rospigliosi (text) and Virgilio Mazzocchi (music), was premiered in 1638, the year when Galileo published his *Discorsi e dimostrazioni matematiche intorno a due nuove scienze*. The libretto contains a new critique of the curiosity of scientists predicated on the theological view of *curiositas* as a vice. Significantly, the opera does not seem to condemn Galileo as a heretic, but rather to rebuke him for forcing the Church to condemn him at a difficult moment in its existence.

It is useful to recall that opera in this period was used as a tool of international propaganda, both *in situ* and *ex situ*. The performance itself was a political event: courts staged operas as part of religious festivals and diplomatic celebrations such as dynastic weddings or visits of princes and ambassadors. The audience of opera was an international one. Virtually all libretti, scores, engravings, and descriptions were conceived in geopolitical terms, and were destined for the entertainment or honouring of specific religious and political personalities; therefore, opera was a meditated genre where all kinds of political, theological and academic messages were implicitly or explicitly included in various subtle ways. So far as the Galileo case was concerned, it is interesting to note that the vast majority of operas from the first half of the seventeenth century were staged in Catholic countries, often produced by popes or cardinals. Galileo's 'appearance' in these operas cannot therefore be separated from contemporary geopolitical concerns of Catholicism.

The Starry Messenger Takes the Stage

The *Sidereus nuncius* (The Starry Messenger; Venice, 1610; see Chapter 10, Ill. 10.2a) earned Galileo a position at the court of his pupil Grand Duke Cosimo II de' Medici. The first edition had an obvious propagandistic character, whereby art and science were deployed for political goals. Its magnificent drawings were designed to astonish the reader. In the *Siderius*

[9]. Hammond 1992.

nuncius, Galileo described the telescope, the mountains of the Moon (some of them, as rightly conjectured, higher than the terrestrial ones; see Chapter 10, Ill. 10.2b), and new stars. In addition, he provided an explanation of the nebulae and the Milky Way. In all, Galileo drew a more expansive universe than hitherto understood. The Grand Duke disseminated the *Sidereus nuncius* with the intention that all of Europe could learn about the achievements of his new protégé.

The discovery of the moons of Jupiter was followed by many other discoveries in the successive months. In the summer of 1610, Galileo informed Belisario Vinta and Giuliano de' Medici of an anomaly he had observed for Saturn, which he could not explain[10]. (Some decades later Christiaan Huygens discovered that the anomaly was caused by rings around the planet.) Galileo's discovery at the end of 1610 of the phases of Venus was even more important. The fact that the phases could not be observed with the naked eye was commonly used as an *a priori* argument against Copernicanism[11]. Galileo had reported his discovery to Giuliano de' Medici using an anagram, although it was necessary to explain the anagram's solution in a subsequent letter to Giuliano de' Medici (and Johannes Kepler — they both resided in Prague) dated 1 January 1611[12]. Galileo not only maintained that the planets had, like the Earth, no light of their own, but also that Venus necessarily moved around the Sun, as did Mercury and other planets. Soon afterwards, Galileo also published the results of his research on sunspots in his *Istoria e dimostrazione intorno alle macchie solari* (History and Demonstration of the Sunspots; Rome, 1613), which features engraved illustrations of his observations (Ill. 6.3a). In a few years, Galileo had contributed many new and convincing arguments in favour of Copernicus's cosmology.

In the same period, Galileo began to defend himself against religious attacks, as is evident in the famous letter to Benedetto Castelli of 21 December 1613[13]. On 20 December 1614, the Dominican Tommaso Caccini spoke against Galileo in a sermon in Santa Maria Novella in Florence. His attack was in keeping with a long tradition, dating back at least to the criticisms against Copernicus by the Dominican Giovanni Maria Tolosani and by Martin Luther in the previous century. In 1612 and 1613, Galileo had already been severely criticized by Martin Horky, Ludovico delle Colombe and his brother the Dominican Raffaello delle Colombe, Francesco Sizzi, Giulio Cesare Lagalla, another Dominican, Niccolò Lorini, and Ulisse Albergotti. There were also Christina of Lorraine's inquiries of 1613 that led to Galileo's letters to Castelli, Piero Dini and Christina herself. The last of these letters followed Bellarmino's own critical letter to Paolo Antonio Foscarini[14], regarding the publication of the latter's *Lettera sopra l'opinione de'*

---

[10]. Galileo to Belisario Vinta, 30 July 1610 (Galilei 1900, pp. 409-410); Galileo to Giuliano de' Medici, 13 November 1610 (*ibidem*, p. 474).

[11]. Galileo to Giuliano de' Medici, 11 December 1610 (*ibidem*, p. 483).

[12]. Galileo to Giuliano de' Medici, 1 January 1611 (Galilei 1901, pp. 11-12).

[13]. Galileo to Benedetto Castelli, 21 December 1613 (Galilei 1895, pp. 281-288; Galilei 1901, p. 610).

[14]. Galileo to Piero Dini, 3 March 1615 (Galilei 1902, pp. 183-185); Roberto Bellarmino to Antonio Paolo Foscarino, 12 April 1615 (*ibidem*, pp. 170-171); Galileo to Christina of Lorraine, 1615 (Galilei 1895, pp. 309-

Pittagorici, e del Copernico della mobilità della terra e stabilità del sole (Letter about the Opinion of the Pythagorians and Copernicus about the Mobility of the Earth and the Stability of the Sun; Naples, 1615), a defence of Galileo's theories[15]. All argued that heliocentrism contradicted the Bible, forcing Galileo to reinterpret some biblical passages to defend the Copernican theory. One of them was Joshua 10:12-13, where it was said that God stopped the moving of the Sun to lengthen the day. It was obvious to everyone that God stopped the revolution of the Sun, but Galileo argued that precisely because the Sun was in the center of the universe, stopping its rotation (not its revolution) was the better way to stop the entire universe without changing all the natural laws[16]. Another argument against Galileo, put forward by Francesco Ingoli in 1616, was that the Hell, the farthest place from Heaven, was in the center of the Earth.

Table 6.3: Partial Family Tree of the House of the Medici

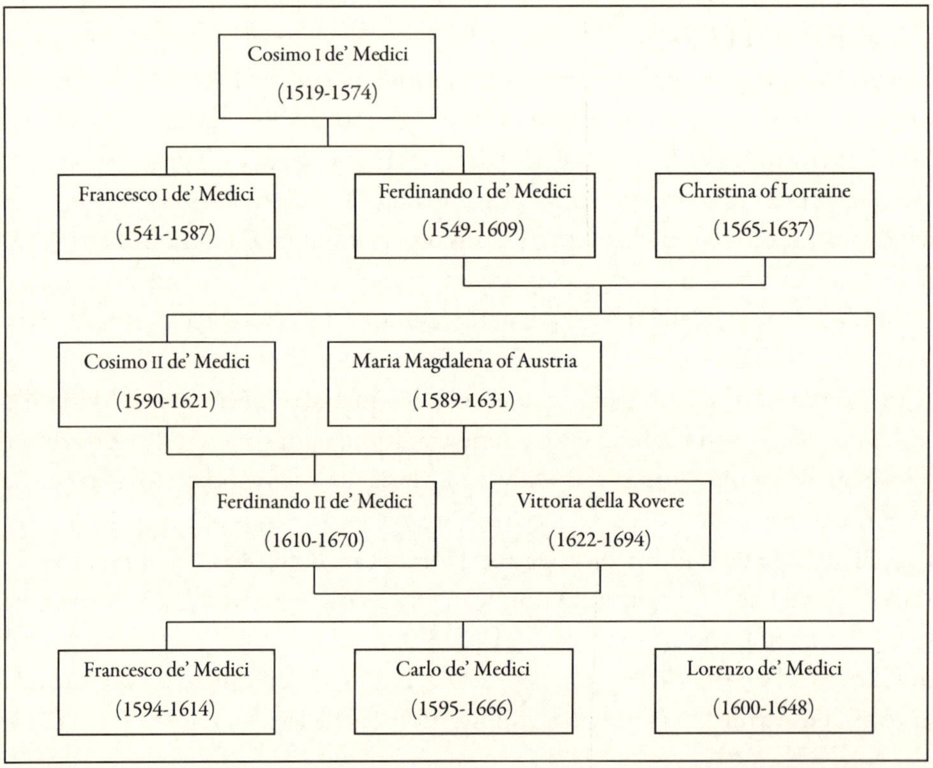

348; modern edition Galilei 2000). See Finocchiaro 1989, Finocchiaro 2010, and Finocchiaro 2019, Feldhay 1995, Guerrini 2009 and Guerrini 2010, Speller 2008, and Donati 2014.

15. In the form of a letter to Sebastiano Fantone, General of the Carmelite Order, dated 6 January 1615.

16. Galilei 1895, p. 286.

On 17 February 1613, occasioned by the visit of Federico Ubaldo della Rovere, Prince of Urbino, to the Medicean court in Florence, a *barriera* (a theatrical entertainment derived from what originally was a danced combat) was created, conceived and set up by Giovanni Villifranchi (1570-1614), in the theatre of the Sala degli Uffizi[17]. Jacopo Cicognini, Alessandro Adimari, Ottavio Rinuccini and Andrea Salvadori collaborated on it; each composed a stage action in addition to the five already written by Villifranchi. Later in 1613 a description of the event by Villifranchi was published in Florence. Political messages are evident from the very start. The stage action contributed by Cicognini tells the story of Cupid, who, tired of the Olympic gods, chose to make his empire in Tuscany. This caused a war between Cupid and other mythological characters, until Jupiter descended from heaven with four companions to bring peace. These four companions were the four moons discovered by Galileo; they represented the power of the Medici. In the *Descrizzione della barriera*, Jupiter's descent is described as follows:

> Jupiter appeared sitting on a very high cloud, all luminous, and Deceptive Love sat next to him. And further down in the clouds appeared the four wandering stars around Jupiter, discovered by Mr. Galileo Galilei od Florence, Mathematician to His Highness, a very rare genius, and unique in our times by the work of the marvelous Telescope. And just as the ancients translated the deserving Heroes to Heaven for their actions, and assigned a Star to them, so, having discovered these Stars, he called them Medicean, assigning the first to His Highness, the second to Lord Prince Don Francesco, the third to Lord Prince Don Carlo, and the fourth to Lord Prince Don Lorenzo[18].

Cicognini transformed his stage action in the *barriera* into a *festino*, that is an opera entitled *Amor pudico* (Chaste Love) and performed in Rome during Carnival of 1614 in the Palazzo della Cancelleria on the occasion of the wedding of Prince Michele Peretti, brother of Cardinal Alessandro Montalto, and Anna Maria Cesi[19]. Music was provided by Cesare Marotta

---

17. See Giovanni Villifranchi, *Descrizzione della barriera, e della mascherata, fatte in Firenze a' XVII & XIX di Febbraio MDCXII* (Florence, 1613). The year MDCXII (1612) in the title of the publication should be read as Old Style, which is, for February, New Style 1613.

18. *Ibidem*, pp. 32-33: «Comparse Gioue sopra una altissima Nube, tutta luminosa, & appresso de lui sedeua l'Inganno amoroso. E più à basso tra le nuuole appariuano le quattro Stelle erranti intorno à Giove, ritrouate dal Signor Galileo Galilei Fiorentino, Mathematico di Sua Altezza, ingegno rarissimo, e singolare a' tempi nostri per opera del maraviglioso Occhiale; E si come gl'antichi traslatauano in Cielo gl'Eroi meritevoli per l'azioni loro, & à quelli assegnauano una Stella, cosi egli havendo ritrouato queste Stelle, l'ha nominate Medicee, assegnando la prima à S. A., la seconda al Signor Principe Don Francesco, la terza al Signor Principe Don Carlo, e la quarta al Signor Principe Don Lorenzo».

19. Giacomo Cicognini, *Amor pudico: Festino e balli danzati in Roma nelle nozze* [...] *Michele Peretti* [...] *e Anna Maria Cesis* (Viterbo, 1614; two editions: Sartori 1990a, nos. 1455-1456, p. 152). See also Franchi 1994, p. 218, and Hill 1997, pp. 279-297.

(*Prima-Seconda hora* [First-Second Hour]), Pellegrino Mutij (*Terza hora* [Third Hour]), Hippolito Macchiavelli (*Quarta-Quinta hora* [Fourth-Fifth Hour]) and Bernardo Nanino (*Cori* [Choruses])[20]. From the libretto (Ill. 6.2a) we know that the new circumstances required some changes in the text. For example, the city where Cupid decided to make his empire was now Rome, instead of Florence, and instead of finishing the action with the appearance of Jupiter and the Medicean Stars, it finished later by an extension of the plot. The scenes of the opera include virtually the entire universe, from Heaven to Hell, from the Elysian Fields to the Sea, and from ancient to new Rome (a reference to the Rome of Sixtus v, from the Peretti family: Felice Peretti da Montalto, pope from 1585 to 1590).

Yet, all this did not seem to pose a problem for the Church. The libretto was dedicated to Cardinal Scipione Borghese, cousin of Pope Paul v (Camillo Borghese). In a long letter by Romolo Paradiso to Giovanni Battista Strozzi, which was published as an appendix to the libretto of *Amor pudico* (Ill. 6.2B), Paradiso underlined that the ceremony was attended by «all the nobles of Rome» (*tutta la nobiltà Romana*), and that they did not have any qualms about praising Galileo. This is how the appearance of the Medicean stars was described:

> Amid so much light sat Jupiter, enthroned well corresponding to his greatness [...]. Around him were four young boys, like custodians, with silver armor, & gilded helmets; from which among many blue plumes a star rose in each one as a crest. I was told, that these four represented the stars that have the name of the Serene House of these Highnesses. It seems to me that in this Mr. Cicognini has not only shown his devotion to his Prince; but also the affection that he has towards Mr. Galileo, who was the first observer of the said stars; and when it was recognized what they were, there were honourous discussions around his person among many scholars[21].

This did not mean that the organizers were unaware of the precariousness of the situation. Federico Cesi, cousin of the bride, was the founder of the Accademia dei Lincei in Rome, to which Galileo belonged since 1611. Cesi was very attentive to the audience's reaction to the Galilean scene. After the celebration, he wrote to Galileo on 1 March 1614 that a small

[20]. See the letter of Romolo Paradiso to Giovanni Battista Strozzi, 15 February 1614, published as *Copia d'una lettera del Sig. Romolo Paradiso, Con la quale dà avviso dell'apparato e grandezza, con che si è rappresentato il Festino dell'Eccellentiss. Sig. Principe Peretti* (Rome, 1614; Sartori 1990b, no. 6642, pp. 229-230), pp. 64-65.

[21]. *Copia d'una lettera del Sig. Romolo Paradiso* (1614), pp. 27-28: «Nel mezzo à tanta luce sedeva GIOVE, in trono ben corrispondente alla sua grandezza [...]. Attorno à lui erano quattro Giovanetti, à guisa di custodi, con armature d'argento, & elmetti dorati; da quali tra molti pennacchi di color turchino sorgeva in ciascuno per cimiere una stella. Fummi detto, che questi quattro, quelle stelle rappresentavano, le quali hanno il nome della Serenissima Casa di coteste Altezze. Parmi, che in ciò il Sig. Cicognino habbia non solo mostrato devotione verso il suo Principe; mà anco l'affetion, che porta al Sig. Galileo, che di dette Stelle è stato il primo osservatore: e riconosciute che furono, tra molti eruditi si tenne ragionamento honorato intorno alla sua persona».

AMOR PVDICO
FESTINO, E BALLI
DANZATI IN ROMA
NELLE NOZZE
De gl'Illuſ.& Ecc.SS.D.Michele Peretti
Principe di Venafro, e Sig.Princi-
peſſa D.Anna Maria Ceſis
Nel Palazzo della Cancellaria l'Anno 1614.
Del Sig. Iacomo Cicognini ne l'Accademia de-
gli Humoriſti di Roma detto il Confidente.
IN VITERBO,
Preſſo Girolamo Diſcepolo. 1614.
Con licenza de' Superiori.

COPIA
D'VNA LETTERA
DEL SIG.
ROMOLO PARADISO.
Con la quale dà auuiſo dell'Appa-
rato, e grandezza, con che ſi è
rappreſentato il Feſtino
Dell'Eccellentiſs. Sig.
PRINCIPE PERETTI.
IN ROMA,
Appreſſo Girolamo Diſcepolo. 1614.
Con licenza de' Superiori.

Ill. 6.2ab: Jacopo Cicognini, *Amor pudico* (Viterbo, ^{2}1614), title pages, and *Copia d'una lettera del Sig. Romolo Paradiso* (Rome, 1614), describing the performance of *Amor pudico*. Copies I-Rn.

part of the Roman audience was upset by the Medicean Stars. Cesi called them «Peripatetic [=Aristotelian] primates»:

> Cicognini certainly satisfied me, since, finding myself at the evening scenic feast in the wedding of Princess Peretti, my cousin, I saw that among the other planets he had, with great grace, placed the Medicean stars as a choir around Jupiter. Everyone liked the show, and the novelty inserted in its place. It is true that I had to speak with some Peripatetic primates, who could not contain themselves from growling, like conservatives and enemies of everything new[22].

In this sense, *Amor pudico* can be connected to other instances of favourable attitudes to Galileo's work, like the reception of his works in 1611 by the Jesuits, the widespread use

[22]. Federico Cesi to Galileo, 1 March 1614 (Galilei 1902, p. 29): «Mi sodisfece certo il Cicognini, poiché, trovandomi alla veglia o festino scenico nelle nozze della Principessa Peretti, mia cugina, vidi che fra l'altri pianeti haveva, con molto garbo, posti i Medicei in choro intorno Giove. Piacque lo spettacolo a tutti, e la novità inserta al suo luogo. Ben è vero ch'io mi feci sentire ad alcuni primati Peripatetici, che non potevano contenersi di ringhiare, come veternosi e nimici d'ogni cosa nuova».

of telescopes by the Society of Jesus, and the fact that Rome allowed the depiction of the mountains of the Moon in Ludovico Cardi detto il Cigoli's fresco *The Virgin of the Immaculate Conception* in the ceiling of the Cappella Paolina in the Basilica di Santa Maria Maggiore in Rome[23]. Cardinal Scipione Borghese was one of the cardinals who received a telescope from Galileo in 1610, along with Cardinals Francesco Maria del Monte and Alessandro Montalto, both of whom were involved from the beginning in the history of Florentine opera[24]. In addition, Galileo gained the respect from religious figures such as Paolo Antonio Foscarini, Tommaso Campanella and Giovanni Ciampoli, along with renowned scientists such as Johannes Kepler. It is not surprising that Galileo, in a letter to Giovanni Battista Baliani of 12 March 1614, rejected some opinions of Tycho Brahe (which seemed derived from Ptolemy) and openly accepted heliocentrism:

> As for Copernicus's theory, I really consider it true, and not just because of the observations of Venus, the sunspots and the Medicean stars, but for his other reasons, and for many other details of mine that seem conclusive to me[25].

But things were about to change. In 1615, the Dominicans Niccolò Lorini and Cosimo (Tommaso) Caccini accused Galileo of heretical views before the Inquisition. On 12 April 1615, Cardinal Roberto Bellarmino wrote a letter to Antonio Paolo Foscarini, in which he stated that Galileo should act with prudence and present his ideas «as hypothesis and not in absolute terms, as I have always believed Copernicus worded the matter», because the opposite could «harm Holy Faith by rendering untrue Holy Scripture»[26]. But the Bible was not the only authority the Church sought to protect. Bellarmino also insisted on the importance of not contradicting the Holy Fathers of the Church. There was a lot at stake at the time of the Thirty Years War and the war against Protestantism.

The Congregatio pro Indice librorum prohibitorum (Congregation of the Index of Forbidden Books) condemned Copernicanism on 26 February 1616 and issued the corresponding decree on 5 March. The decree banned Copernicus's *De revolutionibus orbium coelestium* (On the Revolutions of Heavenly Bodies; Basle, 1543) and other Copernican works,

---

23. Ostrow 1996. See also Chapter 10 of the present volume.

24. Galluzzi 2017, pp. 20-22, and Hill 1997.

25. Galileo to Giovanni Battista Baliani, 12 March 1614 (Galilei 1902, pp. 34-35): «Quanto all'opinione del Copernico, io veramente la tengo sicura, e non per le sole osservazioni di Venere, delle macchie solari e delle Medicee, ma per l'altre sue ragioni, e per molt'altre mie particolari che mi paiono concludenti». See Finocchiaro 2010, p. 57.

26. Roberto Bellarmino to Antonio Paolo Foscarino, 12 April 1615 (Galilei 1902, p. 171): «*ex suppositione* e non assolutamente, come io ho sempre creduto che habbia parlato il Copernico [...] nuocere alla Santa Fede con rendere false le Scritture Sante».

such as *In Job commentaria* (Commentaries on Job; Toledo, 1584; Rome, 1591) by the Spanish scholastic philosopher Diego de Zúñiga and Foscarini's *Lettera sopra l'opinione dei Pitagorici e del Copernico* (1615); it did not explicitly mention Galileo. The latter's social position seems to have prevented him being condemned at the same time. On 26 February 1616, Bellarmino met with Galileo and asked him not to defend Copernicanism. In fact, many believed that Galileo had recanted to avoid condemnation. For this reason, he asked Bellarmino for an official statement denying that he had recanted his views; the cardinal granted this request on 26 May 1616[27].

In a letter of 12 March 1616 to Curzio Picchena, Galileo wrote that he had met with the pope for forty-five minutes. They talked about «the malignity of my persecutors and some of their false slanderous accusations». The pope acknowledged «his integrity and sincerity», safeguarding him from future persecutions[28].

The condemnation of Copernicanism affected the artistic reception of Galileo's work. In the words of Mario Biagioli (1993), «Galileo's discoveries did not continue their career in the Medici mythology they had begun so brilliantly, and they do not seem to have reappeared in Rome either. Their visibility declined even further after 1621 when, following Cosimo II's death, Grand Duchess Christina of Lorraine and her counsellors took over the government of Tuscany and the management of court culture»[29]. As far as opera was concerned, however, the conclusion of Biagioli must be nuanced. During the Duchess's regency, several opera librettos were published in Florence that featured the emblem of the Medicean Stars. In Rome Galileo's discoveries were also acknowledged.

At this point, however, we need to consider Galileo's conflict with the Jesuits.

THE DISPUTE WITH THE JESUITS ABOUT THE COMETS

The pontificate of Paul V, during which heliocentrism was absent from public debate, lasted until 1621. Yet, unforeseen astronomical events had led to new disputes in 1618. Between August and December, three comets were observed in several parts of Europe. The Jesuit Orazio Grassi, in a lecture at the Collegio Romano argued that the comets were, as the absence of parallax demonstrated, removed farther from the Earth than the Moon, whereas Galileo had located them nearer to the Earth. Grassi published his theory in *De tribus cometis anni MDCXVIII* (About the Three Comets of the Year 1618; Rome, 1619), arousing Galileo's criticism. Galileo published his response, titled *Discorso delle comete* (Discourse on the Comets;

---

27. GALILEI 1907, p. 342.

28. Galileo to Curzio Picchena, 12 March 1616 (GALILEI 1902, p. 248): «la malignità de' miei persecutori et alcune delle loro false calunnie [...] l'integrità mia e la sincerità di mente».

29. BIAGIOLI 1993, p. 142.

Florence, 1619), under the name of Mario Guiducci (Ill. 6.3b), although he had not been able to observe the comets himself (due to illness). This marked the start of a long dispute between the Academia dei Lincei, to which Galileo belonged, and the Jesuits of the Collegio Romano, a dispute in which the international prestige of the Collegio was at stake[30].

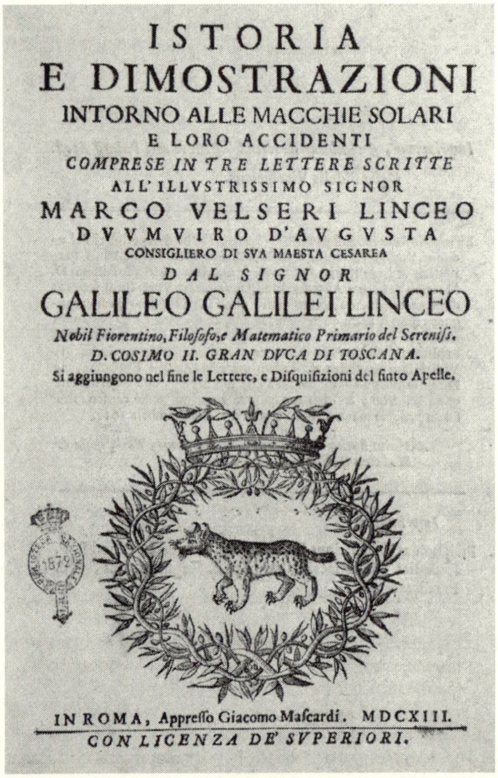

ISTORIA
E DIMOSTRAZIONI
INTORNO ALLE MACCHIE SOLARI
E LORO ACCIDENTI
COMPRESE IN TRE LETTERE SCRITTE
ALL' ILLVSTRISSIMO SIGNOR
MARCO VELSERI LINCEO
DVVMVIRO D'AVGVSTA
CONSIGLIERO DI SVA MAESTA CESAREA
DAL SIGNOR
GALILEO GALILEI LINCEO
Nobil Fiorentino, Filosofo, e Matematico Primario del Serenis.
D. COSIMO II. GRAN DVCA DI TOSCANA.
Si aggiungono nel fine le Lettere, e Disquisizioni del finto Apelle.

IN ROMA, Appresso Giacomo Mascardi. MDCXIII.
CON LICENZA DE' SVPERIORI.

DISCORSO
DELLE COMETE
DI MARIO GVIDVCCI
FATTO DA LVI
NELL'ACCADEMIA FIORENTINA
NEL SVO MEDESIMO CONSOLATO.

IN FIRENZE.
Nella Stamperia di Pietro Cecconcelli, Alle Stelle Medicee. 1619.
CON LICENZIA DE' SVPERIORI.

Ill. 6.3ab: Galileo Galilei, *Istoria e dimostrazioni intorno alle macchie solari* (Rome, 1613), title page, and 'Mario Guiducci' (= Galileo Galilei), *Discorso delle comete* (Florence, 1619), title page, Copies I-Fn and A-Wn, respectively.

In this conflict, Galileo received support from ecclesiastics, the most notable being Cardinal Maffeo Barberini. Barberini included in his *Poemata* (Poems) published in 1620 a poem entitled 'Adulatio perniciosa' (Dangerous Flattery), which was addressed to Galileo and referred to two of his discoveries[31]:

30. Galilei – Grassi 2017.

31. Maffeo Barberini, *Poemata* (Paris, 1620), pp. 46-52. Barberini had sent an autograph copy of the poem to Galileo with his letter to him of 20 August 1620 (Galilei 1903, pp. 48-49; Gattei 2018, p. 281). About the poem, see also Gattei 2018, pp. 281-284 and 304-307.

[...]
Seu Scorpij cor, siue Canis facem,
Miratur alter, vel Iouis asseclas,
Patrisue Saturni, repertos
Docte tuo, Galilaee, vitro.
[...]
Non semper extra quod radiat iubar
Splendescit intra: respicimus nigras
In sole (quis credat?) retectas
Arte tua, Galilaee, labes.
[...]

[...]
Another admires to heart of the Scorpion, or the face of the Dog, or the companions of Jupiter or his father Saturn, discovered, Galileo, by your clever glass.

[...]
Not always light that shines from the outside also glows inside: we see the black spots in the sun (who would believe that?) discovered by your art, Galileo.

[...]

Table 6.4: Partial Family Tree of the House of Barberini

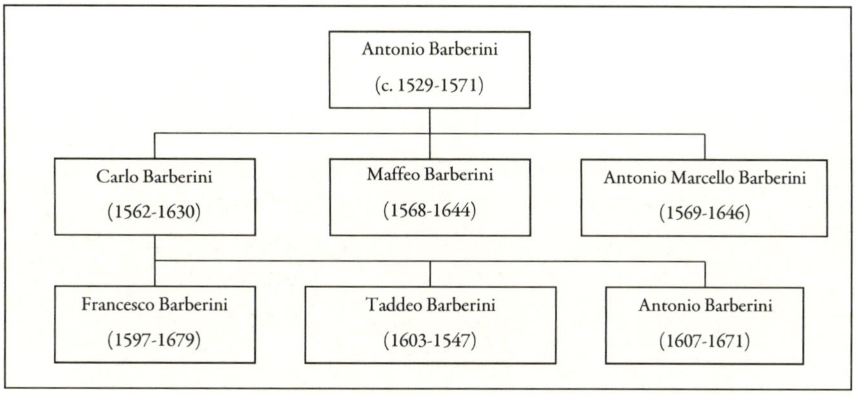

New developments in the Church also seemed to favour Galileo: Paul v and Roberto Bellarmino died in 1621 and the new Pope Gregory xv maintained a neutral position in the conflict. But the situation was far from being unambiguous. It was difficult to separate the controversy on the comets from the condemnation of Copernicanism of 1616. Grassi supported the correction of Copernicus's work that was requested in the decree and that was undertaken by the Cardinal Bonifacio Caetani and later by the priest Francesco Ingoli.

It was in this context that the *Apotheosis sive consecratio SS. Ignatii et Francisci Xaverii*, a homage to Ignatius of Loyola and Francis Xavier, with text by Grassi and music by Giovanni Girolamo Kapsberger, was performed during the celebration of the Jesuit fathers' canonization in 1622 (Ill. 6.4)[32]. Apparently, the Jesuits considered it an appropriate year to honour their

[32]. Two early-nineteenth-century manuscript copies of the libretto and the score are preserved: D-Mbs, MS Coll. Mus. Max. 98, and A-Wn, MS Mus. 16013. An eight-page summary was printed as *Argomento dell'Apotesi o*

founder. Not only was the Thirty Years' War raging in the German-speaking countries, but, in addition, Galilean supporters were appointed on prominent positions during the pontificate of Gregory XV and others dared to defend Galileo. The Dominican monk Tommaso Campanella published his famous *Apologia pro Galileo* in the same year, in Frankfurt. For the Jesuits, it was a critical moment, since Europe's most renowned scientist opposed their opinions, especially those of Grassi, the librettist of the melodrama and an important figure in the creation of the Chiesa di Sant'Ignazio di Loyola in Rome.

Ill. 6.4: Orazio Grassi, *Apotheosis sive consecratio SS. Ignatii et Francisci Xaverii* (1622), on text of Orazio Grassi, set to music by Giovanni Girolamo Kapsberger. Copy D-Mbs, MS Coll. Mus. Max. 98 (early-nineteenth-century copy), title page.

The libretto of the *Apotheosis* has at the same time a clearly mythical and geopolitical character. The globe appears in one scene, both hemispheres governed by the Catholic Monarchy of Spain. On the globe, countries on all continents revolve around the Sun represented by Phoebus. This Sun god is the same as the one in *Amor pudico*, «driving his chariot pulled by

consagratione de' Santi Ignatio Loiola e Francesco Saverio (Rome, 1622; Sartori 1990a, no. 2486b, p. 260); see also Franchi 1994, p. 804.

horses around the world» (*ubi sol rapidum permensus iter | solvit madidos defessus equos*). Several countries are represented by allegorical roles. 'Hispania' represents Spain, the country «where the sun never sets» (*numquam occiduum visura diem*). Hispania is followed by Lusitania (Portugal), who begins an exchange of eulogies with Hispania; together they praise the work of the new saints. Then follow a 'Ballo' and, after another exchange of eulogies, two 'Cori'. The other acts are similarly structured. In Act II there is a dialogue between India and Palestina; in Act III, between Gallia and Japonia; in Act IV, between Italia and Sina (China). This was the worldview that Galileo was confronting, and obviously he was not invited to this 'global theatre'[33]. Only the death of Pope Gregory X in 1623 allowed him to enter the scene again.

FLORENCE RECOVERS JUPITER'S MOONS

The relation between printing and politics is visible in the opera librettos to be discussed next. In 1619, *Il Medoro* by Andrea Salvadori (text) and Jacopo Peri (music) was performed in the Palazzo Pitti, the palace of the Grand Duke of Tuscany, to celebrate the election of Ferdinand II as Emperor of the Holy Roman Empire. That the libretto was not printed until 1623 can be explained by considering the political situation in Rome.

Gregory XV died in 1623. His successor was Maffeo Barberini, the cardinal who had addressed a poem to Galileo three years earlier. Upon his election to the papacy, Barberini took the name of Urban VIII. The Accademia dei Lincei received the news with great enthusiasm. On 12 August 1623, Francesco Stelluti commented to Galileo that «then the creation of the new pope cheered us all up, being of that value and goodness that you know very well, and a particular advocate of the literate, so that we are to have a supreme patron». He also remarked that the new pope had appointed several academicians on key positions, while his nephew Francesco Barberini had become a member of the Accademia[34].

In 1623, Galileo published his *Il saggiatore* (Rome; ILL. 6.5) as a reply to Orazio Grassi's *Libra astronomica ac philosophica* (The Astronomical and Philosophical Scale; Perugia, 1619). Galileo had succeeded in dedicating the work to Urban VIII, just as Copernicus had dedicated *De revolutionibus orbium caelestium* in 1543 to Pope Paul III. The latter sought to confirm his power and influence in this way. The emblems of the Barberini and the Accademia dei Lincei both appeared on the title page of *Il saggiatore*; two telescopes are visible below a figure representing Mathematics, on the opposite side of which appears Natural Philosophy.

---

[33]. COELHO 1997 connect the *Apotheosis* with the colonial politics of the Jesuits rather than cosmological issues.

[34]. Francesco Stelluti to Galileo, 12 August 1623 (GALILEI 1903, p. 121): «la creazione poi del nuovo Pontefice ci ha tutti rallegrati, essendo di quel valore e bontà che V. S. sa benissimo, et fautore particolare de' letterari, onde siamo per havere un mecenate supremo».

Ill. 6.5: Galileo Galilei, *Il saggiatore* (Rome, 1623), title page. Copy D-Mdm.

According to a letter by one the founders of the Accademie dei Lincei, Virginio Cesarini, to Galileo on 28 October 1623, the pope was so content with *Il saggiatore* that «he had it read for him at the table». Cesarini hoped to see Galileo's name «in the possession of immortality, and to raise our age, due to your pen, to such a sign of glory not known to us before, nor to be emulated later»[35]. In the same year, Tommaso Campanella published his *Civitas solis* (City

[35]. Virginio Cesarini to Galileo, 28 October 1623 (*ibidem*, pp. 141-142): «se 'l fa legger a mensa [...] in possesso dell'immortalità, e l'età nostra, mercé la sua penna, alzarsi a tal segno di gloria, che non fu da i primi nostri conosciuta, né sarà da i posteri pareggiata».

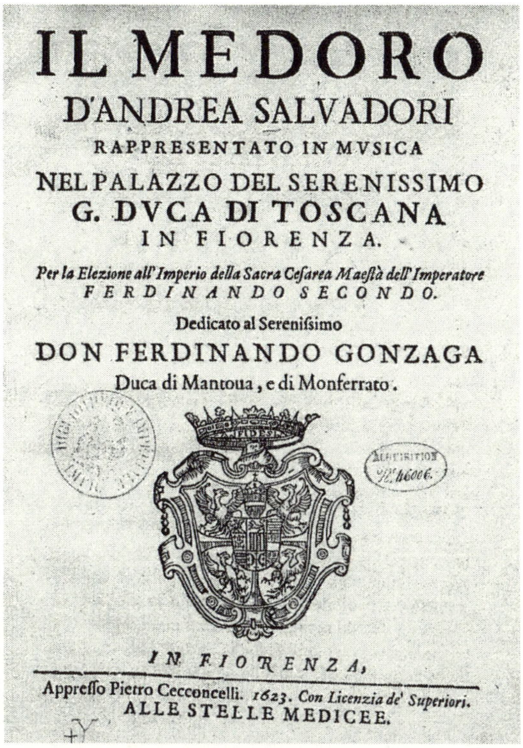

IL MEDORO
D'ANDREA SALVADORI
RAPPRESENTATO IN MVSICA
NEL PALAZZO DEL SERENISSIMO
G. DVCA DI TOSCANA
IN FIORENZA.

Per la Elezione all'Imperio della Sacra Cesarea Maeſtà dell'Imperatore
FERDINANDO SECONDO.

Dedicato al Sereniſſimo
DON FERDINANDO GONZAGA
Duca di Mantoua, e di Monferrato.

IN FIORENZA,
Appreſſo Pietro Cecconcelli. *1623. Con Licenzia de' Superiori.*
ALLE STELLE MEDICEE.

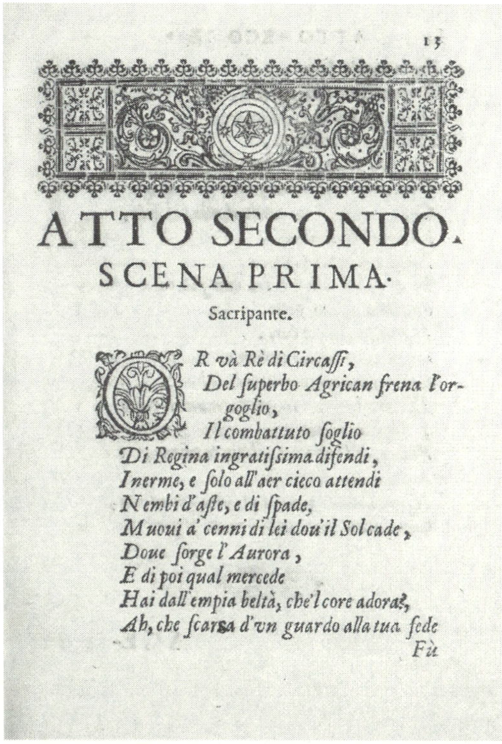

13

ATTO SECONDO.
SCENA PRIMA.
Sacripante.

OR và Rè di Circaſſi,
Del ſuperbo Agrican frena l'orgoglio,
Il combattuto ſoglio
Di Regina ingratiſſima difendi,
Inerme, e ſolo all'aer cieco attendi
Nembi d'aſte, e di ſpade,
Muoui a' cenni di lei dou'il Sol cade,
Doue ſorge l'Aurora,
E di poi qual mercede
Hai dall'empia beltà, che'l core adora?
Ah, che ſcarſa d'vn guardo alla tua fede
Fù

Ill. 6.6ab: Andrea Salvadori, *Il Medoro* (Florence, 1623), title page and page 13, with an ornament containing Jupiter and its four moons. Copy F-Pn.

of the Sun), a utopian text, now incorporating some of Galileo's discoveries[36]. It followed the publication of his *Apologia pro Galileo* (Frankfurt, 1622), which he had written in 1616 and sent to Galileo as «a question, where it was proved theologically that your way of philosophizing is more in conformity with Divine Scripture than the opposite»[37]. In 1624, Galileo spent six weeks in Rome, during which Pope Urban viii received him once a week.

A typographical ornament that includes Jupiter with its four moons is found on several pages in the edition of the libretto of *Il Medoro* published by Pietro Cecconcelli in 1623 (Ill. 6.6ab)[38]. Galileo had a copy of this libretto in his personal library[39]. Jacopo Peri's setting of

---

[36]. *La città del sole* was written first in Italian 1602, then translated into Latin by Campanella in 1613-1614.

[37]. Tommaso Campanella to Galileo, 3 November 1616 (Galilei 1902, p. 287); «una questione, dove si prova theologicamente ch'il modo di filosofare da lei tenuto è più conforme a la Divina Scrittura che non lo contrario».

[38]. The printer's mark in question is also found in other editions published by Cecconcelli, as the *Discorso delle comete* by 'Mario Giudicci' (= Galileo Galilei; 1619; title page), the *Ricettario fiorentino di nuovo illustrato* (1623; at the end), Pietro Accoli's *Lo inganno de gl'occhi* (1625; at the end), and the *Canzoni* by Gabriello Chiabrera (1628; title page). See Martín Sáez 2020b.

[39]. Hall 2013, p. 28.

26

ATTO SECONDO.

SCENA PRIMA.

Oronteo, Feraſpe, Coro di Sacerdoti
di Marte, Iſmano.

Oron. *SE dall'orrida Tana*
Diſceſo armato ſtuol d'Vnni guerrieri,
Fulmin di Tramontana,
Del ſuperbo Occidente arde gl'Imperi;
Nume del quinto Ciel da te dipende:
Tù, quante volte aſcende
Nemiche mura, ò fiere ſquadre aſſale,
Vibri pronta per lui l'Aſta fatale.
Feraſ. *Qual dunque omai preſume*
Colonia ritrouar difeſa, ò ſcampo,
Se tu guerriero Nume
Scendi per gl'Vnni à guerreggiar nel Campo?
Coro. *Dall'oppugnate Mura,*
Ecco,

Ill. 6.7ab: Andrea Salvadori, *La regina Sant'Orsola* (Florence, ²1624), title page, and page 26, with an ornament containing the four moons of Jupiter slightly different from that in *Il Medoro* (Ill. 6b). Copy CDN-Tu.

Salvadori's libretto had premiered in 1619 in the Palazzo Pitti in Florence on the occasion of the election of Ferdinand ii as Holy Roman Emperor. The plot reflects the imperial interest of the celebration. The key concept of the opera is *fedeltà*. Medoro venerates King Dardinello so much that, when he dies in a battle near Paris, «with admirable example of fidelity» he decides to go by night in search of the king's body among the corpses «to give it the due honour of burial». There he is discovered by the enemy, who wounds him in the chest. Moved by pity, Queen Angelica decides to come to his aid. Angelica had received marriage proposals from several kings, but from that moment onwards she was to be faithful to Medoro. Accordingly, the Grand Duchy of Tuscany shows its fidelity to the Empire. Just at the beginning of the Thirty Years' War, this opera rests on dynastic loyalty and Catholic dogma.

Three librettos printed by Cecconcelli have a printer's mark with Jupiter surrounded by its four moons on the title page, with a quotation from Virgil's *Aeneid* (vi, 641): «solemque suum sua sidera norunt» (And they knew their own sun, their own stars). In 1624, *La regina sant'Orsola*, with text by Andrea Salvadori and music by Marco da Gagliano, was performed to celebrate the visit of Charles of Austria, Bishop of Wroclaw, to Florence

ARGOMENTO
DELLA REGINA
SANT'ORSOLA
RAPPRESENTAZIONE
D'ANDREA SALVADORI.

SOLEMQVE SVVM SVA SIDERA NORVNT

IN FIRENZE,
Per Pietro Cecconcelli. 1624. Con Licenza de Superiori.
ALLE STELLE MEDICEE.

Ill. 6.8: Andrea Salvadori, *Argomento della Regina Sant'Orsola* (Florence, 1624), title page. Copy I-Nn.

(Ill. 6.7ab). An *Argomento* (Summary) was published with the same printer's mark on its title page (Ill. 6.8)[40]. The opera was performed again in Florence on the occasion of the visit in 1625 of Ladislaus Vasa, the future king of Poland, an essential ally of the Catholics against Protestants and Muslims[41]. The plot combines the struggle between Heaven and Hell with the war between the Romans and the Huns. In the dedication to the Polish prince, Salvadori compared both kingdoms with the Greeks and the Persians in the ancient world, and the Europeans and the Turks in modern times. After narrating the martyrdom of the saint, the opera culminates with St. Ursula hurling a thunderbolt from the sky that kills the king of the Huns. The work was sponsored by Maria Magdalena of Austria, who is thus presented as a virtuous and powerful ruler[42]. The engraved titled page of the libretto does not have the

[40]. Andrea Salvadori, *Argomento della Regina Sant'Orsola* (Florence, 1624; Sartori 1990a, nos. 2504, p. 261).

[41]. (*Ibidem*; Sartori 1992, no. 19705, p. 27). A new *Argomento* was published the following year (Florence, 1625: Sartori 1990a, no. 2505, p. 261). On the connection between Ladislaus and the history of opera see Martín Sáez 2019.

[42]. Harness 2006.

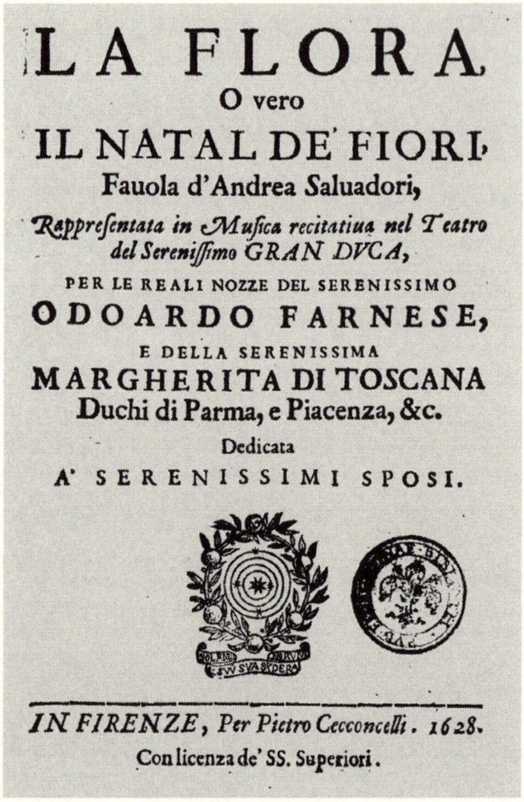

LA FLORA
O vero
IL NATAL DE' FIORI,
Fauola d'Andrea Saluadori,
Rappresentata in Musica recitatiua nel Teatro del Serenissimo GRAN DVCA,
PER LE REALI NOZZE DEL SERENISSIMO
ODOARDO FARNESE,
E DELLA SERENISSIMA
MARGHERITA DI TOSCANA
Duchi di Parma, e Piacenza, &c.
Dedicata
A' SERENISSIMI SPOSI.

IN FIRENZE, Per Pietro Cecconcelli. 1628.
Con licenza de' SS. Superiori.

Ill. 6.9ab: Ferdinando Saracinelli, *La liberazione di Ruggiero dall'isola d'Alcina* (Florence, 1625), title page, and Andrea Salvadori, *La Flora* (Florence, 1628), title page. Copies I-Fn.

printer's mark, but the emblem with Jupiter and its four moons is found on several pages within the booklet. Interestingly, the famous passage from the Book of Joshua (10:12-13) is referred to in the fourth act, when Prince Ireo asks the sun to stop and to inform his father and his servants of his sadness over the martyrdom of Saint Ursula:

Ferma, pietoso Sole,	Stop, merciful Sun,
Là sopra il Regno mio, ferma le rote,	There, above my Reign, stop the wheels,
Al real Genitore, A' Servi miei fa note	Make known to the royal father, to my servants,
L'alte di lei miserie, e 'l mio dolore	The heights of her misery, and my distress.

The second edition of the libretto includes — for the first time in the history of opera — engravings of several scenes signed by Giulio Parigi.

La liberazione di Ruggiero dall'isola d'Alcina (1625) by Ferdinando Saracinelli (text) and Francesca Caccini (music) was performed for Prince Ladislaus Vasa as well. The libretto has an engraved title page, which also includes the image of Jupiter and the four moons (ILL. 6.9A)[43]. It also features engravings representing several scenes. The highlight of the opera was the final *balletto de' cavalli* (horse dance), which was performed in the courtyard of the villa di Poggio Imperiale. The libretto includes a sonnet by Salvadori where the passage from the Book of Joshua is mentioned again. Salvadori stated that the Sun stood in the sky «longer than ever» (*fermossi in Ciel più che mai bello il Sole*), to observe those admirable dances (*l'ammirabili Carole*). The protagonist of the opera is again a woman, the magician Melissa, who saves Ruggiero from the charms of the witch Alcina. The argument stresses the importance of Ruggiero returning to war and abandoning Alcina's siren songs. In the prologue, Neptune recalls how the Polish king had defeated the infidels. Vladislaus is incited to follow the military policy against the heretics.

The last publication in this group is the libretto of *La Flora* (1628), an opera with text by Andrea Salvadori and music by Marco da Gagliano, written to celebrate the accession to the throne of Ferdinando de' Medici. The printed libretto has the printer's mark on the title page (ILL. 6.9B), and an ornament including the four moons of Jupiter at the beginning of the 'Argomento'. The libretto also contains images of the scenes.

THE DANGEROUS CURIOSITY OF THE NEW ENDYMION

Diana schernita (1629) by Giacomo Francesco Parisani (text) and Giacinto Cornacchioli (music) was the first opera in which Galileo (with his telescope) appeared as a character on stage, under the name of Endymion. The opera was first performed in the palace of the German nobleman Johann Rudolph Baron von Hohen Rechberg. No printed libretto is known. The music was published in Rome, with a dedication by Cornacchioli to Taddeo Barberini[44]. As Paolo Fabbri (2003) has suggested, the opera is related to Giambattista Marino's *L'Adone* (1623). In *Diana schernita*, Galileo is presented as the «new Endymion» (*novello Endimione*), eager to watch the Moon with his telescope[45].

Parisiani's libretto tells of Cupid's anger towards Diana because she loves Endymion but pretends to be chaste. In fact, Diana has given Endymion a new device (the telescope) so that he can watch her whenever he wants. Enydmion invokes the Sun asking him to stop its movement in order to enjoy more time with Diana, another clear reference to the biblical passage from the Book of Joshua. In revenge, Cupid decides to make Endymion jealous by assuring him that Diana has a lover. The god of love invites him to seek this out, inciting him to disguise himself

[43]. Ferdinando Saracinello, *La liberazione di Ruggiero dall'isola d'Alcina* (Florence, 1625; SARTORI 1991B, no. 14223, p. 13).

[44]. Giacinto Cornacchioli, *Diana schernita: Favola boscareccia, posta in musica* (Rome, 1629; RISM A/I, C 3938).

[45]. FABBRI 2003, p. 56, and DAOLMI 2006, p. 31. About the libretto, see ANTONICEK 2002.

as the hunter Atteone. When Diana discovers that someone is spying on her, she turns him into a deer and kills him, unaware that he is Endymion. At this moment, Pan proposes to turn the deer into a flower, and three golden lilies appear in its place, on which three golden bees rest, a clear reference to the Barberini emblem. *Diana schernita* was meant to be a comedy, with everyone happy at the end, even Diana. Michele Curnis (2018) has compared the libretto with Marino's poem and has shown significant literary influences. Curnis also elaborated on the connection with Galileo, highlighting the implicit reference to Joshua[46]. There is no doubt that *Diana schernita* is a 'Galilean opera'.

An atmosphere of condemnation hangs over the plot. It is important to remember that Marino's poem was placed on the Index of Prohibited Books in 1624. The message of *Diana schernita* was clearly religious in nature. Endymion represented mortality and science, while Pan stood for immortality and religion. Science could only flourish if it was at the service of religion. This message could be interpreted as an attack against Galileo's curiosity and his uncontrolled love for astronomy. Those who fall in love with the object of their studies, especially in the case of astronomy, are at the risk of being trapped by it. Taddeo Barberini's aim would have been to moderate Galileo's curiosity, transforming the latter's excess into a flower at the service of the Church. In addition to the classic Christian condemnation of curiosity[47], to which I will return later, the message fits with the ambiguous attitude of the Barberini towards Galileo.

In the same year that *Diana schernita* was performed, *De Deo uno* by Agostino Oreggi, a theologian in the service of Pope Urban VIII, was published in Rome. Oreggi articulated the pope's argument about the superiority of God's wisdom and power over human reasoning, according to which any astronomical system could have been created by an omnipotent divinity, with the only condition that it did not involve contradiction. It implied that it was not reasonable to go beyond the realm of hypotheses in astronomy[48].

Opera as a Pontifical Smoke Screen

By 1630, Galileo thought the time was ripe for reconsidering the Copernican theory. He proposed the Pope the publication of a book that would confront the geocentric and heliocentric systems. The book was finished in January 1630[49], but it took several years before it was published as *Dialogo sopra i due massimi sistemi del mondo, Tolomaico e Copernicano* (Florence, 1632; Ill. 6.10a). Urban VIII approved publication; at the same time, he asked Galileo not to

46. Curnis 2018.

47. Blumenberg 1999.

48. Biagioli 1993, p. 310, and Machamer 1998, p. 330, note 26.

49. Machamer 1998, p. 22.

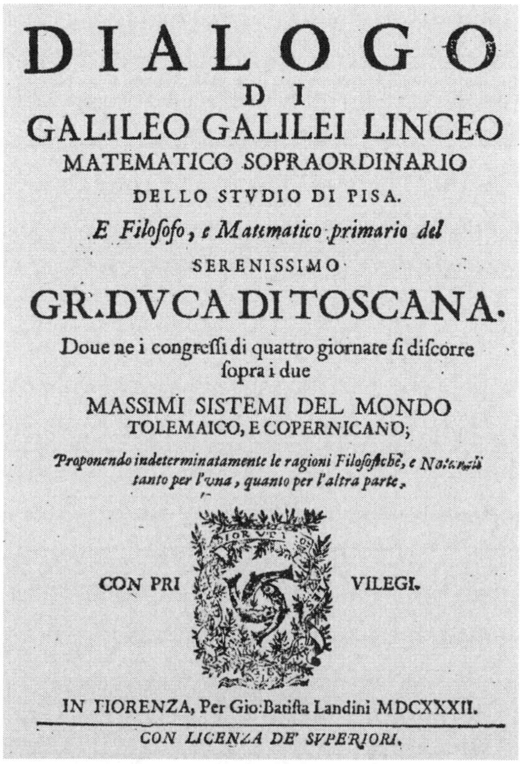
DIALOGO
DI
GALILEO GALILEI LINCEO
MATEMATICO SOPRAORDINARIO
DELLO STVDIO DI PISA.
E Filoſofo, e Matematico primario del
SERENISSIMO
GR.DVCA DI TOSCANA.
Doue ne i congreſſi di quattro giornate ſi diſcorre
ſopra i due
MASSIMI SISTEMI DEL MONDO
TOLEMAICO, E COPERNICANO;
Proponendo indeterminatamente le ragioni Filoſofiche, e Naturali
tanto per l'vna, quanto per l'altra parte.
CON PRI VILEGI.
IN FIORENZA, Per Gio:Batiſta Landini MDCXXXII.
CON LICENZA DE' SVPERIORI.

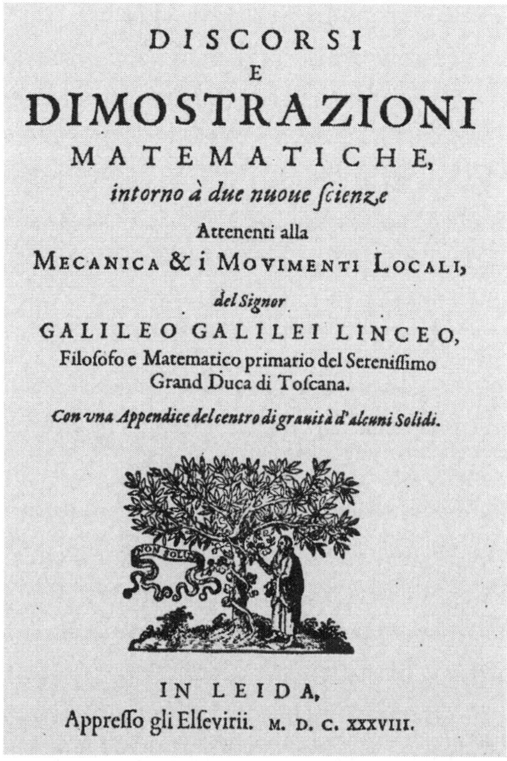
DISCORSI
E
DIMOSTRAZIONI
MATEMATICHE,
intorno à due nuoue ſcienze
Attenenti alla
MECANICA & i MOVIMENTI LOCALI,
del Signor
GALILEO GALILEI LINCEO,
Filoſofo e Matematico primario del Sereniſſimo
Grand Duca di Toſcana.
Con vna Appendice del centro di grauità d'alcuni Solidi.
IN LEIDA,
Appreſſo gli Elſevirii. M. D. C. XXXVIII.

Ill. 6.10ab: Galileo Galilei, *Dialogo sopra i due massimi sistemi del mondo, Tolomaico e Copernicano* (Florence, 1632), title page, and *Discorso e dimostrazioni mathematiche intorno à due nuove scienze* (Leiden, 1638), title page. Copies I-Rn and A-Wn, respectively.

defend the Copernican hypothesis as reality, but to present it as a conjecture and to include a theological argument about the superiority of God. Galileo yielded to this request and the book passed ecclesiastical censorship. The success was immediate. Campanella, Benedetto Castelli, Vincenzo Renieri, Fulgenzio Micanzio and others wrote to Galileo to congratulate him, but the criticism also increased. Galileo had ridiculed the arguments of the famous Jesuit Christoph Scheiner, which undoubtedly contributed to his subsequent conviction[50]. Filippo Magalotti wrote to Mario Guiducci on 7 August 1632 that the Jesuits had taken the decision «to do with pleasure everything to have the book prohibited»[51].

Galileo's position was weakened by the problems with which Pope Urban viii was beset. The Barberini were going through their greatest political crisis. In 1631, Sweden's intervention in the Thirty Years' War, led by the Protestant King Gustavus Adolphus ii, had inflicted great defeats on the Catholic forces, in what is considered the most international period of

[50]. Beltrán Marí 2006, p. 70.

[51]. Filippo Magalotti to Mario Guiducci, 7 August 1632 (Galilei 1904a, p. 370): «lavorar gagliardissimamente perché l'opera sia proibita».

the war, during which alliances began to have an increasingly political character (as the French intervention would confirm in 1636). The opposition of Sweden against Spain, and the French alliances against the Habsburgs, reinforced the pressure of Spain on Rome to clarify its position. Urban VIII had spent a decade closing alliances with France, to diminish the power of Spain in Italy, and the imperial ambassadors began to accuse the pope of favouring the Protestant cause and not daring to declare it a holy war[52]. The Spanish Cardinal Gaspar de Borja (Borgia) y Velasco went so far as to «read (in front of all the other cardinals present) a harsh protest against the pope's lack of support for his king's military efforts against the Protestants in Germany»[53].

This situation likely contributed to the condemnation of Galileo. The mere suggestion that the Church was defending a heretic seemed to be a good excuse to use the astronomer as a scapegoat[54], like the deer in *Diana schernita* that had to be neutralized and was turned into papal lilies. By condemning Galileo, Urban VIII could clean up his image and present himself as a great defender of Catholicism. The use of scapegoats was common in court and consisted of choosing someone well known for the greatest possible impact, something that Galileo's international fame could guarantee. That he was a protégé of Urban VIII fulfilled the requisite needed for the «fall of the favourite», which would allow the pope to present himself as someone who was purifying his own institution[55].

The document of 1616 that recounted the meeting Bellarmine had with Galileo in February of that year appeared at just the right moment. According to this document, Bellarmine had not only forbidden Galileo to defend Copernicanism, but also to teach it *in any way*. Moreover, Urban VIII had expressly asked Galileo to present the Copernican thesis as a hypothesis and, although it was a dialogue, Salviati's character seemed to defend it from a positive position. Finally, the Copernican thesis seemed to be better argued than the Ptolemean, which would break the equilibrium that Galileo was obliged to preserve, not only as a scientist, but also as a courtier. Behind all this, there could also be a personal resentment. In addition to the misgivings of the Jesuits, the inquisitors charged that the pope's theological argument was put in the mouth of a clearly foolish character (*in bocca di un sciocco*)[56].

In this context, the opera of *Sant'Alessio* possibly holds special significance[57]. It was the first opera directly commissioned by Francesco Barberini. The words were written by Cardinal Rospigliosi, future Pope Clement IX, who was well-known as a librettist. Stefano Landi provided the music. *Sant'Alessio* was performed many times between 1629 and 1634, the period in which

[52]. Finocchiaro 2010, p. 147, and Finocchiaro 2019, p. 20.

[53]. Biagioli 1993, p. 335.

[54]. Finocchiaro 1989, pp. 12-13.

[55]. Biagioli 1993, p. 313.

[56]. Galilei 1907, p. 326.

[57]. Murata 1981, Hammond 1994, and Tamburini 2003.

Galileo's *Discorsi*, his condemnation and the propaganda around the affair were conceived and developed. Frederick Hammond (2010) has argued that the performance of 1634, dedicated to Alexander Charles Vasa, brother of Ladislaus iv, now King of Poland, could be interpreted as a reaction to the condemnation of 1633. Convicting the most famous scientist in history had a gigantic impact, and the Church made a great effort to justify its action and to mitigate the discontent it had aroused, which included personalities such as René Descartes, Giovanni Battista Doni, Nicolas-Claude Fabri de Peiresc, and Marin Mersenne[58]. Peiresc's letter of 31 January 1635 to Francesco Barberini, in which he compared the Galileo affair with the trial of Socrates, thereby calling into question the reputation of the pontificate, is perhaps the most eloquent document on this flurry of letters[59]. The comparison was inevitable. Hugo Grotius also compared Galileo to the Greek philosopher in 1636[60].

In fact, despite its numerous performances, it was not until 1634 that it was decided to print the music of the work (Rome; Ill. 6.11a)[61]. *Sant'Alessio* served to present the Church as an advocate of culture through music, arts and literature. At the same time, it exhibited the Church's connection to a Catholic dynasty of great importance — the Vasa kings of Poland — and defended the Jesuits in a new prologue. This was part of a propaganda program that included many other festivities. Hammond (1992) has called attention to the unusual dissemination of the printed score, which was sent not only to various courts, but also to scholars critical of the condemnation. He cited a letter from Fabri de Peiresc to Mersenne of 23 July 1635, in which Peiresc described the opera and asked Mersenne to mention the work in his *Harmonie universelle*, noting the contribution of Cardinal Francesco Barberini[62]. Doni cited *Sant'Alessio* in his 'Trattato della musica scenica', written around 1635 (but published only in 1763), remarking that it was received «always with universal applause»[63].

We can strengthen Hammond's thesis by relating the operatic propaganda to the publicity of the Inquisition about the trial and abjuration of Galileo. In the summer of 1633, all papal nuncios and inquisitors in Europe received copies of Galileo's condemnation and abjuration, as well as orders to publicize them. In a letter to Giacomo Tinti, inquisitor of Modena, Antonio

58. Hammond 1992 and Hammond 2010.

59. Nicolas-Claude Fabri de Peiresc to Francesco Barberini, 31 January 1635 (Galilei 1905, p. 202).

60. Elia Diodati to Galileo, 23 September 1636 (*ibidem*, p. 490).

61. Stefano Landi, *Il S. Alessio: Dramma musicale* (Rome, 1634; RISM A/i L 534). A libretto was never printed, but there is a printed *Argomento del S. Alessio* (Rome, 1634; Sartori 1990a, no. 2480, p. 259). The text is edited in Della Corte 1958, vol. i, pp. 195-265.

62. Nicolas-Claude Fabri de Peiresc to Marin Mersenne, 23 July 1635 (Mersenne 1959, p. 328). See Hammond 1992, p. 88, Hammond 2010, pp. 99-100.

63. Giovanni Battista Doni, *De' trattati di musica tomo secondo* (Florence, 1763), Appendice, 'Trattato della musica scenica', p. 7: «sempre con applauso universale».

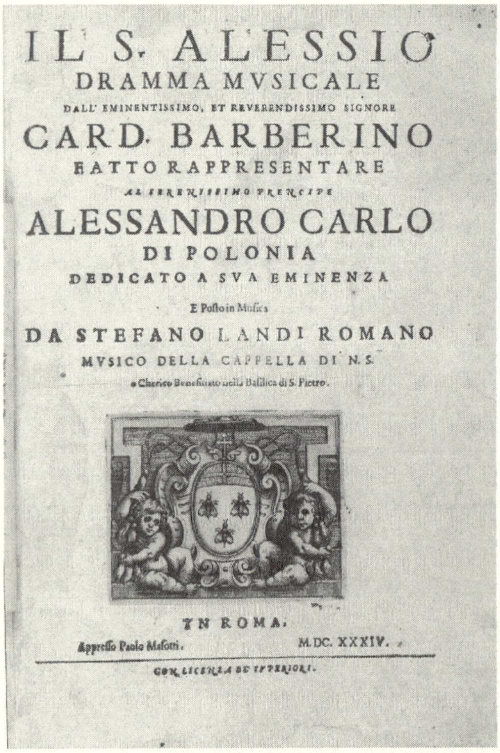
IL S. ALESSIO
DRAMMA MVSICALE
DALL' EMINENTISSIMO, ET REVERENDISSIMO SIGNORE
CARD. BARBERINO
FATTO RAPPRESENTARE
AL SERENISSIMO PRENCIPE
ALESSANDRO CARLO
DI POLONIA
DEDICATO A SVA EMINENZA
E Posto in Musica
DA STEFANO LANDI ROMANO
MVSICO DELLA CAPPELLA DI N. S.
e Chierico Beneficiato nella Basilica di S. Pietro.
IN ROMA.
Appresso Paolo Masotti. M.DC. XXXIV.
CON LICENZA DE SVPERIORI.

Ill. 6.11ab: Stefano Landi, *Il S. Alessio* (Rome, 1634), title page, and Giovanni Carlo Coppola, *Le nozze degli dei* (Florence, 1637), title page. Copies US-NH and US-MAL, respectively.

Marcello Barberini, brother of the pope, ordered the news to be reported to «all the professors of philosophy and mathematics, so that they, knowing how was dealt with the aforementioned Galileo, understand the gravity of the error he committed»[64]. As Maurice Finocchiaro (2010) has noted, «such publicity was unprecedented in the annals of the Inquisition and was never repeated in subsequent inquisitorial practice»[65].

In this sense, the connection of Ladislaus Vasa with this Galilean opera seems significant. The occasion seemed to merit an edition with engravings, the first in the history of Roman opera. Nor should we forget that the King of Poland was an advocate of Galileo[66]. The Church seemed determined to prove that it still admired Galileo, but the situation had forced it to condemn him. On 19 April 1636, the King of Poland asked Galileo for new lenses for his telescope. The king took the opportunity to remind Galileo that «the wish to do him a favour is still with

[64]. Antonio Marcello Barberini to Giacomo Tinti, 2 July 1633 (Galilei 1904b, p. 169): «tutti li professori di filosofia e di matematica, perché, sapendo eglino in che modo si è trattato con il detto Galileo, comprendino la gravità dell'errore da lui commesso».

[65]. Finocchiaro 2010, p. 161.

[66]. *Ibidem*, p. 167.

us»[67]. In fact, on 19 December Roberto Giraldi wrote to Galileo that the king wanted «to talk with you about your interests»[68].

The Church had serious reasons to lessen the impact of the condemnation of Galileo. The very message of the libretto shows how the opera was used for that purpose. The story of Saint Alessio emphasizes the priority of religious life over any other, salvation before science. As the saint himself says in the opera, the truly important things are «beyond the stars» (*sopra le stelle*). We should not forget that Galileo's own *Dialogo* was written, according to the official message the Church included in the Prologue, «to prove non-Catholics that Catholics knew all the arguments» in favour of Copernicanism[69]. Thus, in this case, religion seemed more important than science.

In this context, the performance of Rospigliosi's *Erminia sul Giordano* also merits attention. It was performed in 1633, but the libretto was published only in 1637, with beautiful engravings[70]. This was one year after the publication of Galileo's letter to Christina of Lorraine[71].

The last opera to be published with engravings in the libretto before the opening of the Venetian theatres might be also important from a negative perspective. Giovanni Carlo Coppola's *Le nozze degli dei* (1637; Ill. 6.11b), with music by various composers, was performed in the Palazzo Pitti in Florence to celebrate the wedding of Ferdinando ii de' Medici and Vittoria della Rovere[72]. This opera is from beginning to end a theatre of the world with maritime, earthly, celestial and infernal scenes (Ill. 6.12). Four divine weddings are the pretext for introducing them. The wedding of Proserpina and Pluto takes place in Hell; that of Nettuno and Amphitrite, in the Sea; that of Giove and Giunone, in Heaven; and that of Vulcano and Venere, on Earth. In his dedication, Coppola described the grand ducal couple as «favourable planets» (*favorevoli Pianeti*) and the opera includes all kinds of gods, constellations, and stars. It would have been a perfect excuse to introduce Galileo's discoveries, but nothing suggests a heliocentric influence. The Church had once again achieved its objectives.

---

67. Ladislaus iv a Galileo, 19 April 1636 (Galilei 1905, p. 420): «vive anco con noi volontà di favorirla».

68. Roberto Giraldi a Galileo, 26 December 1636 (*ibidem*, p. 532): «discorrer con lei circa a' sua interessi».

69. Finocchiaro 2019, p. 126.

70. Michelangelo Rossi, *Erminia sul Giordano: Dramma musicale* (Rome, 1637; RISM A/i, R 2743).

71. Galileo Galilei, *Nov-antiqua sanctissimorum patrum, & probatorum theologorum doctrina,* [...] *in gratiam Serenissimæ Christinæ Lotharingæ* (Strasbourg, 1636), with Italian text and Latin translation printed in two columns.

72. The libretto: Giovanni Carlo Coppola, *Le nozze degli dei* (Florence, 1637; Sartori 1991b, no. 16712, p. 252). Other publications related to the performances are the *Argomento delle nozze degli dei* (Florence, 1637; Sartori 1990a, no. 2511, p. 262), the *Descrizione delle feste fatte in Firenze* (Florence, 1637; Sartori 1990b, no. 7646, p. 338), and the *Relazione delle nozze degli dei* (Florence, 1637; Sartori 1992, no. 19749, p. 32).

Ill. 6.12: Giovanni Carlo Coppola, *Le nozze degli dei* (Florence, 1637), engraving representing the first scene of the first act, by Stefano della Bella after a design of Alfonso Parigi. Copy US-MAL.

The Circle Closes: Galileo versus Saint Boniface

In 1638, Galileo published his *Discorsi e dimostrazioni matematiche intorno a due nuove scienze* (Leiden, 1638; Ill. 6.12). In the same year, he reappeared — that is, metaphorically speaking, in an opera — on the Roman scene, in the Palazzo della Cancelleria, where *Amor pudico* had been produced in 1614. *San Bonifatio* was written, again, by Cardinal Rospigliosi. The figure of Saint Boniface had great symbolic meaning during the Thirty Years' War since he was credited for having converted the Germanic people to the Christian faith and having founded the abbey of Fulda. In a context of disputes with Protestant heretics, Galileo's desire to reinterpret the Bible could be a danger, contrasting with the pious attitude of the saints, who in wartime helped to increase the influence of the Church.

One scene in the opera clearly underlines the superiority of religion over scientific discovery. It is done in the most comical passage of the opera, Scene 4 of Act I, where the servant Fagotto complains about those who dedicate themselves to the study of distant things while forgetting human affairs. Specifically, this complaint refers to those who are obsessed with the

Moon, the planets, and the stars. The unpublished libretto is preserved in several manuscripts kept in the Vatican Library. Given that it is little known, it is worth transcribing the entire passage:

Azione 1, Scena 4.	Act 1, Scene 4
Fagotto. Sens'essere indovino	Fagotto. Without being a fortune-teller
Ogn'un distingue il pesce dalla carne	Each one distinguishes fish from meat.
Gli storni dalle starne	Starlings from partridges,
Il pan da sassi, e dal aceto il vino	Stone from bread, and wine from vinegar.
Ogni poca avertenza	Every little alertness
Con chiare distintioni,	With clear distinctions,
riconosce i baiochi da testoni,	Recognizes *baiocchi* from *testoni*,
e pure vi è così poca conoscenza	And yet there is so little knowledge
et è la gente poi si mall'accorta	And people are so little shrewd
quel che tanto importa:	About what matters so much:
che sia pur un cervel sagace e scaltro	Who has yet a smart and sagacious brain,
sapra difficilmente	Will hardly know,
riconoscer ben bene un huom dall'altro,	To recognize one man well from another.
quanti e quanti s'infrascano il cervello	How many and how many break their brain
per conoscer le stelle ad una ad una	To know the stars one by one,
per osservar questo pianeta, e quello,	To observe this planet, and that one,
e saper quanti passi fa la Luna?	And know how many steps the Moon takes?
O diligenze vane?	Or vain diligence?
Che s'ha da far di cose si lontane?	What must you do with things so far away?
Uno studio eccellente	An excellent study
Sarebbe l'imparare	Would be learning
A conoscer la gente	To get to know people
Come si fa dell'un l'altro colore,	How they change from one to another colour,
perche in questo si vede	Because in this you can see
che si piglian de granchi a tutte l'ore:	That they are mistaken all the time.
A me cosi sucede	Therefore, it happens to me
Che possomi à servire un capitano	That I can serve a captain.
Mi credevo per certo	I believed for sure,
Fosse tutto bravura:	That he was all good.
alla fine ho scoperto	I eventually found out
che egli è vile, da poco abbietto e vano,	That he is vile, just abject, and vain,
per professione, per arte, et natura.	By profession, by art, and by nature.
Il peggio è che ogni cosa	The worst is that everything
si finisce in canzoni in dame et armi,	Ends up in songs on ladies and weapons.
ne io posso vantarmi d'haver mangiato	Nor can I boast of having eaten
mai quanto ho bisogno	Ever as much as I need,
e s'ho da dire il vero,	And if I must tell the truth,
non mi satollo mai se non in sogno[73].	I am never satisfied except in my dreams.

[73]. I-Rvat, MS Lat. 10192 (Giulio Rospiglioso, *Sant'Alessio*), pp. 53-54.

As in *Diana schernita*, this scene can be understood through the Christian critique of curiosity. Curiosity could be a vice in at least two basic ways, depending on whether its object was trivial (*negligentia*) or heretical (*haeresis*). In the context of the Counter-Reformation, it was common to oppose the vice of *curiositas* to the virtue of *sancta ignorantia*. Already Saint Augustine had stated that «faithful ignorance is better than reckless science»[74]. For this reason, Campanella distinguished between good and bad curiosity. In his *Discorsi della liberta* (Jesi, 1633), he defined vain curiosity (*curiosità vana*) as «the mother of the arts not founded in natural causes, neither in effects, nor signs, placed by nature, or by God; but in signs placed by human art, and ill-affected will». This kind of curiosity would be the origin of «superstition, idolatry, heresy, schism and ruin of the people». In fact, Campanella immediately referred to the Protestant context and its geopolitical scope: «as seen in France, and Germany, and Greece, and throughout the North; with sedition of peoples, tyranny of princes, and fall of kingdoms, and diminution of obedience, and of strengths and virtues». Yet, there was also the curiosity of the true sciences (*vere scienze*), in which he would undoubtedly include Galileo. Campanella charges the Church for mistaking the curiosity of the true sciences as vain curiosity, thus attacking the papacy in a surreptitious manner. The phrase stating that «the Pope cannot prohibit them [the true sciences], because he is the Vicar of Christ, the Wisdom of God, and every science is the splendour of that» has gone unnoticed, but clearly refers to Galileo[75].

Although the Church accused Galileo of heresy, in *Diana schernita* the scientist seems to be criticized not for his heresy but for his dangerous superficiality. In fact, this fits with the theological argument of Urban VIII: science is as small compared to theology as men are to God, so no reasoning is more than hypothetical and, in a way, irrelevant. Ultimately, science could never damage religion. One can be «a shrewd and cunning brain» in astronomy, but these studies are «vain diligences». It is much better to «know the people» than the moon's steps, as stated in the libretto. Boniface is the best example: he knew people and helped them by converting them, just like the Jesuits extended the Catholic credo across the globe. As a saint (a figure undervalued by Protestants), Boniface achieved something as foreign to science as salvation. With his knowledge of men, he created a world far from those heretics and seditious people of whom Campanella spoke. So maybe the alleged mistake of the scientist was not to be heretical, as stated in the sentence condemning Galileo, but to be a superficial and imprudent person who could favour enemies in wartime. While Galileo gazed at the stars, the pope and

---

74. Saint Augustine, *Sermones*, XXVII, Cap. 3, 4: «[...] *melior est fidelis ignorantia, quam temeraria scientia*». See Vega 2014, p. 172.

75. Tommaso Campanella, *Discorsi della libertà* (Jesi, 1633), p. 11: «madre dell'arti non fondate in cause naturali, né negli effetti, né ne i segni, posti dalla natura, ò da Dio; ma in segni posti dall'arte humana, & volontà mal'affetta [...] come si è visto in Francia, & Alemagna, e in Grecia, & in tutto il Settentrione; con seditione de popoli, tirannide de' Prencipi, e caduta de Regni, e diminutione d'obedienza, e di forze e di virtù [...] non può il Papa prohibirle, perche è Vicario di Christo, Sapienza di Dio, & ogni scienza è splendor di quello».

orders like the Jesuits oversaw dozens of institutions all over the world. The propagandistic message seems to be that Galileo talked too much about irrelevant matters, and someone could use it against the Church; so he had to be controlled. This could explain why the Church refused to see Galileo's arrest as a tragedy in this opera. Galileo's condemnation was, in the end, unimportant. Or that is at least what it wanted the world to believe.

Conclusion

With the advancement of the sciences in the Modern Age, many philosophers have attempted to distinguish major areas of knowledge and to organize them into binary opposites: philosophy *versus* religion, science *versus* letters, humanities *versus* engineering, and so on. This has served to justify the need for various specializations, which were often isolated from each other. The Galileo affair undoubtedly contributed to this phenomenon. Urban VIII defended the need to keep religion out of disputes by arguing that God exceeded human science. Galileo himself — as Descartes or Kant would later do — considered that the Scriptures were not intended to explain nature, just as science was not meant to save souls. Galileo's life and work served rather to stress the opposite. Astronomy, theology, theatre, music, and so on, behaved as communicating vessels, not as watertight compartments.

The history of opera is a unique example of this multidisciplinary porosity. The Galileo affair not only helps us to understand the allegorical meaning of some librettos, their connection with certain academic disputes, and the appearance of new scientific devices such as the telescope, but also to understand the function of the printed sources and the value of operatic materials (scores, librettos, descriptions, etc.) for the propagation of new ideas. Operas are useful for understanding the cultural transition from geocentrism to heliocentrism, the social position of the new scientist, the new idea of curiosity and the modern defense of the freedom to philosophize.

Moreover, the history of opera provides unique examples for understanding the complexity of the disputes at hand. The religious wars between Catholics and Protestants and the existence of several dynasties with diverse geopolitical interests, force us to consider the cultural projects of very diverse European institutions. Although these operas are always performed in Catholic territories, the actors involved dealt with states as different as the Grand Duchy of Tuscany, the Papal States, the Commonwealth of Poland-Lithuania, and the Holy Roman Empire. Moreover, we must not only consider the differences between Church and State, but also between civic dynasties and religious orders, and even between people in the same social group who had opposing worldviews.

All this is only possible from a multidisciplinary approach. We do not need to abandon the existing disciplines, but we can question their limits by letting ourselves be led by the sources. If those sources speak about philosophy, theology, science, mythology, politics, music, etc., why limit us to one of them? Galileo became a heretic because he was not confined to a discipline. His works were as much about philosophy, religion, and literature as they were about physics and mathematics; his activities as a courtier and an engineer were no less important than his work as an astronomer. The very complexity of opera, with its collaboration of artists, writers, musicians, architects, and engineers, reveals something similar. What is essential is not the elements that we can study separately, but that total image of the *theatrum mundi* that both opera and Galileo wanted to understand.

Bibliography

Antonicek 2002

Antonicek, Theophil. 'Giacinto Cornacchiolis *Diana schernita* (Rom 1629)', in: *Album amicorum Albert Dunning*, edited by Giacomo Fornari, Turnhout, Brepols, 2002, pp. 295-310.

Beltrán Marí 2006

Beltrán Marí, Antonio. 'Galileo y Urbano viii. La trama del equívoco', in: *Éndoxa: Series Filosóficas*, no. 21 (2006), pp. 35-73.

Biagioli 1993

Biagioli, Mario. *Galileo, Courtier: The Practice of Science in the Culture of Absolutism*, Chicago, University of Chicago Press, 1993.

Blumenberg 1999

Blumenberg, Hans. *The Legitimacy of the Modern Age*, translated by Robert M. Wallace, Cambridge (MA), MIT Press, 1999.

Coelho 1992

Music and Science in the Age of Galileo, edited by Victor A. Coelho, Dordrecht, Kluwer, 1992.

Coelho 1997

Coelho, Victor A. 'The *Apotheosis* of Francis Xavier and the Conquering of India', in: *The Work of Opera: Genre, Nationhood, and Sexual Difference*, edited by Richard Dellamorra and Daniel Fischlin, New York, Columbia University Press, 1997, pp. 22-47.

Curnis 2018

Curnis, Michele. 'Novelli Endimioni e falsi Atteoni nella *Diana schernita* (Roma, 1629)', in: *Il mito di Diana nella cultura delle corti. Arte, letteratura e musica*, edited by Giovanni Bárberi Squarotti, Annarita Colturato, and Clara Goria, Florence, Olschki, 2018.

Cypess 2016

Cypess, Rebecca. *Curious & Modern Inventions: Instrumental Music as Discovery in Galileo's Italy*, Chicago, University of Chicago Press, 2016.

Daolmi 2006

Daolmi, Davide. 'La drammaturgia al servizio della scenotecnica: le «volubili scene» dell'opera barberiniana', in: *Il saggiatore musicale*, xiii/1 (2006), pp. 5-62.

Della Corte 1958

Della Corte, Andrea. *Drammi per musica dal Rinuccini allo Zeno*, Turin, UTET, 1958.

Donati 2014

Donati, Alberto. *Le motivazioni teologiche della condanna di Galileo Galilei*, Soveria Mannelli, Italy, Rubbettino, 2014.

Fabbri 2003

Fabbri, Paolo. *Il secolo cantante. Per una storia del libretto d'opera in Italia nel Seicento*, Rome, Bulzoni, 2003.

Fabbri 2008

Fabbri, Natasha. *Filosofia, scienza e musica in Mersenne, Descartes e Galileo*, Pisa, Edizioni della Scuola Normale, 2008.

Feldhay 1995

Feldhay, Rivka. *Galileo and the Church: Political Inquisition or Critical Dialogue?*, Cambridge, Cambridge University Press, 1995.

Fenlon 2002

Fenlon, Iain. *Music and Culture in Late Renaissance Italy*, Oxford-New York, Oxford University Press, 2002.

Finocchiaro 1989

Finocchiaro, Maurice A. *The Galileo Affair: A Documentary History*, Berkeley (CA), University of California Press, 1989.

Finocchiaro 2010

Id. *Defending Copernicus and Galileo: Critical Reasoning in the Two Affairs*, London, Springer, 2010.

Finocchiaro 2019

Id. *On Trial for Reason: Science, Religion, and Culture in the Galileo Affair*, Oxford, Oxford University Press, 2019.

Franchi 1994

Franchi, Saverio. *Le impressioni sceniche. Dizionario bio-bibliografico degli editori e stampatori romani e laziali di testi drammatici e libretti per musica dal 1570 al 1800*, Rome, Edizioni di Storia e Letteratura, 1994.

Galilei 1895

Galilei, Galileo. *Opere, Volume v: Delle macchie solari,* [*ecc.*], edited by Antonio Favaro, Florence, Barbèra, 1895.

Galilei 1900

Id. *Opere, Volume x: Carteggio 1574-1610,* edited by Antonio Favaro, Florence, Barbèra, 1900.

Galilei 1901

Id. *Opere, Volume xi: Carteggio 1611-1613,* edited by Antonio Favaro, Florence, Barbèra, 1901.

Galilei 1902

Id. *Opere, Volume xii: Carteggio 1614-1619,* edited by Antonio Favaro, Florence, Barbèra, 1902.

Galilei 1903

Id. *Opere, Volume xiii: Carteggio 1620-1628,* edited by Antonio Favaro, Florence, Barbèra, 1903.

Galilei 1904a

Id. *Opere, Volume xiv: Carteggio 1629-1632,* edited by Antonio Favaro, Florence, Barbèra, 1904.

Galilei 1904b

Id. *Opere, Volume xv: Carteggio 1633,* edited by Antonio Favaro, Florence, Barbèra, 1904.

Galilei 1905

Id. *Opere, Volume xvi: Carteggio 1634-1636,* edited by Antonio Favaro, Florence, Barbèra, 1905.

Galilei 1907

Id. *Opere, Volume xix: Documenti e narrazioni biografiche di contemporanei,* edited by Antonio Favaro, Florence, Barbèra, 1907.

Galilei 2000

Id. *Lettera a Cristina di Lorena,* edited by Franco Motta, Genoa, Marietti, 2000

Galilei – Grassi 2017

Galilei, Galileo – Grassi, Orazio. *Cometas, ciencia y religión. La polémica Galileo-Grassi,* edited by Antonio Beltrán Marí, Madrid, Tecnos, 2017.

Galluzzi 2017

Galluzzi, Paolo. *The Lynx and the Telescope: The Parallel Worlds of Cesi and Galileo,* translated by Peter Mason, Leiden, Brill, 2017.

Gattei 2018

Gattei, Stefano. *On the Life of Galileo: Viviani's Historical Account and Other Early Biographies,* Princeton (NJ), Princeton University Press, 2018.

Guerrini 2009
Guerrini, Luigi. *Galileo e la polemica anticopernicana a Firenze*, Florence, Polistampa, 2009.

Guerrini 2010
Id. *Cosmologie in lotta. Le origini del processo di Galileo*, Florence, Polistampa, 2010.

Hall 2013
Hall, Crystall. *Galileo's Reading*, Cambridge-New York, Cambridge University Press, 2013.

Hammond 1992
Hammond, Frederick. 'The Artistic Patronage of the Barberini and the Galileo Affair', in: Coelho 1992, pp. 67-89.

Hammond 1994
Id. *Music and Spectacle in Baroque Rome: Barberini Patronage under Urban viii*, New Haven (CT), Yale University Press, 1994.

Hammond 2010
Id. *The Ruined Bridge: Studies in Barberini Patronage of Music and Spectacle 1631-1679*, Sterling Heights (MI), Harmony Park Press, 2010.

Harness 2006
Harness, Kelley. *Echoes of Women's Voices: Music, Art, and Female Patronage in Early Modern Florence*, Chicago, University of Chicago Press, 2006.

Hill 1997
Hill, John Walter. *Roman Monody, Cantata and Opera from the Circles around Cardinal Montalto*, Oxford, Clarendon Press, 1997.

Kersten 1997
Kersten, Fred. *Galileo and the Invention of Opera: Study in the Phenomenology of Consciousness*, Dordrecht, Kluwer, 1997.

Machamer 1998
Machamer, Peter. *The Cambridge Companion to Galileo*, Cambridge, Cambridge University Press, 1998.

Mamone 1981
Mamone, Sara. *Il teatro nella Firenze medicea. Problemi di storia dello spettacolo*, Milan, Mursia, 1981.

Mamone 2003
Ead. *Dei, Semidei, Uomini. Lo spettacolo a Firenze tra neoplatonismo e realtà borghese (xv-xvii secolo)*, Rome, Bulzoni, 2003.

Martín Sáez 2019

Martín Sáez, Daniel. 'Ópera y diplomacia entre Italia y Polonia. Del paso de Vladislao Vasa por Florencia hasta la inauguración del teatro de Varsovia', in: *Diplomacy and Aristocracy as Patrons of Music and Theatre in Europe of the Ancien Régime*, edited by Iskrena Yordanova and Francesco Cotticelli, Vienna, Hollitzer Verlag, 2019, pp. 37-62.

Martín Sáez 2020a

Id. 'La Edad Moderna a través de la metáfora del *theatrum mundi*. Cartografía, astronomía, ópera y filosofía de la historia', in: *Anales del Seminario de Historia de la Filosofía*, xxxvii/2 (2020), pp. 247-258.

Martín Sáez 2020b

Id. 'Los melodramas de Gabriello Chiabrera. Favole, favolette, intermedii y vegghie', in: *Revista de Musicología*, xliii/1 (2020), pp. 77-108.

Mersenne 1959

Mersenne, Marin. *Correspondance du P. Marin Mersenne, réligieux minime. 5: 1635*, edited by Cornelis de Waard, Paris, Éditions du Centre National de la Recherche Scientifique, 1959.

Murata 1981

Murata, Margaret. *Operas for the Papal Court, 1631-1668*, Ann Arbor (MI), UMI Research Press, 1981.

Ostrow 1996

Ostrow, Steven F. 'Cigoli's Immacolata and Galileo's Moon. Astronomy and the Virgin in Early Seicento Rome', in: *The Art Bulletin*, lxxviii/2 (1996), pp. 218-235.

Reeves 1996

Reeves, Eileen. *Painting the Heavens: Art and Science in the Age of Galileo*, Princeton (NJ), Princeton University Press, 1996.

Rowland 2003

Rowland, Wade. *Galileo's Mistake: A New Look at the Epic Confrontation between Galileo and the Church*, New York, Arcade Publishing, 2003.

Sartori 1990a

Sartori, Claudio. *I libretti italiani a stampa dalle origini al 1800. Catalogo analitico con 16 indici. 1: A-B*, Milan, Bertola & Locatelli, 1990.

Sartori 1990b

Id. *I libretti italiani a stampa dalle origini al 1800. Catalogo analitico con 16 indici. 2: C-D*, Milan, Bertola & Locatelli, 1990.

Sartori 1991a

Id. *I libretti italiani a stampa dalle origini al 1800. Catalogo analitico con 16 indici. 3: E-K*, Milano, Bertola & Locatelli, 1991.

Sartori 1991b

Id. *I libretti italiani a stampa dalle origini al 1800. Catalogo analitico con 16 indici. 4: L-Q*, Milan, Bertola & Locatelli, 1991.

Sartori 1992

Id. *I libretti italiani a stampa dalle origini al 1800. Catalogo analitico con 16 indici. 5: R-Z*, Milan, Bertola & Locatelli, 1992.

Saslow 1996

Saslow, James M. *The Medici Wedding of 1589: Florentine Festival as Theatrum Mundi*, New Haven (NJ), Yale University Press, 1996.

Speller 2008

Speller, Jules. *Galileo's Inquisition Trial Revisited*, Berlin, Peter Lang, 2008.

Tamburini 2003

Tamburini, Elena. 'Per uno studio documentario delle forme sceniche. I teatri dei Barberini e gli interventi berniniani', in: *Tragedie dell'onore nell'Europa barocca*, edited by Miriam Chiabò and Federico Doglio, Rome, Torre d'Orfeo, 2003.

Vega 2014

Vega, María José. 'El saber como conflicto. Curiosidad herética y saberes inmoderados en la temprana modernidad', in: *Saberes humanísticos*, edited by Cristoph Strosetzki, Pamplona, Universidad de Navarra, 2014, pp. 153-184.

Chapter Seven

«La grandezza del numero sonoro»: Canonic Techniques, Combinatorics, and Early Scientific Thought in Seventeenth-Century Rome

Jason Stoessel[*]
(University of New England, Armidale, Australia)
Denis Collins[**]
University of Queensland, Brisbane, Australia)

Some musical achievements in early seventeenth-century Rome are less well known in today's music histories. Not without justification, historians have concentrated on the development of musical genres like the late madrigal, the monody, the opera, the oratorio, and the new instrumental genres[1]. In part, this situation has arisen out of more recent historiographic biases like the search for the origins of modern music or the quest for musical innovation. This chapter focuses instead on the early seventeenth-century Roman school of contrapuntists who cultivated *artificioso* compositions. Although this learned style originated

*. This research was supported by the Australian Government through the Australian Research Council's Discovery Project funding scheme (project DP180100680). The views expressed herein are those of the authors and are not necessarily those of the Australian Government or Australian Research Council.

**. The first author is responsible for the introduction and first section of this chapter; the second author, for the second section, except the subsection on Valentini to which the first also contributed; both authors contributed equally to the conclusion.

1. See, for example, Bianconi 1987, Butt – Carter 2005.

in the previous century, this generation of composers extended it, principally though the techniques of canon. Here, we speak of canon in the sense of a set of compositional techniques for combining a melodic subject with itself either unaltered or transformed systematically by devices like transposition, contrary motion, retrograde, augmentation and diminution.

The seventeenth-century resurgence of canon at centres like Rome shares some of its motivating forces with Early Modern scientific thought. Without too much difficulty, ingenious canonic techniques can be connected to a renewed interest in combinatorics as a tool for the analysis of the natural world. Canon echoes significant changes in European intellectual society, a musical manifestation of the mathematicization of knowledge of the natural world in writings of new astronomers like Johannes Kepler and Galileo Galilei, wherein natural laws of the sensible world are modelled mathematically, and natural phenomena explored through rational induction. Motivated by an interest in combinatorics, early seventeenth-century composers sought to extend traditional modes of contrapuntal composition as far as possible within the 'laws' of music. Visual representation also plays a part in this playful experimentation, although this element will only be explored here in terms of the 'analytical' use of musical scores[2].

At the epistemological heart of innovative experimentation by early seventeenth-century Roman contrapuntists in musical composition is mechanism, that knowledge system which conceptualizes material structures or objects in terms of their interrelation or interconnection of component parts, and the increased use of mathematics to model those structures in nature. Through the lens of mechanism, historians can begin to understand firstly a renewed emphasis upon combinatorics in the early seventeenth century. Secondly, mechanism, when actualised by artificial machines that crowd seventeenth-century discourse, provides a pervasive thought paradigm that unifies an understanding of the mutually beneficial relationship between art and science in this period.

Various authors have recognised the role that music plays in early scientific thought and its techniques[3]. These scholars have concentrated on music theory, particularly about tuning and consonance. When music is evoked, it is usual to refer to novel genres like monody and early opera, or musical-instrument technologies[4]. The place of canon in this period of scientific revolutions at Rome and its participation in a mechanically motivated combinatorial impulse is the focus of this chapter. The first part will set out the case for recognizing the dual project of mechanism and combinatorics in novel music theory of the early seventeenth century, especially that of Marin Mersenne and the highly derivative Athanasius Kircher. Pointedly, the prominence of the canon in this discourse also extends to the writings of Roman contrapuntist

[2]. For an account of compositional self-representation in terms of early seventeenth-century visual culture at Rome, see Stoessel 2022.

[3]. See, for example, Palisca 1961, Palisca 1992, and Drake 1992.

[4]. See, for example, Coelho 1992 and Cohen 1992.

Pier Francesco Valentini. The second part of this chapter includes specific examples of music composed by Valentini and his fellow Roman composer Paolo Agostini, illustrating a combinatorial impulse not previously discussed in the literature.

Music and Science in Early Seventeenth-Century Rome

Contemporary Italian authors characterized the types of canons examined in this essay as *artificioso*[5]. More than denoting 'artifice' or 'artificiality', the term referred to the idea of composing a piece of music according to strict or at the very least rule-based techniques, such as counterpoint, and the composition itself[6]. Yet, this phenomenon of early seventeenth-century musical creativity is barely known today. Sergio Durante's now classic discussion of *artificioso* compositions (1987) usefully frames a problem in histories of early seventeenth-century music:

> Musical historiography has, on the whole, not paid much attention to these works, whether because of their apparent backwardness or because of their curious but sterile eccentricity. At a time when the most obvious innovations on the musical horizon seem to have been the rise of the *seconda prattica* and of opera, those phenomena not destined for a similar fortune may seem in respect rather opaque and unconnected with the evolving currents, and therefore tend to be neglected by a conception of history concerned with 'determining' events. Furthermore, the rather minimal aesthetic and practical value of these pieces has not helped in arousing interest today; they were in fact composed more to be looked at and studied than to be performed[7].

In addressing this 'teleological vision of history', Durante emphasizes the role of value and value judgement in the historiographic process[8]. That *artificioso* compositions continue to be held in such low esteem indicates the aesthetic gap that separates modern listeners from their seventeenth-century predecessors[9]. While fierce contemporary debates suggest that something like this gap existed in the seventeenth century, the aesthetic and teleological framework that shapes our own responses cannot have motivated it.

5. Durante 1987, p. 195, includes in the category of *artificioso* works polymorphous canons, canons on solmisation syllables, didactic or instructive canons, and strict *obblighi* on a cantus firmus and *oblighi* written for secular or religious purposes.

6. *Ibidem*, p. 196. Also see the useful discussion of the term *artificioso* in Michopoulou 2006, pp. 20-26.

7. Durante 1987, p. 199.

8. Also see the call to revisit the periodisation of seventeenth-century music by recognizing the significant paradigm shift in musical style that occurred in the middle of the century, in Silbiger 1992.

9. See similar judgements in Butt 2005, p. 40.

A possible solution requires shifting our frame of reference from purely musical criteria to an analysis of the intersection of music and early scientific thought. Although the idea of a Scientific Revolution in the seventeenth-century has been debated by historians of science[10], one might still speak of a series of scientific revolutions or at the very least paradigm shifts in systems of knowledge, often arising out of previously dominant systems and transformed by new approaches and interdisciplinary interactions[11]. New applications of mathematics and the pursuit of increasing precise measurement of physical phenomena, often in the conduct of experiments, are just two of the commonly recognised developments in this period. Christophorus Clavius, Professor of Mathematics in the Jesuit Collegio Romano in Rome and the workhorse behind the late sixteenth-century reforms of the Gregorian Calendar, for example, played a significant role in these early transformations by the dissemination of his textbooks on mixed mathematics[12]. Mixed mathematics attempted to synthesize Aristotelian natural philosophy with mathematics. Yet, due to the Aristotelian foundations, mixed mathematics did not play a conceptual role in the geometric and arithmetic (Euclidean) modelling of a heliocentric solar system in the work of Johannes Kepler and Galileo Galilei. The influence of the pursuit of relatively precise measurements and experiments in the tunings of musical instruments by the composer and music theorist Vincenzo Galilei on his famous son has also not escaped the attention of historians of music and science[13]. Galileo's discoveries illustrate the confluence of mathematicization of natural knowledge systems and the more precise measurement and systematic observation of natural phenomena[14]. Although he was not the first to contribute to the transformation of natural philosophy into an early form of physics, Galileo provided an elegant statement concerning the value of mathematics in demonstrating the sensible universe:

> Philosophy is written in this all-encompassing book that is constantly open before our eyes, that is the universe; but it cannot be understood unless one first learns to understand the language and knows the characters in which it is written. It is written in mathematical language, and its characters are triangles, circles, and other geometrical figures; without these it is humanly impossible to understand a word of it, and one wanders around pointlessly in a dark labyrinth[15].

---

[10]. See, for example, Shapin 1996.

[11]. Cohen 2015, p. 94.

[12]. See, for example, Dear 1984, pp. 188-189, and Cohen 2015, pp. 92-93.

[13]. See, for example, Palisca 1961, Palisca 1992, and Cohen 1984.

[14]. Literature on Galileo is vast but see Redondi 1988, Freedberg 2002, and Heilbron 2010.

[15]. Galileo, *Il saggiatore* (Rome, 1623), p. 25: «La Filosofia è scritta in questo grandissimo libro, che continuamente ci stà aperto innanzi à gli occhi (io dico l'vniverso) ma non si può intendere se prima non s'impara à intender la lingua, e conoscer i caratteri, ne' quali è scritto. Egli è scritto in lingua matematica, e i caratteri son triangoli, cerchi, & altri figure Geometriche, senza i quali mezi è impossibile à intenderne vmanamente parola;

The second, and possibly fundamental, concept that impinges on some of the authors discussed here, especially Kircher but also Mersenne (upon whose music treatises Kircher based much of his writings about music) is mechanicism. Mechanism, in the sense of the study of natural and artificial 'machines' is an ancient mode of human thought[16]. To clarify, the term mechanism is used here not in the sense of modern mechanical or corpuscular philosophy that is first clearly demonstrated in Robert Boyle's *The Origine of Formes and Qualities* (Oxford, 1666)[17]. Rather, mechanism is a system of knowledge that stretches back to Antiquity that undergoes multiple transformations that provide the eventual basis of mechanical philosophy in Early Modern science. We can only touch upon the history of mechanism by pointing out the most salient developments that occurred in the late sixteenth and early seventeenth centuries.

Suffice it to observe here that the impetus for a renewed interest in mechanism arose from humanism and its project to recover ancient authors. Unlike the Aristotelian corpus, which was recovered and translated into Latin in the thirteenth century to provide the authority for scholastic thought in the following century, much ancient Greek philosophy from the Alexandrian School, such as, for example, Ptolemy's *Mathematical Arrangement* (later known as the *Almagest*), Euclid's *Elements* and Archimedes's writings, had remained unavailable to European authors until the advent of the humanist philological project of the fifteenth and sixteenth centuries. The recovery of this body of philosophical writings, which included mathematical models of the solar system and descriptions of ancient machines, not to mention thought that contradicted the prevailing Aristotelian paradigm, provided a foundation for early scientific thought[18].

Marin Mersenne was a central figure in early seventeenth-century European intellectual culture whose writings from the 1630s witness his conversion from a previously vociferous adherent to the authority of scholastic Aristotelian thought to a proponent for the synthesis of humanist probabilism and mathematics[19]. Again, humanism played a role in the development of humanist probabilism, or the view that truth might be arrived at by rhetorical argumentation

senza questi è vn'aggirarsi vanamente per vn'oscuro laberinto». English translation from Galileo 2008, p. 183. Nicolaus Copernicus preceded Galileo in proposing the mathematicization of the cosmos in his *De revolutionibus orbium coelestium* (Nuremberg, 1543), Book i, Chapter 10, on the order of the celestial spheres; see Gozza 2002.

16. Dijksterhuis 1950, Dijksterhuis 1961, Westfall 1977, and Garber – Roux 2013. The English edition of Dijksterhuis's book translates the original Dutch terms '*mechanistisch*' and '*mechanicisme*' as 'mechanistic' and 'mechanicism' respectively. In more recent literature, 'mechanical' and 'mechanism' have instead prevailed, despite the fact that they are the English equivalents in meaning to the colloquial Dutch words '*mechanisch*' and '*mechanisme*' that Dijksterhuis sought to avoid.

17. Garber 2013, p. 9.

18. See, for example, Cohen 2015, pp. 79-93.

19. For a discussion of how Mersenne's transformation represents a rejection of philosophical Pyrrhonism, see Dear 1984 and in more detail Dear 1988.

or inductive discourse rather than scholastic demonstration. Motivated by Cicero's theory of oratory, humanist dialectics, beginning with Lorenzo Valla (c. 1406-1457) and continuing with Rudolphus Agricola (1444-1485) and Petrus Fonseca (1528-1599), placed greater emphasis upon cultivating techniques of rhetorical persuasion[20]. In rejecting Aristotle's theory of qualities, Mersenne instead looked to the material relations between entities and components to determine the laws of nature. Pointedly, Mersenne provided the clearest demonstration of his newfound mechanical thought in his 1500-page music treatise, the *Harmonie universelle* (Paris, 1636)[21]. One historian of music theory has succinctly termed the *Harmonie universelle* «a text of musical mechanics», noting that it also included discussions of several problems of pure mechanics recently tackled by Galileo[22].

By reconceptualizing music as a mechanical system with its own natural laws (chiefly the laws of musical consonance) in his *Harmonie universelle*, Mersenne could use combinatorics (*ars combinatoria*) to explore the relationships between music's parts. In doing so, Mersenne drew on the innovations of earlier Italian or Italy-based mathematicians who transformed the science of combinatorics from a linguistic to a mathematical tool[23]. Understood as a system for exploring every possible combination of a series of elements, European combinatorics arose earlier in the thirteenth-century writings of Ramon Llull as an artificial system for exploring the permutations of terms to produce truth propositions[24]. Clavius's treatment of mathematical combinatorics in his *In sphaeram Joannis de Sacro Bosco commentarius* (Rome, 1570), a commentary on *De sphaera mundi* of Johannes de Sacrobosco (written c. 1230), would have been well known to Mersenne from his Jesuit schooling at La Flèche. Through a synthesis of mechanism and combinatorics, Mersenne (in the words of Penelope Gouk, 2005):

> [...] recognised the close links between music's powers and those of speech (especially oratory), compared the structure of musical compositions with the structure of language, and linked both of these to the mathematics of combinatorial analysis. Mersenne used the same combinatorial method in trying to discover the best possible musical composition as in attempting to develop a universal system of rational communication. The underlying assumption was that both music and speech can be broken down into a finite number of elements that can be reconstituted in a variety of ways. He also thought that the elements of speech could be explored through the imitation of the voice by musical instruments, a direct consequence of the assumption that the natural mechanisms of the body (including those responsible for voice, hearing and perception) operate under the same laws as

[20]. Dear 1984, pp. 191-192.

[21]. Lenoble 1971, pp. 370-371.

[22]. Christensen 2013, p. 61.

[23]. Knobloch 2013, pp. 126-132.

[24]. Knuth 2013, pp. 14-18; Knobloch 2013; McKay 2012, pp. 246-256.

Ill. 7.1ab: Cornelis Bloemaert (ii), *Portrait of Athanasius Kircher* (engraving, 28 × 18 cm). Copy NL-Ar, RP-P-1927-677; Athanasius Kircher, *Musurgia universalis, Tomus i* (Rome, 1650), title page. Copy D-Mbs.

> artificial mechanisms and instruments. Of course, this sympathy between bodily and musical instruments was a central tenet of natural magic, but Mersenne distanced himself from this tradition and presented his experimental method as the better alternative[25].

Others have situated the emergence of combinatorics as a universal tool for analysing natural phenomena, including music, alongside Cartesian mechanical philosophy in the second half of the seventeenth century[26], also noting strong precedents in music theory, especially in Book viii of Athanasius Kircher's *Musurgia universalis* (Rome, 1650; Ill. 7.1b) concerning automatic music composition. Kircher, a German Jesuit in seventeenth-century Rome and a central figure in seventeenth-century intellectual history, was an extraordinary polymath who wrote on topics as diverse as the Bible, China, fossils, hieroglyphs, magnetism, music, and

[25]. Gouk 2005, p. 146.

[26]. Kassler 2001.

volcanos. A prolific correspondent, he used his extensive network to collect information on topics from across the globe. Although Kircher clearly got much 'wrong' in the eyes of the next generation of modern scientific thinkers, he was part of an early phase of scientific thought that aimed less at the systematisation of a particular field of knowledge than at bringing together all knowledge into a vast universal system or 'encyclopedia'. Or, as Kircher's biographer Paul Findlen (2004) has stated, «[h]e belonged to an era that combined rather than divided, that took delight in finding unlikely connections in the service of a grand unified theory of absolutely everything»[27].

Kircher's interest in a universal combinatorics and artificial machines is repeatedly evidenced in this letters and published works. Combinatorics was central to his «Universal Lullian-Kircherian Art of Knowing and Examining»[28]. He proposed a linguistic engine in the Llullist fashion in his *Polygraphia nova et universalis* (Rome, 1663), except in this case his system purported to provide a universal translation system across various languages. For his *Musurgia universalis*, Kircher clearly relied on — to say the least — several earlier authors to bring about a synthesis and extension of a theory of musical combinatorics. Foremost among these was Mersenne and his *Harmonie universelle*. Mersenne recorded in a letter that, when he visited Kircher in Rome in 1644, the Jesuit «devoured» a loaned copy of *Harmonie universelle* in four days and soon proposed his own book on universal music[29]. Mersenne's exploration of the permutations and combinations of melodies arising from the notes of the natural hexachord seems to have in part inspired Kircher's theorizing of a machine for the automatic composition of music, the 'Arca musarithmica' (Music-arithmetical box) (ILL. 7.2B)[30].

Concerning counterpoint, Kircher turned in Book V of his *Musurgia universalis* to the writings of Silverio Picerli, translating a part of the latter's *Specchio secondo di musica* (Naples, 1631) from the original Italian into Latin[31]. In the chapter «De noua & admirabili Contrapuncti per varias combinationes instituendi ratione & methodo» (A New and Remarkable Approach and Method for Making Varied Combinations of Counterpoint; Book V, Chapter 18, pp. 328-365), Kircher reproduced Picerli's method of what is now commonplace for teaching double or invertible counterpoint. The complementary inversion of each interval of a subject and its counterpoint is calculated numerically, thereby revealing that some intervals are more suitable for inversion than others. Picerli had included a numerical matrix to aid the composition of invertible counterpoint (p. 63), which Kircher borrowed and labeled in Latin as a *tabula mirifica* (marvellous table; ILL. 7.2A). The result is a near exhaustive method for contrapuntal

27. FINDLEN 2004, p. 7.

28. *Ibidem*, p. 5.

29. Mersenne to Ismaël Boulliaud, 16 January 1645 (MERSENNE 1977, p. 320). See FINDLEN 2004, p. 23.

30. KNOBLOCH 1979, p. 267, and MCKAY 2012, pp. 240 and 254.

31. GERBINO 1995B, pp. 225-228.

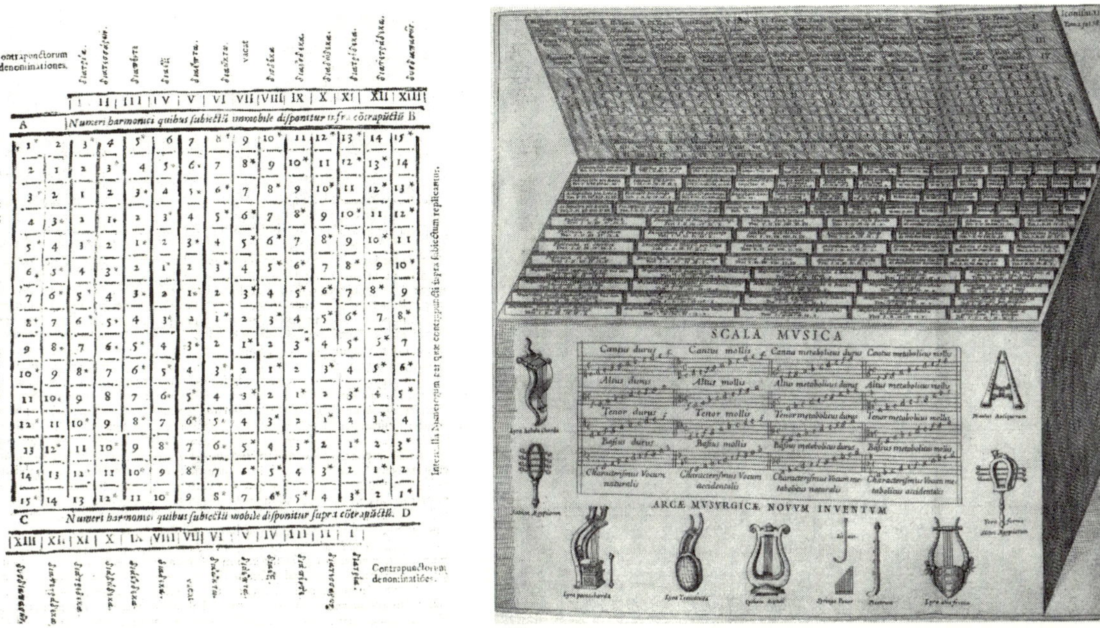

Ill. 7.2ab: 'Tabula mirifica' (marvellous table) and 'Arca musarithmica' (musico-arithmetical box) from Athanasius Kircher, *Musurgia universalis* (Rome, 1650), Tomus i, p. 363, and Tomus ii, p. 188. Copy US-CHH.

composition that sets out all possible invertible combinations of a subject and its counterpoint. Although Kircher briefly acknowledged his source in the *Musurgia*, Eric Bianchi (2017) has aptly remarked that «this fig leaf hardly covers a sixty-page proboscis»[32]. Bianchi, nonetheless, also reads Kircher's borrowing in light of differing attitudes to what today we might call plagiarism.

Kircher extends Book v of the *Musurgia universalis* with a discussion of two extraordinary canons written by the Roman composer, music theorist and noble gentleman Pier Francesco Valentini (1586-1654)[33]. On one hand, these canons provide examples of strict or *observata* composition governed by the so-called natural laws of musical consonance (and implicitly its conceptual antithesis, dissonance). They also embody a musical combinatorics in which one subject can be combined in multiplex ways with itself and its strict transformations. In this sense, the canons selected by Kircher exemplify a specialised and ideal type of musical combinatorics, a theory and practice for the near limitless possibility for combining musical subjects according to the 'laws' of consonance and dissonance[34].

These include Valentini's polymorphous canon *Illos tuos misericordes oculos*. Valentini first published this canon in 1629 with a 155-page demonstration of its 2000 *resolutiones* of the

32. Bianchi 2017, p. 85.

33. On Valentini's life and works, see Cortesi 1983; Gerbino 1995a.

34. McKay 2012, p. 18, suggests that «because of the continued influence of medieval categories of knowledge that placed music theory in the sciences, the new mathematics of combinatorics was first expressed in music».

subject in its prime, contrary, retrograde and retrograde contrary motion as *Canone con le sue resolutioni in piu di duemilia modi* (Rome, 1629). In his preface, Valentini states that his canon invokes the «sonorous number», the true object of musical understanding for a virtuous man (*huomo virtuoso*):

> Of this most honoured science (that is, music) — the fulfilment of virtue in virtuous men — I, without making a profession of it, since boyhood have been so graced by its ornament, to occupy myself [...] in its sweetest study [...] delighting in the speculative diversion that, in the opinion of the Ancients, there was a universal science [...] which [...] encompassed all other sciences. They judged this [universal science] to be Music, not only for the other sciences being proportionally ordered and connected with each other and for one corresponding to the other in same way that many different voices concord with each other in an ensemble (*concerto*). [...] Music is very vast and the grandeur of the sonorous number that is its object is immense, and the counterpoints are most profound. Hence, considering this grandeur, neither to entice nor to give pleasure to hearing, but to assay the profundity of Music in the sonority of human voices, I have composed the canon placed in the frontispiece of this book[35].

Valentini's appeal to Antiquity illustrates the ongoing influence of humanism and appeal of classical erudition in early seventeenth-century Roman musical culture[36]. His juxtaposing of the «most profound counterpoints» with the grandeur of music's «sonorous number», however, reveals an admixture of classical, medieval, and more recent musical thought. «Sonorous number» can be traced to the music theory of Gioseffo Zarlino with which Valentini demonstrates a strong familiarity elsewhere in his writings about music. In Book I, Chapter 19, of *Le istitutioni harmoniche* (1558), Zarlino writes that sonorous number «is artificially revealed in a sounding body, such as in any string, which, containing the ratio of any number in its parts, makes clear the quantity [pitch] of the sounds that it produces, and the quantity of the voice

---

35. Pier Francesco Valentini, *Canone con le sue resolutioni* (Rome, 1629), p. iii: «Di questa honoratissima scienza [musica], compimento di virtù ne'gl'huomini virtuosi, Io senza farne professione, insino da'gl'anni puerili ne so stato vago si per ornamento, come anco per occuparmi fuggendo l'otio ne' dolcissimi studi di essa, & oltre la pratica dilettandomi della speculatiua ritrovo, che fu opinione de'gl'Antichi, che vi fusse vna scienza vniversale chiamata Enciclopedia la quale à guisa d'vn circulo comprendesse in se tutte altre scienze; Questa da essi fu giudicata esser la Musica non solo per esser l'altre scienze proportionalmente ordinate, & connesse insieme, & per corrispondersi l'vna con l'altra in quella maniera che in vn concerto molte voci dispari insieme concordano [...] Non dirò altro; Amplissima è la Musica, & immensa la grandezza del numero sonoro obietto di essa, & profundissimi li Contrapunti. Di qui è, che io considerando questa grandezza non con animo d'allettare, né di dar gusto all'vdito, mà per dare vn saggio della profundità della Musica nella sonorità delle voci humano, ho composto il presente canone posto nel Frontespitio di questo libro».

36. On this phenomenon, see Barker 2015.

parts, referring or applying such sounds to these voice parts»[37]. In this sense, sonorous number as music's object is predicated on the Pythagorean concept of immanent number. In his 'Discorso particolare intorno alla diversità della forme del diapason' (Special Discourse Concerning the Diversity of the Ratios of the Diapason), Vincenzo Galilei was critical of Zarlino's concept of number itself being sonorous, although Galilei himself eagerly reduced phenomena in music to numerical, that is proportional, relationships[38]. Yet, by positing sonorous number, rather than the sense of sound alone, as the object of music, Valentini seems to be lashing himself firmly to Zarlino's mast while also acknowledging the siren call of more recent thought on the experiential and experimental nature of music and musical invention.

Valentini's regard for music as a universal system for understanding all sciences also illustrates a link to contemporary 'encyclopedic' approaches to knowledge. Just as combinatorics lay at the heart of Kircher's approach, firstly to musical knowledge and subsequently to other human knowledge, it was also central to the music of Valentini and several contemporary contrapuntists[39]. For Valentini, sonorous number, most clearly manifest in counterpoint, is the key to this combinatorial encyclopedic project.

More generally, systematic, often mathematicized approaches to music composition could not have occurred without the mathematicization of scientific thought. To be clear, we are not suggesting that canonic musical thought was scientific. Rather, it had embraced mathematical demonstration, resorting to its own system of musical logic principally based on musical consonance, albeit less incontrovertible — at least in the modern mind — than the logic of mathematics. In the pursuit of automated composition, composer-theorists turned to their own symbolic system for demonstrating that certain combinations of canonic subjects were possible. This symbolic system is musical notation, and its 'graph' is the musical score.

Canonic Techniques and Combinatorics: Case Studies from Early Seventeenth-Century Rome

We now present case studies based on the canonic works of two Roman composers, Pier Francesco Valentini and Paolo Agostini (1583-1629). Both were prodigious in their canonic output. Each approached the canonic art from a different direction: Valentini as a member of

[37]. Zarlino, *Le istitutioni harmoniche* (1558), p. 29: «che'l Numero sonore è Numero relato alle voci, & ai soni; il quale si ritroua artificiosamente in vn corpo sonoro, si come in alcuna corda, la qual riceuendo la ragione di alcun numero nelle sue parti, ne fa certi della quantità del suono produtto da essa, & della quantità delle voci, riferendo, ouero applicando essi suoni ad essi voci». Translation, adapted, from Corwin 2008, p. 313.

[38]. Palisca 1989, p. 183, Palisca 1992, p. 149.

[39]. Gerbino 1995a, Klotz 2006, pp. 15-48. More broadly on seventeenth-century encyclopedism and the influence of the *ars combinatoria*, see Sabaino 2013.

the Roman aristocracy and Agostini as a career musician at several Roman institutions. Yet, both are representative of preoccupations amongst a great many contemporary musicians in Rome and throughout Italy with canonic techniques. The surviving sources for this activity have been thoroughly documented in a doctoral dissertation by Michael Lamla (2003)[40]. While Lamla and others have made clear that composers of the time devoted considerable effort to canonic composition, there has been little assessment of specific structural features of the surviving output. The following discussion suggests ways in which a combinatorial impulse found expression in the near-obsessive engagement with canonic techniques amongst Roman composers of the time.

In the corpus of surviving canonic compositions from early seventeenth-century Rome, two techniques stand out for their frequency of occurrence and for their suitability as demonstrations of combinatorial potential. One of these, the polymorphous canon, was known to earlier generations of composers but rarely cultivated before the late sixteenth century. A polymorphous canon can be defined as an original work from which many other versions of the work can be derived. Each of the derivative versions involves alteration to a specific feature (or variable) of the original version. These features may be the interval of imitation, the temporal offset between the canonic voices or the employment of a technique such as imitation by contrary motion, augmentation or retrograde.

The second canonic technique that can be identified as characteristic of Roman sources of the period is what we term the multiple-technique canon. This is where a single work simultaneously employs two or more canonic techniques. For instance, in addition to imitation at a specified interval and temporal offset, the work may require circular repetitions up or down a whole tone — the so-called *canon per tonos*[41]. Other works may be double canons, where each constituent canon involves two or more techniques. In contrast to the polymorphous canon, all the conditions for each canonic technique are combined in just one presentation of a multiple-technique canon. Rarely encountered in sources before the late sixteenth century, multiple-technique canons are abundant in the works of Valentini, Agostini and other composers active in the first decades of the seventeenth century. While Valentini included this canonic technique among the arsenal of compositional procedures presented in his extensive manuscript collections, it was his inventiveness in proposing polymorphous canons with staggeringly large numbers of derivative versions that drew admiration and commentary from Kircher[42].

---

40. Lamla 2003.

41. A *canon per tonos* involves repetition of the entire work up or down by a predetermined interval, usually a second, until the canonic *dux* arrives at a note an octave from the original starting note.

42. For a great variety of canonic techniques, including some polymorphous and multiple-technique canons, see Pier Francesco Valentini, 'Canoni di diversi studi' (V-CVbav, MS Barb. Lat. 4428). The contents of this manuscript have been reviewed, but without a detailed consideration of the specific canonic techniques employed, in Gerbino 1995a and Wuidar 2008.

Kircher, Valentini and the Polymorphous Canon

Athanasius Kircher appears to have been the first writer on music to use a term for the polymorphous canon. In Book v of his *Musurgia universalis*, Chapter 19 («De secretiori canonum methodo») is devoted to two canons by Valentini, the *Illos tuos misericordes oculos* and the *Nodum Salomonis*. Kircher quoted the melody for the first of these works and provided it with the label «Canon Polymorphus». (See Ill. 7.3, where this melody is presented on the title page of Valentini's publication.) The transformations of the melody in inversion, retrograde and retrograde inversion were also quoted by Kircher, who enthused that:

> And this is that marvellous canon, variable in a thousand, even, an infinite number of ways; and it contains ten tempora, one rest, and seventeen notes. First [of all], it can be sung in more than two thousand ways not only by two, three and four voices, but it can also be sung by five voices in more than two thousand ways, and by six voices in more than three thousand ways. Further, by combining each [melody] with each [other melody], it duly multiplies itself into an infinite variety, all of which we would demonstrate gladly and extensively at this point except that the sheer quantity of the combinations would prohibit this[43].

In his *resolutiones* of *Illos tuos*, Valentini moved systematically through the combination of every form of the subject with itself and its transformation, first for two voices, then for three, four, five and six voices. The collection opens with two-part canons based on the melody given on the title page, with following sections of the book systematically exploring canonic possibilities for each of the other three versions of the melody (that is, in inversion, retrograde and retrograde inversion). The range of possibilities also includes examples where one of the versions of the melody is combined with one or more of the other versions. In other examples, voice doublings in parallel thirds yield yet further possibilities. The demonstration of five- and six-voice resolutions is truncated, with Valentini concluding that there are many more than could be provided in the available space. In setting out almost every possible solution for up to six voices in score, Valentini was working in the spirit of early empirical science, providing a demonstration or proof of the near infinite possibilities of his musical subject for combining with itself. Valentini was not using a system of mathematical symbols, geometric diagrams, or the principles of induction, but instead musical notation, scores, and contrapuntal verification.

[43]. Athanasius Kircher, *Musurgia universalis* (Rome, 1650), Tomus i, p. 402: «Atque hic est mirificus ille Canon, mille imò infinitis modis variabilis, continet is decem tempora, unam pausam, & septemdecim notas. Hic primò potest pluribus, quàm bis mille modis cantari, non solùm duabus, tribus, quatuor, sed & quinque vocibus pluries, quàm bis mille modis, & sex vocibus pluries quàm ter millies cantari potest, imò singula cum singulis combinando, meritò is in infinitam varietatem sese extendit, quae omnia fusè hoc loco lubenter extenderemus, nisi moles combiniationum id prohiberet».

Ill. 7.3: Pier Francesco Valentini, *Canone con le sue resolutioni in piu di duemilia modi* (Rome, 1629), title page. Copy I-Bc, CC.74.1.

Kircher shared with his readers his enthusiasm for another of Valentini's canons, the *Nodum Salomonis* (or *Nodo di Salamone*, The Knot of Solomon), published as *Canone nel Nodo di Salomone a novanta sei voci con le sue resolutioni* (Rome, 1631) and *Resolutione seconda del canone nel Nodo di Salomone a cinque cento dodici voci in cento venti otto chori* (Rome, 1631)[44]. He referred to this work as a «Musicus Labyrinthus». This work is a quadruple canon, with the melody of each canonic *dux* based on the notes of the triad built on G (Ill. 7.4). Each *dux* can be imitated by twenty-three *comites* at different time intervals, some of which take the *dux* melody in augmented or diminished note values. After reproducing Valentini's instructions for

[44]. RISM A/i, V 124 and 125.

realizing this polymorphous canon in ninety-six voices, Kircher identified the further possibility of adding choirs up to the grand number of 144,000 parts, a figure corresponding to the number of heaven's elect in the Book of Revelation. This is achieved by varying the notes of each *dux* according to other augmentation values. Kircher further remarked that there need be no limit to the number of voices that can sing this canon:

> [...] and so by augmenting the four said longas to an infinite value, an infinite number of choirs and voices shall also follow, so that hence the huge and unimaginable power of combination underlying the music may become evident[45].

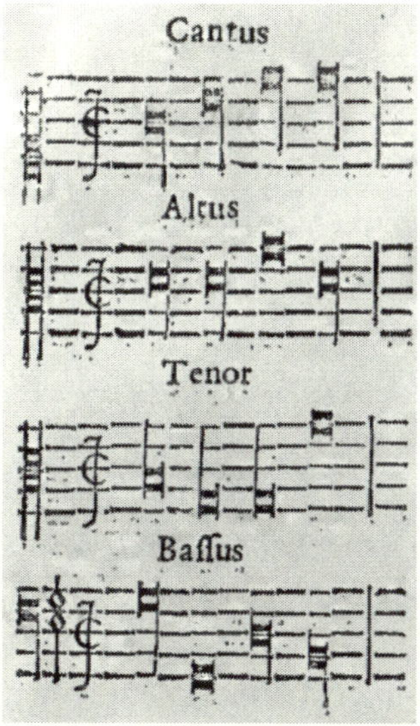

Ill. 7.4: Pier Francesco Valentini, canonic *duces* of the *Nodo di Salomone* quadruple canon, as quoted in Athanasius Kircher, *Musurgia universalis* (Rome, 1650), Tomus i, p. 408. Copy D-Mbs.

The published demonstration by Valentini of his 96-voice canon is partial since the possibility of publishing an illustrative score was probably outside the realms of possibility. Yet the partial demonstration might serve as the basis for the ultimate 'proof of concept' through performance, wherein the judgement of consonance would be exercised not by the eye but by the ear — if one could find ninety-six voices.

Verification through performance seems to have occurred for other canons composed by Valentini. For instance, Valentini himself wrote out parts for his *Gaude Virgo gloriosa* and two versions of his *Cantabimus canticum novum* into small booklets of vocal parts, which also

[45]. Athanasius Kircher, *Musurgia universalis* (Rome, 1650), Tomus i, p. 414: «& sic augmentando 4 dictas longas in infinitum valorem, infinita quoq[ue] consequetur chororum, vocumque multitudo, ut vel hinc appareat maxima, & incomprehensibilis vis combinationis sub musica latens». Kircher continued with a remark that the combined populations of Italy, Spain, France, and Germany would not be enough to sing such a canon.

included some of his non-canonic compositions. All were bound together as MS 369, part of a posthumously assembled collection of manuscripts and prints of Valentini's music and music theory now in the Archivio di Stato of Rome. Giuseppe Gerbino (1995) has suggested that these parts might have been used in the theatre[46], but it seems equally plausible that they were used in a learned academy given the mixture of settings to Latin and vernacular texts that make up its contents. At any rate, they seem to indicate Valentini's interest in the role of the discerning ear in judging the natural order of the sonorous numbers.

Paolo Agostini and the Combinatorial Impulse in Roman Sacred Music

For the remaining case studies, we turn to the canonic compositions of Paolo Agostini, «one of the major representatives of canonic art in Rome in the years 1620-30»[47]. Agostini was employed at several major Roman churches, starting in 1615 at Santa Maria in Trastevere, followed by positions at Santissima Trinità dei Pellegrini, San Lorenzo in Damaso and culminating in his appointment as *maestro di cappella* at the Basilica di San Pietro, a role he held from 1626 to his death in 1629. Agostini published six volumes of music in score in 1627, all printed by Giovanni Battista Robletti:

- *Spartitura delle messe del primo libro* (Rome, 1627)[48].
- *Spartitura del secondo libro delle messe e motetti a quattro voci* (Rome, 1627)[49].
- *Spartitura del terzo libro della messa sine nomine, a quattro* (Rome, 1627)[50].
- *Libro quarto delle messe in spartitura* (Rome, 1627)[51].
- *Spartitura della messa et motetto Benedicam Dominum ad canones, a quattro voci* (Rome, 1627)[52].
- *Partitura delle messe et motetti, a quattro e cinque voci* (Rome, 1627)[53].

These publications comprise settings of the Mass Ordinary, motets and short two-, three- and four-voice counterpoints. Much of the material in these volumes is canonic and reflects Agostini's contemporary reputation as one of the leading contrapuntists of the time[54]. A

46. Gerbino 1995a.

47. *Grove2* 2001, article 'Agostini, Paolo' by Klaus Fischer.

48. Agostini, *Spartitura, Primo libro*: RISM A/i A 413.

49. Agostini, *Spartitura, Secondo libro*: RISM A/i A 414.

50. Agostini, *Spartitura, Terzo libro*: RISM A/i A 416.

51. Agostini, *Libro quarto*: RISM A/i A 415.

52. Agostini, *Spartitura della messa*: RISM A/i A 417.

53. Agostini, *Partitura delle messe*: RISM A/i A 418.

54. For an overview of Agostini's canonic output, see Collins 2018. Details of Agostini's career can be found in Dixon 1981 and Miller 1998, while specific attention to Agostini's tenure at the Saint Peter's is in Rostirolla

posthumously published volume in choirbook format, *Missarum liber posthumus* (Rome, 1630)[55], contains the Mass settings and motets published in 1627 plus two further canonic settings of the Mass Ordinary that were not published previously. This volume does not contain any of the short counterpoints published in the 1627 volumes.

Agostini evidently took pride in his contrapuntal accomplishments: each volume contains a dedication to a member of one of the well-known churches and institutions in Rome where Agostini had found employment earlier in his career. The dedications are followed by instructions for the particularly complex contrapuntal procedures found in certain Mass sections[56]. For instance, after explaining how the canonic Agnus Dei section from his *Messa Si bona suscepimus* (in the *Libro quarto*) could have fourteen solutions involving retrograde motion, Agostini added the comment: «Various canons of all kinds have been made for the curious»[57]. This signals Agostini's preoccupation with providing demonstrations or proofs of the combinatorial potentials of his musical materials. Unusually for the time, the Masses and motets published in 1627 were printed in score format; any partbooks that may have existed for these works are now unfortunately lost. The short counterpoints are notated sometimes in score format, although many of the canons are in 'closed format', that is, only the *dux* is notated, on a single staff. When canons are presented in score format, they often, but not always, include the resolutions of the canonic *dux* on separate staves. Those presented in closed format include signs or letters above certain notes of the *dux* to assist with the resolution of the following canonic voices (the *comites*). That Agostini published most of the materials in the six volumes from 1627 in score points to his choice of this graphical format as a means of verification of his inventiveness with canonic and other contrapuntal techniques. However, retaining the older way of notating canons in closed format allowed Agostini to demonstrate another combinatorial aspect of his creative outlook. Many canons were presented in both formats in the published volumes.

Agostini's music offers a wealth of contrapuntal ingenuity. Of the ten Mass settings published in the 1627 and 1630 volumes, five employ canonic imitation throughout. The remaining five all contain canonic settings of the Agnus Dei, with some Masses also including canonic writing in other sections. Agostini's music surveys the full range of canonic techniques that had been developed during the previous three centuries: imitation at intervals from unison to octave for two or more voices, use of procedures such as contrary motion, retrograde or invertible counterpoint, and simultaneous presentation of two or more canons at the same time, that is, double or triple canons. Even in the noncanonic Mass sections, Agostini employed compositional

2017a. Agostini's music did not, unfortunately, attract the attention of Athanasius Kircher, notwithstanding its many demonstrations of how different contrapuntal combinations can be derived from the same melodic materials.

55. Agostini, *Missarum liber*: RISM A/i A 419

56. Some scholars have viewed the contents of these volumes as comprising a survey by Agostini of his compositional career. See, for example, Dixon 1991, p. 314.

57. Agostini, *Libro quarto* (1627), p. 3: «S'e fatto alcuni Canoni ad ogni facciata per li curiosi».

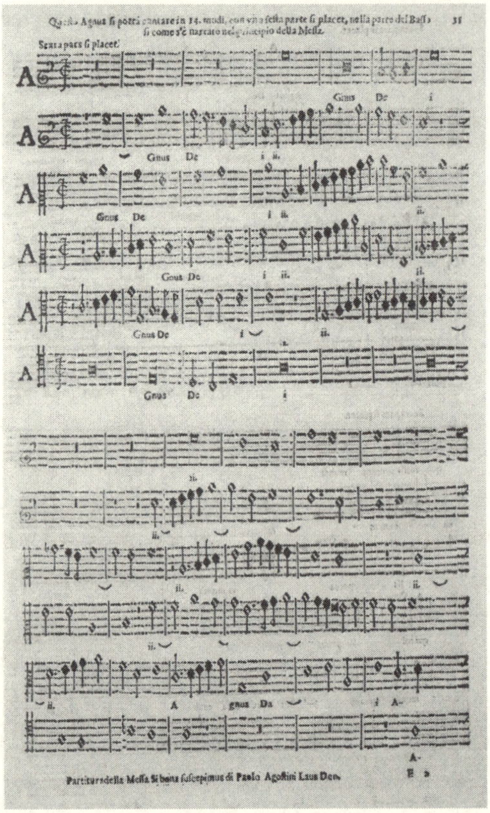

Ill. 7.5: Paolo Agostini, *Messa Si bona suscepimus*, Agnus Dei ii, bars 1-7, a canon with fourteen solutions, from the *Libro quarto delle messe in spartitura* (Rome, 1627), p. 35. Copy I-Bc, V.39.

techniques with long pedigrees: two Masses use the hard hexachord as a cantus firmus or as a freely elaborated motive in all sections; other Mass sections use freely imitative counterpoint, sometimes with an optional additional part leading to two possible versions of a section.

Unlike the polymorphous canons by Valentini, Agostini's canons involve a relatively small number of resolutions. To some extent, this limitation may reflect the liturgical context for his music, where considerations of time and effort may have precluded more extensive polymorphous explorations. The Agnus Dei ii from the *Messa Si bona suscepimus* (mentioned above) contains the most resolutions of any of his polymorphous compositions. The Agnus Dei ii was a site of great contrapuntal experimentation throughout the sixteenth and seventeenth centuries. Early seventeenth-century Roman composers, including Agostini, participated in this tradition through very ingenious canonic settings. Ill. 7.5, for six voices plus continuo, is from the opening of the Agnus Dei ii of the *Messa Si bona suscepimus*[58]. Fourteen versions can be derived from this example through manipulation of the bass line: this voice can proceed

[58]. *Ibidem*, pp. 35-37. The continuo part is located separately at the end of the volume.

from beginning to end or from end to beginning; it may also start from either the beginning or end and turn back in retrograde when it reaches the halfway point. An optional sixth voice is in canon at the twelfth above and follows these retrograde procedures. The other voices remain unaltered in every combination[59].

Other instances of polymorphous writing across Agostini's output typically involve a smaller number of resolutions. By way of example, there are two closely related versions of the Benedictus for three voices from the *Messa Pro feriis, ad canones*. This Mass is in the *Partitura delle messe et motetti*. In one version (Ill. 7.6a), the *dux* is followed by *comites* at the fourth and the octave above; in the second version (Ill. 7.6b), the *dux* is followed by *comites* at the fifth and the octave above. The first version of this Benedictus requires a temporal offset of one semibreve between the *dux* and the first *comes*, while the second version involves a breve between the entries of these two voices. The time distance between the *dux* and the second *comes* is not changed between versions.

Ill. 7.6ab: Paolo Agostini, *Messa Pro feriis, ad canones*, Benedictus, bars 1-8, a canon with two solutions, from the *Partitura delle messe et motetti* (Rome, 1627), pp. 5 (first solution) and 6 (second solution). Copy B-Br, Fétis 1668 C 1.

[59]. A full transcription of this Mass section can be found in Collins 2019, pp. 205-206.

Ex. 7.1: Paolo Agostini, *Messa Pro vigiliis ac feriis*, Benedictus, bars 1-19, a canon two three-voice solutions (Tenore-Canto-Alto and Tenore-Basso-Alto) according to invertible counterpoint at the twelfth, from the *Spartitura delle messe del primo libro* (Rome, 1627), p. 5.

A similar procedure can be observed in the Benedictus from the canonic *Messa Pro vigiliis ac feriis*, published in Agostini's *Spartitura delle messe del primo libro* (Ex. 7.1). In this Benedictus, four voices are presented in score format, with instructions provided by the composer at the bottom of the page:

> In this Benedictus for three [voices], the Tenor is in canon with the Alto and Soprano, and, having had regard for [invertible] counterpoint at the twelfth, the Soprano is silent but [in the second version its melody] is sung a twelfth lower in the Bass[60].

Two versions of the Benedictus are possible in performance, each for only three of the four notated voices. In the first case, the three upper voices form a canon whereby the Tenor (*dux*) is imitated at the octave above by the Soprano and the fourth above by the Alto. By means of invertible counterpoint at the twelfth, the Soprano line becomes the Bass line in the second version of this Benedictus. In this version, the Tenor is imitated at the fifth below by the Bass and (again) at the fourth above by the Alto. In this example overall, Agostini exploits how score format efficiently aligns all four voices vertically in resolution but also allows performers and other viewers to easily select two combinations of three voices that are related through the technique of invertible canon.

In the next example (Ill. 7.7), Agostini seems to adopt the reverse approach in the visual presentation of the musical material: only the canonic *dux* is notated for the short piece at the top of the page. This piece is located above the horizontal decorative border separating it from another work notated in full score on this page. The canon has four clefs (G2, C2, C4, F4) at the beginning of its staff to indicate voice ranges and signs above certain notes to indicate locations of voice entries. In the accompanying textual instructions for this canon, Agostini specifies the addition or replacement of voices to obtain three combinations arising from the canonic melody:

> Canon with three solutions: the first and second solutions are for three voices whereby the first [solution] is without the Canto and the other [solution] is without the Bass. The third solution is sung with four voices. In each case, each voice is a fifth lower [than the previous voice][61].

Two of the solutions involve combinations of three voices and the third solution is for all four voices. Each of these four versions employs a canonic technique known nowadays as the stacked canon[62]. This is where the same temporal distance and the same interval of imitation separate each voice from the next. Consequently, all three solutions involve imitation at the fifth below and the temporal distance of a breve between successive voice entries. By selecting

60. Agostini, *Spartitura* [...] *del primo libro* (1627), p. 5: «Questo Benedictus à 3: il Tenore fa Canon con l'Alto & Soprano, & havendo hauto riguardo al contrapunto alla 12. il Soprano tacera & si potra cantare in Basso 12. più basse».

61. *Ibidem*, p. 11: «Canon in 3. modi. 1. & 2. modo à 3, voci, vna volta senza il Canto, & l'Altro senza il Basso. Il terzo modo canta a 4. & in ciascheduna parte vna quinta più bassa».

62. The term 'stacked canon' was first coined in Gosman 1997.

the appropriate clef and *signum*, the viewer arrives at the following three solutions: (1) Alto, Tenor and Bass; (2) Canto, Alto, Tenor; and (3) Canto, Alto, Tenor, Bass.

ILL. 7.7: On top: canon with three solutions, in Paolo Agostini, *Libro quarto delle messe in spartitura* (Rome, 1627), p. 11. Copy I-Bc, V.39.

In Agostini's publications from 1627, we frequently find canonic melodies at the top of pages with decorative borders separating them from unrelated works below in score format. In the present case (ILL. 7.7), the work below the border is the *Messa Si bona suscepimus*. Most sections of this Mass do not employ canon, as is the case in the excerpt of the Gloria quoted here. In ILL. 7.7 and elsewhere in Agostini's published volumes, it is difficult to see any connection, liturgical or otherwise, between the canon at the top of the page and the work in score beneath it. We here use the term 'garland canon' to refer to a short counterpoint in single-staff notation that shares the same page as another work in score format. Perusal of Agostini's six volumes published in 1627 reveals that most of the garland canons are in fact differently notated versions of sections from canonic Masses presented in full score elsewhere in the same volume or in a different volume (although this is not the case for the canon under consideration here, which does not appear anywhere else)[63]. Many of the garland canons use different texts than their counterparts found in other locations in score format, further obscuring the liturgical contexts for the works that share the same pages. Perhaps Agostini's motivations lay to some extent in taking advantage of the combinatorial possibilities of music notation whereby the same material could appear in different guises at different locations. Works involving canonic techniques are especially well suited to this purpose: Agostini could draw upon the well-established method of notating canons in closed format as well as in the newly emerging score format. The latter format afforded immediate verification of canonic possibilities through visual inspection.

Page 23 of Agostini's second book from 1627, the *Spartitura del secondo libro delle messe e motetti a quattro voci*, has many curious features (Ill. 7.8). The four staves at the top left of the page contain the final bars of the Agnus Dei of the *Missa Ave regina coelorum* in score format. The remainder of the page contains an assortment of short canons, some in score format and others in closed format. These pieces provide demonstrations of what we have termed multiple-technique canon. Agostini crafted the contents of this volume overall so that each section of the *Missa Ave regina cœlorum* is followed by one or more short canons. These canons are based upon melodic phrases drawn from different places in the preceding sections of the Mass. Through this approach, Agostini demonstrated how the same melodic materials could serve as part of a non-canonic texture in the Mass sections but also as the basis for exploration of a variety of canonic techniques.

The canon at the top right-hand side of the page, marked «A5. Canon del Agnus», is in score format. This is a double canon with a fifth free voice (the topmost notated voice). The double barlines with dots on either side indicate that this is a circle canon, that is, the music repeats for as many times as the performers wish. The voices on the second highest staff and the lowest staff form a two-part canon at the octave above after two breves. The staff with two clefs resolves to a unison canon after two breves with a helpfully placed *signum concordantiae* indicating where the *comes* should enter. Comparison to the final bars of the setting of the Agnus Dei immediately to the left of this canon shows that the canon's bass melody is taken from the ending of the bass line of the Agnus Dei. Further, the other canonic *dux* (on the staff with two clefs) is based on the melodic phrase in the final bars of the Alto from the Agnus Dei. These two melodies, or small variations of them, are found at many places in the Mass setting and in other canons. The non-canonic voice in this example (using the treble clef) has a melody that is first heard at the end of Kyrie II. This melody, which we can call the 'Kyrie II' melody, was

63. For instance, the Benedictus from the *Spartitura* [...] *del primo libro* (1627; Ex. 1, discussed above) reappears as a garland canon in the *Spartitura del terzo libro* (1627), p. 4.

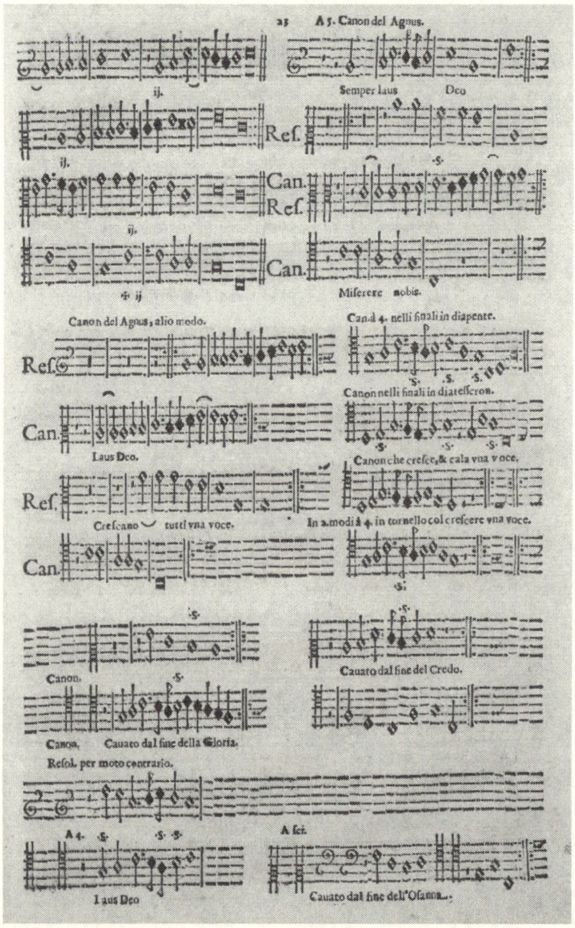

ILL. 7.8: Multiple-technique canons in Paolo Agostino's *Spartitura del secondo libro delle messe e motetti a quattro voci* (Rome, 1627), p. 23. Copy I-Bc, V.39.

particularly favoured by Agostini: it is used many times throughout the Mass and in the remaining canons in ILL. 7.8[64]. It also reappears in several of the garland canons in other volumes.

The other nine canons on page 23 of the *Spartitura del secondo libro* likewise involve multiple techniques, with each canon drawing on melodic material that was heard at different points in the *Missa Ave regina coelorum*. On the centre left-hand side of this page, there is a «Canon del Agnus, alio modo» that is presented in score for four voices, each on a separate staff. This is an alternative version of the first canon on this page: the same two melodies form the basis of another double canon where each pair of voices is in imitation at the fifth above. The

[64]. Laurence Wuidar remarks that the Kyrie II melody seems like a *Leitmotif* (WUIDAR 2008, p. 102): «comme veritable 'leitmotiv'». A transcription of the full Kyrie II section of this Mass can be found in COLLINS 2018, p. 131.

text between the lower two staves reads «*Crescano tutti una voce*» (Let all raise one tone) which indicates that this piece is also a canon *per tonos* at the second above.

By contrast, each of the four staves on the centre right-hand side of the page contains a separate canon. Each of these canons is based on the Kyrie II melody and has its own rubric and *signa concordantiae* to help the user of the volume work out the resolutions. These pieces require either two or three *comites* to follow the *dux* at intervals specified in the rubrics (such as 'diapente' for imitation at the fifth, 'diatessaron' for the fourth). The instructions for the canons notated on the lower two staves in this group of pieces call for canons *per tonos*.

The third and fourth staves from the end of the page contain a further two double canons that are also circle canons. For the canon on the left, the label «*Cavato dal fine della Gloria*» (Taken from the end of the Gloria) indicates that the two melodies come from the end of the Gloria. The canon to the right has a similar label indicating that the melodies were earlier used at the end of the Credo. In each canon, these melodies are clearly small variants of the Kyrie II melody and the bass melody from the end of the Agnus Dei.

The last two canons on this page (on last two staves) introduce further complexity through imitation by contrary motion within double canons. The canon to the left is a double canon whereby the constituent canons use either the original or inverted form of the Kyrie II melody. The last canon, located on the bottom right-hand side of the page, is a six-part double circle canon notated on a single staff with clefs both rightway-up and upside down. The solution involves four voices singing one of the constituent canons by either similar motion or contrary motion according to the four clefs on the left-most side of the staff. Two other voices sing a unison canon according to the clefs on the right-hand side of the staff. The result overall is a succession of alternating chords based on C and G.

Conclusion

At the beginning of this chapter, we identified the confluence of two modes of early scientific thought, namely mechanism and combinatorics, in the early seventeenth century. These found their purest expression firstly in the writings of Mersenne on music. For Kircher, the irresistible urge to incorporate the ideas of his contemporaries into his own voluminous writings on music in the foment of mid-century Rome produced an extension of Mersenne's ideas (while borrowing from other authors like Picerli on the way) to include mechanical processes for musical composition and references to the achievements of Roman exponents of canon. In Kircher, canon, as the epitome of the *artificioso* composition, becomes the site of musical combinatorial experimentation and a means for demonstrating the validity of contrapuntal principles within the 'natural' laws of musical consonance. In other words, canonic techniques served as a means for the multiplex exploration of sonorous number, whose conceptual roots lay in 'consonant' ratios. While the ultimate proof of the combinatorial potential of a canonic subject might ultimately exist in the judgement of hearing, the score format provided a quasi-scientific graph for demonstrating the validity of combining the musical subjects of canons with themselves. Through his synthesis and receptiveness to contemporary developments in musical culture at Rome, Kircher challenges music historians to recognize the canonic endeavours of composers like Valentini and Agostini as sites for experimentation and demonstration.

A close examination of these sites of combinatorial activity suggests ways in which Roman composers responded creatively to the intellectual ferment surrounding them. They not only pushed contrapuntal abstraction in novel directions, but they also found ways to incorporate newly devised combinations of canonic techniques into their daily activities as professional musicians. While Kircher was clearly attracted to the «marvellous power of musical combination»[65] evident in Valentini's polymorphous canons, other approaches to canonic writing were also to the fore of contemporary musical experimentation. A striking feature of surviving sources is the use of multiple-technique canon. Rarely found in earlier repertoires, this aspect of contrapuntal practice fits well with the spirit of experimental and mechanistic artifice. Although Roman composers did not invent any new canonic techniques, their imaginations led to innovative combinations of pre-existing techniques such that their works could demonstrate a multitude of ways to assemble procedures that individually had long histories in compositional practice. It is hardly surprising that new developments in music notation supported canonic experimentation: the emergence of score format allowed ready verification of combinatorial potential through graphical format. At the same time, the persistence of older forms of canonic notation, in particular canons in closed format, allowed a composer such as Agostini to demonstrate (perhaps sometimes even playfully) a subtlety of musical combinatorics whereby the same work could manifest itself in different visual garb, even when setting a different text.

Although it is true, as Sergio Durante (1987) has noted several decades ago, that *artificioso* compositions have generally been overlooked in musical historiography, it may nevertheless be possible to start to gauge some long-term influences of the canonic activities of Roman composers. The canonic outpourings of Valentini, Agostini and others were assembled in collections that are in effect compendia of contrapuntal techniques. Similar volumes, in both print and manuscript form, stretch across the seventeenth and eighteenth centuries, with, for instance, contributions by Giovanni Battista Vitali and by Johann Theile and his circle[66]. Easily

65. Kircher, *Musurgia universalis* (1650), Tomus I, p. 414: «mirificam Musicae combinationis vim».

66. The activities of Theile and the Hamburg school have been documented by Walker 2000. It should be remembered that English musicians at the turn of the seventeenth century were also deeply interested in canonic techniques, most especially those that could be brought to bear on a cantus firmus, in particular the *Miserere* melody. See Collins 2007.

the most well-known contributions from the eighteenth century are those by J. S. Bach where a variety of contrapuntal procedures, including multiple-technique canons, are evident in the great collections, such as the *Musical Offering*, from the end of his life. The possible pathways of influence from Roman canon to later activities are as yet unclear. One of many possible routes may have been the exchanges between Marco Scacchi and Romano Micheli on several aspects of canonic composition that attracted interest from other musicians during the middle decades of the seventeenth century. Scacchi himself left examples of *artificioso* composition that were preserved by Angelo Berardi and perused by later generations of theorists and composers[67]. In any event, it is reasonable to suggest that early seventeenth-century Roman experiments with canon and combinatorics were harbingers of a deep strand of contrapuntal engagement that ran through the work of following generations of composers and theorists.

Bibliography

Barker 2015

Barker, Naomi J. 'Music, Antiquity and Self-Fashioning in the Accademia dei Lincei', in: *The Seventeenth Century*, xxx/4 (2015), pp. 375-390.

Bianchi 2017

Bianchi, Eric. 'Scholars, Friends, Plagiarists. The Musician as Author in the Seventeenth Century', in: *Journal of the American Musicological Society*, lxx/1 (2017), pp. 61-128.

Bianconi 1987

Bianconi, Lorenzo. *Music in the Seventeenth Century*, Cambridge, Cambridge University Press, 1987.

Butt 2005

Butt, John. 'The Seventeenth-Century Musical "Work"', in: Butt – Carter 2005, pp. 27-54.

Butt – Carter 2005

The Cambridge History of Seventeenth Century Music, edited by John Butt and Tim Carter, Cambridge, Cambridge University Press, 2005.

Christensen 2013

Christensen, Thomas. 'The Sound World of Father Mersenne', in: *Structures of Feeling in Seventeenth-Century Cultural Expression*, edited by Susan McClary, Toronto, University of Toronto Press, 2013, pp. 60-90.

Coelho 1992

Coelho, Victor A. 'Musical Myth and Galilean Science in Giovanni Serodine's *Allegoria della Scienza*', in: *Music and Science in the Age of Galileo*, edited by Victor A. Coelho, Dordrecht, Kluwer, 1992, pp. 91-114.

[67]. On the Scacchi-Micheli exchange, see Patalas 2010. Issues around the transmission of contrapuntal theory into eighteenth-century compositional practice are discussed in Collins 2019, pp. 204-214.

Cohen 1984

Cohen, H. Floris. *Quantifying Music: The Science of Music at the First Stage of the Scientific Revolution, 1580-1650*, Dordrecht, Reidel, 1984.

Cohen 1992

Id. 'Beats and the Origins of Early Modern Science', in: *Music and Science in the Age of Galileo*, *op. cit.*, pp. 17-34.

Cohen 2015

Id. *The Rise of Modern Science Explained: A Comparative History*, Cambridge, Cambridge University Press, 2015.

Collins 2007

Collins, Denis. '«Sufficient to Quench the Thirst of the Most Insaciate Scholler Whatsoever»: George Waterhouse's 1,163 Canons on the Plainsong Miserere', in: *Canons and Canonic Techniques, 14th-16th Centuries*, edited by Katelijne Schiltz and Bonnie J. Blackburn, Leuven, Peeters, 2007, pp. 407-420.

Collins 2018

Id. 'Canon in Baroque Italy. Paolo Agostini's Collections of Masses, Motets and Counterpoints from 1627', in: *Music Preferred: Essays in Musicology, Cultural History and Analysis in Honour of Harry White*, edited by Lorraine Byrne Bodley, Vienna, Hollitzer Verlag, 2018, pp. 117-140.

Collins 2019

Id. 'Zelenka and the Combinative Impulse. Contrapuntal Techniques in the Miserere in D Minor, ZWV 56', in: *Musicology Australia*, xli/2 (2019), pp. 199-225.

Cortesi 1983

Cortesi, Mariella Casini. 'Pier Francesco Valentini. Profilo di un musicista barocco', in: *Nuova Rivista Musicale Italiana*, xvii/3-4 (1983), pp. 529-562.

Corwin 2008

Corwin, Lucille. *«Le Istitutioni Harmoniche» of Gioseffo Zarlino, Part 1: A Translation with Introduction*, Ph.D. Diss., New York, City University of New York, 2008.

Dear 1984

Dear, Peter. 'Marin Mersenne and the Probabilistic Roots of «Mitigated Scepticism»', in: *Journal of the History of Philosophy*, xxii/2 (1984), pp. 173-205.

Dear 1988

Id. *Mersenne and the Learning of the Schools*, Ithaca (NY), Cornell University Press, 1988.

Dijksterhuis 1950

Dijksterhuis, Eduard Jan. *De mechanisering van het wereldbeeld*, Amsterdam, Meulenhoff, 1950.

Dijksterhuis 1961

Id. *The Mechanization of the World Picture*, translated by C. Dikshorn, Oxford, The Clarendon Press, 1961.

Dixon 1981

Dixon, Graham Peter. *Liturgical Music in Rome (1605-45)*, Ph.D. Diss., Durham, University of Durham, 1981.

Dixon 1991

Id. 'Tradition and Progress in Italian Mass Settings after Palestrina', in: *Atti del II Convegno internazionale di studi palestriniani. Palestrina e la sua presenza nella musica e nella cultura europea dal suo tempo ad oggi: anno europeo della musica, 3-5 maggio 1986*, edited by Lino Bianchi and Giancarlo Rostirolla, Palestrina, Fondazione Giovanni Pierluigi da Palestrina, 1991, pp. 311-324.

Drake 1992

Drake, Stillman. 'Music and Philosophy in Early Modern Science', in: *Music and Science in the Age of Galileo*, *op. cit.*, pp. 3-16.

Durante 1987

Durante, Sergio. 'On *artificioso* Compositions at the Time of Frescobaldi', in: *Frescobaldi Studies*, edited by Alexander Silbiger, Durham (NC), Duke University Press, 1987, pp. 195-217.

Findlen 2004

Findlen, Paula. 'Introduction. The Last Man Who Knew Everything... or Did He? Athanasius Kircher, S. J. (1602-80) and His World', in: *Athanasius Kircher: The Last Man Who Knew Everything*, edited by Paula Findlen, Abingdon-New York, Routledge, 2004, pp. 1-48.

Freedberg 2002

Freedberg, David. *The Eye of the Lynx: Galileo, His Friends, and the Beginnings of Modern Natural History*, Chicago, University of Chicago Press, 2002.

Galilei 2008

Galilei, Galileo. *Essential Galileo*, translated by Maurice A. Finocchiaro, Indianapolis, Hackett Publishing Company, 2008.

Garber 2013

Garber, Daniel. 'Remarks on the Pre-history of the Mechanical Philosophy', in: Garber – Roux 2013, pp. 3-26.

Garber – Roux 2013

The Mechanization of Natural Philosophy, edited by Daniel Garber and Sophie Roux, Dordrecht, Springer, 2013.

Gerbino 1995a

Gerbino, Giuseppe. *Canoni ed enigmi. Pier Francesco Valentini e l'artificio canonico nella prima metà del Seicento*, Rome, Edizioni Torre D'Orfeo, 1995.

Gerbino 1995b

Id. 'Gli arcani più profondi dell'arte. Presupposti teorici e culturali dell'artificio canonico nel secoli xvi e xvii', in: *Il saggiatore musicale*, ii/2 (1995), pp. 205-236.

Gosman 1997

Gosman, Alan. 'Stacked Canon and Renaissance Compositional Procedure', in: *Journal of Music Theory*, xli/2 (1997), pp. 289-317.

Gouk 2005

Gouk, Penelope. 'Music and the Sciences', in: Butt – Carter 2005, pp. 132-157.

Gozza 2002

Gozza, Paolo. 'La musica', in: *Storia della scienza. 5: La rivoluzione scientifica*, Rome, Istituto della Enciclopedia Italiana, 2002, pp. 556-569.

Grove2 2001

New Grove Dictionary of Music and Musicians, Second edition, edited by Stanley Sadie, 29 vols., Oxford, Oxford University Press, 2001.

Heilbron 2010

Heilbron, J. L. *Galileo*, Oxford, Oxford University Press, 2010.

Kassler 2001

Kassler, Jamie C. *Music, Science, Philosophy: Models in the Universe of Thought*, Aldershot, Ashgate, 2001.

Klotz 2006

Klotz, Sebastian. *Kombinatorik und die Verbindungskünste der Zeichen in der Musik zwischen 1630 und 1780*, Berlin, Akademie Verlag, 2006.

Knobloch 1979

Knobloch, Eberhard. '*Musurgia universalis*. Unknown Combinatorial Studies in the Age of Baroque Absolutism', in: *History of Science*, xvii/4 (1979), pp. 258-275.

Knobloch 2013

Id. 'Renaissance Combinatorics', in: *Combinatorics, Ancient and Modern*, edited by Robin Wilson and John J. Watkins, Oxford, Oxford University Press, 2013, pp. 123-145.

Knuth 2013

Knuth, Donald E. 'Two Thousand Years of Combinatorics', in: *Combinatorics, Ancient and Modern, op. cit.*, pp. 3-40.

Lamla 2003

Lamla, Michael. *Kanonkünste im barocken Italien, insbesondere in Rom*, Berlin, Verlag im Internet, 2003.

Lenoble 1971
Lenoble, Robert. *Mersenne ou la naissance du mécanisme*, Paris, J. Vrin, [2]1971.

McKay 2012
McKay, John Zachary. *Universal Music-Making: Athanasius Kircher and Musical Thought in the Seventeenth Century*, Ph.D. Diss., Cambridge (MA), Harvard University, 2012.

Mersenne 1977
Mersenne, Marin. *Correspondance du P. Marin Mersenne, réligieux minime. 13: 1643-1645*, edited by Cornelis de Waard, Paris, Éditions du Centre National de la Recherche Scientifique, 1977.

Michopoulou 2006
Michopoulou, Aikaterini. *Contrapuntal and Keyboard Idioms in the 'artificioso' Collection (1641) of Giovanni Pietro del Buono*, Ph.D. Diss., York (UK), University of York, 2006.

Miller 1998
Miller, Stephen R. *Music for the Mass in Seventeenth-Century Rome. «Messe piene», the Palestrina Tradition, and the «stile antico»*, Ph.D. Diss., Chicago (IL), University of Chicago, 1998.

Palisca 1961
Palisca, Claude V. 'Scientific Empiricism in Musical Thought', in: *Seventeenth-Century Science and the Arts*, edited by Hedley Howell Rhys, Princeton, Princeton University Press, 1961, pp. 91-137.

Palisca 1989
Id. *The Florentine Camerata: Documentary Studies and Translations*, New Haven (CT), Yale University Press, 1989.

Palisca 1992
Id. 'Was Galileo's Father an Experimental Scientist?', in: *Music and Science in the Age of Galileo*, *op. cit.*, pp. 143-151.

Patalas 2010
Patalas, Aleksandra. '*Ut oratio sit Domina*. The Dispute between Romano Micheli and Marco Scacchi', in: *The Dissemination of Music in Seventeenth-Century Europe: Celebrating the Düben Collection*, edited by Erik Kjellberg, Bern, Peter Lang, 2010, pp. 219-238.

Redondi 1988
Redondi, Pietro. *Galileo Heretic (Galileo eretico)*, London, Allen Lane, 1988.

Rostirolla 2017a
Rostirolla, Giancarlo. 'Il breve magistero di Paolo Agostini (1626-1629)', in: Id. *La Cappella Giulia 1513-2013: Cinque secoli di musica sacra in San Pietro*, 2 vols., Kassel, Bärenreiter-Verlag, 2017 (Analecta Musicologica, 51), vol. i, pp. 430-447.

Sabaino 2013

Sabaino, Daniele. 'Ramificazioni dell' «encyclopaedia». Caramuel, Kircher e l' «ars combinatoria» nella composizione musicale', in: *Steinbruch oder Wissensgebäude? Zur Rezeption von Athanasius Kirchers «Musurgia Universalis» in Musiktheorie und Kompositionspraxis*, edited by Melanie Wald-Fuhrmann, Basel, Schwabe, 2013, pp. 31-58.

Shapin 1996

Shapin, Steven. *The Scientific Revolution*, Chicago, University of Chicago Press, 1996.

Silbiger 1992

Silbiger, Alexander. 'Music and the Crisis of Seventeenth-Century Europe', in: *Music and Science in the Age of Galileo*, *op. cit.*, pp. 35-44.

Stoessel 2022

Stoessel, Jason. 'Pier Francesco Valentini and Musical Canons in the Visual Culture of Early Modern Rome', in: *Belonging and Detachment: Representing Musical Identity in Visual Culture*, edited by Antonio Baldassarre and Arabella Teniswood-Harvey, Vienna, Hollitzer Verlag, 2022.

Walker 2000

Walker, Paul M. *Theories of Fugue from the Age of Josquin to the Age of Bach*, Rochester (NY), University of Rochester Press, 2000.

Westfall 1977

Westfall, Richard S. *The Construction of Modern Science: Mechanism and Mechanics*, Cambridge, Cambridge University Press, 1977.

Wuidar 2008

Wuidar, Laurence. *Canons énigmes et hiéroglyphes musicaux dans l'Italie du 17e siècle*, Brussels, Peter Lang, 2008.

Chapter Eight

Giovanni Battista Doni on the Tonal Space of Chromatic Music: The Case of Gesualdo's *Tu m'uccidi, o crudele*

Carlos C. Iafelice[*]
(São Paulo State University, São Paulo)

(Salviati): When Aristotle would see these things, what do you believe, Mister Simplicio, would he say and do about them[1]*?*

In trying to dissuade Simplicio from his peripatetic convictions, Salviati, spokesman for the Copernican ideas propagated by Galileo, questions whether Aristotle would maintain his convictions on Ptolemaic cosmology after the new discoveries about the movement of celestial bodies. The rhetorical exercise resulting from the question that dislocates Aristotle's *auctoritas* and exposes it to new scientific observations enhances an *a posteriori* method of problem solving and refuting Simplicio's perspective. Salviati implicitly reinforces the significance of observation and experience as a means of demonstrating the effectiveness (or refutation) of a given proposition.

If new evidence could result in questioning postulates acquired by an *intellectus fidei*, similar manifestations may have occurred among Florentine intellectual circles around 1630.

*. The preparation of this chapter was supported by the São Paulo Research Foundation, Doctorate fellowship, grant 2017/18258-5.

1. Galileo Galilei, *Dialogo sopra i due massime sistemi del mondo* (Florence, 1632), p. 43: «(Salviati:) Hora, quando Aristotile vedesse queste cose, che credete voi, signor Simplicio, che'e' dicesse e facesse?».

ILL. 8.1: Gaetano Vascellini (after a drawing by Giuliano Traballesi after an unknown painting), *Giovanni Battista Doni* (engraving, 30 × 20 cm; second half eighteenth century). NL-Ar, RP-P-1909-5446.

An analogy can be seen in the dedication of the *Compendio del Trattato de' generi e de' modi* (Rome, 1635) by the Florentine humanist and music theorist Giovanni Battista Doni (1595-1647; ILL. 8.1), addressed to Cardinal Francesco Barberini:

> That music has great efficacy in tempering the passions of the mind, is something so understood and established among the wise that there is no doubt about it. Not only is this a virtue of the material and the operation of sounds and songs; but also of the noblest, and most excellent music [theory] that joins speculation to practice, and to this adds the historical part: not to rack one's brain with abstract things and to feed vain curiosity, but to find new ways, or renew the old ones, and to illustrate this profession not yet fully returned to its former glory[2].

[2]. Giovanni Battista Doni, *Compendio del Trattato de' modi e de' generi della musica* (Rome, 1635), f. +2r: «Che la musica habbia grandissima efficacia in temperare le passioni dell'animo, è cosa tanto ricevuta, e stabilita trà

In these words, there is a clear reformist intention, in search of a historical truth that had not yet been properly revealed. If, in the field of natural philosophy, the evidence that brought about paradigmatic changes arose from the use of new artifacts (such as Galileo's telescope), Doni's version of such an artifact applied to music could be the careful philological investigation aimed at the proper restitution of Greek musical thought. Therefore, a similar Galilean impetus may be recognized in terms of Doni's re-evaluation of musical knowledge. However, we observe in the case of Galileo a progressive position that differs from that of Doni, whose main purpose was what we could call a 'palingenesis' (to avoid the term 'Renaissance') of the attributes of ancient Greek music and reframing them in current musical practice.

Doni's objective was not only to establish new methods of composition based on knowledge extracted from ancient Greek sources, but also to test these methods on existing compositions. His search was not just a mere mental exercise, but an attempt to change the musical reality that surrounded him[3].

The use of elements from such a different context led Doni to propose a set of adaptations in order to accommodate compositions, both old and new, to the precepts that would provide the 'wonderful effects' (*effetti meravigliosi*) of ancient Greek harmony, until then hidden by the inappropriate way in which, in Doni's view, his predecessors had tried to restore them. In the first chapter of the *Compendio* ('Quanto mal'intesa sia hoggi la materia de' Generi, e de' Modi', How badly understood is today the matter of the genera and the modes), Doni discusses the works of Heinrich Glarean, Nicola Vicentino, Gioseffo Zarlino, Vicente Lusitano, Franchinus Gaffurius, Ercole Bottrigari, and Francisco Salinas, providing an interesting critical overview of the state of knowledge about the musical modes and genera shown by these authors. In the next chapter ('Quanto sia grande la diversità tra i Modi antichi, et i moderni', How big the difference is between the ancient and the modern modes), Doni formulated his own ideas, in line with the proposals made before by Vincenzo Galilei, Giovanni Bardi, and above all, Girolamo Mei.

In order to materialize his ideas, Doni designed several musical instruments that were capable of bringing this theory into practice. Therefore, his theories contain three main elements: the restoration of the modal doctrine of the ancient Greeks, the composition of new musical pieces or the adaptation of existing pieces to this doctrine, and the construction of musical instruments to perform these pieces.

i Savij, che non ha luogo il dubitarne. Ne questa è virtù solo di quella materiale, et operatrice de' Suoni, e de' Canti; ma anco di quella più nobile, et eccellente, che alla pratica unisce la Speculatione, et à questa accompagna la parte Istorica: non per lambiccarsi il cervello in cose astratte, e pascersi di vana curiosità; ma per trovar maniere nuove, ò rinovare le vecchie. Et illustrar questa professione non ancor ritornata del tutto antico splendore».

3. When Doni announced the publication of the *Compendio* in a letter of 8 September 1635 to Marin Mersenne, he stressed his reformist intentions (Mersenne 1959, p. 390): «on y trouvera plein des nouveautés et choses singulières et sera capable de redresser en grand' partie la musique». (One finds there many novelties and singular things and it will be able to redress music considerably.)

COMPENDIO
DEL TRATTATO
DE' GENERI E DE' MODI
DELLA MVSICA.
Di Gio. Battista Doni.
CON VN DISCORSO SOPRA LA PERFETTIONE de' Concenti.
Et vn saggio à due Voci di Mutationi di Genere, e di Tuono in tre maniere d'Intauolatura: e d'vn principio di Madrigale del Principe, ridotto nella medesima Intauolatura.
All'Eminentiss. e Reuerendiss. Sig.
IL SIG. CARDINAL BARBERINO.
IN ROMA, Per Andrea Fei. MDCXXXV. *Con licenza de' Superiori.*

De' Generi, e de' Modi. 33
Secondo i Moderni. *Mutatione di Tuono.*
Terzo Tuono. *Quinto Tuono.*
Mutatione di Modo.
Terzo Modo, secondo altri Quinto *Quarto Modo, ò Sesto.*
Secondo gl'Antichi. *Mutatione di Modo solo.*
Mi, *Tuono, e Modo Dorio.* Re, *Tuono Dorio, Modo Frigio.*
Mutatione di Tuono solo.
Mi, *Tuono, e Modo Dorio,* Mi, *Modo Dorio, Tuono Frigio.*
Ma la vera differenza d'vn Tuono, o Modo reale all'antica, che comprende l'vno, & l'altro; e come scambieuolmente si connettino insieme saluando le loro giuste distãze, si può mostrare nella maniera che segue.
E Dorio

ILL. 8.2AB: Giovanni Battista Doni, *Compendio del Trattato de' generi e de' modi della musica* (Rome, 1635), title page and page 33, with examples of the modern and ancient modes. Copy D-Mbs.

Doni entrusted his ideas to paper several times. In this chapter I will refer to or quote from the following writings:

- 'Trattato dei generi e dei modi della musica' (Treatise on the Genera and Modes of Music; Ms. I-Bc, MS D.143, c. 1630-1634; short: 'Trattato')[4];
- *Compendio del Trattato de' modi e de' generi della musica* (Compendium of the Treatise on the Modes and Genera of Music; Florence, 1635; short: *Compendio*; ILL. 8.2AB);
- *Annotazioni sopra il Compendio del Trattato de' generi e de' modi della musica* (Annotations on the Compendium of the Treatise on the Genera and Modes of Music; Florence, 1640; short: *Annotazioni*; ILL. 8.3AB);
- *De praestantia musicae veteris libri tres* (Three Books on the Excellence of Ancient Music; Florence, 1647; in short: *De praestantia*);

[4]. I thank Professor Massimo Redaelli for access to his database *Manuscripts of Italian Music Theory in Translation* (<http://mimtt.co.uk/>), which contains a transcription of the handwritten text of the 'Trattato'. About the 'Trattato', see PALISCA 1997.

ANNOTAZIONI
Sopra il Compendio de' Generi, e de' Modi
della Musica,
DI GIO. BATTISTA DONI.
DOVE SI DICHIARANO I LVOGHI PIV OSCVRI,
e le Massime più nuove, & importanti si prouano con ragioni, e
testimonianze euidenti d'Autori classici.
Con due Trattati L'vno sopra i Tuoni, e Modi veri, L'altro sopra i Tuoni è
Armonie de gl'Antichi. Et sette Discorsi sopra le materie più princi-
pali della Musica, ò concernenti alcuni instrumenti
nuoui praticati dall'Autore.

IN ROMA, Nella Stamparia d'Andrea Fei. MDCXL.

CON LICENZA DE' SVPERIORI.

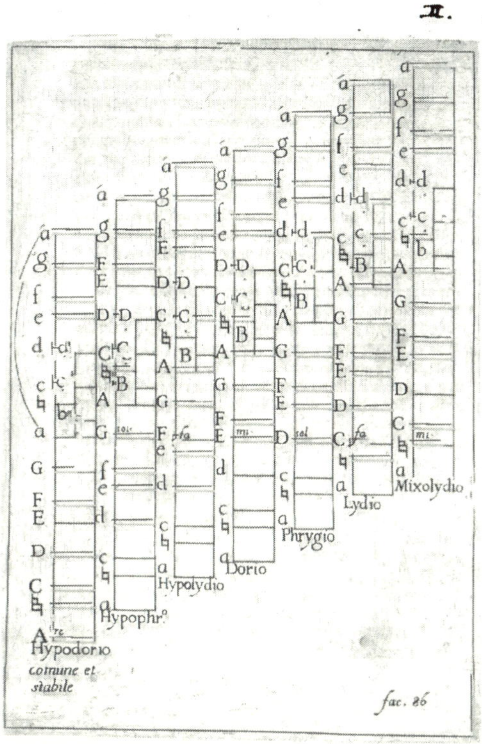

Ill. 8.3ab: Giovanni Battista Doni, *Annotazioni sopra il Compendio del Trattato de' generi e de' modi della musica* (Rome, 1640), title page and the illustration opposite page 86, with the Greek system of tonoi.

• *Lyra Barberina ἀμφίχορδος* [amphichordos], [*Tomus primus*] (The Double-Stringed Barberinian Lyre, First Volume; posthumous: Florence, 1763; short: *Lyra Barberina*);

• *De' trattati di musica* [...] *Tomo secondo* (Treatises on Music, Second Volume; posthumous: Florence, 1763; short: *De' trattati, Tomo secondo*).

The last two publications were prepared by the historian Antonio Francesco Gori (1691-1757) and then realized by Giovanni Battista Passeri (1694-1780). To these publications can be added the edition of his correspondence, edited by Gori:

• *Io. Baptistae Donii, Patricii Florentini, Commercium litterarium* (Florence, 1754; short: *Commercium*).

This chapter investigates how Doni applied his theory to an existing composition (written independently from his doctrine), in particular, bars 1-25 of the madrigal *Tu m'uccidi, o crudele* from the *Madrigali a cinque voci, Libro quinto* (Naples, 1611) by Carlo Gesualdo da Venosa. The excerpt is included in the appendix of the *Compendio* of 1635, first in in score in conventional staff notation, then in tablature for Doni's *viole diarmoniche*. By investigating the characteristics of this peculiar tablature and relating them to Doni's theories, we can gain insight in how Doni interpreted the tonal space of chromatic music, since for writing in tablature an appropriate coding was required in accordance with the tonoi (see below) and modes identified by him in the composition.

We will begin with a summary of the main elements of Doni's modal doctrine. Then an exposition will be given of Doni's conception of chromaticism. And finally, we will investigate the tablature and the tonal space of the excerpt of Gesualdo's *Tu m'uccidi, o crudele* as presented by Doni.

THE ELEMENTS OF DONI'S MODAL DOCTRINE

> [...] *it was like a spark which ignited in me an ardent wish, which subsequently grew more and more by reading the* Dialogo [della musica antica, et della moderna] *by [Vincenzo] Galilei (of which [Girolamo] Mei, about whom I spoke, wrote the larger part) and [still] much more by the treatise 'De modis musicae' by the same author, in manuscript,* [...] *that I had copied to understand it better, and to avail myself with it in these pleasant studies* [...][5].

In this passage, Doni refers to a letter that Girolamo Mei had addressed to Francesco Nori (Bishop of San Maniato in Tuscany) and that was kindly presented to him (Doni) in his youth, and that sparked his interest in the 'ancient harmonies' (*antiche armonie*). This interest was further augmented by reading the *Dialogo della musica antica, et della moderna* (Florence, 1581), written by Vincenzo Galilei (with Mei's collaboration), and above all, by reading the manuscript treatise 'De modis' by Mei himself, which Doni carefully copied, before devoting himself to investigate the relevant Greek sources directly himself[6]. Among Doni's Greek sources was Ptolemy's *Harmonics*, which became, with Mei's treatise, the main foundation of his 'restoration' work. Doni agreed with Mei on the constitution of Greek modes as seven octave species, each of them related to one of the seven tonoi[7]. Here, we will present a summary of the fundamental issue of Doni's modal theory: the difference between tonos and mode.

[5]. Doni, *Lyra Barberina, Tomus I* (Florence, 1763), p. 425: «[...] fu quasi una scintilla, che m'accese un ardente Desiderio il quale viemaggiormente poi si accrebbe dalla lettura del Dialogo del Galilei (dove il Mei, di cui parlo, ebbe la maggior parte) e molto più da un Trattato *de Modis Musicae* dell'istesso Autore, scritto a penna, [...] ch'io feci copiare per bene intenderlo, e valermene in questi ameni studj [...]».

[6]. Doni's copy of Mei's treatise is now I-Fr, MS 815. Mei's autograph is now V-CVbav, MS Vat. Lat. 5323.

[7]. PALISCA 1985, p. 330.

In ancient Greek music theory, a tonos is the Greater Perfect System realized on a certain pitch level or transposition. The number of tonoi may vary: best known are the systems with seven and thirteen tonoi, which are connected with Ptolemy and Aristoxenus respectively. Table 8.1 lists the names and the relative pitches of the tonoi as they occur in these and some other systems.

Table 8.1

Sets of tonoi of various sizes. The first column gives the names of the tonoi; the second, their relative pitches with the Dorian tonos set on A. The names of the tonoi added to the original set of tonoi with diatonic distances between them are indented. The further columns indicate which tonoi belong to the set: (3) The three 'oldest' tonoi; (7) The seven tonoi of Ptolemy; (8) Idem, with an eighth tonos added (to complete the diatonic octave range); (13) The thirteen tonoi of Aristoxenos (the chromatic octave filled); (15) The fifteen tonoi of the later Aristoxenians (with five Hypo, five regular and five Hyper-tonoi). Based on Doni's music example in the *Annotazioni*, p. 208, but with A as fundamental pitch of the Dorian tonos and expanded with the last added fourteenth and fifteenth tonoi.

Tonos	Pitch	3	7	8	13	15
Hypodorian	E		X	X	X	X
Hypoiastian	F				X	X
Hypophrygian	F♯		X	X	X	X
Hypoaeolian	G				X	X
Hypolydian	G♯		X	X	X	X
Dorian	A	X	X	X	X	X
Iastian (Ionian)	B♭				X	X
Phrygian	B	X	X	X	X	X
Aeolian	C				X	X
Lydian	C♯	X	X	X	X	X
Mixolydian (Hyperdorian)	D		X	X	X	X
Hyperiastian	E♭				X	X
Hyperphrygian (Hypermixolydian)	E			X	X	X
Hyperaeolian	F					X
Hyperlydian	F♯					X

For his theoretical considerations, Doni mostly followed Ptolemy's array of seven tonoi and seven octave species or modes (Ex. 8.1); he considered this system as the most effective way to restore the old modal doctrine. For his instrumental practice, however, he limited himself most often to only three tonoi, Dorian, Phrygian, and Lydian, the so-called the principal tonoi (*tuoni principali*)[8].

Ex. 8.1: The Greek modal system, shown in modern notation. Each tonos consists of the Greater Perfect System, a double-octave scale that can be represented by naturals if taken from *A* to *aa*. The tonoi are transpositions of this scale up to a fourth down or up. The octave species from *E* to *e* defines a mode with the name of the tonos in which it occurs.

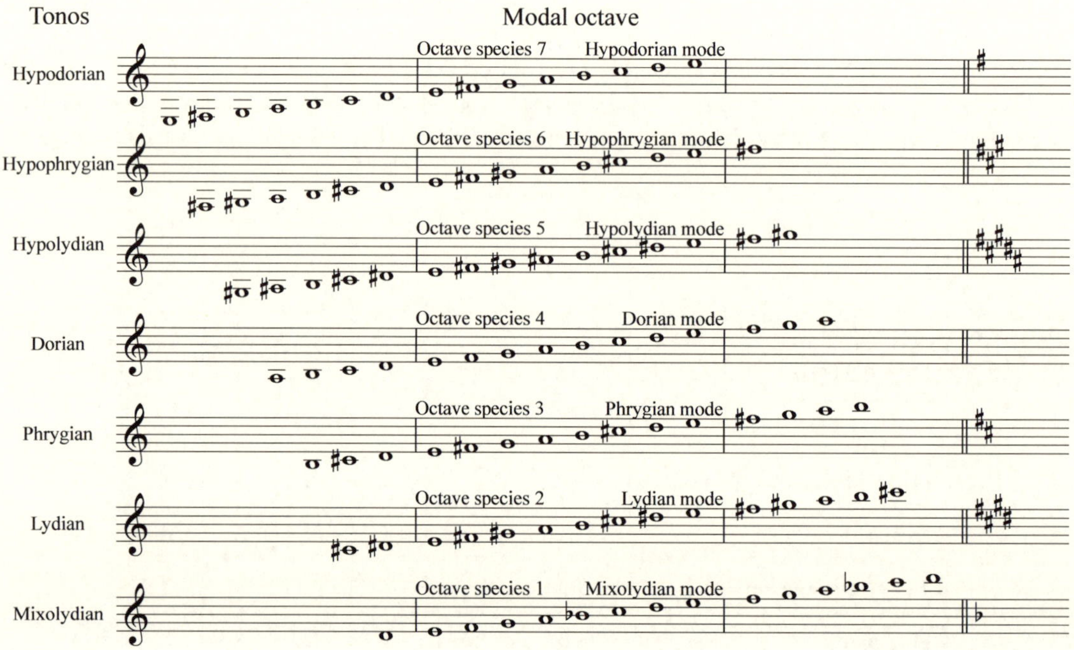

It is important to distinguish carefully between the concepts of tonos (*tuono*) and mode (*modo*), the main elements of Doni's modal theory. The definitions of both concepts were already outlined in the manuscript 'Trattato'. Tonos was defined as «a system or a modulation that is higher or lower in pitch, as when we say to sing lower or higher, even more when proceeding by the same species or mutation (*la medesima specie e mutanza* = mode)»[9].

[8]. These three tones were the oldest tones, according to Doni (*Annotazioni*, p. 104), referring to Ptolemy, *Harmonics*, Book II, Chap. 10 (BARKER 1989, p. 336): «They straighforwardly adopt the three oldest tonoi, called Dorian, Phrygian and Lydian by derivation from the names of the races from which they originated — or for whatever reasons anyone else wants to think up».

[9]. Doni, 'Trattato', [fol. 131v]: «tuono propriamente significa un Sistema o modulatione piu acuta o piu grave come quando diciamo cantare più grave o più acuto; etiandio che si procedi per la medesima specie e mutanza». Quoted from PALISCA 1997, p. 7.

Mode is a «a certain particular mutation or manner (*mutanza o maniera*)», which we would now describe as an 'octave species'[10].

In the *Compendio*, pp. 33-34, Doni presents musical examples that illustrate the three possible relationships between two scales: (1) when the mode is the same, but the tonos is not (Ex. 8.2a); (2) when the tonos is the same, but the mode is not (Ex. 8.2b); and (3) when both tonos and mode are different (Ex. 8.2c).

Ex. 8.2: (a) Two scales with the same mode (Dorian) but with different tonoi (Dorian and Phrygian). (b) Two scales with the same tonos (Dorian) but with different modes (Dorian and Phrygian). (c) Scales with different tonoi (Dorian and Phrygian) and different modes (Dorian and Phrygian). After Doni, *Compendio*, pp. 33-34.

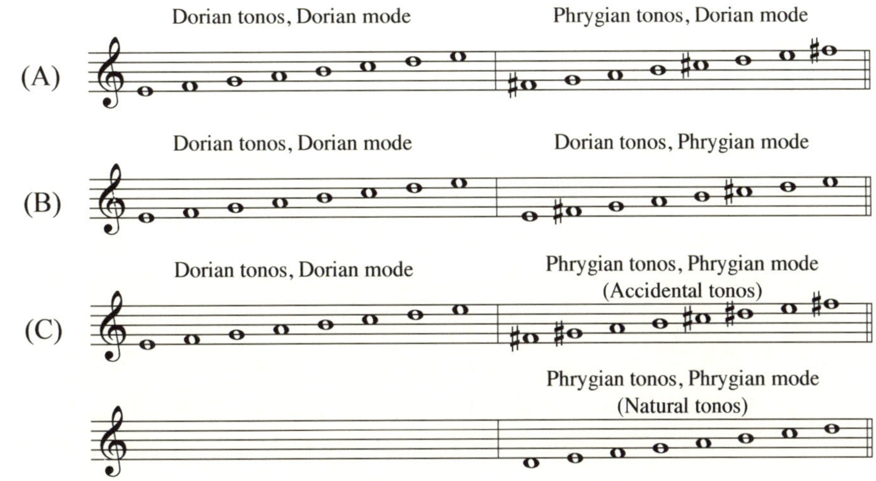

The impact of these operations becomes particularly clear when we generalize them over all modes and tonoi. Keeping the mode the same, but applying it to another tonos means selecting the octave span within that tonos that corresponds with the mode or octave species. The Dorian mode in the Dorian tones spans the octave from the fifth to the twelfth note from below. The Dorian mode in other tonoi equally spans the octave from the fifths to the twelfth note in the tonos. Keeping the tonos the same, but changing the mode results in filling in the modal octave, from *E* to *e*, with a different mode.

Changing tonos and mode at the same time has a peculiar effect. The change of tonos results in a displacement of the modal octave, which (at least in Doni's doctrine) always runs from the fifth to the twelve note of the tonos (and not just between *E* and *e*). The new modal octave can be filled in with any mode and this will for any mode other than the Dorian result in new (accidental) pitches in the tonos in question. Ex. 8.2c concerns the Phrygian tonos with

[10]. Doni, 'Trattato', [fol. 131v]: «modo propriamente significa quella tale o tal mutanza o maniera». Quoted from Palisca 1997, p. 7.

the Phrygian mode. Choosing the Phrygian tonos means the adoption of the octave F♯-f♯ as modal octave. This is filled in by the octave species of the Phrygian mode or T S T T T S T (T = whole tone, S = semitone), and this results in the scale F♯ G♯ A B C♯ D♯ E f♯. The displacement of tonos and mode with the same name is in fact a special case of the more general class of displacements with any tonos and any mode.

Doni calls the scale F♯ G♯ A B C♯ D♯ E f♯ the 'accidental tonos' (*tuono accidentale*). It is in fact the Phrygian tonos and mode presented in the notation of the Dorian tonos. To this accidental tonos, Doni adds a 'natural tonos' (*tuono naturale*), which presents the same octave species or (mode) with only naturals. In the case of the scale F♯ G♯ A B C♯ D♯ E f♯ it is the natural scale from D to d. To align the natural Phrygian tonos with the Dorian tonos, notes must sound a major third higher than notated.

The application of the Lydian mode in the Lydian tonos can be approached in a same way. The Lydian tonos implies the octave from G♯ to g♯ as modal octave. This is filled in by the octave species of the Lydian mode, which is T T S T T T S. The resulting scale is G♯ A♯ B♯ C♯ D♯ E♯ F× g♯, which defines the accidental Lydian tonos. This corresponds with the natural scale C D E F G A B c, functioning as the natural Lydian tonos. To align the natural Lydian tonos with the Dorian tonos, notes must sound an augmented fifth higher than notated.

It becomes clear from Ex. 8.1 and Ex. 8.2 that, when presented in modern musical notation, most tonoi have accidentals. Only one tonos can be fully written out without accidentals; in the notation chosen for Ex. 8.1 this is the Dorian tonos, being the middle tonos of the set of seven. Referring to Ptolemy's *Harmonics*, Book II, Chapter 11, Doni remarked that the Dorian tonos, because of that middle position, would be most suitable to be the reference of the whole system[11]. He called the ancient system of tonoi the 'mobile system' (*sistema mobile*), in contrast to the modern system of modes, which he termed the 'stable system' (*sistema stabile*), because all modes were derived from the same series of notes (in fact corresponding to the Dorian and Lydian tonoi, for modes without or with a flat in the scale, respectively)[12]. According to Doni, the modern system had arisen from the equivalence of the terms mode (*modus*), trope (*tropus*), and tone (*tonus*) that was established by Boethius when defining all three terms as octave species in Chapter 15 of Book IV of his *De institutione musica*[13].

---

[11]. Doni, *Trattato*, f. 145r. For the passage of Ptolemy referred to, see BARKER 1989, p. 338.

[12]. In Doni's words (*Compendio*, p. 10): «Haueuano dunque gl'antichi Tuoni, ciascuno la sua propria scala, o sistema, in modo, che uno non si continuava con l'altro; et non come i nostri, che si servono tutti d'una medesima scala, o Gamma». (Therefore, each of the old tones had its own scale or system so that one did not continue in the other; and not as our modern [modes], which are all served by one and the same scale.)

[13]. Doni, *Trattato*, [f. 140r], quoting Boethius (*De institutione musica*, BOETHIUS 1867, p. 341.19-21): «Ex diapason igitur consonantiae speciebus existunt, qui appelatur modi, quos eosdem tropos vel tonos nominant». This passage is also quoted and commented in PALISCA 1997, p. 10.

Doni designed a number of musical instruments, notably stringed instruments and keyboard instrument, on which it was possible to play several tonoi[14]. The *cembalo triarmonico* combined three tonoi (Dorian, Phrygian, and Lydian), the *violino diarmonico* or (better) *viola diarmonica* two tonoi (Dorian and Phrygian; Ill. 8.4).

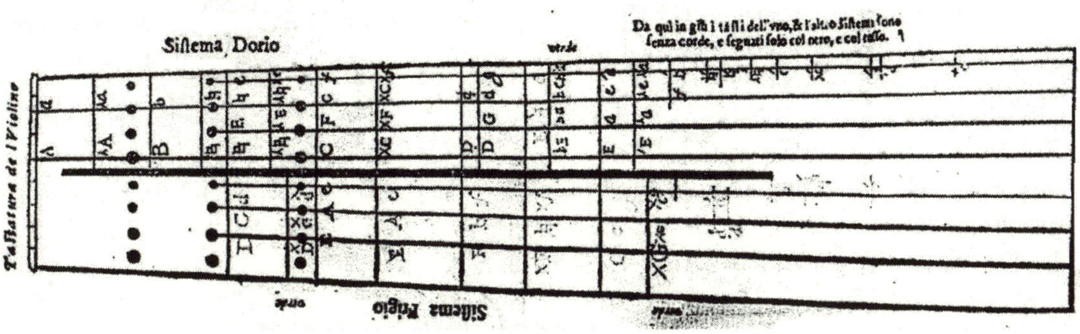

Ill. 8.4: Diagram of the *violino diarmonico* (bi-harmonic violin), from Doni, *Compendio*, opposite page 30. The Dorian system (*Sistema Dorio*) is implemented on the first four strings (at the top of the diagram), the Phrygian system (*Sistema Frigio*) has only three strings, so that the instrument has seven strings. Despite the name, it must have been a viol rather than a violin.

However, when several tonoi are combined in a single instrument, there will be differences of a comma interval (81:80) between the intonation of certain corresponding notes between tonoi. This arises because the tonoi are tuned internally according to just intonation, with unequal whole tones (larger ones of 9:8 and smaller ones of 9:10). We may remind ourselves here that the scales of the Greater Perfect System consist of tetrachords. The tetrachords consist of a semitone, a greater whole tone, and a lesser whole tone, in ascending order. They are tuned (again, in ascending order) as (16:15) × (9:8) × (10:9) = (4:3). This means that the D in the natural Phrygian tonos is a comma higher than the F# in the Dorian system, to which it corresponds. Similarly, there is a comma difference between the F# in the natural Phrygian tonos and the D in the natural Lydian tonos. In the notation of musical compositions, the lower note of such a comma pair is identified by Doni by a dot below the note (or the name of the note in tables, and so on). The pitch relations between the tonoi are depicted in Ex. 8.3, which is based on an illustration in Doni's *Compendium*, p. 41.

For the *viola diarmonica*, Doni designed a tablature notation, consisting of letters in different forms for the two octaves of five tonoi (Dorian, Phrygian, Lydian, Aeolian, and Iastian). These letter forms are tabulated in Table 8.2, using Doni's illustration in the *Compendio*, p. 127. Doni needed the Iastian tonos, in pitch between the Dorian and Phrygian tonoi, for

[14]. On Doni's instruments, see Barbieri 2008, pp. 221-276, and Kirnbauer 2013, *passim*.

the analysis of Gesualdo's *Tu m'uccidi, o crudele*, for which reason he will have included it in his tablature notation, just as he also included the other 'intermediate' tonos, the Aeolian, between Phrygian and Lydian.

Ex. 8.3: The relationship between the Dorian, Phrygian, and Lydian tonoi in their natural and accidental versions, after Doni, *Compendio*, p. 41. The straight horizontal dotted lines in Doni's illustration (dashed vertical lines in the music example) identify unisons between the systems. The curved dotted lines connect pitches with a comma difference between them. The letter name of the lower note of a 'comma pair' has a dot below it.

Table 8.2: The Letter Forms Designed by Doni for the Five Tonoi in the Tablature of the *Viola diarmonica*

System	A	B♭	B♮	C	D	E	F	G	a	b♭	b♮	c	d	e	f	g
Dorian	A	b	B	K	Δ	∈	F	Ƃ	α	β	♭	x	∂	ϵ	ſ	ℽ
Phrygian	A	B	♮	C	D	E	F	G	a	b	♮	c	d	e	f	g
Lydian	A	B	B	C	D	C	F	G	a	b	b	c	d	e	f	g
Aeolian	A	B	B	‹	D	E	F	G	a	b	b	(	∂	ε	f	g
Jastian	A	B	B	C	D	C	F	G	a	b	b	c	∂	ε	f	g

Chromatic Music

> *Francesco [Viola]: Do you believe therefore that the songs that those modern composer compose for chromatic [pitches], that they are chromatic?*
> *Adrian [Willaert]: I do not believe so*[15].

Before presenting his own opinion in the *Compendio* on compositions with 'many sharps and flats' (*diesi e b molli*), Doni quotes Zarlino, who, in the dialogue of 'Proposta XI' of the 'Ragionamento quarto' in *Le dimostrationi harmoniche* (Venice, 1571), rejected the presence of the chromatic genus in pieces designated as having 'many sharps and flats'[16]. Although Doni supported (and praised) the Zarlinian position, he also stated that Zarlino did not know «the form and essence of the true and ancient tonoi» (*la forma, & essenza de' Tuoni veri, & antichi*). According to Doni, Zarlino had disregarded one of the main points of his doctrine: the understanding that compositions and passages characterized by accidentals pointed to the use of different tonoi, and not to the use of different genera[17]. In Doni's words:

> But what is most important to know, and comprises the principal secret of this new or, rather, renewed doctrine, is this: that modern compositions marked with those accidental notes with sharps and flats are not, as wrongly believed up to now, mixtures of genera, but of tonoi[18].

The 'chromatic mixture' (*mistura di chromatico*) can normally be considered a simple 'thickening' of the diatonic genus, when a whole tone (for example, E-F♯) is divided into a larger semitone (E-F) and a smaller semitone (F-F♯). Doni explains these notes as indicating a mixture of the diatonic geus (because of the larger semitone E-F) and the chromatic genus (because of the smaller semitone F-F♯)[19]. If, however, an accidental changes the order of whole

---

15. Zarlino, *Dimostrationi* (1571), pp. '276' [=236]: «Fran[cesco Viola]: [...] Credete adunque che quelle canzoni, le quali compongono questi Moderni Compositori per Chromatiche [corde], elle siano Chromatiche? Adri[ano Willaert]: Credo che non siano».

16. Doni, *Compendio*, p. 17. See also the *Annotazioni*, p. 11.

17. As also asserted, for example, by Nicola Vicentino in his *L'antica musica ridotta alla moderna prattica* (Rome, 1555), f. 95r.

18. Doni, *Compendio*, pp. 16-17: «Ma quello, che più importa a sapersi, et in che consiste il principal segreto di questa nuova o più tosto rinovata dottrina, è questo; che le compositioni moderne segnate con quelle note accidentali di diesis, e *b* molli, non sono, come sin hora con notabile errore s'è creduto, mescolanza di Generi, ma di Tuoni». Translation, adapted, from Barbieri – Hurry 2007, pp. 92-93.

19. An interval classification that agrees, for example, with that practiced by Vicentino, *L'antica musica ridotta alla moderna prattica* (Rome, 1555), ff. 13v-15r.

tones and semitones, then the melodic quality should not be classified as 'chromatic', but rather as 'metabolic' (modulating, that is, modulating to a different tonos) because it results from a «mixture of notes from accidental modes or tonoi» (*mescolata con corde di Modi, ò Tuoni accidentali*[20]; Ex. 8.4).

Ex. 8.4: 'Metabolic' passages or «mixtures with notes of accidental modes and tonoi» (*mescolata con corde di Modi o tuoni accidentali*). After Doni, *Annotazioni*, p. 12.

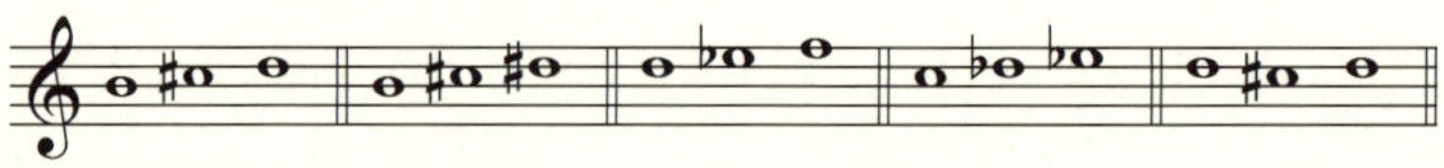

Ex. 8.5: Chromatic passages interpreted as mixtures of tonoi. The fundamental tonos is the natural Phrygian tonos. The black notes are derived from the Dorian and Lydian tonoi, respectively. When the Dorian tonos is notated to align it with the pitches of the natural Phrygian tonos, it has a key signature of four flats so that it can provide the E♭. When the Lydian tonos is notated to align it with the pitches of the natural Phrygian tonos, it has a key signature of four sharps so that it can provide the D♯. After Doni, *Annotazioni*, p. 14.

Ex. 8.6: Chromatic rising fourth interpreted as a mixture of genera and tonoi. Doni does not specify the system: the example is valid if the notes from the added tonos come from a tonos that is a whole tone higher than the fundamental tonos. Here, we have adopted the Dorian tonos as fundamental tonos. The black notes are derived from the accidental Phrygian tonos. The C♯ can also be seen as belonging to the chromatic genus of the Dorian tonos. After Doni, *Annotazioni*, p. 14.

Doni does, but for a different reason, also speak of chromaticism when a smaller semitone divides one of the whole tones D-E or G-A. These were intervals that were not normally considered appropriate places for chromaticism in the chromatic genus, the latter having C-D and F-G as the first choices to split into a chromatic and a diatonic semitone (because of their positions within the tetrachords B-C-D-E and E-F-G-A). For Doni, the division of D-E or G-A could make the composition chromatic not because of a mixture of genera, but because the accidental notes came from an accidental tonos (*tuono accidentale*) that differed from the fundamental tonos (*tuono fondamentale*). The following quotation explains this aspect:

[20]. Doni, *Annotazioni*, p. 12.

> If the system thickens not in the ordinary places of the chromatic genus, that is (to speak in more intelligible terms) between the *fa* and the *sol*, but elsewhere, as between the *sol* and the *la*, or below *mi*, then this composition can [also] be called chromatic, although it does not change by notes of the chromatic genus, but by the accidental notes from other tonoi[21].

In Ex. 8.5, Doni presents two examples of chromaticism as mixtures of tonoi. Both examples have the Phrygian tonos as fundamental tonos; in the first example, the Eb comes from the accidental Dorian tonos, and in the second, the D♯ from the accidental Lydian tonos.

Another example presents a chromatically filled ascending fourth (Ex. 8.6). There, Doni recognizes mixtures of both genera and tonoi, as can be concluded from his comment:

> [this passage] can absolutely be said to be chromatic; but considering it exactly, it will be found to be a mixture of genera and tonoi, because the first, second, and sixth notes are notes common to the chromatic and diatonic [genera], but the fifth [note] is diatonic in a tonos that is higher than the fundamental tonos, and the third is similarly diatonic in the same accidental tonos, and at the same time it is chromatic, belonging to the fundamental tonos[22].

Given these chromatic examples, the chromatic quality of a passage is related to the presence of accidental notes either from the chromatic genus of the fundamental tonos or from an accidental tonos. Doni repeated this view in his treatise *De præstantia musicæ veteris* (Florence, 1647):

> [...] (which the ignorant calls 'with sharps and flats', whereas you, being more educated, call it extra-harmonic and metabolic, as I believe) are not [changes] of genus, but of tonos: so that anyone who in the past hundred years began to utilize them (which more than anyone else the Prince of Venosa was wont to do) obtained not the sounds and intervals of the chromatic genus, except in a few cases, and still less those of the enharmonic, but instead modulated melodies (*metabolici cantus*), composed of notes belonging to the most varied and dissimilar tonoi[23].

---

21. *Ibidem*, p. 14: «Se il Sistema s'inspesserà non ne' luoghi ordinarij del Cromatico; cioè (per parlare in termini più intelligibili) trà il *fa* e 'l *sol*; ma altrove; come tra il *sol*, & il *la*; ouero sotto il *Mi*; tal compositione si potrà chiamare Cromatica; benche non s'alteri con le corde Cromatiche; mà con le accidentali d'altri Tuoni».

22. *Ibidem*: «si potrà dire assolutamente, che sia Cromatico: mà considerando esattamente, si troverà misto di Genere, e di *tuono*: perche la prima, seconda, e sesta nota, sono voci, o corde comuni al Cromatico, e Diatonico: La quinta è Diatonica; mà d'un tuono più acuto del fondamentale; e la terza è similmente Diatonica del medesimo *tuono* accidentale; et insieme Cromatica propria del *tuono* fondamentale».

23. Doni, *De præstantia*, p. 24: «(quos diesatos et bemollatos barbari vocant, vos politiores exharmonios, et metabolicos, ut opinor) non generis esse, sed toni: ac proinde quicumque intra hos centum annos in suis

Doni's reference to Gesualdo is particularly important for us when proceeding with the next step of our study, since an excerpt of Gesualdo's madrigal *Tu m'uccidi, o crudele* is found in conventional score and in viol tablature in the *Aggiunta* of the *Compendio* (pp. 151-160).

The term '*stylus metabolicus*' was also used by Athanasius Kircher in Chapter 9 (*De mutatione modi, sive toni, sive stylo metabolico*) of Book VII of the *Musurgia universalis, Tomus primus* (Rome, 1650), pp. 672-675. Kircher, «who clearly took the term from Doni», as Patrizio Barbieri (2007) states[24], quotes Giacomo Carissimi, Domenico Mazzocchi, Pietro Eredia, Pietro Della Valle, Benedetto Narducci, as well as Gesualdo, as composers in that style. In the same chapter, Kircher mentions Doni, although only to praise his erudition concerning the ancient tonoi and to point to the presence of a composition by Eredia (*Passa la vita all'abbassar d'un ciglio*) in the *Compendio*. In a letter of 17 December 1641 to Doni, Kircher mentions the preparations for the publication of his *Ars magna lucis et umbrae* (which would only be published in 1645) and his *Musurgia universalis* (published in 1650). When writing about the *Musurgia*, Kircher says that he will frequently quote Doni and he asks him for some 'chromatic compositions', to incorporate in the volume. In the words of Kircher:

> And I prepare another work at the instigation of the new descendant of the Pope [Alexander VI], Cardinal [Giambattista] Pamphilj [later Pope Innocent X], with is called *Musurgia curiosa*, in which I promise, with help of the *ars combinatoria* or the rabdological theory of music[25], to teach in little time all the artifice to compose music, also to one who does not know music. I will quote you often in the work, and I would wish that you send me some chromatic compositions, which I will include with your name as example[26].

melodijs usurpare eos ceperunt (quod prae ceteris Venusinus Princeps factitauit) non chromatici generis sonitus atque intervalla (si pauca quaedam loca exceperis) multoque minus enarmonia; sed metabolicos potius cantus, ex peregrini cuiusdam atque extranei toni chordis compositos». Translation, adapted, from Barbieri – Hurry 2007, p. 93.

24. *Ibidem*, p. 93.

25. 'Rabdology' is a term introduced by John Napier, in his *Rabdologiae, seu Numerationis per virgulas libri due* (Edinburgh, 1617), for methods to simplify arithmetic operations such as multiplication and division.

26. Doni, *Commercium*, cc. 155-156: «& preparo un'altra operetta al istanza del novo nipote del Papa, Cardinal Panfilio, che si chiama *Musurgia curiosa*; nella quale io promesso per l'arte combinatoria, o vero Rabdologia musurgica. insegnar tra poco tempo tutto l' artificio di comporre la Musica, ancora a uno che non sappi la Musica. Cito V[ostra] S[ignoria] spesso nella detta opera, et desidererei, che V[ostra] S[ignoria] mi mandasse qualche Composizione cromatica, la quale io col suo nome inserirei *paradigmatis loco*».

The Tonal Space of Gesualdo's *Tu m'uccidi, o crudele*

But, to sweeten the taste of the curious reader, embittered perhaps by my badly composed music, I have put here at the end the beginning of the most artful madrigal Tu m'uccidi, o crudele, by the Prince of Venosa, really the prince of the modern composers[27].

Doni brought his doctrine into practice in the appendix ('Aggiunta') of the *Compendio*, which includes three musical compositions, each preceded by analytical comments that explain the tonoi found in them. The first is an instrumental Duo composed by Doni himself, with passages in the Dorian and Phrygian tonoi in the diatonic and chromatic versions (pp. 139-150); the second, an excerpt (bars 1-25) from Gesualdo's madrigal *Tu m'uccidi, o crudele* (pp. 151-160); and the third, the sonnet *Passa la vita all'abassar d'un ciglio* by Maffeo Barberini (Pope Urban VIII) set to music by Pietro Eredia (pp. 161-171).

The first two compositions are presented in two notational formats: first in conventional staff notation in score, then in tablature for Doni's special instruments, also in score. Eredia's composition is presented only in conventional staff notation. In the comments that precede the notation of Gesualdo's madrigal, Doni presents, in detail, justifications for his tonal interpretation, providing a set of considerations that allow us to investigate this piece (at least part of it) from Doni's perspective.

The versions (score and tablature) of bars 1-25 of the madrigal *Tu m'uccidi crudele* (Fig. 22) as found in the *Compendio* (pp. 153-160; Ill. 8.5) largely follow the edition of the complete madrigal in Gesualdo's *Madrigali a cinque voci, Libro quinto* (Naples, 1611, and later editions) and in the *Partitura delli sei libri de' madrigali a cinque voci* (Genoa, 1613). The brevis bars of the edition in score are replaced by semibrevis bars. In the text incipit, the interjection «o» is omitted, and after the text incipit no further text is given. In his commentary, Doni explains that the tablature for the viols contains only the four lines that correspond with the strings of the Dorian partition of the viols. Five such instruments are needed, of three types, with the string tuned *E – A – e – a* (for the Basso part), *A – e – a – e1* (for the Tenore and Alto parts), and *e – a – e1 – a1* (for the Quinto and Canto parts). There is no tablature for the strings of the Phrygian partition of the instrument. Yet, the Dorian tablature contains a few notes derived from other tonoi, notably the Phrygian, Lydian, and Iastian tonoi. They are distinguished by their letter forms.

27. Doni, *Compendio*, p. 151: «Ma per raddolcire il gusto del curioso Lettore, amareggiato forse da queste mie mal composte Modulationi, ho posto in vltimo luogo vn principio di quell'artificiossimo Madrigale, *Tu muccidi crudele. &c.* Del Principe di Venosa; veramente Principe de' Compositori moderni».

Ill. 8.5a: Carlo Gesualdo da Venosa, *Te m'uccidi, o crudele*, bars 1-12, in Giovanni Battista Doni, *Compendio del trattato de' generi et de' modi* (Rome, 1635), pages 153-154 (staff notation) and pages between pages 156 and 157 (tablature notation).

Doni considers only a few notes in bars 1-25 of Gesualdo's *Tu m'uccidi crudele*, individually regarding their tonal qualities (Ex. 8.7). The Dorian tonos is understood to be the fundamental tonos. The Bb can be seen as belonging to the 'conjunct tetrachord' A-B♭-C-D, the notes F♯ and C♯ as belonging to the chromatic tetrachords E-F-F♯-A and B-C-C♯-E, respectively, within the Dorian tonos.

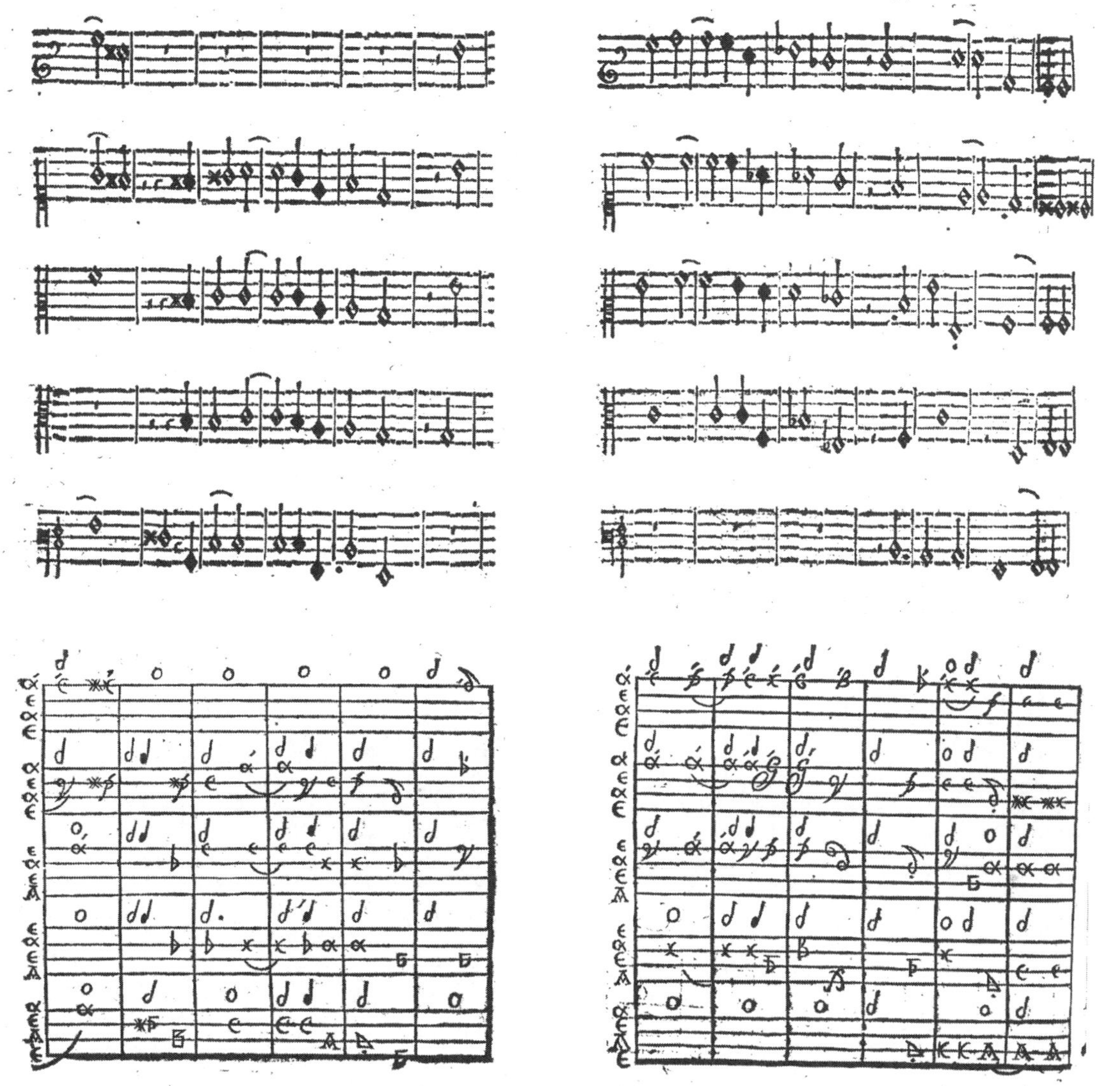

Ill. 8.5b: Carlo Gesualdo da Venosa, *Te m'uccidi, o crudele*, bars 13-24, in Giovanni Battista Doni, *Compendio del trattato de' generi et de' modi* (Rome, 1635), pages 155-156 (staff notation) and pages between pages 156 and 157 (tablature notation).

In bar 4 of the Quinto, Doni allots the D♯ to the (accidental) Phrygian tonos, corresponding to B♮ in the natural Phrygian tonos. The justification presented by Doni is essentially based on the impossibility of presenting D♯ as a chromatic note in the Dorian tonos. In the tablature, however, the note is included with the letter for the Bn in the Dorian tonos. This is probably an engraving error. The G♯ in bar 3 can equally be allotted to the Phrygian tonos, but the corresponding tablature letter rather looks (again) like a Dorian E.

Ex. 8.7: Carlo Gesualdo da Venosa, *Tu m'uccidi, o crudele*, from the *Madrigali a cinque voci, Libro quinto* (Naples, 1611; Genoa, 1613), bars 1-25, with Doni's interpretation of the notes not belonging to the diatonic Dorian mode. Chrom. = Chromatic; Iast. = Iastian; Phryg. = Phrygian.

In bar 5 of the Alto, Doni identifies the A♯ as a note of the accidental Lydian tonos, corresponding to D in the natural Lydian tonos. In his commentary he states that he has not included this note on the fingerboards of his viols, «to avoid the confusion of so many notes and not to multiply in so many tonoi, in addition to the aspect of reserving that fret for the enharmonic note A× [in the enharmonic conjunct tetrachord]». Yet, he continues, «here you see the note written with its own and natural sign»[28]. The tablature gives, however, a Phrygian E, which would sound as G♯ in the Dorian tonos.

Bars 6-20, with no comments by Doni, can be read in the Dorian tonos for the larger part, with a few chromatic notes. Two notes, a D♯ in bar 15 of the Alto and the G♯ in bar 16 of the Quinto, can be related to the Phrygian tonos. They are misrepresented in the tablature in the same way as these notes in bars 3-5. The (Dorian) C in bars 19-20 of the Quinto is misrepresented in the tablature by the letter for the Dorian A.

Four notes in bars 20-21 (A♭, D♭, and Ev in two different octaves) are identified by Doni in his comments as belonging to the Iastian tonos. They are correctly notated in the tablature by the corresponding Iastian letter forms for G, C and D. That Doni called the Iastian tonos here for help, was certainly a consequence of the absence of an A♭, D♭, and E♭ in the three basic tonoi, the Dorian, Phrygian and Lydian tonoi.

Ex. 8.8: Tonal space of bars 1-25 of Gesualdo's *Tu m'uccidi, o crudele*, according to Doni's approach of the modes, tonoi and genera in 'chromatic music'.

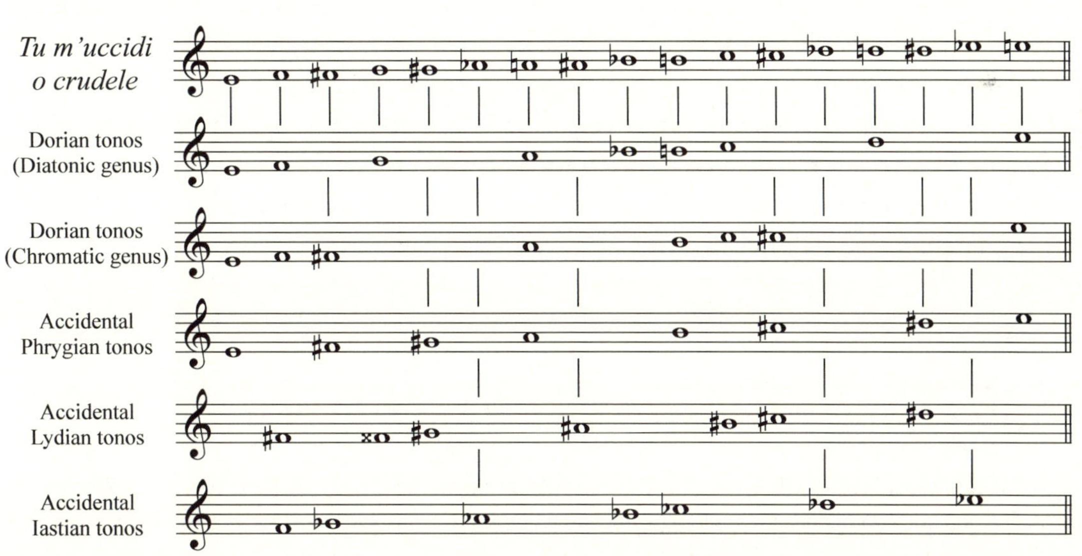

[28]. *Ibidem*: «per euitare la confusione di tante Voci, e non multiplicare in tanti Tuoni, oltre il rispetto di saluare quel tasto per la Voce Enarmonica XA, tuttavia qui si vede notata col suo proprio, e natural carattere».

Despite the errors found in the tablature, it is clear that the excerpt from *Tu m'uccidi, o crudele* represents the diatonic Dorian tonos as the predominant tonos (with a few notes using the chromatic genus), enriched by notes from three accidental tonoi — Phrygian, Lydian and Iastian, all diatonic. This is summarized in Ex. 8.8.

Doni's Musical Legacy: a '*maniera nuova*'

Follow the few, and not the common mass[29].

Doni's decision to include Gesualdo's madrigal among the pieces to illustrate the practical applicability of his restored doctrine may be more than just to «sweeten the taste of the curious reader», as claimed in the comments to the madrigal[30]. Perhaps more significantly, addressing a piece that, in the eyes of the author, was composed by one of the main exponents of modern music, could be seen as a subtle means of proving the applicability of his theory. This could establish a theoretical consistency that intended to lead the prevailing tradition of the time towards a 'new manner' (*maniera nuova*) or a 'very old now renewed manner' (*maniera antichissima rinovata*). These were the words of Pietro della Vale when he defined Doni's aims in the dedication of his *Dialogo per la Festa della Santissima Purificazione* to Girolamo Rosini[31].

In this way, Doni's comments on the excerpt of *Tu m'uccidi, o crudele*, which was intended to be performed by adapted viols, act as a means to validate his restoration plan, so that his objective would not remain solely speculative. Although it was only *Tu m'uccidi, o crudele* that received prominence in the *Compendio*, other pieces and composers are cited and briefly commented on throughout the *Compendio* and in other writings. These references show a genuine interest on the part of the author to be a reformist agent trying to transform the understanding of the science of harmony as a result of his new discoveries.

Therefore, Doni's tireless effort as a 'restorer of antiquity' (*ristoratore dell'antichità*), to use Giovanni Battista Martini's words (1762)[32], or as a 'musicological activist', following

---

29. *Ibidem*, p. [172]: «Seguite i pochi; e non la volgar gente». A motto printed on the last page of the *Compendio*. From Petrarca, *Sonnetto in vita di M. Laura* 67 (73).

30. *Ibidem*, p. 151.

31. I-Rn, Fondi Minori, MS Mus. 123. When commenting on the music of the *Dialogo*, Della Valle confirms his dedication to Doni's doctrine, although until that moment (1640) it was little practiced (Solerti 1905, p. 280): «La musica, conforme al mio solito, è di quella maniera nuova, ò per dir meglio, antichissima rinovata dal Sig[nore] Gio[vanni] Battista Doni, che sola io tratto, e che infin' ad hora da altri poco si pratica, cioè con mutatione di varij tuoni al modo de gli antichi».

32. From Giovanni Battista Martini's letter of 12 June 1762 to Giovanni Battista Passeri, in Doni, *De' trattati, Tomo secondo*, p. 265.

Claude Palisca's characterisation (1997)[33], although not fully understood by the musicians of his own time, reached a high degree of theoretical formulation and erudition which certainly cannot be overlooked in the history of music theory. His understanding of tonal space opens an important theoretical perspective for the analysis not only of Gesualdo's 'chromatic', or better, 'metabolic' music, but also of related repertoire, composed mainly in Rome during the first half of the seventeenth century.

Bibliography

Barbieri 2008

Barbieri, Patrizio. *Enharmonic Instruments and Music 1470-1900: Revised and Translated Studies*, Latina, Levante, 2008.

Barbieri – Hurry 2007

Barbieri, Patrizio – Hurry, Ken. 'Pietro della Valle. The 'Esthèr' Oratorio (1639) and Other Experiments in the «stylus metabolicus». With New Documents on Triharmonic Instruments', in: *Recercare*, xix/1-2 (2007), pp. 73-124.

Barker 1989

Greek Musical Writings. 2: Harmonic and Acoustic Theory, translated and edited by Andrew Barker, Cambridge, Cambridge University Press, 1989.

Boethius 1867

Boethius, Anicius Manlius Torquatus Severinus. *De insititutione arithmetica libri duo, De institutione musica libri quinque*, edited by Gottfried Friedlein, Leipzig, Teubner, 1867.

Kirnbauer 2013

Kirnbauer, Martin. *Vieltönige Musik. Spielarten chromatischer und enharmonischer Musik in Rom in der ersten Hälfte des 17. Jahrhunderts*, Basel, Schwabe, 2013.

Mersenne 1959

Mersenne, Marin. *Correspondance du P. Marin Mersenne, réligieux minime. 5: 1635*, edited by Cornelis de Waard, Paris, Éditions du Centre National de la Recherche Scientifique, 1959.

Palisca 1985

Palisca, Claude. *Humanism in Italian Renaissance Musical Thought*, New Haven (CT), Yale University Press, 1985.

---

33. Palisca 1997.

Palisca 1997

Id. 'Giovanni Battista Doni's Interpretation of the Greek Modal System', in: *The Journal of Musicology*, xv/1 (1997), pp. 3-18.

Solerti 1905

Solerti, Angelo. 'Lettere inedite sulla musica di Pietro della Valle a G. B. Doni ed una Veglia drammatica-musicale del medesimo', in: *Rivista Musicale Italiana*, i/4 (1905), pp. 271-338.

Chapter Nine

Me veux-tu voir mourir: Joan Albert Ban versus Antoine Boësset (1640-1641)

Rudolf Rasch
(Utrecht University)

In the history of music and its theory, every century has known its controversies. In the sixteenth century, for example, there originated disputes between Giovanni Spataro and Pietro Aaron on the rules of counterpoint, Nicola Vicentino and Vicente Lusitano on the interpretation of ancient genera, Vincenzo Galilei and Gioseffo Zarlino on a variety of topics, Giovanni Maria Artusi and Ercole Bottrigari mainly about authorship issues of Bottrigari's *Il Desiderio* (Venice, 1594), and Artusi and Claudio Monteverdi on the admissibility of a counterpoint freer than what was described by Gioseffo Zarlino[1]. Friedrich Wilhelm Marpurg's criticisms of Johann Philipp Kirnberger on several themes date from the eighteenth century[2], and the disputes that Eduard Hanslick aroused with his *Vom Musikalisch-Schönen* (Leipzig, 1854) from the nineteenth.

1. On the Spataro-Aaron dispute, see Blackburn – Lowinsky – Miller 1991; on the Vicentino-Lusitano dispute, Kaufmann 1966, pp. 22-32; on the Galilei-Zarlino controversy, Walker 1974, Palisca 1985a, pp. 265-279, and Goldberg 2011; on the Artusi-Bottregari dispute, Maniates 1993; and on the Artusi-Monteverdi dispute, Palisca 1985b and Carter 1992.

2. Serwer 1970.

Table 9.1: Overview of the Galilei-Zarlino Controversy

Vincenzo Galilei	Gioseffo Zarlino (Giovanni Maria Artusi)
	Zarlino, *Le istitutioni armoniche* (Venice, 1558)
Galilei, *Dialogo della musica antica, et della moderna* (Florence, 1581)	
	Artusi, pamphlets (lost) Zarlino, *Sopplimenti musicali* (Venice, 1588)
Galilei, *Discorso intorno all'opere di Messer Gioseffo Zarlino da Chioggia* (Florence, 1589)	
	Artusi, *Trattato apologetico in difesa del* [...] *Zarlino da Chioggia* (Venice, 1590; lost)
Galilei, Critique of Zarlino (I-Fn, MS Gal. 5)	

Table 9.2: Overview of the Artusi-Bottrigari Controversy

Giovanni Maria Artusi	Ercole Bottrigari
	Alemanni Benelli [=Bottrigari], *Il Desiderio* (Venice, 1594) Bottrigari, *Il Desiderio* (Bologna, 1599)
Artusi, *L'Artusi, overo Delle imperfettioni della moderna musica* (Venice, 1600) Artusi, Letter of dedication and preface to Annibale Melone [=Bottrigari], *Il Desiderio* (Milan, 1601)	
	Bottrigari, *Lettera di Federico Verdicelli* (Milan, 1602; I-Bc, MS B.146) Bottrigari, 'Ant-Artusi' (lost)
Artusi, 'Considerationi musicali', appendix to his *Seconda parte dell'Artusi, overo Delle imperfettioni della moderna musica* (Venice, 1603)	
	Bottrigari, 'Aletelogia di Leonardo Gallucio' (I-Bc, MS B.43; 1604)

Table 9.3: Overview of the Artusi-Monteverdi Controversy

Giovanni Maria Artusi	Claudio Monteverdi (L'Ottuso; Giulio Cesare Monteverdi)
	Monteverdi, *Cruda Amarilli*, *Anima mia, perdona*, *Era l'anima mia*, *Ecco Silvio*
Artusi, *L'Artusi overo Delle imperfettioni della moderna musica* (Venice, 1600)	
	'L'Ottuso', letters (lost)
Artusi, *Seconda parte dell'Artusi, overo Delle imperfettioni della moderna music* (Venice, 1603)	
	Monteverdi, 'Studiosi lettori', in: *Il quinto libro de' madrigali* (Venice, 1605)
Artusi, *Discorso musicale di Antonio Braccino da Todi* (lost)	
	Giulio Cesare Monteverdi, 'Dicharatione', in: Claudio Monteverdi, *Scherzi musicali* (Venice, 1607)
Artusi, *Discorso secondo musicale di Antonio Braccino da Todi per la dicharatione posta ne' Scherzi musicali del Sig. Claudi Monteverdi* (Venice, 1608)	

Often these controversies took the form of a dialogue, not in the manner of a conversation, but in the form of an exchange of writings. Something written or published on a certain topic was the starting point. An opponent wrote a rebuttal to the opinions presented in writing. That rebuttal was usually followed by a written rejoinder by the first author, to use the legal term. The opponent might have written a response to the rejoinder, and so on. Thus, there was often a whole series of written exchanges. Tables 1, 2 and 3 summarize the exchanges in the controversies between Zarlino and Galilei, Bottregari and Artusi, and Artusi and Monteverdi.

The seventeenth century also was full of controversies. One of the most interesting is certainly the dispute between the Dutch priest and amateur composer Joan Albert Ban and the French professional composer Antoine Boësset, a controversy complicated by the support that Boësset received from his fellow-countrymen Marin Mersenne and René Descartes, who acted like 'solicitors' on behalf of Boësset. (Boësset himself entered relatively late in the exchange of writings, as we will see below. Before that, the controversy was instead 'about him' than 'with him'.) This controversy is the subject of the present chapter. One could — *cum grano salis* — describe the unfolding of the controversy as a tragedy (or perhaps a comedy) in five acts with prologue and epilogue. That is what we will do here, before discussing the arguments put forward by both parties.

There are five main characters in this tragedy, two Dutchmen and three Frenchmen. The French characters, the *père minime* and universal scholar Marin Mersenne (1588-1648), the composer Antoine Boësset (1586-1643), and the philosopher René Descartes (1596-1650), are no strangers to those interested in seventeenth-century music and its theory. By constrast, the two Dutchmen may need a few words of explanation.

Ill. 9.1ab: Theodor Matham, *Vanitas Still Life with Musical Instruments* (Engraving; 23 × 32 cm; 1622). NL-Ar, RP-P-1904-30. The musical composition shown is a two-part setting of *Omnia hominum tenui pendentia filo* (All things human hang by a thin thread; after Ovid, *Epistulae ex Ponto*, iii, 4, 35-36) by Joan Albert Ban.

Joan Albert Ban (Joannes Albertus Bannius) (1597/8-1644) was a Catholic priest and amateur musician. Born in Haarlem (Holland), he studied law in Louvain and Cologne[3]. After his studies he became a parish priest in his native town. In the United Provinces (to which the province of Holland and Haarlem belonged), Calvinist faith was the official religion, but Catholicism was tolerated in private spheres. Ban had developed a theory of vocal music in which

[3]. On Ban, see Huygens 1882, pp. xxxvi-cxlv, and Rasch 1983.

Ill. 9.2: Paulus Pontius (after Anthony van Dyck), *Constantijn Huygens* (engraving, 24.5 × 17.5 cm). NL-Ar, RP-P-OB-16.462.

every significant element of the text had to be expressed in the melody as well as in the harmony. In his opinion, the performance of Gregorian chant during the Catholic liturgy needed such a theory, as well as vocal music in general, because melody was indispensable for conveying the meaning of the text to the audience. Accordingly, he developed an *Affektenlehre* for musical intervals: to each interval he assigned a certain *Affekt*. To express this *Affekt* well, the interval had to be tuned perfectly. To this end, Ban designed a perfect intonation system with nineteen notes per octave. Such a system could not be realized on a normal keyboard with twelve keys but required a 'perfect keyboard' with eighteen keys[4]. Apart from melody and harmony, the prosody of the text (that is, the sequence of stressed and unstressed syllables) also had to be reflected carefully in the composition, both by the pitches of the notes and by their durations.

The other Dutchman in this tragedy, Constantijn Huygens (1597-1687), was part of the cultural and political elite of the United Provinces. He was secretary to the Stadtholder Prince Frederick Henry in The Hague. He wrote poems in several languages (Dutch, French, Latin,

[4]. Rasch 1983.

English, etc.) and was an avid amateur musician[5]. He had learned to play the lute, the viola da gamba and keyboard and composed pieces for these instruments. He corresponded with many distinguished figures in the arts, letters and sciences, in Holland, and abroad. The famous physicist Christiaan Huygens was his son.

At the core of the controversy between Ban and Boësset were differing views on the nature of vocal musical composition. Ban believed that vocal music should be entirely dependent on the text, in all respects, with an almost deterministic relationship between the meaning of the text and the musical composition. For Boësset, the most important composer of *airs de cour* of the time, musical composition was more a matter of musical rules. The musical settings of the French poem *Me veux-tu voir mourir* that Ban and Boësset composed, Ban according to his deterministic rules, Boësset in his own musical style, occupied a principal place in the controversy. Ban defended his own composition and criticized that of Boësset. Boësset's composition was defended not only by the composer himself, but also by Mersenne, by a 'Treasurer-General' (Jacques de Gouy), and by René Descartes.

Most of the combatants in this controversy never met in person. Mersenne and Boësset, who both lived in Paris, never met the two Dutchmen, Ban and Huygens. It is unknown if Descartes has met Boësset in Paris, but certainly he knew Mersenne. In Holland, Descartes met with Huygens and Ban multiple times. Due to the geographical separation of the main characters, the controversy itself was an epistolary one, rather than one conducted through printed pamphlets or more extensive publications; consequently, very few people in the seventeenth century knew of the controversy[6].

In the vein of a tragedy, we can discern in the Ban-Boësset controversy a prologue, five acts and an epilogue.

PROLOGUE: 1636-1640

The prologue covers the period from 1636 until the beginning of 1640. Ban had first developed his theories without communicating them by word or writing to other people. In 1636 Huygens visited the city of Haarlem and met Ban there. In a few letters after this meeting Ban exposed his theories to Huygens, who was the first to hear about them[7].

In 1637 André Rivet, a French Calvinist theologian who had settled in Holland, introduced Ban as a potential correspondent to Marin Mersenne. Mersenne's first letter to Ban was probably one of 23 November 1637 (now lost). In this letter Mersenne asked Ban four questions about the vibrations of a tensioned string. Ban replied amply, although his letter arrived in Mersenne's hands in a roundabout way. Ban sent his answers first to William Boswell,

[5]. On 'Huygens and music', see HUYGENS 1882, NOSKE 1987, HUYGENS 2007, pp. 118-134.

[6]. The first publication of the documents concerning this controversy is HUYGENS 1882, pp. lxix-cxx.

[7]. Ban to Huygens, 11 August 1636 (HUYGENS 2007, pp. 272-278), 18 August 1636 (*ibidem*, pp. 280-287).

the English ambassador to The Hague[8]. After receiving a reply from Boswell, he then sent his letter (now lost) to Boswell[9], asking him to give it to Huygens to forward it to Mersenne.

On 17 April 1639, Ban sent another letter to Mersenne[10]. The letter itself is very short but its shortness is amply compensated by the appendices, which include a series of compositions by Ban for two to seven voices, an 'Anatomia quarundam cantionum musicarum Parisiensium' (Analyses of Some Compositions by French Composers), an analysis of some of his own compositions titled 'Anatomia musicae', a little treatise on music written by him, and the table of contents of a book that he intended to write[11].

The 'Anatomia quarundam cantionum musicarum Parisiensium' contains the analysis of three compositions, *O crux ave spes unica* for four voices by Fleury (the text is an excerpt from the hymn *Vexilla regis*), *Est Deus pastor* (Psalm 23) for four voices by Jacques Mauduit (published in Mersenne's *Quaestiones celeberrimae in Genesim* [Paris, 1623], pp. 1633-1634) and *Odi profanum vulgus* for two voices by Du Chemin (published in Mersenne's *Harmonie universelle* [Paris, 1636], *L'art de bien chanter*, p. 395). In each composition, Ban found errors against his rules, for which he proposed corrections. Such errors were missing, it goes without saying, in the compositions of his own hand.

Here we come across one of Ban's peculiarities when discussing music: the reworking of other composer's pieces of music that he did not find entirely satisfactory, that is, not completely compliant to his rules for vocal music. A reworking of Pomponio Nenna's madrigal *Fili mentre ti miro* had been sent by Ban to Huygens with a letter of 11 August 1636. Ban's letter to Mersenne of 17 April 1639 contained a two-part version of *O crux ave spes unica* as an improvement of Fleury's four-part *O crux ave*. With the same letter Ban sent a corrected version of Mauduit's *Est Deus pastor* to illustrate his analysis of this piece. He also sent a new version of the chanson *Blessé d'une playe unhumaine* (Wounded by an inhuman stroke), originally composed by Claudin Le Jeune and published in Le Jeune's *Livre de mélanges* (Antwerp, 1585)[12]. The revision of this composition by Ban led Huygens to add to his letter to Mersenne of 26 August 1639, in a postscript, the following words:

---

8. Ban at William Boswell, 14 December 1637 (Huygens 1882, pp. lxiii-lxvi).

9. Ban to William Boswell, 15 January 1638 (Mersenne 1962, no. 641, p. 3).

10. Ban to Mersenne, 17 April 1639 (Mersenne 1964, pp. 384-386).

11. Several of the appended documents are edited in Mersenne 1962, pp. 205-209, and Mersenne 1964, pp. 256-262, 343-346, and 386-401. The compositions sent with the letter are *Ego dormio* [*et cor meum vigilat*] (text from the Song of Songs, 5:2) à 2; *O Domine Jesu Christe*, à 3; *Jesu dulcedo*, à 4; *Hodie apparuit*, à 5; *O crux ave*, à 4; *Blessé d'une playe inhumaine*, à 4; *Libera Domine*, à 7. None of these compositions are preserved, except the treble part of *Jesu dulcedo*, which was attached to Ban's letter to Boswell of 14 December 1637 (Huygens 1882, pp. lxv-lxvi).

12. *Blessé d'une playe inhumaine* is the first piece in Claudin Le Jeune's *Livre de mélanges* (Antwerp, 1585). It consists of three parts, à 5, à 3 and à 5, respectively.

> You are of the same opinion as me so far as the comparison of Claudin's *La playe inhumaine* (The inhuman wound) and that of Bannius is concerned, and who judges it differently seems to me to be struck at the brain by an even more inhuman wound[13].

So, with this prelude onwards to the first act!

THE FIRST ACT: APRIL-MAY 1640

The first act covers the months of April and May of 1640. In his letter to Mersenne of 1 April, Constantijn Huygens sowed — probably unknowingly — the seed of controversy when writing about Ban and

> [...] his discourses on music, according to which he only has never composed any good air, which is an insolent paradox[14].

As an answer, Mersenne must have sent the text of *Me veux-tu voir mourir* with the suggestion that Ban would put this text to music, but it is unknown when or how.

We know that Ban, on 5 May, received the text of *Me veux-tu voir mourir* with the request to set it to music, thus allowing the comparison of his music with a French composition of the same text. Ban immediately set himself to work and finished his composition on the same day (see Ex. 9.1)[15]. Then, he sent it to Huygens, to be forwarded to Mersenne[16]. Huygens sent the composition to Mersenne, who must have received it at the end of May or early June.

THE SECOND ACT: THE FRENCH FIRST RESPONSE, JUNE 1640

The second act contains the arrival in Paris of Ban's composition, the comparison of his composition with that by Boësset on the same text there, the judgment of the two

---

[13]. Mersenne to Huygens, 26 August 1639 (MERSENNE 1964, p. 494; HUYGENS 2007, p. 307): «Vous estes bien de mesme advis que moy en ce qui est de la comparaison de *La playe inhumaine* de Claudin, et de Bannius, et qui en juge autrement me semble bien frappé au cerveau d'une playe plus inhumaine».

[14]. Huygens to Mersenne, 1 April 1640 (MERSENNE 1965, pp. 259-260; HUYGENS 2007, p. 325): «ses discours de musique, suivant lesquels jamais il n'a esté faict aucun bon air seulement, qui est un insolent paradoxe».

[15]. Earlier editions of Ban's composition are found in HUYGENS 1882, pp. lxxvi-lxxvii, WALKER 1976, p. 242 (= 1978, p. 92), STOLL 1978, p. 241, and probably elsewhere.

[16]. Ban to Huygens, 19 May 1640 (MERSENNE 1965, pp. 321-323; HUYGENS 2007, pp. 331-333).

Ex. 9.1: Joan Albert Ban, *Me veux-tu voir mourir*, from his letter to Anna Maria van Schurman, 20 August 1640. See Mersenne 1967, p. 36, and Huygens 2007, pp. 365-366.

compositions by 'the French', and the dispatch of a letter by Mersenne to Ban (sent to Huygens, to be forwarded to Ban), containing the French composition and the judgment to Ban's setting.

The letter that Mersenne wrote to Ban is lost but later Ban quoted from it the brief judgment made by the French of his composition:

> We ['the French'] sang your [Ban's] parts, the bass and the soprano, the simple version and the version with diminutions, several times, and then compared them to Boësset's air. You will judge for yourself which of the two is best suited to the text, to the ears, and to the mode. It is believed that your version with diminutions is more appropriate for instruments than for the voice, and that your simple version is not so pathetic (*patheticus*) as ours, which you have already noticed yourself, I think. Farewell[17].

[17]. Ban to Anna Maria van Schurman, 20 August 1640 (Mersenne 1967, p. 20; Huygens 2007, p. 351): «Tuas partes, nempe Bassum et Cantum simplicem ac varium, non semel cani, et Boësseti cantu comparari curavimus. Ipse judicabis uter ad literam, aures et circulum accommodatius et convenientius. Tua diminutio

Ex. 9.2: Antoine Boësset, *Me veux-tu voir mourir*, à 2, from the letter by Joan Albert Ban to Anna Maria van Schurman, 20 August 1640. See MERSENNE 1967, p. 31, and HUYGENS 2007, pp. 361-362. Barlines original; time signatures **2** and **3** modernized.

aptior instrumentis quam voci judicata est, neque simplex cantus adeò patheticus ac noster, quem (uti credo) jam pervidisti. Vale».

Ex. 9.3: Antoine Boësset, *Me veux-tu voir mourir*, à 4, from Antoine Boësset, *IX Livre d'airs de cour à quatre et cinq parties* (Paris, 1642), all partbooks, ff. 28v-29r. Barlines added and time signatures **2** and **3** modernized to *alla breve* and 3/2 respectively.

Me veux-tu voir mourir, trop aymable inhumaine,
Viens donner à tes yeux ce funste plaisir.
L'excez de mon amour & celuy de ta haine,
S'en cont en vn moment contenter ton desir.
Mais au moins souuiens toy cruelle,
Si je meurs malheureux, que j'ay vescu fidelle.

Do you want to see me die, you, too lovely inhuman being? Come and give your eyes this fatal pleasure. The excess of my love and that of your hate, will in a moment satisfy your desire.

But at least remember, you cruel one, that, if I die unhappy, I have lived faithful.

Tes extremes rigueurs qui secondent tes charmes
Ont blessé tous mes sens, & me priuent du jour:
Ie sens couler mon sang aussi bien que mes larmes,
Et mesprisant la mort, je veux mourir d'amour.
Mais au moins.

Your extreme rigours that accompany your charms have hurt all my senses and deprive me of the light of day. I can feel my blood as well as my tears, And despising death, I want to die of love.

But at least.

The air submitted by the French had been composed by Antoine Boësset already, in 1639[18]. Boësset may have composed his air originally for voice and lute, but such a version is not known[19]. The version Ban eventually received was a two-part vocal setting, for treble and bass,

---

[18]. In his letter to Ban, Mersenne had remarked that Boësset's composition had been composed six or eight months before, so by the end of 1639, as appears from Ban's letter to Anna Maria van Schurman. See MERSENNE 1967, p. 24, and HUYGENS 2007, p. 355.

[19]. Bars 1-6 of the Basse-contre of the four-part versions do not have text and the remark «Le luth», and this could point to an original setting for voice and luth. But the two-part setting that Ban received has text with both parts, which seems to point to a derivation from a setting with a vocal bass throughout.

though we do not know if it was really Boësset who produced this version (see Ex. 2)[20]. It seems instead that this version was a two-part reduction of the four-part vocal setting, for Dessus, Haute-contre, Taille and Basse-contre that was published later in Boësset's *IX Livre d'airs de cour à quatre et cinq parties* (Paris, 1642; see Ex. 3). Although Boësset's full authorship is therefore not assured for the version sent to Ban, in all subsequent discussions Boësset was named as the composer. Some seventeenth-century manuscripts transmit the piece for voice and an unfigured instrumental bass[21]. In the 1642 publication, a second stanza was added.

The text of *Me veux-tu voir mourir* is traditionally ascribed to the French poet Germain Habert de Cérisy (1610-1654), but the first publication connecting him with the text is the *Poësies choisies, Cinquiesme partie* (Paris, 1660), p. 322[22].

Mersenne had sent Boësset's air to Huygens, who then forwarded it to Ban with his letter of 26 June 1640[23]. Huygens did not mention in this letter to Ban the judgment of the French on his composition, but it seems likely that this judgment arrived in Haarlem by the same way and on the same occasion.

The Third Act: Ban's First Response, August 1640

The third act is Ban's response to Mersenne, particularly as related to the judgment of his and Boësset's composition. Ban was enraged, furious, angry. What stinged him most was the total absence of an argument from the French side: the French gave their judgment without comment, without explanation. Ban had, on the contrary, a whole musical theory to justify his approach. Furthermore, he had produced a new musical composition according to his well-defined principles, while the French had sent an already existing composition and had declared it superior without explaining why.

At first, Ban did not respond directly to Mersenne. At Huygens's instigation he formulated a long letter, addressed to the learned woman Anna Maria van Schurman, in which

[20]. Earlier editions of Boësset's composition as Ban received it are found in Huygens 1882, pp. lxx-lxxii, Walker 1976, p. 243 (= 1978, p. 93), Stoll 1978, p. 240, and probably elsewhere.

[21]. Antoine Boësset, *Me veux-tu voir mourir*: D-W, Cod. Guelf. 52 Noviss. 80 (Nr. 114), no. 113; F-Pn, VMA MS-854, p. 195; F-Pn, VM7-501, f. 2v (in C minor); GB-Lbl, MS Stowe 1081, f. 5v; GB-Lbl, Add. MS 17991, f. 47v (melody only).

[22]. In the *Poesies choisies, Cinquième partie*, of 1660, the second line of the second stanza differs from Boesset's: «*M'ayant osté l'espoir, me vont priver du jour*». The poem is cited, as second and third stanza, in a sacred poem by Simon Morin, entitled 'Reproche que Iesus mourant en Croix fait à la nature humaine', in his *Pensées* (1647), pp. 149-150. Habert's first stanza is quoted verbatim, his second stanza is heavily paraphrased.

[23]. Huygens to Ban, 26 June 1640 (Mersenne 1965, pp. 439-440; Huygens 2007, pp. 336-339).

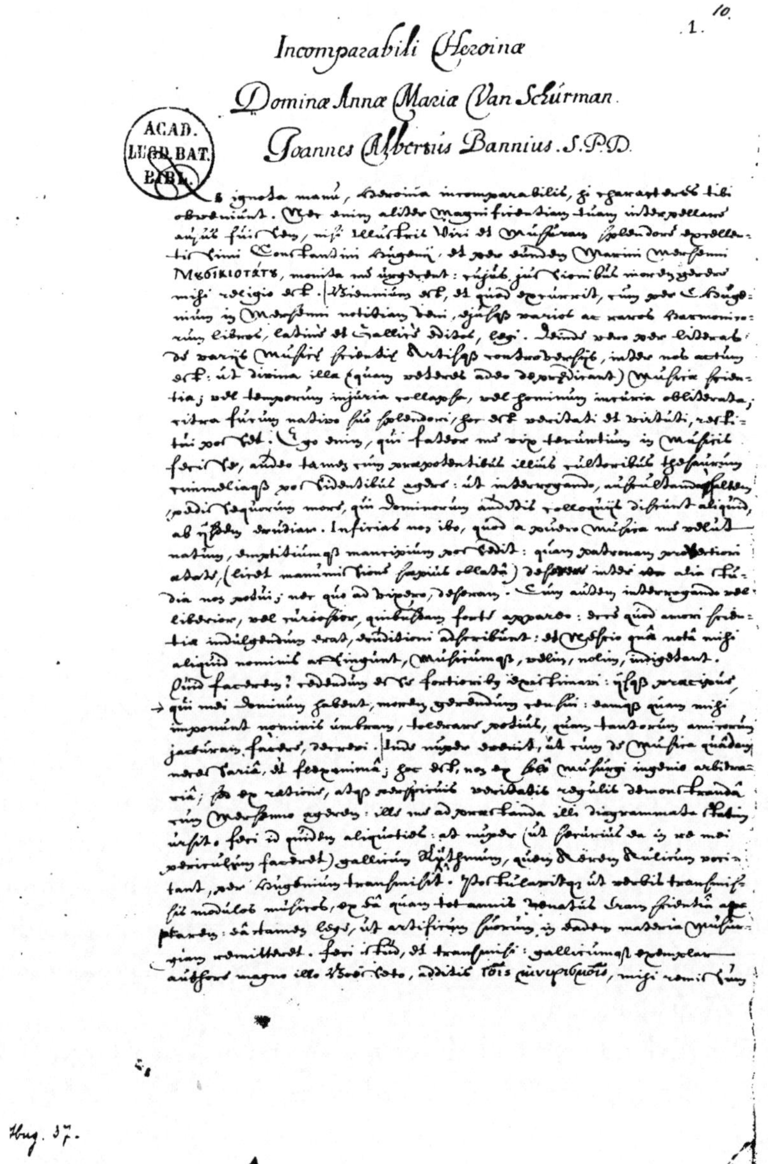

ILL. 9.3: Joan Albert Ban to Anna Maria van Schurman, 20 August 1640 (copy written by Ban's copyist), first page. NL-Lu, Cod. Hug. 27.

he had to temper his rage (ILL. 9.3)[24]. After an introduction of a few pages, he discussed and assessed Boësset's composition in accordance with his own rules. It goes without saying that

[24]. The letter itself (that is, the document sent to Anna Maria van Schurman) is not preserved. Two copies are known in Ban's hand, which are in NL-Lu (Cod. Hug. 37 and Cod. Bibl. Publ. 2212); a third copy, by Mersenne, is in F-Pm (Ms. 3696). For modern editions, see note MERSENNE 1967, pp. 18-46, and HUYGENS 2007, pp. 348-394.

the composition only partially complied with these rules, and that it should be rejected for that reason. Then he explained, bar by bar, his own air, why he had composed it as he had done. In this way, his composition was the perfect one, in his opinion almost the only possible way to set the submitted text to music.

The letter to Anna Maria is dated 20 August 1640. A copy was sent to Huygens, to be sent on to Mersenne. This copy must have arrived in Paris in September. It is not preserved. Mersenne produced a summary in French for Boësset, who seems not to have had enough Latin to understand the original letter. This is the 'Examen du Sieur Bannius de l'air *Me veux-tu voir mourir* composé par Monsieur Boësset' (Examination by Mr. Ban of the air *Me veux-tu voir mourir* composed by Mr. Boësset)[25]. Mersenne sent a copy of his abbreviated translation to Huygens with his letter of 3 November 1640[26].

The Fourth Act: The French Second Responses, November-December 1640

The fourth act contains responses of the French to Ban's letter to Anna Maria van Schurman, of which there were no fewer than four. All of them were disastrous for Ban. The arguments proposed by him to defend his own composition were rejected without exception as was his critique of Boësset's air.

Mersenne was first to attack Ban and to defend Boësset, first in his letter of 3 November, then again in a more detailed letter of a few weeks later, 14 November[27]. The second letter was annexed by a document entitled 'Jugement d'un Trésorier-Général sur la lettre de Monsieur Bannius' (Judgment of a Treasurer-General on the Letter of Mr. Ban). The «Trésorier-Général» can be identified with the French composer Jacques de Gouy[28].

Boësset now entered the discussion himself, with a letter to Huygens of 15 November[29].

The fourth response was that of René Descartes, living in Holland at the time and a close friend of both Ban and Huygens. He wrote a letter in Latin which is preserved as an undated copy that Descartes made for himself. The letter itself must have been written in December 1640[30].

---

25. For the 'Examen', see Huygens 1882, pp. lxx-lxxix, Meresenne 1967, pp. 836-844, and Huygens 2007, pp. 406-410.

26. Mersenne to Huygens, 3 November 1640 (Mersenne 1967, pp. 205-207; Huygens 2007, pp. 402-416).

27. Mersenne to Huygens, 14 November 1640 (Mersenne 1967, pp. 236-249; Huygens 2007, pp. 419-437.

28. See Ban's letter to Mersenne of 13 January 1641 (Mersenne 1967, p. 403). For the 'Jugement', see Huygens 1882, pp. xci-xciv, Mersenne 1967, pp. 258-262, and Huygens 2007, pp. 428-230.

29. Boësset to Huygens, 15 November 1640 (Mersenne 1967, pp. 250-253; Huygens 2007, pp. 440-444).

30. Descartes to Ban, December 1640 (Descartes – Huygens 1926, pp. 293-298; Descartes 1975, pp. 829-834).

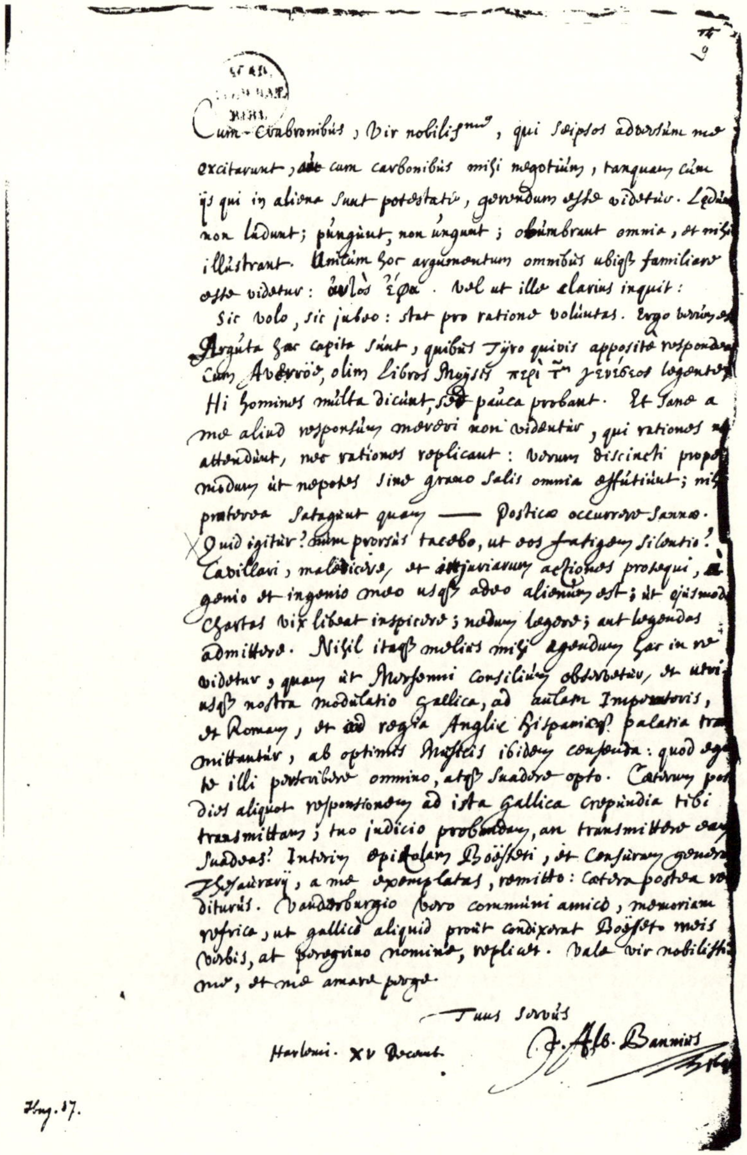

Cum Crabronibus, Vir nobilissime, qui seipsos adversum me excitaverunt, cum carbonibus mihi negotium, tanquam cum iis qui in aliena sunt potestate, gerendum esse videtur. Laedunt non laedunt; pungunt, non ungunt; obumbrant omnia, et mihi illustrant. Unicum hoc argumentum omnibus ubique familiare esse videtur: αὐτὸς ἔφα. Vel ut ille clarius inquit:
Sic volo, sic jubeo: stat pro ratione voluntas. Ergo ver...
Arguta haec capita sunt, quibus Tyro quivis apposite responde[re] cum Averroe, olim libros Magistri περὶ τῆς γενέσεως legente[s]. Hi homines multa dicunt, sed pauca probant. Et sane a me aliud responsum moveri non videntur, qui rationes n[on] attendunt, nec rationes replicant: verum discincti prop[e] modum ut nepotes sine grano salis omnia effutiunt; ni[hil] praeterea satagunt quam — postico occurrere sannae.
Quid igitur? Num prorsus taceo, ut vos fatigem silentio? Cavillari, maledicere, et injuriarum actiones prosequi, a genio et ingenio meo usque adeo alienum est; ut ejusmodi chartas vix libeat inspicere; nedum legere; aut legendas admittere. Nihil itaque melius mihi agendum hac in re videtur, quam ut Mersenni consilium observetur, et utri[us]que nostra modulatio gallica, ad aulam Imperatoris, et Romam, et ad regia Angliae Hispaniaeque palatia trans[-]mittantur, ab optimis Musicis ibidem censenda: quod ego ve illi perscribere omnino, atque suadere opto. Caeterum post dies aliquot responsionem ad ista gallica crepundia tibi transmittam; tuo judicio probandam, an transmittere eam suadeas? Interim epistolam Boësseti, et censuram generosi Thesaurarii, a me exemplatas, remitto: Caetera postea ve[ni]ditures. Vanderburgio vero communi amico, memoriam refrica, ut gallice aliquid prout condixerat Boësseto meis verbis, at peregrino nomine, replicet. Vale vir nobilissime, et me amare perge.

Tuus servus

Harlemi. XV Decemb. J. Alb. Bannius

Hug. 27.

ILL. 9.4: Joan Albert Ban to Constantijn Huygens, 15 December 1640 (autograph). NL-Lu, Cod. Hug. 27.

Huygens received the letters from Mersenne and Boësset in late November or early December 1640. He immediately forwarded them to Ban, with a letter by himself, dated 9 December[31]. Ban responded with a furious letter, dated 15 December (ILL. 9.4)[32].

---

[31]. Huygens to Ban, 9 December 1640 (HUYGENS 2007, pp. 447-448).

[32]. Ban to Huygens, 15 December 1640 (*ibidem*, pp. 448-451).

What was the quintessence of responses of the French? Mersenne's letter of 14 November can be summarized as follows: (1) Ban did not understand the French text correctly; (2) all modes could be used to express any emotion or passion; (3) the prosodic accent that Ban considered very important was not so strong in the French language; (4) it was wrong to express all the emotional elements of the text in the music; and (5) Boësset's music was admirable. De Gouy paid attention to the use of the mode. He also noticed that the abundance of details too greatly constrained a composer. Boësset, on the other hand, had probably not intended his letter to be seen by Ban, but Huygens was vile enough to forward it to Ban anyhow. Boësset wrote about (1) the interpretation of the text; (2) the choice of mode; and (3) details in the text. At the end there are some very denigrating judgements of Ban's work and his personality. Descartes, finally, dealt with (1) the interpretation of the text; (2) the use of accents; (3) the relationship between polyphony and intelligibility; and (4) the expression of the semantic elements of the text. At the end he mentioned that music should not be considered mathematically, nor physically, but morally:

> And let them die, all these jokers, if they like it. Let us laugh in the meantime. You must know that I must express myself as I do so, not to contradict you, who are serious, but to show that the considerations of this matter, which depend more on the interpretation of French verses than on the science of music, seem to me to be neither mathematical nor physical, but only moral, and of such a kind that I can, with their help, easily contradict not only someone else, but also myself[33].

The Fifth Act: Ban's Second Responses, January 1641

After the Fourth Act, Ban began to defend himself. He wrote extended letters to Huygens (12 January 1641) and Mersenne (13 January 1641), in which he complained about the tone of the French[34]. He replied to Boësset's letter and then, point by point, to De Gouy's 'Jugement'. Huygens replied to Ban (on 16 January) that one should not get so carried away and that he himself did not completely agree neither with him nor the French[35]. Huygens obviously wanted

[33]. Descartes to Ban, December 1640 (Descartes – Huygens 1926, p. 298; Descartes 1975, p. 834): «At moriantur profectò isti nugatores si ipsis placet, nos verò interim rideamus. Et scias hæc me ludibundum effudisse, non ut tibi serio contradicam, sed ut tester istiusmodi rationes, quæ non tam à musicæ scientiâ quam ab interpretatione carminis Gallici pendebant, nec mathematicas, nec physicas, sed morales tantum mihi videri, talesque ut illarum ope non solum alteri, sed etiam mihi ipsi contradicere facile possim. Vale».

[34]. Ban to Huygens, 12 January 1641 (Huygens 2007, pp. 452-472); Ban to Mersenne, 13 January 1641 (Huygens 1882, pp. cviii-cxii; Mersenne 1967, pp. 400-404).

[35]. Huygens to Ban, 16 January 1641 (Huygens 2007, pp. 475-479).

to remain friends with both Ban and Mersenne. Ban continued his expositions in a second long letter to Huygens dated 18 January[36].

Curiously, after this exchange of letters the controversy disappears more or less from the correspondence between Huygens-Mersenne-Boësset. They had all said what they wanted to say, and no one wanted to prolong this fire of pro and contra arguments. End of the controversy and of the tragedy itself.

Epilogue: 1642-1647

Finally, some echoes of the controversy during the years that followed deserve to be considered.

With his letter of 29 April 1642, Huygens sent Ban a new Parisian commentary on the letter to Anna Maria van Schurman and a new composition of *Me veux-tu voir mourir*, both by a «Parisian Councillor» (*Consiliarius Parisiensis*)[37]. Ban replied with another letter full of indignation and fury (1 May 1642)[38]. And another French composition of *Me veux-tu voir mourir* was sent by Huygens to Ban on 19 August 1642. Ban's answer reveals that this piece was by the French violinist Guillaume Dumanoir[39].

Ban died on 27 July 1644. This news was passed on to Huygens — who was with the army on campaign — by their mutual friend Jacob van der Burgh, who wrote to Huygens on 6 August 1644:

> I do not know if you know that God has taken poor Ban from this world.
> I hope he hears better music [in heaven] than he imagined when he was alive.
> Although he had his errors, he had a good soul[40].

Huygens wrote a Latin epigraph and epitaph on Ban's death[41]. He shared the news with Mersenne in his letter of 16 August, and concluded with the following words:

[36]. Ban to Huygens, 18 January 1641 (*ibidem*, pp. 479-488).

[37]. Huygens to Ban, 19 April 1642 (Mersenne 1970, pp. 133-134; Huygens 2007, pp. 639-642).

[38]. Ban to Huygens, 1 May 1642 (Huygens 2007, pp. 642-646).

[39]. Huygens to Ban, 19 August 1642 (*ibidem*, pp. 658-661). Dumanoir is mentioned as the composer of the piece sent by Huygens in Ban's letter to Huygens of 16 September 1642 (*ibidem*, p. 662). Dumanoir had been, from 1636 to 1640, dancing master at the court of Stadtholder Frederick Henry in The Hague; he may thus have been a personal acquaintance of Huygens.

[40]. Jacob van der Burgh to Huygens, 6 August 1644: (Huygens 1915, p. 27): «Je ne sçay si vous sçavez que Dieu a retiré de ce monde le pauvre Bannius. J'espère qu'il entendra meilleure musique [dans le ciel] qu'il ne s'est imaginée estant en vie. S'il avoit des erreurs, il avoit l'âme bonne».

[41]. Constantijn Huygens, 'In obitum Joannis Alberti Banii' and 'Epitaphium [Joannis Alberti Bannii]'. See the Appendix to this chapter.

> And the sciences lose a lot, especially the one [music] he embraced so strongly. For, as I have always said, there were considerable things in his principles, but the practice did not succeed in his hands. For it is one thing to know prosody well and another to be a good poet. Finally, an honest soul has gone, but God did it[42].

Late 1646 and early 1647, Huygens and Descartes discussed Ban's musical theories on several occasions. Descartes sent with his letter of 30 November 1646 a copy of his letter to Ban of December 1640. He wrote:

> You will see here the musical letter about which I spoke to you. I had enough trouble copying it from a bad draft that had remained with me, but you will have even more to guess what it means, if you do not compare it with the letter that Mr. Ban wrote to Mademoiselle Schurman, of which I suppose you have a copy. If you are patient enough, you may take pleasure in seeing what a man who has never learned to sing *ut re mi fa sol la*, nor to judge whether someone else sang it right, has conjectured on a subject that depends only on the judgment of the ear. And if Monsieur your second son [Christiaan Huygens] wants to be exercised in this matter, he can have the opportunity to criticize both Mr. Ban and me, and to show that we have understood nothing, none of us. For, because our reasons are neither mathematical nor physical but only moral, as I said in my letter, it is easy to find other reasons that are contrary to them. If he writes something on this subject, I would like to see it[43].

Huygens replied to Descartes on 7 January 1647:

---

[42]. Huygens to Mersenne, 16 August 1644 (Mersenne 1977, p. 196): «Et les sciences y perdent beaucoup, nommément celle qu'il embrassoit si fort. Car, comme j'ay tousjours dit, il y avoit des choses considérables en ses principes, quoyque la prattique n'en réuscissoit pas entre ses mains. Car autre chose est de bien sçavoir la prosodie et autre d'estre bon poète. Enfin, un honeste amy perdu, mais Dieu l'a faict».

[43]. Descartes to Huygens, 30 November 1646 (Descartes – Huygens 1926, pp. 247-248; Descartes 1976, pp. 787-788): «vous verrez icy la letre musicale dont je vous avois parlé. J'ay eu assez de peine à la tirer d'un méchant brouillon qui m'estoit demeuré, mais vous en aurez encore plus à deviner ce qu'elle veut dire, si vous ne la conférez avec celle de Monsieur Bannius escrite à Mademoiselle Schuermans, dont je suppose que vous avez copie. Si vous estes assez patient pour cela, vous prendrez peut-estre plaisir à voir ce qu'un homme qui n'a jamais sceu apprendre à chanter *ut re mi fa sol la*, ny à juger si un autre le chantoit bien, a conjecturé touchant un sujet qui ne dépend que du jugement de l'oreille. Et si Monsieur vostre second fils [c'est-à-dire, Christiaan Huygens] se veut exercer en cete matière, il en peut avoir occasion en nous reprenant Bannius et moy, et monstrant que nous n'y avons rien entendu, ny l'un ny l'autre. Car, nos raisons n'estant ny mathématiques ny physiques mais seulement morales, comme j'ay dit là, il est aysé d'en trouver d'autres qui leur soient contraires. S'il escrit quelque chose sur ce sujet, je seray bien ayse de le voir».

> I thank you, early or late, most humbly for the excellent refutation of poor Bannius. He needed a censor of your authority, whom only he would recognize among us other little dogs, whom he thought were only barking. However, I thought I had a say in this matter, and I am in favour of it, but it would be too long for this paper. On some occasion we will discuss it[44].

Descartes responded to this letter on 4 February:

> However, I cannot refrain from writing these lines, to ask you to continue your willingness to inform me of your considerations on the subject of my dispute with Bannius, for I have no doubt that they are of much use for my instruction, and if I die only of old age, I still want some day to write about the theory of music[45].

And, finally, as a posthumous tribute, Mersenne inserted the brief musical treatise that Ban had sent him almost ten years ago, with the letter of 17 April 1639, in the second edition of his *Harmonicorum libri XII* (Paris, 1648)[46].

After this publication, the controversy between Ban and Boësset fell into oblivion, until Jonckbloet and Land described it amply in the preface to their edition of Constantijn Huygens's musical correspondence, published in 1882, based on documents they had found in the Library of the University of Leiden and in the Royal Library in The Hague[47]. Since then, the controversy has not stopped attracting the attention of musicologists and other researchers. Substantial discussions have been provided by Jan Pieter Nicolaas Land (1891), André Pirro (1907), Geneviève Rodis-Lewis (1972), Daniel Walker (1976), Georges Lote (1978), Rudolf Rasch (2006) and Steven Nadler (2013)[48]. Other commentators have mentioned the controversy in passing[49].

---

[44]. Huygens to Descartes, 7 January 1647 (DESCARTES – HUYGENS 1926, p. 249; DESCARTES 1976, p. 789): «Je vous rends donq, tost ou tard, mes remerciemens très-humbles pour la très-bonne réfutation du pauvre Bannius. Il avoit besoin d'un censeur de vostre autorité, que seule il recognoissoit parmi nous autres petits chiens, qu'il croyoit ne faire qu'aboyer. Cependant, j'ay creu avoir le mot à dire en ceste matière, et en suis encor d'advis, mais cela seroit long pour ce papier. A quelqu'entreveue nous nous en entretiendrons».

[45]. Descartes to Huygens, 4 February 1647 (DESCARTES – HUYGENS 1926, pp. 250-251; DESCARTES 1976, pp. 790-791): «Cependant je ne puis m'abstenir d'escrire ces lignes, pour vous prier de continuer en la volonté de me faire part de vos considérations sur le sujet de ma dispute avec Bannius, car je ne doute point qu'elles ne servent beaucoup à mon instruction, et si je ne meurs que de vieillesse, j'ay encore envie quelque jour d'escrire de la théorie de musique».

[46]. Marin Mersenne, *Harmonicorum libri XII* (Paris, 1648), pp. 153-154.

[47]. HUYGENS 1882, pp. xl-xlvii.

[48]. LAND 1891, PIRRO 1907, pp. 109-120, RODIS-LEWIS 1971, WALKER 1976 (= 1978, pp. 81-110), LOTE 1988, pp. 150-156, RASCH 2005, and NADLER 2013.

[49]. See, for example, ADAM 1937, pp. 79-80, and GOULET 2004, p. 349.

Argumentation

After this brief history of the controversy between Ban and Boësset, it is time to examine the arguments put forward by the combatants. Every controversy must have at least one topical focal point apart from possible circumstantial points. The Ban-Boësset controversy can be said to have a whole series of focal points, both topical and circumstantial. Topical arguments concern the interpretation of the text, the role of accents in the French language, the choice of tone or mode, the expression of the elements of the text in the music, and the choice of the time signature, particularly the use of triple time. In addition to these topical arguments there are non-topical arguments, the *ad hominem* arguments, concerning the superior status of Boësset as a composer (versus Ban's insignificance in this respect), the position of the French that they did not want to be taught by Ban how to use their language, the denigrating remarks of the French towards Ban, and that Ban did not feel considered by the French an honest, serious and sincere man. Let's look at these arguments one by one.

Topical Arguments

1. *The interpretation of a text and its consequence for a musical composition.* Both sides, Ban and his opponents, agreed on the principle that the character of a text should be reflected in its musical setting. For a joyful text one should compose music different than for sad words. Unfortunately, Ban interpreted the text of *Me veux-tu voir mourir* differently than the French. For Ban, the dominant feelings were anger, fury, and indignation. The French, on the contrary, took the text rather as a hyperbole, as the language of a lover who put into play the metaphor of wanting to die for the object of his affection simply to describe the depth of his love. From this point of view the text was flattering and lovely rather than angry and resentful.

In his opposition to the French interpretation, Ban remarked that the text that had been sent to him differed in a few words from the text set by Boësset: in his (Ban's) text, the initial sentence ended with the words «insensible Climaine»; in the text that Boësset had set one reads «trop aymable Inhumaine». Indeed, it cannot be denied that this difference has an influence on the interpretation of the general tenor of the text. The French, subsequently, did not touch upon this detail.

2. *The choice of mode.* In Ban's musical thought, the intonation of the musical scale was just intonation, so with steps of different size, with larger and smaller whole tones. Accordingly, each mode had a scale with a different intonations of the intervals within the scale[50]. With

[50]. For this period of transition between modal and tonal conceptions of the musical scale, it is difficult to decide what word to use for the musical scale: mode, key, tone? Bann called the scale 'circulus musicus', Mersenne

these differences in intonation and the differences between the intervals *per se* (as the difference between the major and the minor third, and so on), each mode also had its own character and its own associated emotion. Therefore, in the case of a text with emotional content, the choice of the mode was not arbitrary. Ban defended his choice for *Me veux-tu voir mourir* — F major — in his letter to Anna Maria van Schurman of 20 August 1640 with the following words:

> In the first place I chose the most appropriate mode for indignation, that is, the mode that goes from F by c to f, of which the strong fifth [F-c] is the basis:
>
> 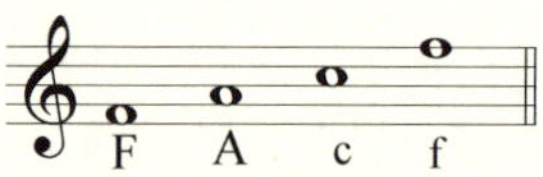
>
>
> The very excited fourth c-f completes the octave. The impetuous third F-A forms the basis of the fifth. The strongest kind of the minor third A-c makes up the upper part of the fifth. In this way, everything is ready for indignation. At the same time, the tone A is appropriate to moderate the anger[51].

It is not difficult to understand why, for a text like *Me veux-tu voir mourir*, Ban rejected the mode of D minor, Boësset's choice.

It seems, from several seventeenth-century French writings on the music of this period, that the doctrine of the character of the modes, as developed earlier by Zarlino, was not yet forgotten[52]. Boësset discussed the character of the modes quite extensively in his letter of 15 November 1640 to Huygens. He began by agreeing with Ban regarding the choice of mode, if the interpretation of the text was as Ban assumed:

> It is true that, if the meaning of the words was such as he [Ban] explains them, I confess that I would have made a bad choice with the mode which I have employed myself [D minor], which is full of sweetness and more apt to express pain and complaint than anger, its octave being divided in its cadence by the minor third,

---

and Boësset used 'mode' or 'modus'. In this chapter, we have, somewhat arbitarily, chosen to use the word 'mode', but to specify them by their final and by the quality of the third of the scale, as us usual for keys.

[51]. Ban to Anna Maria van Schurman (MERSENNE 1967, p. 37, and HUYGENS 2007, p. 367): «Primò quidem circulum harmonicum delegi, qui indignationi aptissimus est, nimirum ex F in C procedens ad F, ubi potentissima quintæ species basin octavæ constituit:

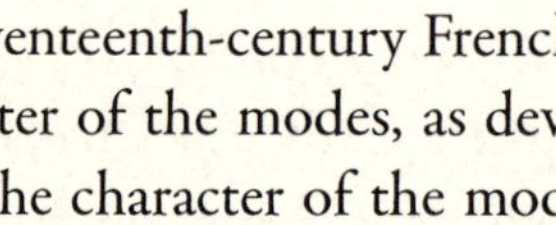

Incitatissima etiam quartæ species C-F, ut-fa, culmen octavæ habet. Imò vehementissima ditoni species F-A, fa-la, quintæ basin occupat. Fortissima etiam semiditoni species A-C, re-fa, acutiorem locum quintæ tenet. Omnia ergo ad indignationem disposita sunt. Una etiam hæc sedes in A ad temperiem iræ aptissima est».

[52]. See, for example, Antoine Parran, *Traité de la musique théorique et pratique* (Paris, 1639).

> and its compound [that is, its octave enlargement] being pathetic and soft. [...] If the meaning of the words would have been as the said Mr. Ban imagined it, I would have employed the mode in which he composed his own composition, which is the seventh mode [= F major], proper to fury, cries and despair, because of the division by the major third, which is as virile that the minor third is soft[53].

But here Boësset changes the discourse and stresses that an able composer could, by the use of accidentals, apply any mode for any emotion and passion:

> [...] although, when I feel like it, I can express all kinds of passions equally well in one mode as in another. And it is a mistake to believe differently about accidentals, which can be used with propriety, to make them [the modes] all equal[54].

These remarks were followed by a commentary on Zarlino's theories of modal character.

3. *The influence of the emphasis of the text on the musical composition*. For Germanic languages, including Dutch, and for languages such as Italian and Spanish, the prosodic accents in a text must be observed in musical composition: an accented syllable of the text must be placed on an accented note of the music. For French the relationship between textual and musical accents is much less coercive (except in cadences at the end of a verse), to the point that the French can say that the role of the textual accent is almost negligible. The Genevan Psalter as well as the *air de cour* repertoire are full of illustrations of the limited role of the prosodic accent in musical composition. In the words of Mersenne, in his letter of 3 November 1640[55]:

> Our language and our words have almost no accent in France, and therefore [...] the musician can apply them according to his design. For singing and speaking

[53]. Boësset to Huygens, 15 November 1640 (Mersenne 1967, pp. 251-252; Huygens 2007, p. 441): «Il est vray que, si le sens des parolles estoit tel qu'il [Ban] les explique, j'advoue que j'aurois faict un mauvais cheoix du mode dont je me suis servy, qui est plain de douceur et plus propre à exprimer la douleur et la plainte que la cholère, son diapason estant divisé dans la cadence par la tierce mineure, et sa replique estant pathétique et molle. [...] Que sy la signification des parolles eussent esté telles que ledict Sieur Bannius se l'est imaginé, je me serois servy du mode dont il a composé le sien, qui est du 7e mode, propre à la furie, aux crys et au désespoir, à cause de la division de la tierce majeure, qui est aussy virile que la tierce mineure est molle».

[54]. Boësset to Huygens, 15 November 1640 (Mersenne 1967, p. 252; Huygens 2007, p. 441): «bien que, quand il me plaira, j'exprimeray toute sorte de passion aussy bien en un mode qu'en l'aultre. Et c'est une erreur de croire le contraire des accidentz, desquels l'on se peut servir avec addresse, les rendant tous esgaulx».

[55]. Mersenne to Huygens, 3 November 1640 (Mersenne 1967, p. 295; Huygens 2007, p. 404): «Nostre langue et nos paroles n'ont quasy point d'accent en France, et partant [...] le musicien leur en peut donner suivant son dessein. Car le chant et la parole sont si différents, que lorsqu'un parle comme les Italiens, nous nous en moquons et disons qu'il chante».

> are so different that, when a person speaks like the Italians, we laugh at it and say that he is singing.

Nonetheless, Ban believed that he heard a difference in accent between, for example, «[il] parle» (with the accent on the first syllable of «parle») and «[nous] parlons» (with the accent on the second syllable of «parlons») and wanted these differences to be observed in the musical composition. This had to be done in every conceivable way, that is, melodically (by placing accented syllables on higher notes), rhythmically (by placing accented syllables on longer notes), and metrically (by placing accented syllables on a strong position in the bar). It is not difficult to understand that the French found these demands exaggerated if not downright ridiculous.

4. *The role of semantic elements of the text in musical composition.* Ban wanted every semantically significant element of the text to be expressed in the music. For example, for the text of *Me veux-tu voir mourir*, he wanted a musical expression of the following words:

- The whole first line, as an expression of indignation.
- From the first line more specifically: «mourir» (to die), and «insensible» (insensitive).
- From the second line: «yeux» (eyes), «funeste» (fatal), «plaisir» (pleasure).
- From the third line: «excès» (excess), «amour» (love), «haine» (hate).
- From the fourth line: «contenter» (to please), «souviens» (remember).
- From the fifth line: «cruelle» (cruel).
- From the sixth line: «je meurs» (I die), «malheureux» (unhappy), «fidelle» (faithful).

In his letter to Anna Maria van Schurman of 20 August 1640, he provided ample explanations of his choices for expressing the semantic content of these words in music. He believed that he had found not only an adequate musical expression for each element, but also the best possible musical expression, if not the only acceptable musical expression for each element.

In Ban's theories there were two ways of achieving musical expression with musical intervals: by the size of the interval and its melodic direction. For positive feelings he recommended major intervals (including the larger whole tone) in upward direction, for negative feelings the minor intervals (including the smaller whole tone and the semitone) in downward direction.

The French objected, and I think we would agree, that such recipes would kill musical creativity and would lead to an undesirable and reprehensible schematism and predictability, and, furthermore, to similar motion in the polyphony without interruption, almost impossible to maintain with more than two voices. Consider the following words from the De Gouy's 'Jugement':

> He [Ban] always attaches himself too much to wanting to express the meaning of a word or diction taken alone and isolated, and detached from all the rest of the discourse, without considering in general what passion, or affection, or what substance results and appears from the whole period, or from the verse, or from the whole discourse. As, for example, to express this word "*mourir*" (to die) or "*cruelle*" (cruel), he wants the first to be expressed by low notes, swallowed, brief, and short, for he says that a dying man has a low voice, and failing, so that he does not want to make any repetition, as if death would come suddenly, and that at one moment one would lose all strength and breath[56].

Circumstantial Arguments

Perhaps more interesting than the topical arguments are the non-topical or circumstantial arguments, which are, of course, not arguments *per se*. Rather they are non-arguments that disturb and confuse the discussion. There are several of these disruptive facts in the Ban-Boësset controversy (as there are probably in most controversies).

1. *Ban alone against the French collective*. Ban felt alone against what he felt as a French block: Mersenne, Boësset, Gouy, Descartes. He sought Huygens's support, for example, but he did not find it. On the contrary, Huygens refused active support, certainly fearful that too strong a connection with Ban would harm his relations with Mersenne, Boësset and perhaps other Frenchmen who might take part in the debate. On 26 June 1640 Huygens wrote to Ban:

> You want to know what I think of your composition. Rather I urge you to teach me with what rule, with what words, newly made for this science or already existing, I must express myself when I speak of any piece of music. Certainly, for the 'Compendium musicae' that you promised, I am waiting eagerly, so that I can express my opinion with the right words and sentences, and with confidence[57].

---

[56]. 'Jugement d'un Trésorier-Général' (Huygens 1882, p. xciii, Mersenne 1967, p. 261, and Huygens 2007, p. 429): «Il [Ban] s'attache toujours par trop à vouloir exprimer la signification d'un mot ou diction prise toute seule en son particulier, et destachée de tout le reste du discours, sans considérer en général quelle passion, ou affection, ou quelle substance résulte et paroist de toute la période, ou du vers, ou du discours entier. Comme, par exemple, pour exprimer ce mot "mourir" ou bien "cruelle", il veut que le premier soit exprimé par des notes graves, et ravalées, briéves, et courtes, pour ce qu'il dit qu'un homme mourant a la voix basse, et défaillante, et pour ce ne veut que l'on y face aucune répétition, comme si la mort venoit tout à coup, et que deslors au premier instant l'on perdist toute force et halaine».

[57]. Huygens to Ban, 26 June 1640 (Huygens 2007, p. 337): «Expectas tuâ de cantione quid senserim; ego verò magis expecto, ut me doceas ad quam amussim, quibus ad scientiam factis natisve vocalibus disserere me de

What was worse, Huygens advised Ban to contact Anna Maria van Schurman. It was advice that would only aggravate the Ban's problems. The relationship between Huygens and Anna Maria had been troubled for some time. Huygens now employed Ban for his own interests. He sought to restore his relationship with Anna Maria by asking Ban to send her some documents on which he wanted her opinion. Ban walked right into the trap and wrote his long letter of 20 August 1640 to Anna Maria. As noted above, the letter, by its length alone, made him extremely vulnerable to his opponents. It is no surprise that Anna Maria never replied, a silence that Ban deeply regretted, if not resented.

Ban also sought the support of Descartes, whom he considered unreservedly a good friend, as indeed he was. Yet, Descartes sided with Ban's opponents in his letter to Ban in December 1640.

2. *As a Dutchman*, from a small nation, *Ban felt powerless against the French*, from a large and dominant nation in the fields of culture and music. His nationality and modest origin are recurring elements in his letters. He accentuated that he was descended from a family that had lived in Haarlem for several generations, a guarantee for an honest, solid, serious, impartial, and unprejudiced judgment.

The French also used the nationalist argument. According to Mersenne, the superiority of Boësset's air and Boësset's method were simply guaranteed by the composer's fame. On 3 November 1640, Mersenne wrote to Huygens that:

> [...] the King, and the whole court, and the whole of France with it, and even *Italy, which admires the airs of Monsieur Boësset as well as we do*, have never complained that he observes the accents badly, and have not found that it hurted either their ears or their mind[58].

On 14 November 1640 Mersenne described Boësset as «our Orpheus» (*nostre Orphée*). Boësset himself called his composition something «which all of France admired» (*ce que toute la France admire*)[59].

Furthermore, Descartes opened his letter by noting that, being French, he was obliged to take the side of his compatriots, just as Ban had sought the support of Anna Maria van Schurman.

---

omni modulatione deceat. Compendium, inquam, quod promiseras, avidè expecto, mox animi mei sensa congruis verbis ac sententiis et bonâ item fide expositurus».

[58]. Mersenne to Huygens, 3 November 1640 (Mersenne 1967, p. 206, Huygens 2007, p. 405): «le Roy, et toute la cour avec toute la France, et mesme l'Italie qui admire aussy bien les airs de Monsieur Boësset comme nous, ne se sont jamais plaintz qu'il observast mal les accentz, et n'en ont point trouvé leur oüye ny leur esprit blessez».

[59]. Mersenne to Huygens, 14 November 1640 (Mersenne 1967, p. 249, Huygens 2007, p. 427). and Boësset to Huygens, 15 November 1640 (Mersenne 1967, p. 251, Huygens 2007, p. 441).

3. *The combatants did not shy away from* ad hominem *attacks*. Perhaps the most extreme was Boësset's judgment on Ban, expressed in his letter to Huygens of 15 November 1640:

> And the said Mr. Ban studied philosophy too much to have been able to acquire the title of great musician; and, in order to make the teachers of France to believe that, he did not have to send the 'Examen' of my work, and even less his air, by which he believes to have triumphed and gained the victory, which is worthy of compassion more than of envy, since there is no choirboy who cannot make a better one. To do a review of it is not what I claim. Suffice it to say that he imitated those miserable painters who, beginning to scribble, want to design an important painting. Therefore, I will say that to do something good with his composition it would be necessary to compose again the beginning, the middle, and the end, both for the meaning of the words and for the order of the consonances, which he has misunderstood completely. And I would consider my time to correct the enormous faults he committed as badly used as the time he has lost in making a work so imperfect. Make therefore, please, Sir, that he studies this science for a dozen years, and, if I am still in this world, I will oblige him to examine his works, which will perhaps be worthy of my attention[60].

Huygens forwarded this letter to Ban, who was furious after reading, which is easy to understand.

Descartes's words, in his letter of December 1640 to Mersenne, written probably soon after his letter to Ban, were somewhat more polite than those of Boësset:

> As for Mr. Ban's music, I believe it differs from Boësset's song, as the exercise of a student trying to apply all the rules of rhetoric differs from a speech by Cicero where it is difficult is to recognize those rules. I told him this, and I think he sees it now[61].

[60]. Boësset to Huygens, 15 November 1640 (Mersenne 1967, pp. 252-253, Huygens 2007, p. 442): «Et ledict Sieur Bannius a trop estudié en philosophie pour avoir peu acquérir le tiltre de grand musicien; et pour le faire croire aux professeurs de la France, il ne devoit pas envoyer l'examen de mon ouvrage, et encores moins son air, dont il croit avoir triomphé et gaigné la victoire, qui est digne de compassion plustost que d'envye, puisqu'il n'y a point d'enfant de chœur qui n'en puisse faire un meilleur. D'en faire un examen, c'est ce que je ne prétends pas. Je me contenteray de vous dire qu'il a imité ces misérables peintres qui, commenceant à griffonner, veullent desseigner un tableau d'importance. Ainsy je diray que, pour faire quelque chose de bien de sondict ouvrage, il faudroit refaire le commancement, le milieu et la fin, tant pour le sens des parolles que pour l'ordre desdictes consonances, qu'il a très-mal entendues. Et je tiendrois mon temps aussy mal employé à corriger les énormes faultes qu'il a commises, que celluy qu'il a perdu à faire un ouvrage sy imparfaict. Faictes donc, s'il vous plaist, Monsieur, qu'il estudie en cette science une douzaine d'années, et, sy je suis en ce monde, je l'obligeray d'examiner ses ouvrages, qui seront peut-estre dignes de m'en donner la peyne».

[61]. Descartes to Mersenne, December 1640 (Descartes 1899, p. 255; Mersenne 1967, p. 325): «Pour la musique de Monsieur Ban, je croy qu'elle diffère de l'air de Boësset, comme la creye d'un escolier qui a voulu

In addition, he let these words follow by some friendly remarks:

> But that does not prevent him from being a good musician and a sincere person and a good friend of mine, nor does it prevent the rules from being good, either in music or in rhetoric[62].

Ban, however, could show an equal lack of courtesy, as, for example, in his letter of 15 December 1640 to Huygens. There he wrote the following words about the French:

> I must deal, sir, with horseflies [*crabrones*], which are pricking me, or with scribblers [*carbones*], as possessed of a foreign power. They hurt, they don't play fair, they sting, they don't caress, they obscure everything, they illuminate nothing. Only one argument is known to them, which they use all the time: "the master said it himself", or, as he says more clearly: "So I want it, and so I command it", or "wishing prevails over reason". That is why it is true. These heads are cunning, to which a beginner would aptly respond with Averroës, when he read Moses's books on the creation: "These gents say many things but prove very few". And indeed, these people do not deserve to receive any answer from me, these people who do not observe reason and do not respond to reasonable proposals. They behave recklessly, almost like little children chatting without any restraint, and when they pay attention to the matter, the only result is a lot of farting[63].

Huygens tried to keep an intermediary position in this avalanche of insinuations. In his letters to Ban he tried to calm down Ban by showing his disapproval of the behaviour of the French. But he defended French music in a general sense, as, for example, in his letter to Ban of 26 June 1640:

---

pratiquer toutes les règles de sa réthorique diffère d'une oraison de Cicéron, où il est malaisé de les reconnoistre. Je luy en ay dit la mesme chose, et je croy qu'il l'avoue à présent».

[62]. Descartes to Mersenne, December 1640 (DESCARTES 1899, p. 255; MERSENNE 1967, pp. 325-326): «Mais cela n'empesche pas qu'il ne soit très-bon musicien, et d'ailleurs fort honneste homme, et mon bon amy, ny aussi que les règles ne soient bonnes, aussi bien en musique qu'en réthorique».

[63]. Ban to Huygens, 15 December 1640 (HUYGENS 2007, p. 449): «crabronibus, vir nobilissime, qui seipsos adversum me excitarunt, aut carbonibus, mihi negotium, tanquam iis qui in alienâ sunt potestate, gerendum esse videtur. Lædunt, not ludunt; pungunt, not ungunt; obumbrant omnia, and nihil illustrating. Unicum hoc argumentum omnibus ubique familiare esse videtur: "αὐτός ἔφα". Vel ut ille clarius inquit: "Sic volo, sic jubeo"; "stat pro ratione voluntas". Ergo verum est. Arguta hæc capita sunt, quibus tyro quivis appositè respondeat Averröe, olim libros Moyses περὶ τῆς γενέσεος legente: "Hi homines multa dicunt, sed pauca probant". And sanè à me aliud responsum mereri non videntur, qui rationes non attendunt, nec rationes replicant. Verum discincti propemodum ut nepotes sine grano salis omnia effutiunt, nihil præterea satagunt quam — posticæ occurrere sannæ».

> It must be said that these mortals [that is, the French] are by far the most versed in this matter [music], both by nature and habit[64].

In his letters to Mersenne and Boësset, Huygens avoided at all costs being associated too strongly with Ban's thoughts and ideas, as shown in the following excerpt from the letter of 19 January 1641 to Boësset:

> You will see how he feels pricked that he was sent back to school for twelve years. He is a knowledgeable man and as far as the theory of harmonic tones and intervals is concerned, as well versed as I have ever seen, so that I have always hoped that he would clarify these hidden matters, which the Ancients have treated obscurely in writings that moderns have only feigned to understand well. But concerning the application of an art, and in particular concerning the true genius that is not taught to anyone, that is the soul of the practice, he understands as little of it as you, Sir, understand much, and to the delight of everyone. The rules which, moreover, he claims to prescribe to the composer of an air in agreement with the words are, in my opinion, so far removed from reason, that, [even] when I had not seen the bad attempt that he has sent you, I would not have refrained, with you, from rejecting them. There will be a lot to say about the judgments of the best musicians in Europe, on whom he is happy to rely. But, however condemned he may be, he will never give up his imagination, if I know him well[65].

4. *In addition, Ban felt deceived and trapped*, because the French had not bothered to compose a new piece but had taken something already composed. And Ban was irritated by the fact that the French did not provide a theory on which their judgments were based, while he had produced long theoretical explanations about how to proceed with musical composition. And indeed, most of the considerations of the French were reactive: they said that the views of Ban were wrong and those of Boësset right.

---

[64]. Huygens to Ban, 26 June 1640 (Mersenne 1965, p. 439, Huygens 2007, p. 337): «Cujus rei fatendum est mortales illos longè peritissimos esse, naturâ simulatque habitu».

[65]. Huygens to Boësset, 19 January 1641 (Mersenne 1967, p. 416, Huygens 2007, pp. 489-490): «Vous verrez comme il se sent picqué de ce qu'on l'a renvoyé à l'escole pour 12 ans. Il est homme sçavent et pour ce qui est de la théorie des tons et intervalles harmoniques autant versé que j'en aye encor veu, de sorte que j'ay tousjours espéré qu'il rendroit ces matières esclarcées, que les Anciens ont traictés obscurément en des escrits que des modernes n'ont faict que la mine de bien entendre. Mais pour ce qui est de l'application de l'art, et nommément de ce vray génie que ne s'enseigne à personne, et que fait l'âme de la prattique, il y entend aussi peu que vous, Monsieur, en possédez amplement, et au ravissement de tout le monde. Les règles d'ailleurs qu'il prétend de précrire au compositeur d'un air à l'advenant de la lettre, sont, à mon advis, si esloignées de raison, que, quand je n'auroy pas veu le mauvais essay qu'il vous en a envoyé, je ne lairroy pas de les rejetter aveq vous. Il y aura du p[l]aisir à veoir là-dessus les arbitrages des meilleurs musiciens de l'Europe, auxquels il est content de s'en remettre. Mais, tout condamné qu'il sera, il ne demordra jamais de son imagination, si je le cognoy».

5. *Discussions took place on different levels.* Ban wanted to develop a complete theory, from the first principles to practical applications, whereas the French produced judgments based on practice. Ban expected something from the French that they could not and did not want to give him.

The Role of Huygens

Actually, Huygens's role as an intermediary and the communication 'hub' in this controversy was not totally honest and without self-interest. It introduced an additional layer of complexity into the controversy. On several occasions Huygens tried to confuse the discussion by introducing pieces of text and music from his own hand, always anonymously, making the receiver believe that they were French poetry, French composition or French commentary. This was all done with the sole purpose coercing answers (if not praise) from the combatants for his own benefit. There are three instances of this behaviour.

1. Huygens sent with his letter of 26 June 1640 to Ban a text with the request that he set it to music:

> You see here how Boësset set to music the same poem [*Me veux-tu voir mourir*] by accident or by habit. Added is another entirely sad and sorry poem, so that you two [Ban and Boësset] could, if you like, take another challenge, like noble athletes. A copy of the French version with the music is in my possession, so you should not be afraid of further deceit[66].

I am convinced that Huygens sent Ban a French text that he had written himself and, subsequently, already set to music[67]. Apparently, he wanted to play the same game with his own composition with Ban as Mersenne had done with Boësset's music! Ban seems not to have taken up the challenge.

2. Mersenne also fell victim to Huygens's intrigues. With his letter of 31 August 1640, Huygens sent Mersenne an air. Huygens stated that he had found it somewhere and that it was

---

66. Huygens to Ban, 26 June 1640 (Mersenne 1965, p. 439, Huygens 2007, p. 337): «Ut Boëssettius eosdem versiculos casu vel consuetudine expresserit, hic habes, subjunctis aliis merè tristibus ac luctuosis, ut, si libet, alterum hoc vestri periculum inter vos faciatis nobilissimi athletæ. Gallicum enim exemplar manuscriptum, ne fucum rursus extimescas, jamnunc in potestate meâ est, unâ μινυρισμῷ».

67. In these years, 1639-1640, composed the first French and Italian airs that were a few year later published in his *Pathodia sacra et profana* (Paris, 1647). See Rasch 1997.

anonymous. In truth, he had composed the air himself. He wondered if perhaps Mersenne could tell him who the author was:

> I like it [the air sent] very much, and it is evident that it is not of the hand of good Ban. After the final "*ne gouste plus*" the bass of the lute descends by beautiful steps all along towards the last cadence. But maybe I send the water back to the fountain, and you know it already. If you encounter other airs of the same kind, you could oblige me a lot with them. If Mr. Boësset is not the author, I would like to hear his opinion about it[68].

From the reference to the words «*ne gouste plus*» and the descending bass line, it appears that the composition was Huygens's own *Graves tesmoins*, later published in his *Pathodia sacra et profana* (Paris, 1647), ff. 37v-38r. No response from Mersenne is known.

3. Huygens would play the game yet a third time. With his letter of 16 January 1641 to Ban, he sent the 'Jugement d'un Trésorier-Général' for the second time, in a slightly retouched version, with the question of whether Ban could agree with it[69]. According to Huygens, the revision was done by the De Gouy, but the chronology of the controversy makes this very unlikely. Therefore, it is more likely that Huygens himself had produced the retouched version of the 'Jugement'. Ban did not notice the duplicity in Huygens's request. No response by Ban is known.

Despite these slight deceptions, Huygens still managed to please everyone.

Conclusion

The Ban-Boësset controversy has, as perhaps all controversies do, two levels. The first level is that of real and objective (topical) arguments. The second level is that of circumstantial and subjective (personal) arguments. It goes without saying that the protagonists themselves state in every respect that only the topical arguments counted. Yet, from beginning to end the choice of a topical argument is determined by the circumstances in which the combatants find themselves. It seems that no controversy can be studied and judged on the basis of topical facts

---

68. Huygens to Mersenne, 31 August 1640 (Mersenne 1967, p. 92; Huygens 2007, p. 398): «Il est fort à mon gré, et paroist bien que le bon Bannius ne l'a pas engendré. Depuis le dernier "*ne gouste plus*" la basse du luth descend par des beaux degrez tout du long vers la dernière cadence. Mais peut-estre que je renvoye l'eau à la fontaine, et que vous l'aurez veu. Que si vous en rencontrez d'autres de mesme veine, vous m'en pourriez beaucoup obliger. Si Monsieur Boësset n'en est pas l'auteur, je seray bien-ayse d'en sçavoir son sentiment».

69. Huygens to Ban, 16 January 1641 (Mersenne 1967, p. 412; Huygens 2007, p. 476).

alone. It is always necessary to understand the circumstantial facts as well: the characters of the combatants, their methods, their interests, and, last but not least, their personalities.

Postscript 1

In his letter of 3 November 1640 Mersenne made a remarkable proposal:

> [...] and if this would be refused under the pretext that we are prejudiced, we will send the two compositions to Rome, and to the courts of the Emperor, and the Kings of Spain and England, so that the best musicians who are found there will judge this dispute in the last resort[70].

In his letter of 13 January 1641 to Huygens, Ban asked Huygens to remind Mersenne of his promise[71]. Huygens replied that Mersenne was waiting for Ban to provide him with copies of the two pieces to send them to the courts of Rome (the pope!), the empress (in Vienna) and the kings of Spain (at war with the United Provinces!) and England. The plan was never executed at the time.

Yet there was a more recent opportunity to fulfil Ban's wishes. During a concert on the occasion of a congress on Descartes in Utrecht, March 2005, the ensemble Camerata Trajectina, under the direction of Louis Peter Grijp, executed the two compositions on *Me veux-tu voir mourir*, by Ban and Boësset respectively, without informing the public which composition was by which composer. After the performance of the two compositions, the audience was provided an opportunity to vote for which of the two compositions they preferred. A clear majority of the audience voted for Ban's composition as the most successful. Listening to the two compositions, we indeed realize that Boësset's is a little abstract and unsurprising, almost too perfect. Ban's composition, on the other hand, arouses more interest by its clumsiness and its unexpected turns. Posthumous rehabilitation?

Postscript 2

I have discussed two compositions of *Me veux-tu voir mourir*: (1) the one by Ban (a two-part version for voice and bass; Ex. 9.1) and (2) the one by Boësset (for two voices; Ex. 9.2). Another version by Ban, (3) 'with diminutions', was mentioned in the quote drawn from

---

[70]. Mersenne to Huygens, 3 November 1640 (Mersenne 1967, p. 207; Huygens 2007, p. 405): «et si nous sommes recusez soubz prétexte que nous sommes préocupez, nous envoyerons les deux airs à Rome, et à la cour de l'Empereur, du Roy d'Espagne et d'Angleterre, afin que les maistres de musique qui s'y trouveront, jugent de ce différent en dernier ressort».

[71]. Ban to Huygens, 13 January 1641 (Huygens 2007, p. 474).

Ex. 9.4: Nicolas Renier, *Me veux-tu voir mourir*, in: *Recueil d'airs sérieux et à boire de différents auteurs* (Paris, 1708), p. 66.

Mersenne's letter to Ban of June 1640 in Ban's letter to Anna Maria van Schurman of 20 August 1640; this version is lost[72]. Boësset's composition also existed in other version: (4) one for four voices that was published in 1642 (Ex. 9.3).

But the correspondence of Mersenne, Huygens and Ban contains references to no fewer than *seven* more compositions on the same text, all now lost. (5) A composition by «a physician in Rennes» is mentioned in a letter by Christophe Villiers, a physician in Sens, to Mersenne of 7 July 1640[73]. Mersenne had sent this version to Villiers as well as the compositions of Ban and Boësset. Villiers wrote that he found Boësset's composition to be the best, that of the physician from Rennes the worst, with Ban's composition in an intermediate position. With his letter, Villiers sent Mersenne two compositions of *Me veux-tu voir mourir*: (6) one from his own hand,

[72]. Ban to Anna Maria van Schurman, 20 August 1640 (Mersenne 1967, p. 20; Huygens 2007, p. 351): «Tuas partes, nempe Bassum et Cantum simplicem ac varium, [...]. Tua diminutio aptior instrumentis quam voci judicata est, [...]».

[73]. Villiers to Mersenne, 7 July 1640 (Mersenne 1965, pp. 456-458).

Ex. 9.5: Anonymous, *Veux-tu me voir mourir*, in: *Recueil d'airs sérieux et à boire de différents auteurs* (Paris, 1718), p. 20.

and (7) one by Le Riche, music master in Sens. Villiers's letter provides descriptions of some features of these compositions. In his letter of 21 August 1640, Villiers sent (8) a new version from his own hand, now based on the text as sent to Ban[74].

(9) A further composition on *Me veux-tu voir mourir*, by Jacques de Gouy, is mentioned by Ban in his letter to Huygens of 13 January 1641[75]. Two more versions were composed in France. With his letter of 29 April 1642, Huygens sent Ban (10) a *Me veux-tu voir mourir* by a certain *Consiliarius Parisiensis*[76]. Finally, with his letter of 19 August 1642, Huygens sent Ban (11) a composition on the same words by the French violin player Guillaume Dumanoir[77].

The words of *Me veux-tu voir mourir* would also attract the interest of eighteenth-century composers. A setting for voice and figured bass by Nicolas Renier is found in the *Recueil*

---

[74]. Villiers to Mersenne, 21 August 1640 (Mersenne 1967, pp. 48-50).

[75]. Ban to Huygens, 13 January 1641 (Huygens 2007, p. 473).

[76]. Huygens to Ban, 19 April 1642 (Mersenne 1970, p. 133; Huygens 2007, p. 640).

[77]. Huygens to Ban, 19 August 1642 (Huygens 2007, p. 659) and Ban to Huygens, 16 September 1642 (Mersenne 1970, p. 180; Huygens 2007, p. 662).

d'airs sérieux et à boire de différents auteurs pour l'année 1708 (Paris, 1708), p. 66 (Ex. 9.4), an anonymous setting for voice and unfigured bass in the corresponding volume for 1718, p. 20 (Ex. 9.5). These settings would have pleased Ban more than Boësset's composition: the match of textual and musical accents is much better and the phrase «*si je meurs malheureux*» is clearly musically translated into a falling melodic line with several *Seufzer*.

Appendix: Huygens's Poems on the Death of Joan Albert Ban

Text from Huygens 1893, p. 319. Both poems were published in Constantijn Huygens, *Momenta desultoria* (The Hague, 1655), p. 328. In the publication of the first poem, the word «tonos» in the second line was replaced by «modos».

In obitum Joannis Alberti Bannii,
Jurisconsulti Musici

Bannius, humanos moderatus lege magistrâ,
Correctum superos ivit ad astra tonos.
Posthâc, Astronomi, Cœlis confidite. Tandem
Vestra solœcismum sidera non facient.

Ad Saxum-Gandavi Assenedâ, 16 Augusti [1644].

On the death of Joan Albert Ban,
musical lawyer

Ban, who first ruled human tones with his master law, has now travelled to the stars to correct the celestial melodies. From now on, you, astronomers, can trust the sky, for your stars will not sin against the rules!

Assenede, Before Sas-van-Gent, 16 August 1644

Epitaphium

Bannius hic silet. O tristes epicedia Musæ
Dicite, nec Charitum de tribus ulla sile.
Quid verò? Tacet illa cohors, nec præfica magnis
Funeribus magno in funere plectra movet.
Hæsitat; hæc ratio est, post Banni fata choralæ
Nescia quâ lessum lege modoque canat.

Ibidem, 17 Augusti.

Epitaph [for Joan Albert Ban]

Ban is silent here. O sorrowful muses, sing a lament. And let no one of the three Graces be silent. But what? They are silent, and neither does the leader of the great funerals use her plectrum at this great funeral. She stops; this is the reason: after the death of the musician Ban, she no longer knows by what law and manner to sing her mournful song.

At the same place [=Assenede, before Sas-van-Gent], 17 August [1644]

Bibliography

Adam 1937

Adam, Charles. *Descartes. Ses amitiés féminines*, Paris, Boivin,1937.

Blackburn – Lowinsky – Miller 1991

Blackburn, Bonnie J. – Lowinsky, Edward E. – Miller, Clement A. *A Correspondence of Renaissance Musicians*, Oxford, Clarendon Press, 1991.

Carter 1992

Carter, Tim. 'Artusi, Monteverdi, and the Poetics of Modern Music', in: *Musical Humanism and Its Legacy: Essays in Honor of Claude V. Palisca*, edited by Nancy Kovaleff Baker and Barbara Russano Hanning, Stuyvesant (NY), Pendragon Press, 1992, pp. 171-194.

Descartes 1899

Descartes, René. *Correspondance. 3: janvier 1640-juin 1643*, edited by Charles Adam and Paul Tannery, Paris, Léopold Cerf, 1899.

Descartes 1975

Id. *Correspondance. 3: janvier 1640-juin 1643*, edited by Charles Adam and Paul Tannery, Paris, Vrin, [2]1975.

Descartes 1976

Id. *Correspondance. 4: juillet 1643-avril 1647*, edited by Charles Adam and Paul Tannery, Paris, Vrin, 1976.

Descartes – Huygens 1926

Descartes, René – Huygens, Constantijn. *Correspoandence of Descartes and Constantyn Huygens 1635-1647*, edited by Leon Roth, Oxford, Clarendon Press, 1926.

Goldberg 2011

Goldberg, Randall E. *Where Nature and Art Adjoin: Investigations into the Zarlino-Galilei Dispute, Including an Annotated Translation of Vincenzo Galilei's «Discorso Intorno all'opere di Messer Gioseffo Zarlino»*, Ph.D. Diss., Bloomington (IN), Indiana University, 2011.

Goulet 2004

Goulet, Anne-Madeleine. *Poésie, musique et sociabilité au xviie siècle. Les «Livres d'airs de différents auteurs» publiés chez Ballard de 1658 à 1694*, Paris, Honoré Champion, 2004.

Huygens 1882

Huygens, Constantijn. *Musique et musiciens au xviiie siècle. Correspondance et œuvre musicales de Constantin Huygens*, edited by Jan Pieter Nicolaas Land and Willem Joseph Andries Jonckbloet, Leiden, Brill, 1882.

Huygens 1893

Id. *De gedichten. Derde deel, 1636-1644*, edited by Jacob Adolf Worp, Groningen, Wolters, 1893.

Huygens 1915

Id. *Briefwisseling. Deel 4: 1644-1649*, edited by Jacob Adolf Worp, Den Haag, Nijhoff, 1915.

Huygens 2007

Id. *Driehonderd brieven over muziek van, aan en rond Constantijn Huygens*, translated and edited by Rudolf Rasch, Hilversum, Verloren, 2007.

Kaufmann 1966

Kaufmann, Henry W. *The Life and Works of Nicola Vicentino*, Rome, American Institute of Musicology, 1866.

Land 1891

Land, Jan Pieter Nicolaas. 'Joan Albert Ban en de theorie der toonkunst. ii', in: *Tijdschrift der Vereeniging voor Noord-Nederlands Muziekgeschiedenis*, iii/4 (1891), pp. 204-218.

Lote 1988

Lote, Georges. *Histoire du vers français. Tome iv: Le xvie et xviie siècles. 1: Les éléments constitutifs du vers. La déclamation*, Aix-en-Provence, Presses Universitaires de Provence, 1988.

Maniates 1993

Maniates, Maria Rika. 'The Cavalier Ercole Bottrigaro and His Brick-bats', in: *Music Theory and the Exploration of the Past*, edited by Christoph Hatch and David W. Bernstein, Chicago (IL), University of Chicago Press, 1993.

Mersenne 1962

Mersenne, Marin. *Correspondance du P. Marin Mersenne, réligieux minime. 7: janvier-juillet 1638*, edited by Cornelis de Waard, Paris, Éditions du Centre National de la Recherche Scientifique, 1962.

Mersenne 1964

Id. *Correspondance du P. Marin Mersenne, réligieux minime. 8: août 1638-décembre 1639*, edited by Cornelis de Waard, Paris, Éditions du Centre National de la Recherche Scientifique, 1964.

Mersenne 1965

Id. *Correspondance du P. Marin Mersenne, réligieux minime. 9: Du 2 janvier 1640 au 6 août 1640*, edited by Cornelis de Waard, Paris, Éditions du Centre National de la Recherche Scientifique, 1965.

Mersenne 1967

Id. *Correspondance du P. Marin Mersenne, réligieux minime. 10: Du 6 août 1640 à fin décembre 1641*, edited by Cornelis de Waard, Paris, Éditions du Centre National de la Recherche Scientifique, 1967.

Mersenne 1970

Id. *Correspondance du P. Marin Mersenne, réligieux minime. 11: 1642*, edited by Cornelis de Waard, Paris, Éditions du Centre National de la Recherche Scientifique, 1967.

Mersenne 1977

Id. *Correspondance du P. Marin Mersenne, réligieux minime. 13: 1643-1645*, edited by Cornelis de Waard, Paris, Éditions du Centre National de la Recherche Scientifique, 1977.

Nadler 2013

Nadler, Steven. *The Philosopher, the Priest, and the Painter: A Portrait of Descartes*, Princeton (NJ), Princeton Unversity Press, 2013.

Noske 1987

Noske, Frits. 'Huygens de musicus. Enkele aspecten', in: *Veelzijdigheid als levensvorm. Facetten van Constantijn Huygens' leven en werk*, edited by Arie Theodorus van Deursen, Elidius Klaas Grootes, and Petrus Eduardus Ludovicus Verkuyl, Deventer, Sub Rosa, 1987, pp. 129-140.

Palisca 1985a

Palisca, Claude V. *Humanism in Italian Musical Thought*, New Haven (CT), Yale University Press, 1985.

Palisca 1985b

Id. 'The Artusi-Monteverdi Controversy', in: *The New Monteverdi Companion*, edited by Denis Arnold and Nigel Fortune, London, Faber & Faber, 1985, pp. 127-156.

Pirro 1907

Pirro, André. *Descartes et la musique*, Paris, Fischbacher, 1907.

Rasch 1983

Rasch, Rudolf. 'Ban's Intonation', in: *Tijdschrift van de Vereniging voor Nederlandse Geschiedenis*, xxxiii/1-2 (1983), pp. 75-99.

Rasch 1997

Id. 'Waarom schreef Constantijn Huygens zijn «Pathodia sacra et profana»?', in: *Constantijn Huygens 1596-1996. Lezingen van het tweede Groningse Huygens-symposium*, edited by Nanne Frederik Streekstra, Groningen, Passage, 1997, pp. 95-124.

Rasch 2005

Id. 'Descartes en de Ban-Boësset-controverse', in: *Née Cartésienne / Cartesiaansch Gebooren. Descartes en de Utrechtse Academie 1636-2005*, edited by Willem Koops, Leon Dorsman, and Theo Verbeek, Assen, Van Gorcum, 2005, pp. 178-195.

Rodis-Lewis 1971

Rodis-Lewis, Geneviève. 'Musique et passions au xviie siècle', in: *xviie siècle*, no. 92 (1971), pp. 81-89.

Serwer 1970

Serwer, Howard. 'Marpurg versus Kirnberger. Theories of Fugal Composition', in: *Journal of Music Theory*, xiv/2 (1970), pp. 206-236.

Stoll 1978

Stoll, Albrecht D. *Figur und Affekt. Zur höfischen Musik und zur bürgerlichen Musiktheorie der Epoche Richelieu*, Tutzing, Schneider, 1978.

Walker 1974

Walker, Daniel Pickering. 'Some Aspects of the Musical Theory of Vincenzo Galilei and Galileo Galilei', in: *Proceedings of the Royal Musical Association*, c (1973-1974), pp. 33-47 (= Walker 1978, pp. 14-26).

Walker 1976

Id. 'Joan Albert Ban and Mersenne's Musical Competition of 1640', in: *Music & Letters*, lvii/3 (1976), pp. 233-255.

Walker 1978

Id. 'Mersenne's Musical Competition of 1640 and Joan Albert Ban', in: Id. *Studies in Musical Science in the Late Renaissance*, London, Courtauld Institute, 1978, pp. 81-110.

Chapter Ten

La Missa *La luna piena* (1657-1658) di Giuseppe Corsi e le teorie galileiane

Galliano Ciliberti
(Conservatorio di Musica 'Nino Rota', Monopoli)

Il Sig. Cigoli si è portato divinamente nella Cupola Cappella di S[ua] S[anti]tà a Santa Maria Maggiore, e come buon amico leale ha, sotto l'immagine della beata Vergine, pinto la Luna nel modo che V[ostra] S[ignoria] è stata scoperta, con la divisione merlata e le sue isolette[1].

Così il 23 dicembre 1612 il fondatore nonché 'principe' dell'Accademia dei Lincei Federico Cesi comunicava a Galileo Galilei la fine dei lavori nella cupola della grandiosa Cappella Borghese (o Paolina) della Basilica di Santa Maria Maggiore (o Basilica Liberiana) a Roma. Era stato lo stesso papa Paolo v (Camillo Borghese) a commissionare l'affresco della cupola della cappella al pittore Lodovico Cardi detto 'il Cigoli' raffigurante *L'assunzione della Vergine* (anche chiamato *La donna del Sole*) coronata da dodici stelle, in piedi sopra una luna dipinta similmente al «primo paesaggio lunare» descritto attraverso il telescopio da Galileo[2] (Ill. 10.1). Il Cigoli già da tempo stava lavorando per la famiglia Borghese e nella Basilica di Santa Maria Maggiore[3]. Gli affreschi con l'Assunta (opera tarda del pittore),

1. Federico Cesi a Galileo, 23 dicembre 1612 (Galilei 1901, no. 814, p. 449; Tognoni 2009, pp. 120-121).

2. Drake 1960, pp. 58, 60, Dame 1966, Ottavi Cavina 1976, Edgerton 1984, Winkler – Helden 1992, Helden 1996, e Spiller 2000.

3. Schraven 2001.

Ill. 10.1: Lodovico Cardi detto Il Cigoli, *L'assunzione della Vergine* (1610-1612). Roma, Basilica di Santa Maria Maggiore, Cappella Borghese (Paolina), cupola, con due particolari.

nonostante le inevitabili ostilità, piacquero al pontefice e al cardinale nipote Scipione Borghese che si adoperarono per insignire l'artista della nomina di cavaliere dell'ordine militare di Malta[4].

4. Tongiorgi Tomasi 2009, p. 20.

Galileo e Cigoli: tra musica 'assoluta', pittura e astronomia

Lodovico Cardi (1559-1613) — comunemente chiamato 'Il Cigoli' dal nome del suo paese natìo — era stato molto amico di Galileo Galilei (1564-1642; *cfr.* Ill. 6.1). I due intellettuali (pressoché prossimi di età) dovettero incontrarsi giovanissimi a Firenze dove frequentarono le lezioni del matematico granducale Ostilio Ricci che impartì a entrambi gli insegnamenti della geometria. Cigoli e Galileo si scambiarono numerose missive.

Tra le più importanti vi è senza dubbio la lettera (di cui purtroppo abbiamo solo una copia della seconda metà del XVII secolo ma ritenuta autentica dagli studiosi) inviata dallo scienziato in Firenze al pittore in Roma il 26 giugno 1612 dove Galileo esprime chiaramente le sue idee sulla musica:

> Non ammireremo noi un musico, il quale cantando e rappresentandoci le querele e le passioni di un amante ci muovesse a compassionarlo, molto più che se piangendo ciò facesse? E questo, per essere il canto un mezzo non solo diverso, ma contrario ad esprimere i dolori, e le lagrime et il pianto similissimo. E molto più l'ammireremmo, se tacendo, col suo strumento, con crudezze et accenti patetici musicali, ciò facesse, per esser le inanimate corde meno atte a risvegliare gli affetti occulti dell'anima nostra, che la voce raccontandole[5].

Come già ebbe a notare Erwin Panofsky (2008), la presa di posizione di Galileo in favore della musica strumentale quale capace di influenzare meglio l'animo umano rispetto a quella vocale[6], proprio per la sua astrattezza semantica, poneva lo scienziato su un piano di speculazione razionale e umanistica molto avanzato e nuovo. Galileo divergeva su questo punto dal padre Vincenzo che nel suo *Dialogo della musica antica, e della moderna* del 1581, citando *Sulla musica* dello Pseudo-Plutarco, asseriva che una canzone «per accomodarla al soggetto delle parole con giuditio grandissimo, e tutti gli affetti che in essa erano stati dal Poeta spiegati» doveva essere espressa «con arte maravigliosa; della qual cosa importantissima e principale dell'arte musica[le], non è fatto conto alcuno da prattici d'hoggi»[7]. Galileo supera la già innovativa teorizzazione del rapporto musica/testo, specchio della *seconda prattica* monteverdiana: la preminenza della musica strumentale quale unica capace di risvegliare «gli affetti occulti dell'anima», risulta infatti un concetto che anticipa ciò che avrebbe detto Jacob Burckhardt (sulla scia dell'estetica antiwagneriana di Eduard Hanslick) circa duecentocinquant'anni dopo e cioè che «la musica, se vogliamo penetrare la sostanza del suo essere, deve essere considerata

5. Galileo a Cigoli, 26 giugno 1612 (Galileo 1901, no. 713, p. 342; Tognoni 2009, pp. 93-94).
6. Panofsky 2008, pp. 28-29.
7. Galilei 1581, p. 79.

come musica strumentale, disgiunta dalle parole e, soprattutto, separata dalla rappresentazione drammatica»[8]. Per Panofsky:

> L'insistenza di Galileo per una chiara e nitida distinzione di valori e di procedimenti che, al suo tempo, erano comunemente considerati inscindibili, è la testimonianza di un purismo critico che può essere considerato la vera sigla del suo genio. Così come preferiva la musica pura senza parole alla canzone, figuriamoci al canto inframezzato da singhiozzi o da risate, analogamente insisteva sulla necessità di separare la quantità dalla qualità, la scienza dalla religione, dalla magia, dal misticismo e dall'arte[9].

Galileo in questa lettera, dunque, procede — secondo Nathalie Heinich (2016) — a un vero e proprio lavoro epistemologico:

> [...] 'di abilitazione' (e non di 'riabilitazione' che supporrebbe l'esistenza di una precedente età dell'oro) della musica: sottolineando ciò che le appartiene propriamente, al contrario di ciò che invece rientrerebbe nelle componenti eteronome, non specifiche, come i riferimenti testuali che, analogamente alla pittura, assumevano un po' agli occhi del mondo erudito il ruolo di garanti della 'dignità' dell'arte e della 'liberalità' dei suoi praticanti[10].

Questa modernità epistemologica avanzata da Galilei nell'ambito della musica e delle arti non è senza rapporto alla sua concezione di rigore scientifico, obbligando — per dirla con la Heinich — a «separare ciò che era ritenuto indistinto, a isolare i generi, a dissociare l'attività poetica e l'attività scientifica, la finzione e la realtà, la favola e dati della esperienza, leggenda e l'anatomia, il misticismo e l'algebra, la filosofia e la fisica, la religione e l'astronomia»[11]. Galileo, infatti, rifiutava la numerologia (sia essa biblica che pitagorica)[12] e

[8]. Burckhardt 1905, p. 228: «[...] die Musik, die man, um auf den Grund ihres Wesens zu kommen, ohne Verbindung mit Texten und vollends ohne Verbindung mit der dramatischen Darstelling als Instrumentalmusik betrachten muß». Traduzione inglese in Burckhardt 1943, p. 321, e citato in Panofsky 1956, p. 5. Inoltre si veda Canguilhem 1992 e Cypess 2012, pp. 453-454.

[9]. Panofsky 2008, p. 33.

[10]. Heinich 2016, p. 8 «'d'habilitation' (et non de 'réhabilitation' qui supposerait l'existence antérieure d'un âge d'or) concernant la musique: soulignant ce qui lui appartient en propre, au contraire de ce qui relèverait de composantes hétéronomes, non spécifiques, telles les références textuelles qui, comme en peinture, jouaient un peu aux yeux du monde savant le rôle de garants de la 'dignité' de l'art et de la 'libéralité' de ses praticiens».

[11]. *Ibidem*, p. 8: «à séparer ce qui était tenu pour indistinct, à isoler les genres, à dissocier activité poétique et activité scientifique, fiction et réalité, fable et données de l'expérience, légende et anatomie, mystique et algèbre, philosophie et physique, religion et astronomie».

[12]. Panofsky 1956, p. 6; Panofsky 2008, p. 34.

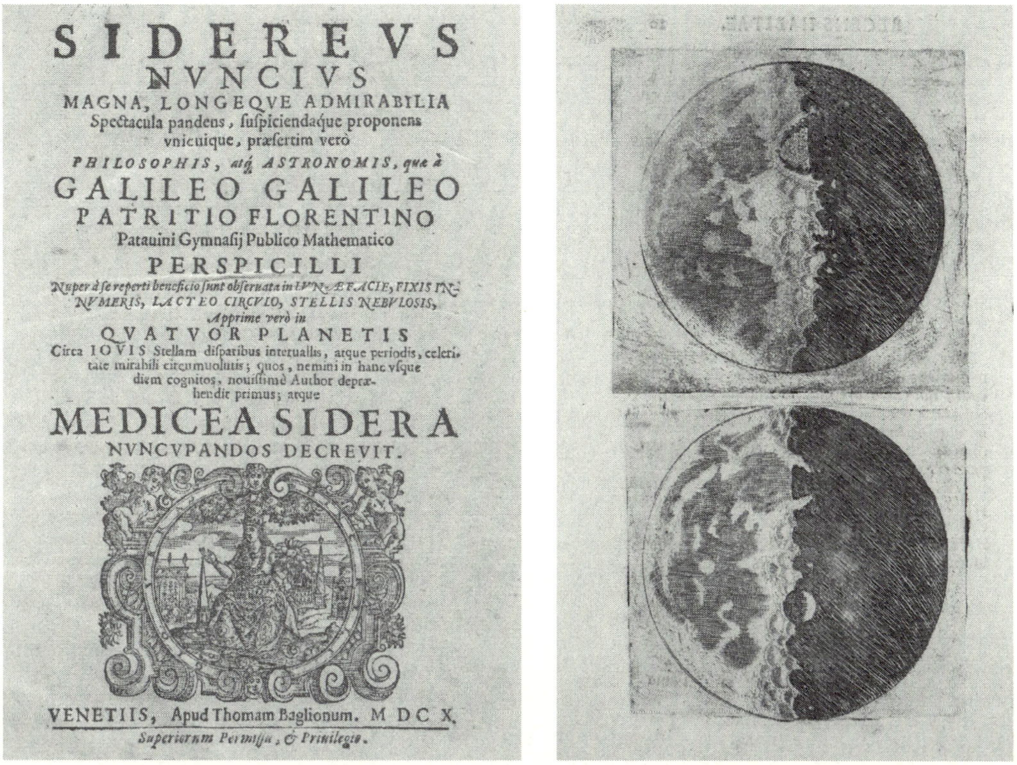

SIDEREVS
NVNCIVS
MAGNA, LONGEQVE ADMIRABILIA
Spectacula pandens, suspiciendaque proponens
vnicuique, praesertim verò
PHILOSOPHIS, atque ASTRONOMIS, quae à
GALILEO GALILEO
PATRITIO FLORENTINO
Patauini Gymnasij Publico Mathematico
PERSPICILLI
Nuper à se reperti beneficio sunt observata in LVNAE FACIE, FIXIS INNVMERIS, LACTEO CIRCVLO, STELLIS NEBVLOSIS,
Apprime verò in
QVATVOR PLANETIS
Circa IOVIS Stellam disparibus interuallis, atque periodis, celeritate mirabili circumuolutis; quos, nemini in hanc vsque diem cognitos, nouissimè Author depraehendit primus; atque
MEDICEA SIDERA
NVNCVPANDOS DECREVIT.
VENETIIS, Apud Thomam Baglionum. M DC X.
Superiorum Permissu, & Priuilegio.

Ill. 10.2ab: Galileo Galilei, *Sidereus nuncius* (Venezia, 1610), frontespizio e le incisioni che mostrano montagne e crateri sulla luna. Copy US-CAh.

lo stesso simbolismo (soprattutto nei poemi eroici come la *Gerusalemme liberata* di Torquato Tasso) perché costringeva il lettore «a interpretare ogni cosa come riferita seconditamente a qualcos'altro»[13].

Egli, inoltre, conosceva le teorie di Johannes Kepler[14] e probabilmente anche le sue idee sulla musica secondo cui Dio avrebbe attribuito ad ogni pianeta «non una 'nota' unica, ma una frase musicale propria e a formare, partendo da queste frasi, un'armonia polifonica e contrappuntistica svolgentesi nel tempo»[15]. Teorie che secondo Panofsky sarebbero state considerate da Galileo come una sorta di «astronomia manierista»[16].

Se la lettera del 26 giugno 1612 era stata prontamente inviata a Cigoli da Galilei per aiutare il pittore «in una di quelle tediose discussioni sui rispettivi meriti della pittura e della scultura»

[13]. Panofsky 2008, p. 36. Panofsky 1956, p. 6: «forcing the reader to interpret everything as a reference to something else».

[14]. Panofsky 1956, pp. 13-14, Panofsky 2008, p. 57.

[15]. Gozza 1989, pp. 27-28.

[16]. Heinich 2016, p. 9: «astronomie manièriste».

e quindi a rispondere ai «minacciosi argomenti dei suoi oppositori» conosciuti tramite Andrea Cioli (segretario della madre del Granduca di Toscana nonché legato a Galileo)[17], così anche lo scienziato si era valso «ampiamente dell'aiuto dell'amico pittore»[18]. Tale collaborazione era avvenuta non solo «nella realizzazione delle tavole a corredo dell'*Istoria e dimostrazioni intorno alle macchie solari* (Roma, 1613)»[19], ma anche nelle sedute di ben 26 osservazioni celesti che Cigoli effettuò proprio dall'alto della cupola della Basilica di Santa Maria Maggiore[20]. È questa la ragione per cui, nelle raffigurazioni della Cappella Borghese, Cigoli aveva voluto dipingere la figura della Vergine sopra una Luna galileiana con i suoi crateri, ispirandosi chiaramente alle tavole del *Sidereus nuncius* (Venezia, 1610; Ill. 10.2) che era stato pubblicato solo sei mesi prima dell'inizio dei lavori[21]. I due amici si erano del resto incontrati a Roma proprio nel 1611 e quindi non è da escludere che essi abbiano avuto modo di conferire concretamente sul progetto tale «rivoluzionaria iconografia»[22].

À la rencontre di Galilei: Roma

Roma costituiva il punto di approdo finale di molti seguaci di Galileo. Nell'Urbe si recavano per incontrare lo scienziato che ebbe a soggiornarvi nel corso di sei viaggi: 1587; 29 marzo - 4 giugno 1611; 10 dicembre 1615 - 4 giugno 1616; 23 aprile - 16 giugno 1624; 3 maggio - 26 giugno 1630; 16 febbraio - 6 luglio 1633[23]. Roma però poteva costituire soprattutto per coloro che venivano dalla Francia anche la tappa finale di un 'viaggio galileiano' che prevedeva una sosta a Firenze, città dove lo scienziato si era stabilito. Tra questi accoliti e sostenitori è da ricordare Jean Tarde, amministratore ecclesiastico ma anche uomo di scienze (astronomo e matematico), che compì il suo secondo viaggio in Italia (21 settembre 1614 - 3 febbraio 1615) proprio per incontrare Galileo a Firenze, visita dalla quale trarrà conoscenze fondamentali per la costruzione del telescopio che realizzerà al suo ritorno in Francia[24].

Durante il soggiorno a Roma, Tarde si recò proprio alla Basilica di San Maria Maggiore (novembre 1614) per visitare la Cappella Borghese e nel giorno della festa di Santa Cecilia (22 dello stesso mese) andò a visitare la chiesa omonima «che è presso la Ripa Grande» dove

17. Si veda Panofsky 2008, p. 24, ma anche Chappel 1975.
18. Tongiorgi Tomasi 2009, p. 15.
19. *Ibidem*.
20. Cigoli a Galileo, 23 marzo 1612 (Galilei 1901, n. 666, pp. 286-289; Tognoni 2009, pp. 81-84).
21. Ostrow 1996.
22. Tongiorgi Tomasi 2009, p. 19.
23. Huemer 1983; Topper – Gillis 1996; Cole 2005; Konečný 2005; Shea – Artigas 2009.
24. Tarde 1984, pp. 60-66.

ebbe modo di ascoltare «una bella musica a quattro cori, ciascun coro con il suo organo». Egli, inoltre, «per diversi giorni» poté udire «la musica nella [Basilica di] Sant'Apollinare» oltreché le prediche al Gesù e alla Chiesa Nuova[25].

La Cappella Liberiana, la Cappella Borghese in Santa Maria Maggiore e Giuseppe Corsi

Il 27 gennaio del 1613 la Cappella Borghese fu solennemente consacrata da papa Paolo v attraverso una grandiosa cerimonia a cui parteciparono i cantori pontifici e dove furono eseguiti due mottetti a 8 voci in due cori (*Alma redemptoris mater* e *Sancta Maria succurre*) di Arcangelo Crivelli, tenore della Cappella Sistina[26]. La possibilità di cantare a due cori anche all'interno di questo spazio era resa possibile dalla presenza di altre due piccole cappelle (una dedicata a San Carlo Borromeo e l'altra a Santa Francesca Romana). Come testimoniò nel 1655 l'oratoriano francese Nicolas de Bralion: «vi sono su ciascuna di queste due cappelle due Cori per collocare i musicisti, che hanno all'interno di quella grande balaustre e gelosie»[27].

Nel 1615 Scipione Borghese approvò gli statuti che ne organizzavano il funzionamento:

> La fondazione di Paolo v per il servizio di questa cappella prevede la presenza di un priore, di dodici cappellani (uno dei quali è sacrestano ed un altro camerlengo, cioè esattore), di quattro chierici, di un Corpo di Musica e di molti altri ufficiali che sono sotto la giurisdizione e la protezione di un cardinale[28].

Il «Corpo di Musica» prevedeva il servizio della 'Salve' da parte di dodici musicisti, un maestro di cappella e un organista (questi due ultimi già in servizio nella compagine stabile della cappella musicale della basilica). Essi avevano il compito di cantare tutti i sabato sera nonché tutte le feste e la vigilia delle feste della Santa Vergine sia le compiete a due cori che le litanie e soprattutto la *Salve Regina* (da qui si spiega perché i contemporanei chiamassero questi uffizi semplicemente 'la Salve')[29].

---

25. *Ibidem*, pp. 70-72: «qui est près de Ripa Grande [...] une belle musique à quatre chœurs, chaque chœur ayant son orgue [...] par divers jours [...] la musique à l'Apollinaire».

26. Lionnet 2018, p. 175.

27. Nicolas de Bralion, *Les curiositez de l'une et de l'autre Rome* (Parigi, 1655), p. 396: «Il y a sur chacune de ses deux chapelles deux Chœurs pour placer des Musiciens, lesquels ont au-dedans de la grande des balustres & des ialousies».

28. *Ibidem*, p. 398: «La fondation de Paul v pour le service de cette Chapelle est d'un Prieur, douze Chapelains, dont l'un est Sacristain, & un autre Camerier, c'est-à-dire Receveur; quatre Clercs; d'un Corps de Musique, & plusieurs autres Officiers, lesquels sont sous la iurisdiction & protection, d'un Cardinal comme on voie».

29. Lionnet 2018, pp. 175-176.

Nel corso del Seicento tutti i maestri di cappella di Santa Maria Maggiore dovettero organizzare tali funzioni liturgiche. Toccò quindi anche al ventiseienne Giuseppe Corsi (Tavola 10.1) eletto nel settembre del 1657 nel magistero musicale della Basilica Liberiana a dirigere ed organizzare la cappella grazie al probabile interessamento di Antonio Maria Abbatini che era stato direttore di questa istituzione per quasi sedici anni (1640-1646 e 1649-1657)[30].

Tavola 10.1: biografia di Giuseppe Corsi[31]

1631-1632	Nasce a Celano (Abruzzi) dal ramo nobile dei Corsi-Vangelisti. Grazie ai Peretti-Montalto si stabilisce a Roma
1646	Dal 10 febbraio al 30 novembre è Soprano «puer» nella Cappella Giulia sotto il magistero di Virgilio Mazzocchi di cui ne diventò allievo. Dopo la morte del maestro (6 ottobre) fu probabilmente presso i padri Gesuiti e allievo di Giacomo Carissimi.
Ante 1653	È a Gallese (Lazio) al servizio di Pietro Altemps.
1654-1655	Il 2 marzo del 1654 è eletto maestro di Cappella della Cattedrale di San Florido di Città di Castello (Umbria) carica che mantiene sino alla fine di settembre 1655.
1655-1657?	A Napoli al servizio di un principe.
1657-1659	Nel settembre del 1657 viene nominato maestro di cappella della Basilica di Santa Maria Maggiore a Roma magistero che terrà sino al gennaio 1659.
1658-1684	Il suo nome compare nei verbali della *Congregazione dei musici di Roma sotto l'invocazione di Santa Cecili*a il 24 gennaio 1658 come maestro di cappella della chiesa di Santa Maria Maddalena.
1661-1665	Dal gennaio 1661 al giugno 1665 dirige la cappella della Basilica di San Giovanni in Laterano.
1666-1667	Il 13 marzo è maestro di cappella nella Cattedrale di San Giovenale di Narni (Umbria).
1668-1675	Dal 28 febbraio 1668 al 28 febbraio 1674 (ma ufficialmente rimane in carica sino al 30 settembre 1675) è maestro di cappella nella Basilica della Santa Casa di Loreto (Marche).
1676-1677	È attivo nella Confraternita del SS. Crocifisso e quindi dirige due oratori nell'Oratorio di San Marcello e in Santa Maria in Vallicella (Chiesa Nuova) sempre come compositore di oratori.
1677	Il 29 luglio viene arrestato per il «defloramento di una zitella». Torturato è incarcerato a Narni (Umbria) nella Rocca Albornoziana.
1681-1689	Entra nella corte di Ranuccio ii Farnese divenendo maestro di cappella della Basilica di Santa Maria della Steccata di Parma.
1689	Forse è ad Ancona a dirigere nella Cattedrale di San Ciriaco. Muore subito dopo.

30. Burke 1984, p. 18; Simi Bonini 2010; Ciliberti – Tribuzio 2014, pp. 13-20.

31. Burke 1984, p. 18, Simi Bonini 2010, Ciliberti – Tribuzio 2014, Rostirolla 2017, pp. 466, 476, 498, Luisi 2021, pp. 240-243.

All'interno del complesso sistema istituzionale romano, la Basilica di Santa Maria Maggiore costituiva uno degli organismi ecclesiastici più rilevanti dopo San Pietro in Vaticano, ambedue chiese collegiali e capitolari assieme a San Giovanni in Laterano, San Lorenzo in Damaso e Santa Maria in Trastevere. «Nel Monte Esquilin dove è Maggiore, la Gran Madre di Dio»[32], per dirla con i versi autobiografici di Abbatini, si operava, dunque, solo in quanto artista capace e di credito. Tutto ciò si evince dalla lunga tradizione di maestri di cappella di indiscusso prestigio che, prima di Corsi, erano stati chiamati a reggervi la musica (tra i quali Palestrina).

Il lavoro in questa basilica risultava assiduo e logorante tanto che qualche decennio più tardi sempre l'Abbatini testimonierà che «per l'impiego continuo» che qui svolgeva quale insigne maestro di cappella, rischiava addirittura di contrarre la tisi[33]. Quindi trovava ampia giustificazione la presenza di un giovane pieno di energie e di entusiasmo quale era Giuseppe Corsi. Il coro negli anni 1657-1658 era formato da sedici elementi (quattro soprani, quattro contralti, quattro tenori e quattro bassi), più l'organista Francesco Mutij, importante strumentista attivo in altre significative cappelle romane[34].

L'alacre attività di Corsi alle feste più rilevanti celebrate con musica nella basilica quali quelle di Natale (negli anni 1657-1658 si svolsero però con un organico ridotto probabilmente a causa della pestilenza), di Pasqua, di San Marco con la processione (25 aprile), del *Corpus Domini* con la processione, della Madonna della Neve (5 agosto), dell'Assunta (15 agosto), della natività della Madonna (8 settembre) e di Sant'Ildefonso di Toledo (23 gennaio), dove era consuetudine fare musica a più cori[35]. Si trattava di celebrazioni solenni e fastose che necessitavano di imponenti esecuzioni straordinarie per l'impiego sostanzioso di musicisti provenienti da altre chiese e necessari per rinforzare l'organico della musica regolare (ovvero i salariati costituenti la compagine abituale della cappella).

Tra gli incarichi di Corsi anche l'organizzazione della musica (febbraio-novembre 1658) nella Cappella Borghese (a spese del principe Giovanni Battista Borghese) con la sua liturgia specifica concernente — come detto — l'esecuzione delle Litanie e della *Salve Regina* ogni sabato sera[36]. Il giovane musicista abitava in una casa riservata dal capitolo al maestro di cappella collocata guarda caso proprio «incontro alla Cappella Paolina [= Borghese]» con la moglie Anna (di 20 anni), con la domestica Margherita D'Amati (di 40 anni), e con due soprani: Agostino Calderi (di 13 anni) e Giovanni Battista Cofani (di 12 anni). Il suo stipendio era di scudi 12 al mese più altri 10 per il mantenimento dei due soprani[37].

---

32. Ciliberti 1986, p. 429.

33. *Ibidem*, p. 429.

34. Simi Bonini 2010, p. 548; Silbiger 1980, pp. 47-49, 52, 55-58, 64, 72; Annibaldi 1990, pp. 402, 405, 425.

35. Simi Bonini 2010, pp. 548-549; Ciliberti – Tribuzio 2014, pp. 15-16.

36. Simi Bonini 2010, p. 549.

37. Burke 1984, p. 18; Simi Bonini 2010, pp. 548, 565; Ciliberti – Tribuzio 2014, p. 14.

Gli *Avvisi di Roma* ricordano come il 15 settembre 1658 Corsi diresse nella Cappella Borghese della Basilica una messa solenne «a 4 cori di Musica» fatta celebrare dal cardinale Girolamo Colonna «protettore del regno d'Aragona» nell'ottava della Natività della Beata Vergine[38]. Negli archivi della Basilica Liberiana non esiste alcuna messa del musicista a 4 cori ma ne esistono custodite due a 8 voci ripartite in due cori. Si tratta in realtà dello stesso componimento, scritto su gradi o modalità diverse: la Missa *La luna piena* e la Missa *Divo Juvenali dicata*[39]. Il titolo della prima versione fa esplicito riferimento alla 'Luna piena' ovvero all'immagine della 'Donna del sole' (l'Assunta), affrescata dal Cigoli nella Cappella Borghese. La seconda versione (più tarda e risalente al magistero di Corsi nella Chiesa Nuova tra il 1676-1677) possiede sempre un nesso con la Luna. Nella *Vita* scritta da don Pietro Giacomo Bacci del padre filippino nonché vescovo e insigne compositore Giovenale Ancina (1545-1604) — «Divo Juvenali» recita, appunto, il titolo del brano del Corsi[40] — si narra che questi:

> [...] andò su la meza notte à destare i Padri, acciò che venissero seco à vedere l'eclissi di Luna: e perché uno di loro in particolare si sentì svegliare nel meglio del sonno (il che non si suol fare, se non per grave causa) quando intese, che Giovenale non voleva altro, che invitarlo a veder l'eclissi, gli rispose: "Se fa l'eclissi non me ne curo": il servo di Dio incapace di quella (pareva à lui) insensibilità, sospirò & ammirato partì, dicendo: "come? Fa l'eclissi, e tu non te ne curi"[41].

La Missa *La luna piena*, l'affresco del Cigoli e le teorie galileiane

Non è dato sapere se la Missa *La luna piena* fosse stata composta da Corsi per l'eclissi lunare del 20 dicembre 1657 (in quell'anno ve ne furono tre ma il maestro prese servizio alla Basilica di Santa Maria Maggiore nel settembre)[42] o per l'anniversario della consacrazione della Cappella Borghese (27 gennaio) o per festeggiare il protettorato aragonese del cardinale Girolamo Colonna (15 settembre)[43]. Certo è che essa fu pensata in rapporto al luogo (la Cappella Borghese), alla sua liturgia specifica (la Salve) e, visto il titolo ammiccante *La luna piena*, all'affresco del Cigoli che rappresentava la Luna galileiana.

---

[38]. V-CVasv, *Segr. Stato, Avvisi*, 28, c. 514, edito in Franchi 2004, p. 573.

[39]. Tribuzio 2014, pp. 151-152, n. 2.

[40]. E non *San* Giovenale come indica Simi Bonini 2010, pp. 554-558, prima e Luisi 2021, p. 243, poi.

[41]. Pietro Iacomo Bacci, *Vita del v[enerabile] servo di Dio Gio: Giovenale Ancina* (Roma, 1671), pp. 217-218.

[42]. Antonio Carnevale, *Gli arcani delle stelle* (Firenze, 1657); Francesco Barzini, *Astrologiche precognizioni* (Firenze, 1657), ff. [3v-4r]; Giacinto Martij, *Cognetture astronomiche dell'eclisse della Luna nel 1657, alli 25. di Giugno* (Todi, 1657); Rutilio Benincasa, *Almanacco perpetuo* (Ancona, 1658), p. 22.

[43]. La fonte più antica della Missa *La luna piena* si trova in I-Rsm, MS 73/12. Per un elenco completo delle fonti (per la maggior parte del Sette- e Ottocento), si rimanda a Tribuzio 2014, pp. 151-152, n. 2.

Es. 10.1: Giuseppe Corsi, Missa *La luna piena*, Gloria, battute 1-5.

Riguardo al primo aspetto, ovvero la relazione con il contesto liturgico in cui la Missa venne concepita, è possibile constatare quanto Giuseppe Corsi impieghi sostanzialmente quattro temi principali che vengono sviluppati in diversi punti (Tavola 10.2). Ognuno di questi temi corrisponde sia all'*incipit* di brani liturgici preesistenti (Temi **1**, **2**, **4**) che a delimitate sezioni interne di essi (Tema **3**).

Tavola 10.2: temi nella Missa *La luna piena* di Giuseppe Corsi

Cantus firmus	Corsi, Missa *La luna piena*
Tema **1**	
Kyrie «Salve» A (*Missel grégorien* 1985, p. 129)	Kyrie: *Kyrie* I Gloria: *Et in terra pax* | *Qui tollis peccata mundi* Credo: *Factorem cœli et terrae* Sanctus: *Sanctus* Agnus Dei: *Agnus dei*

TEMA 2	
Nativitas est hodie (Antifona II dei vespri della natività BMV, *Liber usualis* (1961), p. 1625)	Kyrie: *Christe* Credo: *Cuius regni*
TEMA 3:	
Salve Regina: «misericordes oculos ad nos» (Antifona BMV, *Liber usualis* (1961), p. 279)	Kyrie: *Kyrie* II Gloria: *Gratias agimus tibi* Credo: *Amen*
TEMA 4	
Maria Virgo assumpta est (Antifona II dei vespri dell'Assunzione BMV, *Liber usualis* (1961), p. 1605)	Gloria: *Domine Deus* | *In gloria Dei Patris. Amen* Credo: *Qui propter nos* Agnus Dei: *Qui tollis peccata*

In alcuni casi il compositore si accinge ad elaborare contemporaneamente anche diversi canti fermi. È il caso dell'inizio del Gloria dove al Coro Primo (ed in particolare al Tenore) (ES. 10.1) è affidata l'intonazione dell'antifona (diffusa soprattutto in area iberica) *Sancta Maria Virginum piissima* per l'assunzione della Vergine Maria:

[44].

mentre le altre voci elaborano ad imitazione la melodia del Kyrie «Salve» (tema **1**):

[45].

A Santa Maria Maggiore si celebrava la 'Messa di Spagna' il giorno della festa di Sant'Ildefonso alla quale partecipava solennemente tutto il corpo diplomatico e la nobiltà spagnola presente a Roma. La presenza di un brano specifico della liturgia appartenente a questa zona geografica quindi non è da considerare un'eccezione.

Il processo compositivo tendente ad intersecare melodie preesistenti dissimili si evince anche in due casi molto importanti: il Kyrie II e la parte conclusiva del Gloria: «in gloria Dei Patris. Amen». Qui si entra però nel secondo aspetto quello del possibile legame della musica con le relative descrizioni galileiane del *Sidereus nuncius*.

Analizzando la struttura generale della Missa *La luna piena* (TAVOLA 10.3) risalta la medesima architettura compositiva di questi brani: in ambedue le sezioni, infatti, le parti

44. P-BRs, MS 028, f. 153r.

45. *MISSEL GRÉGORIEN* 1985, Kyrie «Salve» A, p. 129.

di Soprano del Coro Primo e del Coro Secondo procedono omoritmicamente e ad unisono evidenziando una melodia costituita da una lunga successione di brevi legate tra loro che formano ampi gruppi di battute nelle quali l'elaborazione imitativa di tutte le altre voci è costruita su valori più stretti.

Tavola 10.3: struttura generale ed indicazioni dei tempi nella Missa *La luna piena* di Giuseppe Corsi

Kyrie

Sezione	Tempus	Annotazioni
Kyrie I a 8	c	
Christe a 4 (C I-II, A I-II, Org)	c	
Kyrie II a 8	3/1	*Color*; C I e II: 36 brevi di seguito (la 27ª longa bianca su *color* delle altre voci, eccetto T II che presenta minime bianche concluse da una *ligatura cum opposita proprietate*)

Gloria

Sezione	Tempus	Annotazioni
Et in terra pax a 8	c	Organum: 9 semibrevi + 6 brevi di seguito
Gratias agimus a 8	c 3/1	*Color*
Domine Deus Rex a 8	¢	
Qui tollis peccata a 8 *Suscipe* *Deprectationem*	c 3/1 c	 C I-II 45 brevi di seguito (*Gloria Dei Patri. Amen*)

Credo

Sezione	Tempus	Annotazioni
Patrem omnipotentem a 8 *Et ex patre*	c 3/1	 *Color*
Qui propter nos a 4 (T I-II, B I-II, Org)	c	
Et incarnatus a 8	c	*Color*, Org e B II con molte brevi di seguito

Crucifixus a 4 (C, A, T, B, Org) *Et resurrexit* *Cuius regni*	c 3/1 ¢	 *Color*
Et in Spiritum *Simul adoratur* *Confiteor* *Mortuorum* *Amen*	3/1 ¢ 3/1 ¢ 3/1	*Color* *Color* *Color*

SANCTUS

Sezione	Tempus
Sanctus	c

AGNUS DEI

Sezione	Tempus
Agnus Dei	c

Questa scrittura molto particolare (anche dal punto di vista visivo) dove le voci superiori dei cori intonano sequenze di note lunghe legate per la durata di diversi insiemi di battute, sembra collegarsi per la sua conformazione seriale proprio ad alcune descrizioni del *Sidereus nuncius*.

LE COSTELLAZIONI

A proposito della Costellazione delle Pleiadi Galilei asserisce «ché presso di esse si affollano più di quaranta invisibili [...]; di queste ne abbiamo segnate trentasei soltanto»[46]. E più oltre descrivendo la Nebulosa del Presepe lo scienziato osserva quanto questa «non è una stella soltanto, ma una congerie di quaranta stelline; ne abbiamo segnate oltre gli Asinelli, trentasei»[47].

[46]. Galileo Galilei, *Sidereus nuncius* (Venezia, 1610), f. 16bis-recto: «quibus aliae, plures quam quadraginta invisibiles adiacent; [...]; harum nos tantum triginta sex adnotavimus». Traduzione italiana da GALILEI 1993, pp. 126-127.

[47]. *Ibidem*, f. 16ter-verso: «quae non una tantum Stella est, sed congeries Stellularum plurimum quam quadraginta: nos praeter Asellos, trigintasex notavimus». Traduzione italiana da GALILEI 1993, pp. 128-129.

Nel Kyrie II la successione di 36 brevi (collocate sopra l'elaborazione contrappuntistica del tema **3** tratto da un frammento dell'antifona *Salve regina*) è ricavata dall'*incipit* dell'inno *Jesu corona Virginum*[48]. Gesù è corona gloriosa delle Vergini, concepito da Maria, ma l'allusione è soprattutto alla corona delle dodici stelle situata sopra l'immagine dell'Assunta. La sequenza delle 36 brevi (12 × 3) (Tavola 10.4) è costituita da una serie di sette note che procedono per gradi congiunti disposte secondo una piccola scala ascendente (le prime quattro: Sol, La, Si e Do) e discendente (le restanti tre: Si, La e Sol) (Es. 10.2) come se fossero astri che sorgono e tramontano nella volta celeste secondo quanto testimonia lo stesso Galileo nel descrivere il movimento di tali corpi celesti.

Tavola 10.4: la sequenza delle 36 brevi nel Kyrie II della Missa *La luna piena* di Giuseppe Corsi

Numero di battute	Valore delle note	Nota
6	Brevi	Sol
5	Brevi	La
5	Brevi	Si
7	Brevi	Do
3	Brevi	Si
2	brevi = 1 longa	La
8	Brevi	Sol
Totale: 36		

Es. 10.2: Giuseppe Corsi, Missa *La luna piena*, Gloria, Cantus I-II.

[48]. Stäblein 1956, n. 16.

Ill. 10.3ab: Galileo Galilei, *Sidereus nuncius* (Venezia, 1610), ff. [16b-recto] e [16b-verso].

Circa la Costellazione delle Pleiadi (Ill. 10.3a) Galilei dichiara: «ho disegnato le sei stelle del Toro, dette Pleiadi (dico sei, in quanto che la settima non appare quasi mai), racchiuse in cielo entro limiti angustissimi»[49].

Nel Kyrie ii alla nota di partenza e base della sequenza, il Sol (base modale della messa dove nome del suono potrebbe essere inteso anche quale abbreviazione di *sol*e e quindi di astro), sono affidate sei brevi che delimitano con esatta precisione anche le prime sei battute della serie formata da valori aumentati. Le sei stelle della costellazione del Toro corrispondono nel disegno di Galileo agli astri più luminosi quelli collocati in primo piano.

Nella parte conclusiva del Gloria la sezione contrappuntistica del versetto finale «in gloria Dei Patris. Amen» è basato sull'antifona *Maria Virgo assumpta est* che presenta nel testo l'esplicito riferimento al «stellatum solium» [trono stellato] dove siede Dio. Dopo sedeci battute in cui il processo imitativo investe tutte le otto voci delle due formazioni, i Soprani del Coro Primo e del Coro Secondo fuoriescono dall'intreccio contrappuntistico cambiando il loro *ductus*. Analogamente al caso del Kyrie ii, essi procedono con omoritmia e all'unisono nell'enunziare, attraverso l'impiego di brevi e semibrevi legate in nuclei diversificati, due sequenze (Tavola 10.5): la prima di tre note ma con sei figure per ognuna nota costruita sull'*incipit* della melodia del Sanctus ix utilizzato appositamente nella liturgia per la festa della Vergine[50]; la seconda una sequenza di sei suoni come le stelle del Toro.

[49]. Galileo Galilei, *Sidereus nuncius* (Venezia, 1610), f. 16bis-recto: «sex Stellas Tauri, Pleidas dictas, depinximus (dico autem sex, quandoquidem septima fere nunquam apparet) intra angustissimos in cœlo». Traduzione italiana da Galilei 1993, pp. 126-127.

[50]. *Missel grégorien* 1985, p. 108.

Tavola 10.5: i nuclei melodici della parte conclusiva del Gloria della Missa *La luna piena* di Giuseppe Corsi

1	Re	6
2	Si	6
3	Sol	6

1	La
2	Si
3	Do
4	Si
5	La
6	Sol

Il disegno della Nebulosa del Presepe (Ill. 10.3b) pubblicato nel *Sidereus nuncius* è delimitato da due stelle collocate una in basso di poco più piccola dell'altra posta in alto. La sequenza melodica del Kyrie ii si apre e si chiude con lo stesso suono: Sol. Quello iniziale (sei brevi) è di poco più piccolo di quello finale (otto brevi).

La luna piena

L'ultimo aspetto di questa trattazione concerne il rapporto tra la musica e l'immagine stessa della Luna così come appare nell'iconografia del Cigoli (Ill. 10.1) che nelle rappresentazioni di Galileo (Ill. 10.2b):

> E di queste macchie, sempre più scure sono quelle che sono contermini al confine tra luce e tenebre, mentre le più remote appaiono e minori e meno scure; cosicché quando finalmente la Luna diventa piena nell'opposizione, l'ombra delle cavità si distingue dal candore delle parti rilevate per un modico e assai tenue distacco[51].

Il riferimento a questa descrizione della Luna e alla conseguente immagine fornita dal Cigoli risulta più che evidente nella Missa di Corsi poiché l'utilizzazione abbondante di note bianche e di *color* (si veda Tavola 10.3) nelle sezioni ternarie della composizione acquisisce un parallelo valore visivo inequivocabile. Le *notae nigrae* apposte dal musicista con sapiente retorica

[51]. Galileo Galilei, *Sidereus nuncius* (Venezia, 1610), f. 11r: «quarum quidem macularum semper nigriores sunt illae, quae confinio luminis et tenebrarum conterminae sunt, remotiores vero tum minores, tum obscurae minus apparent; ita ut tandem, cum Luna in oppositione totum impleverit orbem, modico admodumque tenui discrimine cavitatum opacitas ab eminentiarum candore discrepet». Traduzione italiana da Galilei 1993, pp. 100-101.

semiografica a contrasto con i valori notazionali bianchi, ricordano alla vista (più che all'udito) proprio quel giuoco di luci e di ombre presenti nei crateri della Luna. In particolare verso la fine del Kyrie II un passo sembra l'espressione esatta di queste rappresentazioni extramusicali del nostro satellite naturale. Alle parti di Canto I e di Canto II è affidata una longa bianca (Es. 10.2) che sovrasta sia alcune *notae nigrae* collocate nelle voci di Contralto, Tenore, Basso e organo del Coro Primo sia 12 *tactus* al Tenore del Coro Secondo (come le dodici stelle dell'Assunta) formati da otto minime e da una *ligatura cum opposita proprietate* composta da due brevi (= due semibrevi) (Es. 10.3).

Es. 10.3: Giuseppe Corsi, Missa *La luna piena*, Kyrie II, battuta 107.

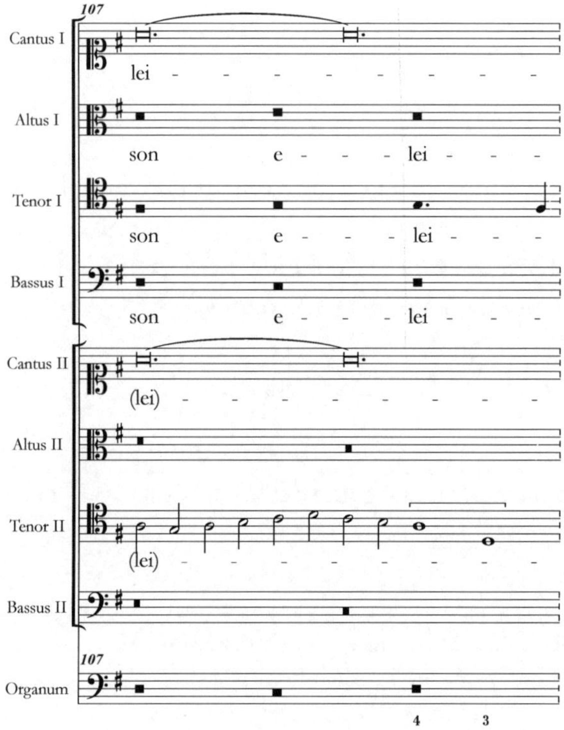

Nella Missa naturalmente vi sono altri passi che fanno riferimento alla similitudine della corona stellata dell'affresco del Cigoli (Ill. 10.1c) come, ad esempio, alla fine del Christe (Contralto I ed Organo):

Un riferimento al passo dell'*Apocalisse* dove si fa riferimento esplicito ad «Un segno grandioso» che «apparve nel cielo: una donna vestita di sole [ovvero di Dio], con la luna sotto i suoi piedi e, sul capo, una corona di dodici stelle»[52].

Giove

La parte finale del *Sidereus nuncius* è dedicata all'osservazione di Giove e alla scoperta delle sue quattro lune medicee. La scoperta che questo pianeta fosse circondato da quattro satelliti «fu salutata con grida di orrore da coloro che pretendevano che Dio non avrebbe mai permesso che gli elementi del sistema planetario superassero il sacro numero di sette»[53]. Ancora nel 1657 (l'anno della composizione della messa di Giuseppe Corsi) il padre minore osservante fra Giovanni D'Avella legava astrologicamente la teoria musicale dei modi ai «sette Pianeti, ò Luna, ò Mercurio, ò Venere, ò Sole, ò Marte, ò Giove, ò Saturno», collocati tra il «Cielo stellato» e la Terra (Ill. 10.4) dove le eclissi costituiscono gli elementi di alterazione[54].

Ill. 10.4: Giovanni D'Avella, *Regole di musica divise in cinque trattati* (Roma, 1657), p. 78.

---

52. *Apocalisse* 12,1.

53. Panofsky 2008, p. 33. Panofsky 1956, p. 5: «His discovery of the four Jupiter satellites was greeted with cries of horror by those who claimed that God would never have permitted the elements of the planetary system to exceed the sacred number of seven».

54. Giovanni D'Avella, *Regole di musica divise in cinque trattati* (Roma, 1657), pp. 78, 88-108.

Altri accolsero invece la scoperta di Galilei «with cries of triumph» in quanto scorgevano nei quattro nuovi satelliti «la quadruplice essenza di Dio, dell'universo e dell'uomo» nonché le qualità dell'intelletto umano (Saggezza, Scienza, Arte e Prudenza)[55].

Tavola 10.6: motivi ritmici nella Missa *La luna piena* di Giuseppe Corsi basati su disegni di costellazioni delle lune di Giove come mostrate nel *Siderus nuncius* (1610) di Galileo Galilei

	Osservazione	Costellazione	Motivo corrispondente nella Missa *La luna piena*
(1)	8 gennaio 1610		Altus II 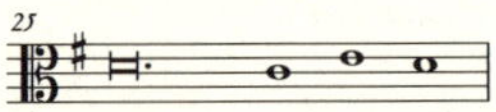 (Kyrie II)
(2)	10 gennaio 1610		Bassus II  (Gloria)
(3)	11 gennaio 1610		Cantus I  (Kyrie I)
(4)	12 gennaio 1610		Bassus II 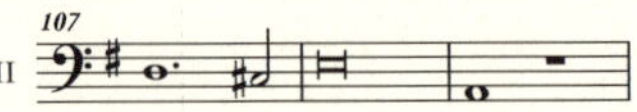 (Gloria)

Galileo, che come si è detto era completamente distaccato dalle fedi numerologiche, aveva scoperto le nuove quattro lune di Giove, attraverso precise osservazioni dello spazio riportate e descritte con oggettività scientifica nel suo *Sidereus nuncius*. Lo scienziato propone infatti una serie di disegni in cui colloca Giove e i relativi satelliti molto simili a schemi ritmici.

In tal senso e in rapporto alla Missa *La luna piena* di Giuseppe Corsi si possono avanzare quattro considerazioni (Tavola 10.6):

(1) Nell'osservazione dell'8 gennaio 1610 Giove si presenta seguito da 3 stelle. Se il pianeta assume un valore più grande e le tre stelle uno più piccolo, come del resto la raffigurazione suggerisce, si può evincere una scansione ritmica simile in un passo del Kyrie II nella parte di Contralto del Coro Secondo.

(2) Nell'esplorazione del 10 gennaio Giove appare preceduto da due stelle. Seguendo il medesimo procedimento di prima è possibile riscontrare una congruente analogia nel Gloria nella parte di Basso del Coro Secondo.

(3) Nell'indagine dell'11 gennaio Giove è preceduto da due stelle una più grande e una più piccola. La struttura è accostabile alle prime due battute del Kyrie I dove tutte le voci procedono omoritmicamente.

[55]. Panofsky 2008, pp. 33-34. Panofsky 1956, p. 5: «with cries of triumph by those who felt that Galileo's discovery had showed forth once more the metaphysical importance of the number four».

(4) Nell'osservazione del 12 gennaio Giove è preceduto da due stelle di dimensioni diverse e seguito da una di dimensioni intermedie rispetto le altre due: si hanno quindi 4 valori differenti. Una similitudine rilevabile nel Gloria nella parte di Basso del Coro Secondo.

L'aver sposato le idee rivoluzionarie di Galileo portò Giuseppe Corsi ad essere ingiustamente accusato nel 1677 di aver deflorato una fanciulla. Il musicista fu per questo arrestato, torturato e imprigionato a Narni (Umbria) nella Rocca di Albornoziana ed espulso per sempre dalla vita musicale romana.

Bibliografia

Annibaldi 1990

Annibaldi, Claudio. 'Musical Autographs of Frescobaldi and His Entourage in Roman Sources', in: *Journal of the American Musicological Society*, xliii/3 (1990), pp. 393-425.

Burckhardt 1905

Burckhardt, Jacob. *Weltgeschichtliche Betrachtungen*, a cura di Jakob Oeri, Berlino, Spemann, 1905.

Burckhardt 1943

Burckhardt, Jacob. *Force and Freedom: Reflexions on History*, New York, Pantheon Books, 1943.

Burke 1984

Burke, John. 'Musicians of S. Maria Maggiore Rome, 1600-1700. A Social and Economic Study', in: *Note d'archivio per la storia musicale*, n.s., ii (1984), Supplemento.

Canguilhem 1992

Canguilhem, Philippe. 'Tel père, tel fils? Les opinions esthétiques de la famille Galilei', in: *International Review of the Aesthetics and Sociology of Music*, xxiii/1 (1992), pp. 27-42.

Chappel 1975

Chappel, Miles. 'Cigoli, Galileo, and Invidia', in: *The Art Bulletin*, lvii/1 (1975), pp. 91-98.

Ciliberti 1986

Ciliberti, Galliano. *Antonio Maria Abbatini e la musica del suo tempo (1595-1679). Documenti per una ricostruzione bio-bibliografica*, Perugia, Regione dell'Umbria, 1986.

Ciliberti – Tribuzio 2014

Ciliberti, Galliano – Tribuzio, Giovanni. '«Un buon virtuoso agitato dalla fortuna, dalla quale sortì vari accidenti». Giuseppe Corsi: un maestro marsicano nel Seicento europeo', in: *«E nostra guida sia la Stravaganza». Giuseppe Corsi da Celano musicista del Seicento*, a cura di Galliano Ciliberti e Giovanni Tribuzio, Bari, Florestano, 2014, pp. 2-117.

COLE 2005
COLE, Janie. '«Se di fuori è dorata, dentro è d'oro». Maffeo Barberini, Michelangelo il giovane e Galileo', in: *Belfagor*, LX/1 (2005), pp. 1-26.

CYPESS 2012
CYPESS, Rebecca. 'Instrumental Music and «Conversazione» in Early Seicento Venice: Biagio Marini's «Affetti Musicali» (1617)', in: *Music & Letters*, XCIII/4 (2012), pp. 453-478.

DAME 1966
DAME, Bernard. 'Galilée et les taches solaires (1610-1613)', in: *Revue d'Histoire des Sciences et de leurs Applications*, XIX/4 (1966), pp. 307-370.

DRAKE 1960
DRAKE, Stillman. 'Galileo Gleanings VIII, The Origin of Galileo's Book on Floating Bodies and the Question of the Unknown Academician', in: *Isis*, LI/1 (1960), pp. 56-63.

EDGERTON 1984
EDGERTON, Samuel Y. Jr. 'Galileo, Florentine «Disegno», and the «Strange Spottednesse» of the Moon', in: *Art Journal*, XLIV/3 (1984), pp. 225-232.

FRANCHI 2004
FRANCHI, Saverio. 'Musiche per le feste e spettacoli a Roma durante il pontificato di Alessandro VII Chigi (1655-1667)', in: *Florilegium Musicae. Studi in onore di Carolyn Gianturco. 2*, a cura di Patrizia Radicchi e Michael Burden, Pisa, Edizioni ETS, 2004, pp. 549-593.

GALILEI 1581
GALILEI, Vincenzo. *Dialogo* [...] *della musica antica et della moderna*, Firenze, Giorgio Marescotti, 1581.

GALILEI 1901
GALILEI, Galileo. *Opere, Volume XI: Carteggio 1611-1613*, a cura di Antonio Favaro, Firenze, Barbèra, 1901.

GALILEI 1993
ID. *Sidereus nuncius*, traduzione di Maria Timpanaro Cardini, a cura di Andrea Battistini, Venezia, Marsilio, 1993.

GOZZA 1989
GOZZA, Paolo. 'Introduzione', in: *La musica nella rivoluzione scientifica del Seicento*, a cura di Paolo Gozza, Bologna, Il Mulino, 1989, pp. 9-66.

HEINICH 2016
HEINICH, Nathalie. 'Panofsky épistémologue', in: PANOFSKY, Erwin. *Galilée critique d'art*, a cura di Nathalie Heinich, Rennes, Les Impressions Nouvelles, 2016, pp. 5-13.

Helden 1996

Helden, Albert van. 'Galileo and Scheiner on Sunspots: A Case Study in the Visual Language of Astronomy', in: *Proceedings of the American Philosophical Society*, cxl/3 (1996), pp. 358-396.

Huemer 1983

Huemer, Frances. 'Rubens and Galileo 1604: Nature, Art, and Poetry', in: *Wallraf-Richartz-Jahrbuch*, xliv (1983), pp. 175-196.

Konečný 2005

Konečný, Lubomír. 'Peter Paul Rubens, Galileo Galilei und die Schlacht am Weißen Berg', in: *Artibus et Historiae*, xxvi/52 (2005), pp. 85-91.

Lionnet 2018

Lionnet, Jean. *«Parve che Sirio [...] rimembrasse una florida pimavera». Scritti sulla musica a Roma nel Seicento con un inedito*, a cura di Galliano Ciliberti, Bari, Florestano, 2018.

Luisi 2021

Luisi, Maria. 'Integrazioni alla storia della cappella musicale della cattedrale di Narni tra xvii e xviii secolo', in: *Symposium Musicae. Saggi e testimonianze in onore di Giancarlo Rostirolla per il suo 80° genetliaco*, a cura di Federica Nardacci e Benedetto Cipriani, Faleria, Associazione culturale Recercare; Palestrina, Fondazione Giovanni Pierluigi da Palestrina, 2021, pp. 233-246.

Missel grégorien 1985

Missel grégorien des dimanches noté en chant grégorien par les moines de Solesmes, Solesmes, Abbaye Saint-Pierre, 1985.

Ostrow 1996

Ostrow, Steven F. 'Cigoli's Immacolata and Galileo's Moon: Astronomy and the Virgin in Early Seicento Rome', in: *The Art Bulletin*, lxxviii/2 (1996), pp. 218-235.

Ottani Cavina 1976

Ottani Cavina, Anna. 'On the Theme of Landscape: ii, Elsheimer and Galileo', in: *The Burlington Magazine*, cxviii/876 (mazo 1976), pp. 139-145.

Panofsky 1956

Panofsky, Erwin. 'Galileo as a Critic of the Arts: Aesthetic Attitude and Scientific Thought', in: *Isis*, xlvii/1 (1956), pp. 3-15 [traduzione italiana in Panofsky 2008].

Panofsky 2008

Id. *Galileo critico delle arti*, a cura di Maria Cecilia Mazzi, Milano, Abscondita, 2008 (Carte d'artisti, 98).

Rostirolla 2017

Rostirolla, Giancarlo. *La Cappella Giulia 1513-2013. Cinque secoli di musica sacra in San Pietro*, 2 voll., Kassel, Bärenreiter, 2017.

Schraven 2001

Schraven, Monou. 'Giovanni Battista Borghese's Funeral «Apparato» of 1610 in S. Maria Maggiore, Rome', in: *The Burlington Magazine*, cxliii/1174 (2001), pp. 23-28.

Shea – Artigas 2009

Shea, William R. – Artigas, Mariano. *Galileo a Roma. Trionfo e tribolazioni di un genio molesto*, Venezia, Marcianum Press, 2009.

Silbiger 1980

Silbiger Alexander. 'The Roman Frescobaldi Tradition, c. 1640-1670', in: *Journal of the American Musicological Society*, xxxiii/1 (1980), pp. 42-87.

Simi Bonini 2010

Simi Bonini, Eleonora. 'Alcuni aspetti della vita di Giuseppe Corsi da Celano', in: *Musica tra storia e filologia. Studi in onore di Lino Bianchi*, a cura di Federica Nardacci, Roma, Istituto Italiano per la Storia della Musica, 2010, pp. 547-565.

Spiller 2000

Spiller, Elizabeth A. 'Reading through Galileo's Telescope: Margaret Cavendish and the Experience of Reading', in: *Renaissance Quarterly*, liii/1 (2000), pp. 192-221.

Stäblein 1956

Stäblein, Bruno. *Hymnen i. Die mittelalterlichen Hymnenmelodien des Abendlandes*, Kassel-Basel, Bärenreiter, 1956.

Tarde 1984

Tarde, Jean. *Deux voyages en Italie. À la rencontre de Galilée*, a cura di François Moureau, François Moureau e Marcel Tetel, Ginevra, Slatkine, 1984.

Tognoni 2009

Il carteggio Cigoli-Galileo 1609-1613, a cura di Federico Tognoni, Pisa, Edizioni ETS, 2009.

Tongiorgi Tomasi 2009

Tongiorgi Tomasi, Lucia. *Cigoli e Galilei. Cronaca di un'amicizia*, in: Tognoni 2009, pp. 13-26.

Topper – Gillis 1996

Topper, David – Gillis, Cynthia. 'Trajectories of Blood: Artemisia Gentileschi and Galileo's Parabolic Path', in: *Woman's Art Journal*, xvii/1 (1996), pp. 10-13.

Tribuzio 2014

Tribuzio, Giovanni. 'Catalogo della produzione musicale di Giuseppe Corsi', in: Ciliberti – Tribuzio 2014, pp. 145-189.

Winkler – Helden 1992

Winkler, Mary G. – Helden, Albert van. 'Representing the Heavens. Galileo and Visual Astronomy', in: *Isis*, lxxxiii/2 (1992), pp. 195-217.

The (Long) Seventeenth Century

Science

Chapter Eleven

René Descartes and Isaac Beeckman: The Philosopher and the Schoolmaster

Rudolf Rasch
(Utrecht University)

For researchers in the field of the history of music and science, the *Compendium musicae*, written by René Descartes (1596-1650; Ill. 11.1), is a well-known text[1]. Several books and numerous articles deal with its contents[2]. Descartes wrote this text in 1618; it is, in fact, his first known writing. He wrote it in Breda, in the Netherlands, where he had become friends with Isaac Beeckman (1588-1637), a young Dutchman, from Middelburg in

[1]. Descartes's autograph manuscript is lost. Early manuscript copies are in NL-MIzb, MS 6471 (in Isaac Beeckman's Journal; Ill. 11.4), NL-Lu, Cod. Hug. 29a (from Constantijn Huygens), NL-Gu, MS 108, ff. 60-83 (from the Dutch mathematician Frans van Schooten junior), and GB-Lb, Add. MS 4388 (from the English mathematician John Pell). The first edition is *Musicae compendium* (Utrecht, 1650; Ill. 11.2). Reprints appeared in Amsterdam (1656, 1683) and Frankfurt (1695). Early translations are *Excellent Compendium of Musick* (London, 1653; English translation by Walter Charleton), *Kort Begryp der Zangkunst* (Amsterdam, 1661; Dutch translation by Jan Hendrik Glazemaker) and *Abrégé de musique* (Paris, 1668; French translation by Nicolas Poisson). The Latin text is available in Descartes 1908, pp. 89-150, and in a critical edition with French translation prepared by Frédéric de Buzon (Descartes 1987). Other recent translations include Descartes 1961 (English translation by Walter Robert), Descartes 1978 (German translation by Johannes Brockt), Descartes 1979 (Italian translation by Patrizia Iandolo), Descartes 1990a (French translation by Pascal Dumont), Descartes 1990b (Italian translation by Luisa Zanoncelli), Descartes 1992 (Spanish translation by Primitiva Flores and Carmen Gallardo), and Descartes 2010, pp. 57-49 (Dutch translation by Rudolf Rasch). We refer to the *Compendium musicae* as 'Descartes, *Compendium*' with the page numbers as found in Descartes 1908. These page numbers are also used for reference in most modern editions, translations, and studies.

[2]. Just a small selection: Mercadier 1901, Pirro 1907, August 1965, Seidel 1970, Gozza 2000, Jorgensen 2012, and Psychoyou 2018.

ILL. 11.1: Cornelis A. Hellemans, *René Descartes in his Study* (engraving, 21 × 14 cm; c. 1690). NL-Ar, RP-P-OB-55.341.

Zeeland, trained in a variety of disciplines and who would embark on a career as schoolmaster, first in Utrecht, then in Rotterdam and finally in Dordrecht. Descartes wrote the *Compendium musicae* for his new friend and handed the manuscript to him on the last day of the year 1618, as the final paragraph of the text tells us.

RENATI
DES-CARTES
MUSICÆ
COMPENDIUM.

TRAJECTI AD RHENUM,
TYPIS *Gisberti à Zyll, & Theodori ab Ackersdyck,*
cIↄ Iↄ c L.

Ill. 11.2: René Descartes, *Musicae compendium* (Utrecht, 1650), title page.

Isaac Beeckman has also received due attention from scholars, although more from those working in the field of the history of science proper than from musicologists[3]. During his lifetime, Beeckman did not publish a single word[4], but he did leave us a comprehensive scientific journal in manuscript form, titled 'Loci communes', now preserved in the Provincial Library of Zeeland in Middelburg (Ill. 11.3)[5]. It contains, on several hundreds of folios, notes and

3. About Beeckman, see Van Berkel 1983, Van Berkel 1990. About Beeckman's musical interests, see Cohen 1984 and Rasch 1992.

4. Except Beeckman's dissertation for the degree in medicine, defended in Caen in France in 1618: *Theses de febre tertiana intermittente* (Caen, [1618]). A number of entries from his Journal were published posthumously by his younger brother Abraham Beeckman as *Mathematico-physicarum meditationum, quaestionum, solutionum centuria* (Utrecht, 1644).

5. NL-MIzb, MS 6471. See Beeckman 1939, Beeckman 1942, Beeckman 1945, Beeckman 1953. We refer to Beeckman's Journal by the volume and page numbers of Beeckman 1939, Beeckman 1942, Beeckman 1945.

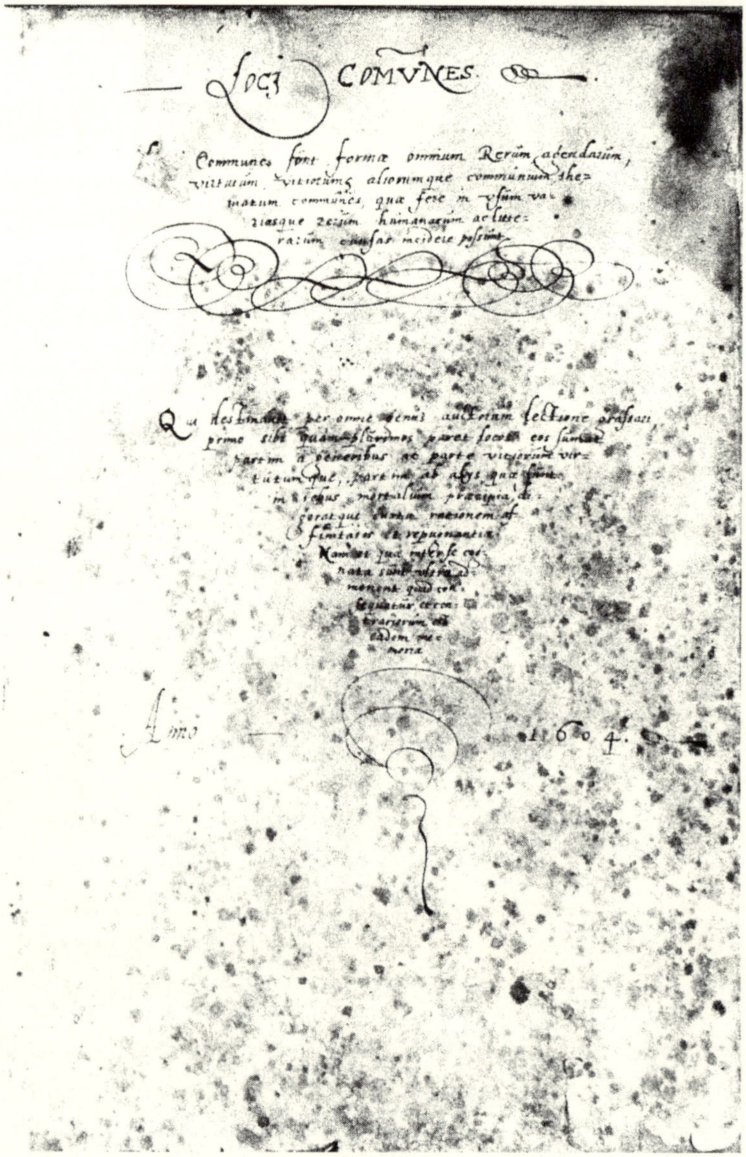

Ill. 11.3: Isaac Beeckman, 'Loci communes', better known as Beeckman's Journal, title page. The horizontal line above the 'v' [= u] and 'm' 'comvnes' indicates that the 'm' must be doubled, so that the word is read as 'communes'. NL-MIzb, MS 6471. The manuscript suffered severe water damage when the isle of Walcheren, on which Middelburg is located, was flooded in 1944-1945, at the end of the Second World War.

observations about all branches of contemporary science and knowledge, including many on physical and mathematical problems, engineering questions, diseases, moral and ethical topics, and, last but not least, music. They were written from 1604 to 1634.

When Descartes and Beeckman got to know one another by the end of 1618, they discussed music on more than one occasion. Otherwise, there would have been no reason for

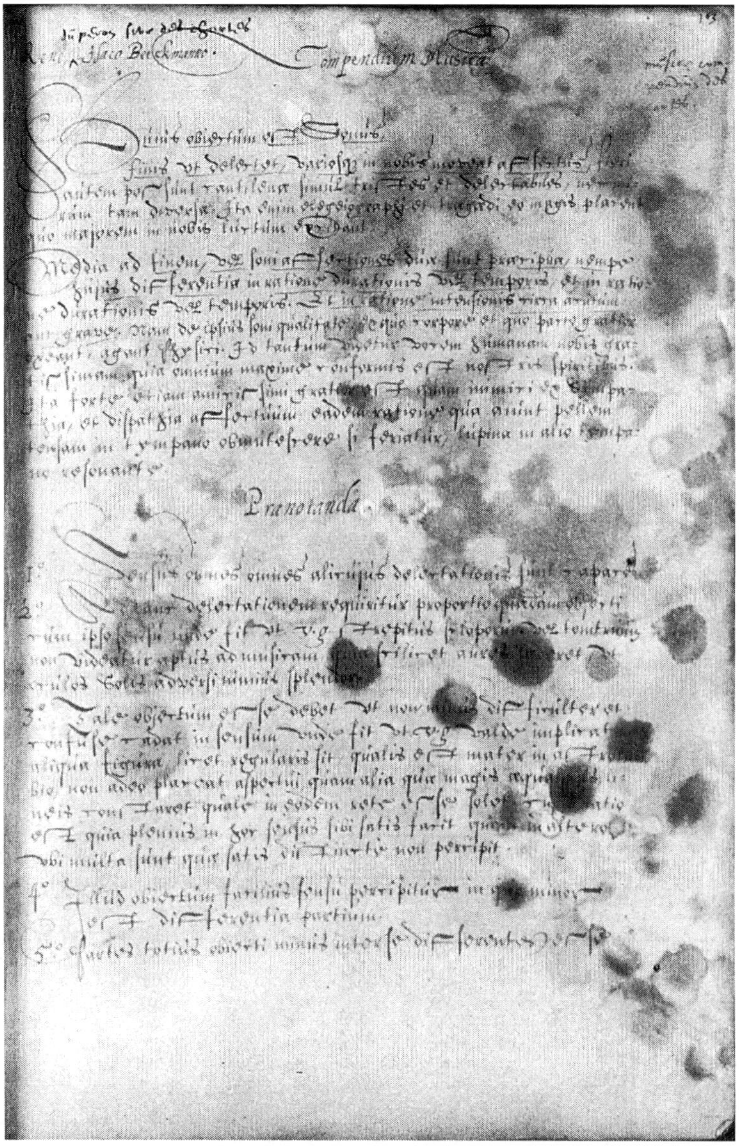

Ill. 11.4: René Descartes, *Compendium musicae*, first page, as copied in Isaac Beeckman's Journal, f. 163r. The copy is not in Beeckman's hand.

Descartes to dedicate his *Compendium musicae* to Beeckman. And indeed, Beeckman's Journal contains several entries on music written down in these months, some of which explicitly mention Descartes.

Descartes and Beeckman exchanged a few letters in the early months of 1619. Then Descartes left Breda and apparently the two men did not have any contact with one another for several years.

Rudolf Rasch

Descartes and Beeckman: 1628-1631

In 1628 Descartes went to the Dutch Republic again. One of his first actions was to visit Beeckman. Since he was unaware of Beeckman's appointment as Principal of the Latin School in Dordrecht, he first went to Middelburg. There he found out that Beeckman was in Dordrecht. On 28 October 1628 the two friends met again, a most hearty reunion, if we may believe Beeckman's description of it in his Journal[6]. During the year that followed they exchanged several letters, which are, unfortunately, lost today. At the same time, however, a shadow was also cast over their relationship, as an unintentional result of Beeckman's correspondence with Marin Mersenne in those years.

Today Mersenne is best known for his monumental *Harmonie universelle*, published in 1636. This work can also be seen as a kind of second edition of a similarly titled but relatively modest publication, the *Traité de l'harmonie universelle*, published in 1627. Right after the publication of the *Traité*, Mersenne had begun to prepare the more ambitious version, for which he initiated a comprehensive correspondence with scholars inside and outside France. These correspondents were asked to give their opinions about a multitude of problems. From their answers Mersenne hoped to be able to compile a more prestigious text than the *Traité* had been. One of his correspondents was the Dutch professor of theology André Rivet (1572-1651), who must have mentioned Beeckman's name in one of his answers to Mersenne. Mersenne formulated a few questions for Beeckman, which were first sent to Rivet in order to be forwarded to Beeckman[7]. But on 28 February 1629 he wrote directly to Beeckman, a letter now lost[8]. From Beeckman's preserved answer, however, which is undated but probably written in March 1629[9], it appears that Mersenne had asked, among other things, a question about the number of vibrations of a string in a given time. Beeckman wrote that he did not consider this problem very urgent. In the same letter, Beeckman referred to his friendship with Descartes of ten years before:

> He is the one, I say, to whom I communicated ten years ago what I had written about the causes of the sweetness of the consonances, and who has given you, I believe, the opportunity of this question[10].

---

6. Beeckman 1945, p. 94 (8 October 1628).

7. Mersenne to André Rivet, 30 October 1628 (Mersenne 1945, p. 107).

8. As mentioned in Mersenne's letter to André Rivet of 28 February 1629 (*ibidem*, p. 208).

9. Beeckman to Mersenne, March 1629 (Beeckman 1953, pp. 141-143; Mersenne 1945, pp. 217-224).

10. Beeckman to Mersenne, March 1629 (Beeckman 1953, p. 142; Mersenne 1945, p. 218): «Ipsus, inquam, is est, cui ante decem annos ea quae de causis dulcedinis consonantiarum scripseram, communicavi, quemque tibi quaestionis hujus occasionem dedisse putabam».

Mersenne must have communicated this information at some point to Descartes, probably in a now lost letter of September 1629, to which Descartes answered on 8 October, with thanks to Mersenne for telling him that Beeckman had written to him (Mersenne) that he (Beeckman) had been Descartes's teacher ten years ago, an idea that Descartes did not like very much, to put it mildly[11]. Descartes decided to ask Beeckman to return the manuscript of his *Compendium musicae*. On 18 December he wrote to Mersenne that he had received the manuscript one month prior[12]. Apparently, Descartes had used toward Beeckman the argument that he needed it (which indeed is probable)[13] and had not indicated his discontent over Mersenne's suggestion that Beeckman had been Descartes's teacher[14].

Things would get worse, however, the next year, 1630. In that year Mersenne travelled to the Netherlands and visited Beeckman in Dordrecht[15]. Beeckman showed him his Journal and Mersenne spent considerable time going through it. Beeckman proudly wrote the following words about this visit to Descartes:

> And when your friend Mersenne looked through my manuscript book entire days and saw in it several things that he thought were yours, and wondered, from the dates attached to those notes, rightly about their author, I explained what had happened to him perhaps more frankly than would please you or him[16].

Now Descartes got really angry and wrote to Beeckman — late September or early October (the letter is undated) — that it was out of the question that he would have learned anything from him and that his (Beeckman's) behaviour was offensive and totally unacceptable (Ill. 11.5)[17]. Beeckman's answer did not calm Descartes, who replied with a long and angry letter only

[11]. Descartes to Mersenne, 8 October 1629 (Descartes 1897, p. 24; Mersenne 1945, pp. 301-302).

[12]. Descartes to Mersenne, 18 December 1629 (Descartes 1897, p. 100; Mersenne 1945, p. 350).

[13]. Descartes referred to the *Compendium musicae* in his letter to Mersenne of 18 December 1629 (Descartes 1897, p. 100; Mersenne 1945, p. 349) when applying to the fifth a reasoning found in the *Compendium musicae* in relation to the major third (Descartes 1908, p. 109).

[14]. Descartes to Mersenne, 8 October 1629 (Descartes 1897, p. 24; Mersenne 1945, p. 301): «je ne luy en manderay rien» (I will say nothing about it to him).

[15]. Beeckman 1945, pp. 160-161 (12 August 1630). See also Sassen 1964.

[16]. The letter is lost but Descartes quoted from it in his letter to Mersenne of 4 November 1630 (Descartes 1897, p. 171; Mersenne 1945, pp. 543): «Cumque Mersennus tuus totas dies in Libro meo manuscripto versaretur, atque in eo pleraque, quæ tua esse existimabat, videret, et ex tempore illis addito, de illorum authore meritò dubitaret, id quod res erat, illi liberius fortassis, quam tibi aut illi placuit, aperui».

[17]. Descartes to Beeckman, September-October 1630 (Descartes 1897, pp. 154-156; Beeckman 1953, pp. 194-195).

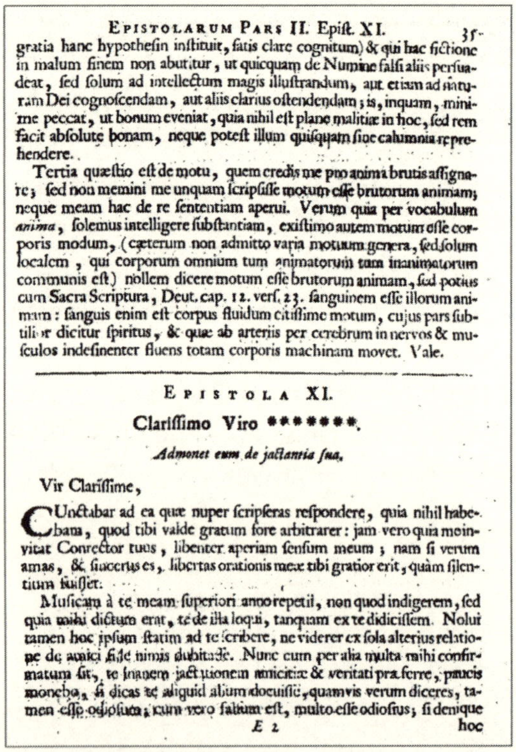

EPISTOLARUM PARS II. Epiſt. XI. 35

gratia hanc hypotheſin inſtituit, ſatis clare cognitum) & qui hac fictione in malum finem non abutitur, ut quicquam de Numine falſi aliis perſuadeat, ſed ſolum ad intellectum magis illuſtrandum, aut etiam ad naturam Dei cognoſcendam, aut aliis clarius oſtendendam; is, inquam, minime peccat, ut bonum eveniat, quia nihil eſt plane malitiæ in hoc, ſed rem facit abſolute bonam, neque poteſt illum quiſquam ſine calumnia reprehendere.

Tertia quæſtio eſt de motu, quem credis me pro anima brutis aſſignare; ſed non memini me unquam ſcripſiſſe motum eſſe brutorum animam; neque meam hac de re ſententiam aperui. Verum quia per vocabulum *anima*, ſolemus intelligere ſubſtantiam, exiſtimo autem motum eſſe corporis modum, (cæterum non admitto varia motuum genera, ſed ſolum localem, qui corporum omnium tum animatorum tum inanimatorum communis eſt) nollem dicere motum eſſe brutorum animam, ſed potius cum Sacra Scriptura, Deut. cap. 12. verſ. 23. ſanguinem eſſe illorum animam: ſanguis enim eſt corpus fluidum citiſſime motum, cujus pars ſubtilior dicitur ſpiritus, & quæ ab arteriis per cerebrum in nervos & muſculos indeſinenter fluens totam corporis machinam movet. Vale.

EPISTOLA XI.

Clariſſimo Viro *******.

Admonet eum de jactantia ſua.

Vir Clariſſime,

CUnctabar ad ea quæ nuper ſcripſeras reſpondere, quia nihil habebam, quod tibi valde gratum fore arbitrarer: jam vero quia me invitat Corrector tuus, libenter aperiam ſenſum meum; nam ſi verum amas, & ſincerus es, libertas orationis meæ tibi gratior erit, quàm ſilentium fuiſſet.

Muſicam à te meam ſuperiori anno repetii, non quod indigerem, ſed quia mihi dictum erat, te de illa loqui, tanquam ex te didiciſſem. Nolui tamen hoc ipſum ſtatim ad te ſcribere, ne viderer ex ſola alterius relatione de amici fide nimis dubitaſſe. Nunc cum per alia multa mihi confirmatum ſit, te inanem jactationem amicitiæ & veritati præferre, paucis monebo, ſi dicas te aliquid alium docuiſſe, quamvis verum diceres, tamen eſſe odioſum; cum vero falſum eſt, multo eſſe odioſius; ſi denique

E 2 hoc

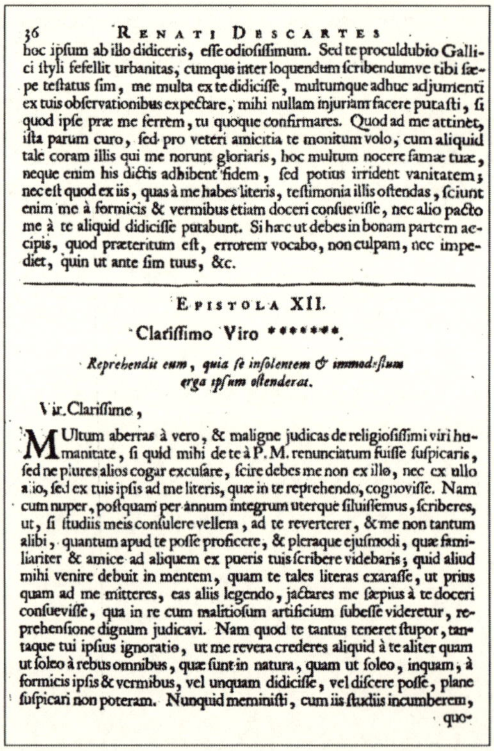

36 RENATI DESCARTES

hoc ipſum ab illo didiceris, eſſe odioſiſſimum. Sed te proculdubio Gallici ſtyli fefellit urbanitas, cumque inter loquendum ſcribendumve tibi ſæpe teſtatus ſim, me multa ex te didiciſſe, multumque adhuc adjumenti ex tuis obſervationibus expectare, mihi nullam injuriam facere putaſti, ſi quod ipſe præ me ferrem, tu quoque confirmares. Quod ad me attinet, iſta parum curo, ſed pro veteri amicitia te monitum volo, cum aliquid tale coram illis qui me norunt gloriaris, hoc multum nocere famæ tuæ, neque enim his dictis adhibent fidem, ſed potius irrident vanitatem; nec eſt quod ex iis, quas à me habes literis, teſtimonia illis oſtendas, ſciunt enim me à formicis & vermibus etiam doceri conſueviſſe, nec alio pacto me à te aliquid didiciſſe putabunt. Si hæc ut debes in bonam partem accipis, quod præteritum eſt, errorem vocabo, non culpam, nec impediet, quin ut ante ſim tuus, &c.

EPISTOLA XII.

Clariſſimo Viro *******.

Reprehendit eum, quia ſe inſolentem & immodeſtum erga ipſum oſtenderat.

Vir Clariſſime,

MUltum aberras à vero, & maligne judicas de religioſiſſimi viri humanitate, ſi quid mihi de te à P. M. renunciatum fuiſſe ſuſpicaris, ſed ne plures alios cogar excuſare, ſcire debes me non ex illo, nec ex ullo alio, ſed ex tuis ipſis ad me literis, quæ in te reprehendo, cognoviſſe. Nam cum nuper, poſtquam per annum integrum uterque ſiluiſſemus, ſcriberes, ut, ſi ſtudiis meis conſulere vellem, ad te reverterer, & me non tantum alibi, quantum apud te poſſe proficere, & pleraque ejuſmodi, quæ familiariter & amice ad aliquem ex pueris tuis ſcribere videbaris; quid aliud mihi venire debuit in mentem, quam te tales literas exaraſſe, ut prius quam ad me mitteres, eas aliis legendo, jactares me ſæpius à te doceri conſueviſſe, qua in re cum malitioſum artificium ſubeſſe videretur, reprehenſione dignum judicavi. Nam quod te tantus teneret ſtupor, tantaque tui ipſius ignoratio, ut me revera crederes aliquid à te aliter quam ut ſoleo à rebus omnibus, quæ ſunt in natura, quam ut ſoleo, inquam, à formicis ipſis & vermibus, vel unquam didiciſſe, vel diſcere poſſe, plane ſuſpicari non poteram. Nunquid meminiſti, cum iis ſtudiis incumberem,

quo-

ILL. 11.5: Descartes's letter to Beeckman of September-October 1630 as published in Latin translation in *Renati Descartes Epistolae, Pars secunda* (Amsterdam, 1683), no. XI, pp. 35-36. The text is either Descartes's original Latin text or a Latin back translation of the French translation of Descartes's original Latin published by Claude Clerselier in *Correspondance de Mr Descartes, Tome second* (Paris, 1659), no. XI, pp. 55-57. Epistola XII is Descartes's letter to Beeckman of 17 October 1630.

to confirm the meanness of Beeckman's actions[18]. Apparently Beeckman thought it pointless to respond further to Descartes's allegations: no further letters about this question are known.

Remarkably, the two men were on speaking terms again rather soon after the dispute. On 7 October 1631 Beeckman could write to Mersenne that he had had dinner with Descartes in Amsterdam just a few days before[19]. And in August 1634 Descartes could borrow — albeit for a very short time — Beeckman's copy of Galileo's recently published *Dialogo sopra i due massimi sistemi del mondo, Tolemaico e Copernicano* (Florence, 1632)[20].

Without the availability of Beeckman's Journal it would be very difficult to establish whether Descartes's anger was justified or not, but with help of the Journal it is possible to shed more light on this episode. Let us go back to when it all began, December 1618.

18. Descartes to Beeckman, 17 October 1630 (DESCARTES 1897, pp. 156-170; BEECKMAN 1953, pp. 195-202).

19. Beeckman to Mersenne, 7 October 1631 (BEECKMAN 1953, p. 207; MERSENNE 1969, p. 203).

20. Descartes to Mersenne, 14 August 1634 (DESCARTES 1897, pp. 303-304; MERSENNE 1955, p. 297).

Descartes and Beeckman: 1618

Beeckman received Descartes's *Compendium musicae* on 31 December 1618. Two days later, 2 January 1619, he wrote in his Journal:

> After he had seen [what I had written] about the strokes of sounds and the four non-sweet modes because of the augmented fourth, and about the six notes [of the hexachord], Mr. Descartes included this in his [*Compendium*] *musicae*. It means that also those considerations of mine had pleased him[21].

Thus, according to Beeckman, Descartes had included text in his *Compendium musicae* that was derived from his ideas regarding three topics: the 'strokes of sounds', the 'non-sweet modes' and the 'six notes of the hexachord'.

The 'strokes of sound' referred to Beeckman's theory of musical sound. In an entry written down in 1614 or 1615 he had described «how sound could be divided in several strokes» («Sonus in plures ictus divisibilis»):

> One must not believe that the sound that our ears perceive is one and indivisible, because the pause between [one] sound and [the next] sound is not heard. The sound that we hear is composed of as many sounds as there are returns of the strings to their place[22].

This was, in the mid-1610s, a novel idea. Together with a statement made a little later, namely that the number of strokes in a unit of time was inversely proportional to the length of the string producing the sound, all other features (thickness, tension) being equal[23], it opened the way to a theory that could explain the consonance of intervals with low-integer string-length ratios such as the octave and the fifth. This is the so-called 'coincidence theory'. By describing musical sounds as a series of pulses or strokes that hit the ear, consonances could be distinguished from dissonances. In consonances, many pulses or strokes of one tone coincide in time with pulses or strokes of the other tone; in dissonances, on the contrary, many or most pulses or strokes of one tone do not coincide with pulses or strokes of the other tone. Viewed from modern acoustic and psycho-acoustic conceptions this is a problematic theory: the description

---

[21]. Beeckman 1939, p. 269 (2 January 1619): «Quae de ictibus sonorum et quatuor modis non dulcibus propter falsam quartam, deque sex notis, Mr. Duperon, cum vidisset, *Musicae* suae interseruit. Significat et meas illas cogitationes placuisse».

[22]. *Ibidem*, p. 53 (1614): «Non oportet existimare sonum quem percipiunt aures nostrae, unum et individuum esse, quia pausa inter sonum et sonum non est perceptibilis. Componitur enim sonus quem audimus ex tot sonis, quot sunt reditus chordarum ad locum suum».

[23]. *Ibidem*, pp. 54-55 (1614): 'Chordarum musicarum trepidatio' (The vibrations of musical strings).

of tones in general as a series of pulses or strokes is not correct, while the theory also presupposes a synchronization of the two series of strokes, a problem that — remarkably enough, in fact — did not occur to Beeckman's mind, nor to others who propagated the theory.

The strokes of sound appear in Descartes's *Compendium musicae* in the chapter 'About the Major Third, the Minor Third, and the Sixths' (*De ditono, tertia minore et sextis*). After explaining consonance in terms of mere small-integer ratios, he wrote

> In the same way it [consonance] will be understood if someone would say that a sound hits the ears with many strokes, and the faster the strokes come, the higher the sound is[24].

Note the indefinite pronoun 'someone' (*quis*): this probably is to be read as referring to Beeckman. After this sentence, the intervals with proportions 5:1 (two octaves plus major third or seventeenth) and 5:2 (octave plus major third or tenth) are compared with respect to their coinciding strokes. In the rest of the *Compendium musicae* there are no further references to the 'strokes of sound'.

Descartes returned to the 'strokes of sound' in his letter to Mersenne of January 1630, where he answered a question from Mersenne, who was, in the meantime, informed by Beeckman about the coincidence theory[25]. In an undated letter by Descartes to Mersenne that must have been written early in 1631, there is a relatively comprehensive exposition of the coincidence theory and its relationship to consonance and dissonance[26]. To illustrate his ideas, he inserted a diagram in the letter, in which the horizontal lines must be understood as time axes, and the small vertical strokes as the 'strokes of sound' for a reference tone (line C), the lower octaves (lines B and A), and a fifth (line D), a fourth (line E), and a major third (line F) upward (ILL. 11.6). Finally, there is a description of the coincidence theory when sound and hearing are discussed by Descartes in his *Traité de l'homme* (Treatise on Man), written around that time (early 1630s) but published only posthumously[27].

The second topic in the *Compendium musicae* where Beeckman discerned his inspiration can be described as 'the four non-sweet modes'. This refers to the authentic and plagal modes on E and F, which have the augmented fourth F-B in the lower part of their scales and therefore

24. Descartes, *Compendium*, p. 110: «Eodem pacto illud concipietur, si quis dixerit sonum aures ferire multis ictibus, idque eò celerius quò sonus acutior est».

25. Descartes to Mersenne, January 1630 (DESCARTES 1897, p. 108; MERSENNE 1945, pp. 370-371).

26. Descartes to Mersenne, early 1631 (DESCARTES 1897, pp. 223-226; MERSENNE 1969, pp. 24-27).

27. Descartes, *L'homme* (Paris, 1664), pp. 35-37 (DESCARTES 1909, pp. 149-150). A Latin translation, *De homine*, had been published in Paris already in 1662.

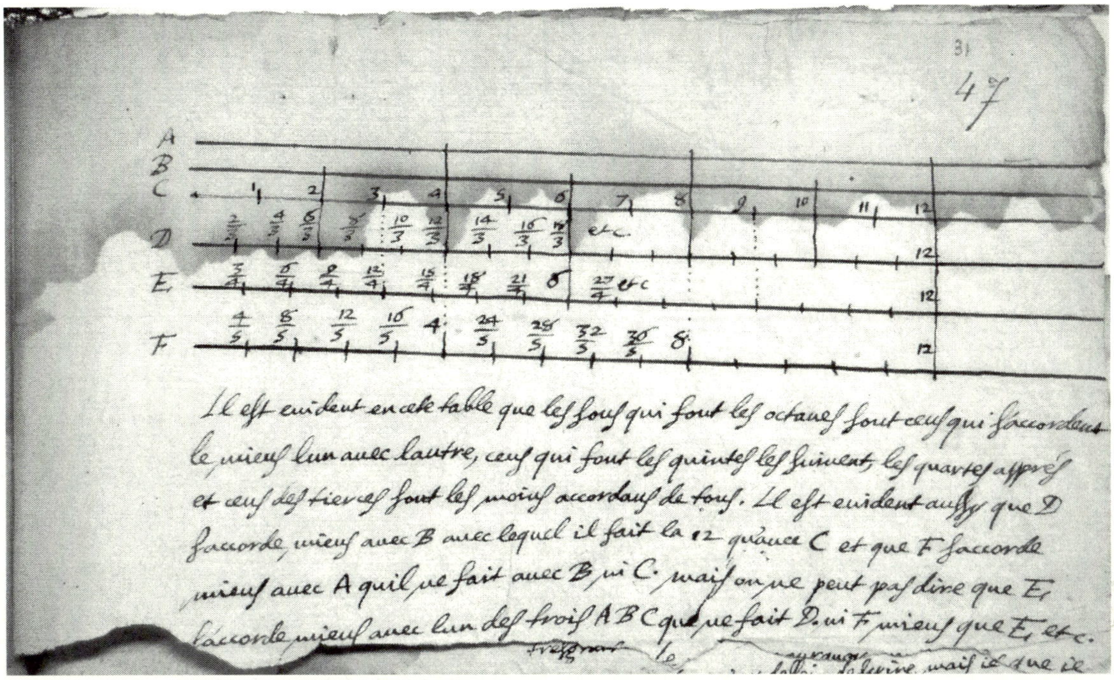
Il est euident en cete table que les sons qui font les octaues sont ceus qui s'accordent le mieus l'un auec l'autre, ceus qui font les quintes les suiuent, les quartes apprés et ceus des tierces sont les moins accordans de tous. Il est euident aussy que D s'accorde mieus auec B auec lequel il fait la 12 qu'auec C et que F s'accorde mieus auec A qu'il ne fait auec B ni C, mais on ne peut pas dire que E s'accorde mieus auec l'un des trois A B C qu'il ne fait D ni F, mieus que E, etc.

Ill. 11.6: Fragment of Descartes's letter to Mersenne of January 1630, with a diagram showing coinciding pulses in consonant intervals. F-Pn, MS Nouv. Acq. Franç. 5160, f. 47r.

were rejected as usable modes by Beeckman in an entry written down in 1616[28]. In Descartes's *Compendium musicae* this idea is found in the discussion of the modes, with the following words:

> From this, only twelve modes arise, four of which are less elegant because an augmented fourth occurs within their fifths. Because of this they cannot rise or fall with the main fifth (the fifth for which the whole melody seems to have been composed) with tone steps without necessarily creating an augmented fourth or a diminished fifth[29].

The third topic in the *Compendium musicae* that Beeckman related to his own ideas is that of the 'six notes', that is, the hexachord. The hexachord is discussed both in Beeckman's notes and in Descartes's *Compendium musicae* a number of times, most often in a rather conventional way. Beeckman probably had in mind an entry from 1614-1615 in which he explained that four

[28]. Beeckman 1939, pp. 90-91 (1616).

[29]. Descartes, *Compendium*, p. 139: «Unde orti sunt tantum duodecim modi, ex quibus etiam quatuor sunt minus elegantes, ex eo quod in horum quintis tritonus reperiatur, ita ut non possint à quintâ principali, et cujus gratiâ tota cantilena videtur componi, per gradus ascendere vel descendere, quin necessariò occurrat falsa relatio tritoni aut falsæ quintæ».

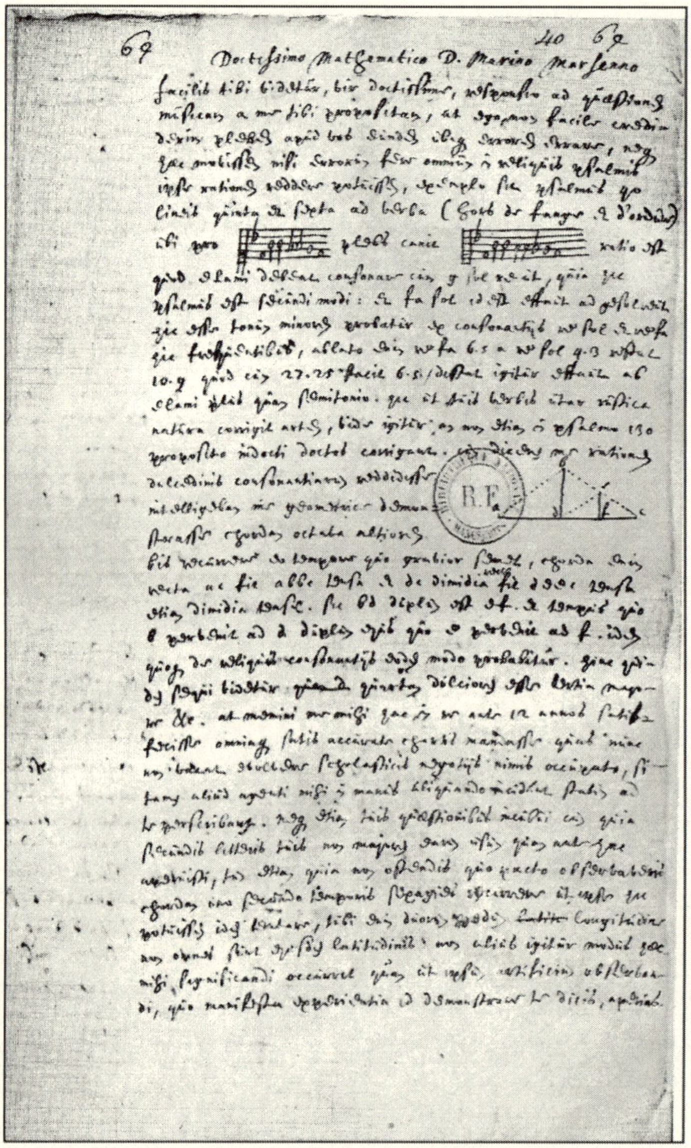

Ill. 11.7: First page of Beeckman's letter to Mersenne of June 1629. F-Pn, MS Nouv. Acq. Franç. 6202, f. 40r (p. 69).

note names could suffice in musical practice[30]. This somewhat unusual topic is also found in Descartes's *Compendium musicae*[31].

Therefore, we may conclude that Beeckman's entry in his Journal of 2 January 1619, that Descartes borrowed or at least repeated some ideas from his Journal, is justified, at least from Beeckman's point of view.

30. Beeckman 1939, p. 52 (1614).

31. Descartes, *Compendium*, p. 124.

On the other hand, however, Beeckman's Journal for the last months of 1618 makes clear that Descartes also brought up some original ideas on his part which were then written down in Beeckman's Journal. There are five entries about music in the Journal that describe ideas explicitly attributed to Descartes. They are the following ones, of which four of which correspond to remarks included in the *Compendium musicae*: (1) Lower strings can make higher strings resonate, but not vice versa[32]; (2) A recorder gives a pitch an octave higher when overblown[33]; (3) Instruction how to tune a lute[34]; (4) A lower string can make a higher string resonate when the interval is a fifth, not when it is a fourth[35]; and (5) The bisection of strings to generate musical intervals[36]. In addition, there is one entry in which Descartes's name is connected with a mathematical problem[37].

Descartes and Beeckman: 1630 again

With the observations just made, we are in a better position to judge the conflict between Beeckman and Descartes that arose in 1630. The first stumbling block for Descartes was the following sentence in Beeckman's letter of March 1629 to Mersenne[38]:

> He is the one, I say, to whom I communicated ten years ago what I had written about the causes of the sweetness of the consonances, [...].

Note that Beeckman uses the rather neutral word «communicated» (*communicavi*) here; he referred to the coincidence theory rather than to the strokes of sounds. We do not know in which words Mersenne transferred to Descartes what Beeckman had written, but when Descartes responded, on 8 October 1629, to Mersenne's letter with the words «when he wrote to you that he was my teacher ten years ago», it looks like a distortion[39]. Descartes also described Beeckman's behaviour as a sign of ungratefulness (*ingratitude*).

The second, and in fact much more serious stumbling block was a passage in a letter to Descartes by Beeckman himself, from which we have already quoted above but which we can now assess better:

32. Beeckman 1939, p. 244. Compare Descartes, *Compendium*, p. 99.
33. Beeckman 1939, p. 246. Compare Descartes, *Compendium*, p. 99.
34. Beeckman 1939, p. 246. Not found in Descartes's *Compendium musicae.*
35. Beeckman 1939, p. 247. Compare Descartes, *Compendium*, p. 103.
36. Beeckman 1939, p. 258. Compare Descartes, *Compendium*, p. 102.
37. Beeckman 1939, pp. 255-256.
38. Beeckman to Mersenne, March 1629 (Beeckman 1953, p. 142; Mersenne 1945, p. 218): «Ipsus, inquam, is est, cui ante decem annos ea quæ de causis dulcedinis consonantiarum scripseram, communicavi [...]».
39. Descartes to Mersenne, 8 October 1629 (Descartes 1897, p. 24; Mersenne 1945, pp. 301-302): «s'il vous écrivoit qu'il a esté mon maistre il y a dix ans».

> And when your friend Mersenne looked through my manuscript book entire days and saw in it several things that he thought were yours, and wondered, from the dates attached to those notes, rightly about their author, I explained what was at stake to him perhaps more frankly than would please you or him[40].

This could be read as if Beeckman claimed — albeit implicitly — *authorship* of certain ideas in the *Compendium musicae*, and this truly unleashed Descartes's rage. The latter's first response was a still relatively friendly Latin letter of reasonable length, 265 words. Descartes raised the following points[41]:

(1) 'that he did not ask back the *Compendium musicae* because he needed it, but because he was told that Beeckman spread the idea that he (Beeckman) had taught him (Descartes) about music.' It seems that Descartes is not really sincere here. In his letter of 8 December 1629 to Mersenne he inserted text from the *Compendium musicae* so that it seems that he indeed needed it in order to be able to quote from it. Descartes used the phrases 'I was told' (*mihi dictum erat*) but in his letter to Mersenne there are no references to any other source of information than Mersenne's letter.

(2) that it was hateful (*odiosum*) to say you have taught someone, more hateful (*odiosius*) if it was not true and most hateful (*odiosissimum*) if it was the other way around.

(3) that he had indeed declared to Beeckman that 'he had learned a lot from him' (*me multa ex te didicisse*), but that should be understood as 'French-style courtesy' (*Gallici styli urbanitas*), not as fact.

(4) that he did not care for himself but rather for Beeckman's good name in the eyes of people who also knew him (Descartes).

Descartes ended the letter by saying that they could remain friends as before. As previously noted, we do not know Beeckman's answer. But now, really, Descartes's anger knew no bounds, prompting him to write a letter of almost 2300 words, almost ten times the length of the previous letter. It is remarkable to see how eloquent and verbose a man in anger can become. The letter is too long to be discussed here in all its detail, but some points cannot go unnoticed.

First, Descartes begins by saying that it was not through Mersenne that he had heard about Beeckman's misconceptions, an idea apparently raised by Beeckman in his last letter, but

---

40. Descartes to Mersenne, 4 November 1630 (DESCARTES 1897, p. 171; MERSENNE 1945, pp. 543), citing Beeckman's letter: «Cumque Mersennus tuus totas dies in Libro meo manuscripto versaretur, atque in eo pleraque, quæ tua esse existimabat, videret, et ex tempore illis addito, de illorum authore meritò dubitaret, id quod res erat, illi liberius fortassis, quam tibi aut illi placuit, aperui».

41. Descartes to Beeckman, September-October 1630 (DESCARTES 1897, pp. 154-156; BEECKMAN 1953, pp. 194-195).

rather by Beeckman's own letter of August-September 1630, apparently referring to the passage describing Mersenne's visit to Dordrecht and the latter's reading of Beeckman's Journal. But there was more in that letter that Descartes disliked. Apparently Beeckman had proposed that Descartes would visit him again so that they could continue their discussions, and this would be to Descartes's benefit. Descartes felt belittled, for here Beeckman seemed to adress him like a student.

Descartes then goes on to explain various modes of learning with the conclusion that in no way could there exist a teacher-pupil relationship between Beeckman and him. This is followed by similar expositions about discoveries, with the parallel conclusion that none of Beeckman's ideas could possibly be called discoveries.

Then two concrete cases are discussed: the strokes of sound and the hyperbola. Concerning the strokes of sound, Descartes remarks that Beeckman could not call himself the inventor of this idea, because already Aristotle had described sound as strokes. This is true, but only to a certain extent. In *On Things Heard*, formerly ascribed to Aristotle but no longer considered to be his work, there is a description of sound as consisting of 'many and separate strokes', each corresponding to a single vibration of a string, a description that approximates the idea of the 'strokes of sound':

> The impacts [strokes] made on the air by strings are many and separate, but because of the smallness of the time between them the ear is unable to detect the gaps, and hence the sound seems to us single and continuous[42].

In the lines that follow this idea is expanded to cover the perception of concords:

> For in all concords the impacts [strokes] of the air belonging to the higher notes occur more frequently, because of the swiftness of the movement; but the last of the sounds strikes our hearing simultaneously with the sound from the slower movement[43].

It seems that «the last of the sounds» (τὸν δὲ τελευταῖον τῶν ἤχων) must be read as 'the last of the sounds that belong to a sequence of sounds of the higher note with the same duration as the simultaneous sound of the lower note' and then it is indeed a description of the 'coincidence theory'. The same idea is found in Pseudo-Aristotelian *Problems*[44].

---

42. Pseudo-Aristotle, *On Things Heard*, 803b34-39: «αἱ δὲ πληγαὶ γίνονται μὲν τοῦ ἀέρος ὑπὸ τῶν χορδῶν πολλαὶ καὶ κεχωρισμέναι, διὰ δὲ μικρότητα τοῦ μεταξὺ χρόνου τῆς ἀκοῆς οὐ δυναμένης συναισθάνεσθαι τὰς διαλείψεις, μία καὶ συνεχὴς ἡμῖν ἡ φωνὴ φαίνεται, καθάπερ καὶ ἐπὶ τῶν χρωμάτων· καὶ γὰρ τούτων τὰ διεστηκότα δοκεῖ πολλάκις ἡμῖν συνάπτειν ἀλλήλοις, ὅταν φέρωνται ταχέως». Translation from Barker 1989, p. 107.

43. Pseudo-Aristotle, *On Things Heard*, 804a1-6: «πλεονάκις μὲν γὰρ ἐν πάσαις ταῖς συμφωνίαις ὑπὸ τῶν ὀξυτέρων φθόγγων αἱ τοῦ ἀέρος γίγνονται πληγαὶ διά τὸ τάχος τῆς κινήσεως· τὸν δὲ τελευταῖον τῶν ἤχων ἅμα συμβαίνει προσπίπτειν ἡμῖν πρὸς τὴν ἀκοήν καὶ τὸν ἀπὸ τῆς βραδντέρας γιγνόμενον». Translation from Barker 1989, p. 107.

44. Pseudo-Aristotle, *Problems*, Book xix, Problem 39, 921a15-20 (Barker 1989, p. 95).

The idea was kept alive by Boethius, in his *De institutione musica libri quinque* (early sixth century):

> One should not think that, when a string is struck, there is only one sound, or that there is only one stroke (*percussio*) in these sounds. Rather, the air is struck as often as the vibrating string will have struck it. But since the rapid motions of the sounds are connected, no interruption is sensed by the ears, and a single sound, either low or high, impresses the sense, although either consists of many sounds, a low sound of slower and less frequent sounds, a high sound of faster and more frequent ones[45].

This was picked up by Early Modern Italian scientists such as Francesco Maurolico (*Opuscula mathematica*, 1575) and Giovanni Battista Benedetti (*Diversarum speculationum mathematicarum et physicarum liber*, 1585)[46]. We do not know if Beeckman was, in 1614, aware of these descriptions. Probably not: references to Aristotle and his works are found the Journal in small number, none of them dealing with the passages — rather by a Pseudo-Aristotle — quoted above[47]. Direct references to Boethius are missing at all. Nor do we know for certain if Descartes knew, in 1618, the relevant passages from *On Things Heard* and Boethius, but it could well be so. If not, he could have encountered them in the meantime[48].

Looking back at the conflict between Descartes and Beeckman after nearly four hundred years we can conclude that Descartes was touchy, to put it mildly, while Beeckman was naive, having no idea his words could be so offensive to Descartes.

---

[45]. Boethius, *De institutione musica*, Liber I, Cap. 3 (Boethius 1867, p. 190.8-15): «Neque enim quotiens chorda pellitur, unus edi putandus est sonus aut unam in his esse percussionem, sed totiens aer feritur, quotiens eum chorda tremendabunda percusserit. Sed quoniam iunctae sunt velocitates sonorum, nulla intercapedo sentitur auribus et unus sonus sensum pellit, vel gravis vel acutus, quamvis uterque ex pluribus constet gravis quidem ex tardioribus et rarioribus, acutus vero ex celeribus ac spissis.». English translation from Boethius 1989, pp. 11-12 (adjusted).

[46]. Francesco Maurolico, *Opuscula mathematica* (Venice, 1575) pp. 149-150, and Giovanni Battista Benedetti, *Diversarum speculationum mathematicarum et physicarum liber* (Turin, 1585), p. 283, as quoted in Barbieri 2020, pp. 237-238.

[47]. Beeckman 1938, pp. 96, 319, and 220, and 2, pp. 11 and 15. Problem 33 of Book XIX of the *Problems* is discussed in an entry from 1619, citing the Greek text (1, p. 319). The catalogue of Beeckman's library as made up after his death (*Catalogus variorum et insignium librorum* [...] *Isaaci Beeckmanni*, Dordrecht, 1637) lists the complete works of Aristotle in Greek (*Opera*, Frankfurt, 1584-1596; five volumes) and in Latin (*Opera omnia*, Leiden, 1580; eight volumes).

[48]. The hyperbola refers to a problem proposed by Descartes to Beeckman during his visit of 8 October 1628. Beeckman solved the problem a few days later, but Descartes had already included the solution in his manuscript 'Algebra' of which he sent a copy to Beeckman. Beeckman may have thought that also regarding this topic Descartes had used ideas developed by him. See Beeckman 1945, pp. 98-99.

Now I will tackle the question of how in general Descartes and Beeckman compare to one another where musical matters are concerned. I will deal with this question from several points of view, namely (1) musical ability; (2) musical education; (3) inspirations; and (4) the nature of their interest.

Descartes and Beeckman: Musical Capabilities

If we may believe his own words, Descartes apparently had no musical capabilities whatsoever. Twice he admitted this in his letters, first in a letter of 15 April 1630 to Mersenne, where he wrote:

> As in fact I myself would not know how to distinguish the fifth from the octave, while there are people who distinguish the larger semitone from the smaller semitone[49].

and then, sixteen years later, in a letter of 30 November 1646 to Constantijn Huygens, where he described himself as:

> [...] a man who has never been able to learn to sing *ut re mi fa sol la*, nor to judge whether someone else sings it well, [...][50].

In fact, he was proud to be able to say so much about music without any capabilities. But he must have had some familiarity with the lute and with the recorder, perhaps through acquaintances. Both instruments are mentioned several times in the *Compendium musicae* and in later letters to Mersenne[51]. When in Breda, he explained the tuning of the lute to Beeckman[52].

---

49. Descartes to Mersenne, 15 April 1630 (Descartes 1897, p. 142; Mersenne 1945, p. 429): «[...] comme de fait je ne sçauroit distinguer la 5 de l'8, et il y en a qui distinguent le demi-ton majeur du mineur».

50. Descartes to Huygens, 30 November 1646 (Descartes – Huygens 1926, p. 248): «[...] un homme qui n'a jamais sceu apprendre à chanter *ut re mi fa sol la*, ny à juger si un autre le chantoit bien, [...]».

51. The lute (French *luth*, Latin *testudo*) is mentioned in the *Compendium musicae*, pp. 97, 99, 103, and 109, and in letters to Mersenne of 8 October 1629 (Descarte 1897, p. 29; Mersenne 1945, p. 306) and 18 December 1629 (Descartes 1897, p. 96; Mersenne 1945, p. 346), January 1630 (appendix: Descartes 1901, p. 686, in Latin translation; Mersenne 1945, p. 605), 15 April 1630 (Descartes 1897, p. 143; Mersenne 1945, p. 429), and 25 November 1630 (Descartes 1897, p. 180; Mersenne 1945, p. 562). The recorder (French *fluste*, Latin *fistula*) is mentioned in the *Compendium musicae*, p. 99, and in letters to Mersenne of 18 December 1629 (Descartes 1897, p. 103; Mersenne 1945, p. 353), January 1630 (appendix: Mersenne 1945, p. 622), 24 February 1630 (Descartes 1897, p. 117; Mersenne 1945, p. 393), and 18 March 1630 (Descartes 1897, p. 134; Mersenne 1945, pp. 418-419).

52. Beeckman 1939, p. 246 (1618).

Beeckman, on the other hand, must have possessed rather well-developed musical capabilities. He was a performer on the cittern[53] and was able to write down song tunes by ear[54]. In his Journal there are several entries that deal with keyboard instruments such as the organ, the harpsichord and clavichord, and other instruments such as the recorder and the Jew's harp[55].

Descartes and Beeckman: Musical Education

Descartes received, as we know, his secondary education at the Jesuit College in La Flèche in the Pays de la Loire[56]. There, music, in the form of *musica speculativa*, that is, the theory of music, was a permanent component of the teaching of mathematics. Jesuit teaching programs of the sixteenth and seventeenth centuries organize the teaching of mathematics following the disciplines of the medieval quadrivium, which consisted of arithmetic, geometry, astronomy and music. It must be said, however, that the time allowed for music was much less than that devoted to the other disciplines. The teaching programs of Jerónimo Nadal (1507-1580; 'De studii generalis dispositione et ordine', 1552)[57] and Christophorus Clavius (1538-1612; 'Ordo servandus in legendis operibus', 1581)[58] both refer to *De musica libris demonstrata quattuor* by Jacobus Faber Stapulensis (Jacques Le Fèvre d'Étaples; Paris, 1496, reprints until 1552) as basic text for the study of music, but this may have changed by the time Descartes received his training in mathematics in La Flèche. His mathematics teacher was Jean François (1582-1668), of whom, however, no particular interest in music is documented[59].

---

[53]. *Ibidem*, pp. 57 (1614) and 193 (1618).

[54]. *Ibidem*, pp. 12 (1612), 50 (1614), and 190 (1618).

[55]. Organ: Beeckman 1942, p. 374 (1626), and 1953, pp. 32-33 (1615). Harpsichord (Dutch *claversyne*, Latin *clavicymbalum*): 1, pp. 88-89 (1616), 309-310 (1619), and 320 (1619), and 2, pp. 11 (1620) and 16-17 (1620), p. 374 (1626). Clavichord (Dutch *clavechorde*, Latin *clavechordum*): 2, p. 5 (1619). Recorder (Dutch *fleute*, Latin *fistula*): 1, pp. 54 (1614-1615) and 312-313 (1619), and 2, pp. 278 (1624) and 374 (1626). Jew's harp (Dutch *tromp*): 2, p. 223 (1623). Lute (Dutch *luyte*): 1, p. 320 (1619), and 2, p. 374 (1626). Cittern (Latin: *cythara*): 1, p. 57 (1614), p. 320 (1619), 2, p. 374 (1626).

[56]. Rochemonteix 1889, Tome quatrième, pp. 50-53.

[57]. Published in Lukács 1965, pp. 134-163. On p. 149: 'Secunda lectio complectitur musicen speculativam et perspectivam. Haec legetur, vel communis vel Vitellionis, illa Fabri Stapulensis, vel alterius si quis commodior videatur.' See also Gatto 2006, p. 237.

[58]. Published in Gatto 2006, pp. 249-252: 251: 'Musica speculative Fabri Stapulensis, quam tamen praecedet Arithmetica Jordani, si prius non percepta [= praecepta?] fuit. Haec tradita est Fabro Stapulensi.' The 'Arithmetica' by Jordanus Nemorarius (13th century) were published by Faber Stapulensis together with the latter's own 'Musica'.

[59]. Dainville 1954, p. 110, and Dainville 1978, p. 326. Later, Descartes asked Pierre Bourdin, in a letter of October 1644 (Descartes 1901, p. 144), to give a copy of his recently published *Principia philosophiae* (Amsterdam, 1644) to 'R. P. F., mon ancien maître' (among several other French scientists). One assumes this is

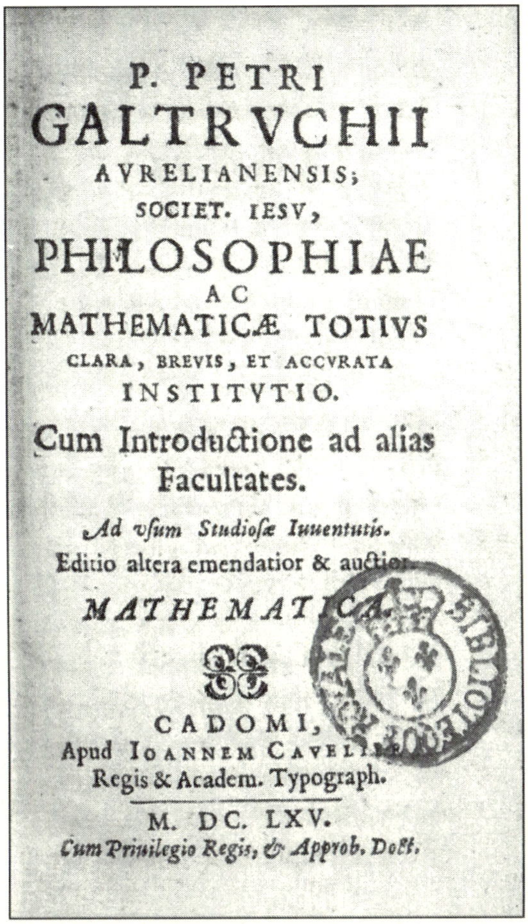

P. PETRI
GALTRVCHII
AVRELIANENSIS;
SOCIET. IESV,
PHILOSOPHIAE
AC
MATHEMATICÆ TOTIVS
CLARA, BREVIS, ET ACCVRATA
INSTITVTIO.
Cum Introductione ad alias
Facultates.
Ad vsum Studiosæ Iuuentutis.
Editio altera emendatior & auctior.
MATHEMATICA.
CADOMI,
Apud IOANNEM CAVELIER
Regis & Academ. Typograph.
M. DC. LXV.
Cum Priuilegio Regis, & Approb. Doct.

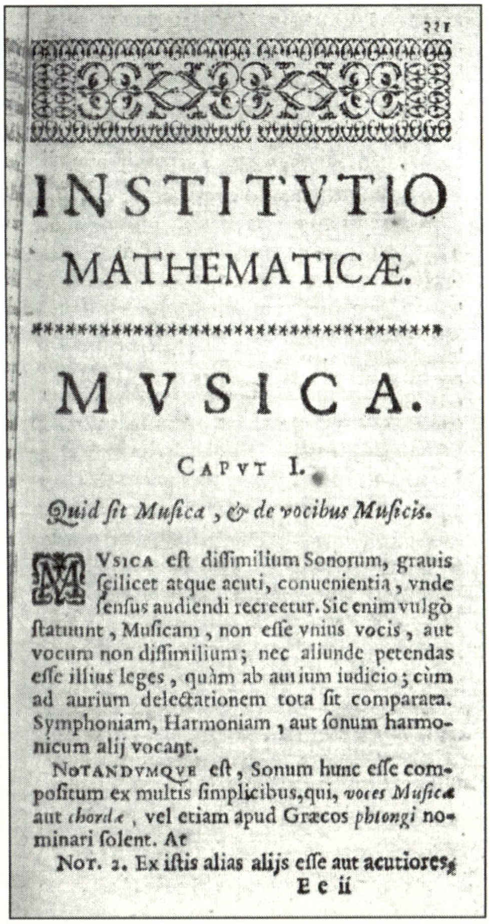

INSTITVTIO
MATHEMATICÆ.
MVSICA.
CAPVT I.
Quid sit Musica, & de vocibus Musicis.

MVsica est dissimilium Sonorum, grauis scilicet atque acuti, conuenientia, vnde sensus audiendi recreetur. Sic enim vulgò statuunt, Musicam, non esse vnius vocis, aut vocum non dissimilium; nec aliunde petendas esse illius leges, quàm ab aurium iudicio; cùm ad aurium delectationem tota sit comparata. Symphoniam, Harmoniam, aut sonum harmonicum alij vocant.

NOTANDVMQVE est, Sonum hunc esse compositum ex multis simplicibus, qui, *voces Musicæ* aut *chordæ*, vel etiam apud Græcos *phtongi* nominari solent. At

NOT. 2. Ex istis alias alijs esse aut acutiores,
Ee ii

Ill. 11.8ab: Pierre Gautruche, *Philosophiae ac mathematicae totius clara, brevis et accurata institutio* (Caen, ²1665), title page and first page of the section on music (p. 331). Copy F-Pn.

Camille de Rochemonteix, in his comprehensive classic study of the Jesuit College in La Flèche, published in 1889, considered the texts published by Pierre Gautruche (1602-1681; also known as Petrus Galtruchius) in several volumes as *Philosophiae ac mathematicae totius institutio* (Caen, 1655, 2/1665; Ill. 8ab) as a reasonable reflection of what was taught also in Descartes's times at the college[60]. Volume 5 of Gautruche's work, with the subtitle *Mathematica*, is devoted to mathematics. It shows the traditional division into the four disciplines of the quadrivium, with the fewest number of pages devoted to music, at the end of the volume[61].

R[évérend] P[ère Jean] F[rançois], in particular in view of François's letter to Mersenne of 28 September 1647 (Mersenne 1983, p. 451), where he asks Mersenne to send his regards to Descartes.

60. Rochemonteix 1889, Tome quatrième, pp. 30-49.

61. Pierre Gautruche, *Philosophiae ac mathematicae totius institutio* (Caen, 1655), pp. 331-359: 'Musica'.

Gautruche's text on music is a well-organized and very effective introduction to the theory of music as it stood in the decades around 1600. Its eight chapters discuss notes, intervals, consonances, proportions, modes, genera, basic compositional theory, and singing, mainly solmisation and notation. The style of its Latin writing is not very different from Descartes's Latin in the *Compendium musicae*. Many descriptions found in Descartes's text also occur in Gautruche's works, but it must be said immediately that they probably can be found in many more introductory texts about the theory of music written in Latin in the decades around 1600. Nevertheless, a training according to a text like Gautruche's would very well explain Descartes's expertise regarding the theoretical aspects of music as visible in the *Compendium musicae*.

In addition, Descartes must have studied Gioseffo Zarlino's *Istitutioni harmoniche* (Venice, 1558), considering the text of the chapter on composition in the *Compendium musicae*, in which he also refers to Zarlino[62]. In the discussion of rhythm, he uses the Italian word for beat or measure, *battuta*, presumably also after Zarlino[63].

That he had indeed received training in the theory of music also becomes clear from sentences in the *Compendium musicae* about things he had forgotten, such as:

> I do not remember well enough to be able to explain how and why that happens [the addition of a sharp to a note], nor why one always attaches a flat to the tone above *la*[64].

One would say that if someone says he cannot remember something well, he must first have learned it.

Beeckman's education was a different one. He attended the Dutch School of Middelburg from 1595 to 1601 and then the 'Latin school' of Arnemuiden, a small town not far from Middelburg, from 1601 to 1607[65]. No particulars about music lessons at these schools are known, but at the Dutch School he certainly had some music instruction, namely what was needed for singing the psalms in church. Dirck Adriaenszoon Valcoogh's *Regel der Duytsche Schoolmeesters* (Rule of the Dutch Schoolmasters), published in 1591, provides this instruction in print and adds a few more topics from the theory of music[66]. There are also printed psalm books of the time that contain a few pages that explain the very basics of the theory of music.

We do not know to what extent music belonged to the curriculum of the Latin schools in the Dutch Republic in the early decades of the seventeenth century. But we may mention

62. Descartes, *Compendium*, p. 134.

63. *Ibidem*, p. 93.

64. *Ibidem*, p. 124: «Quod quomodo et quare fiat, jam satis non memini, ut possim explicare. Neque item quare, dum una duntaxat nota supra *la* elevatur, illi = solet affigi».

65. Beeckman 1939, pp. iv-v.

66. Dirck Adriaenszoon Valcoogh, *Den Regel der Duytsche Schoolmeesters* (Amsterdam 1591), pp. 60-71.

a publication such as the anonymous *Brevia musicae rudimenta Latino-Belgicae, Corte Onderwijsinghe van de Musike int Latijn ende Duyts*, published in Leiden in 1605 for use at the Latin School of Harderwijk in the province of Gelderland; it contains the basics of music theory both in Latin and Dutch. It may be interesting to note that, at the end of this text, two publications of polyphonic music are recommended for further study, namely the anthology *Livre septième des chansons vulgaires* (many editions from 1560 onward, in Louvain, Antwerp and Amsterdam) and Abraham Speth's *Psalmorum Davidis Prophetae Regii, paraphrasis metrorhythmica* (Heidelberg, 1591; Frankfurt, 1612). References to these works are found in Beeckman's notes of 1616-1618 and 1620 respectively[67].

Beeckman, however, must have expanded his knowledge of the theory of music considerably after his years at school. From his notes it becomes clear he possessed and studied a fair number of printed works on the theory of music, among them Heinrich Glarean's *Dodecachordon* (Basel, 1547), Gioseffo Zarlino's *Istitutioni harmoniche* (Venice, 1558), Pierre Maillart's *Les tons, ou Discours sur les modes de musique* (Tournai, 1610), Johannes Nucius's *Musices poeticae sive de compositione cantus* (Neisse, 1613) and Marin Mersenne's *Traité de l'harmonie universelle* (Paris, 1627). He was especially interested in what these men had to say about the theory of the modes and the intonation of the hexachord.

Additionally, while teaching at the Latin School of Utrecht, he took singing and clavichord lessons with Evert Verhaer, precentor of the Reformed Dom Church in Utrecht[68]. Verhaer had been a pupil of the Amsterdam organist and composer Jan Pieterszoon Sweelinck[69].

Interest in Music

Going through Descartes's writings on music from the earliest text — the *Compendium musicae* — to the last remarks — in his correspondence with Constantijn Huygens in 1646-1647 — it is noteworthy that he is always responding to questions posed by others. The *Compendium musicae* was written for Beeckman, while the numerous remarks about musical topics in his correspondence from 1628 to 1635 are answers to questions from Mersenne. In 1640 he got involved in the so-called Ban-Boësset controversy, which prompted him to write a long letter to Joan Albert Ban (December 1640)[70], which is a commentary on a letter by Ban to Anna Maria van Schurman (of 20 August 1640)[71].

67. Beeckman 1939, p. 116, and 1942, pp. 23-24, respectively.

68. Beeckman 1942, pp. 4-5 (1619) and 15-16 (1620).

69. *Ibidem*, p. 16 (1620).

70. Descartes to Joan Albert Ban, [December 1640] (Descartes – Huygens 1926), pp. 293-298. About Descartes's role in the controversy, see Rasch 2005, pp. 178-195, and Chapter 9 of this volume.

71. Joan Albert Ban to Anna Maria van Schuurman, 20 August 1640 (Mersenne 1967, pp. 18-46; Huygens 2007, pp. 348-394).

Beeckman's notes on music, on the other hand, are clearly written down from an intrinsic motivation. The large majority of his notes on music are based on observations he made and thoughts that came to his mind. Some notes are responses to statements he read in books. Mersenne posed him some questions, but Beeckman actually avoided answering them.

Descartes's interest in music had a character entirely different from Beeckman's. Descartes was apparently trained in *musica speculativa* and had no experience in *musica practica* himself. Therefore, his remarks on music have a strictly theoretical character. Beeckman, on the other hand, was an amateur musician and was interested in the music around him, including the psalms sung in church. Moreover, he was interested in the physical aspects of musical sound and all sorts of phenomena that had a relation to music.

The differences between Descartes and Beeckman in background and interest lead to different profiles regarding the topics covered by the two men, although it must be remarked that there is a large overlap in their musical interests as well.

Descartes's *Compendium musicae* is a rather general text about music. The letters exchanged with Mersenne and Ban add a number of specific topics such as the nature of sound, its propagation, the rules of counterpoint, the comparative sweetness of consonances, intonation systems, as well as aesthetic and ethical questions related to music.

Most of these topics were also discussed by Beeckman in his *Journal*. But Beeckman also spent many words on topics not touched by Descartes, such as modal theory (more specifically as applied to the Genevan psalm tunes), the tuning of organs and harpsichords, remarks on other instruments, the occurrence of leading notes in psalm and song tunes, performance practice of the psalms, the relation between text and music in vocal music, and so on. Among the entries there is even a design for a metronome based on the periodic movement of a pendulum[72]. In Beeckman's notes there are many references to identifiable musical repertoire, the Genevan psalms in the first place, but also to sacred and secular songs and vocal polyphony, such as the *Livre septième* and the four-part psalms by Andreas Speth, as mentioned before.

The amount of text devoted to music also differs markedly between the two men. The *Compendium musicae* has about 9,400 words; the next longest 'musical text' by Descartes is his letter to Ban of December 1640, which has about 1,200 words. Together with the passages on music in other letters and the few remarks in published texts, I estimated the total number of words devoted by Descartes to music and related subjects at about 15,000. About the size of Beeckman's musical writing I cannot yet give precise data. But I assume that the number of entries on music in his Journal is in the order of 500 or more and with an average length of 200 words this amounts to 100,000 words or more taken all together.

---

[72]. BEECKMAN 1945, pp. 183-184 (1631).

Conclusion

Comparing Descartes's and Beeckman's contributions to what could be called the science of music as we have done in the preceding observations must lead, I believe, to the conclusion that Beeckman's contributions to musical questions far surpass Descartes's. Was Descartes aware of this, or, even, was he willing to be aware of this? In his letter of January 1630 to Mersenne, Descartes depicted Beeckman as a near-total ignoramus concerning musical matters:

> But I swear to you that at the time that this person [Isaac Beeckman] boasted to have written such fine things about music, he did not know more of it than what he had learned from Faber Stapulensis, and considered it a big secret that the fifth was the ratio 2:3 and the major third 4:5, and he had never gone further than that. And he found this so beautiful that, although it was entirely not at issue, he inserted that in his thesis [for the degree] of Medicine[73], which he had defended shortly before that time[74].

This is unfair, I believe, but perhaps we must see it as self-defence.

There are also certain similarities between Descartes's and Beeckman's remarks on music. Both men disliked the idea that they would just have to accept that music was as it was. They wanted to understand *why* music was as it was. Therefore, both of them gave numerous explanations about the causes of musical phenomena, from the vibration of strings and the sound production of recorders to the rules of counterpoint and the existence of leading tones. Many of these explanations — perhaps even most of them — appear to us as fallacies: often they are vague or impossible to check or they are based on statements that themselves need explanation. Beeckman's famous theorem that the number of vibrations of a musical tone is inversely proportional to string length is in principle correct, but his proof is not, because it assumes constant speed of the moving string[75]. (Let us remember here that Beeckman and Descartes lived in a pre-Newtonian era.) Descartes's statements about the generation of higher

[73]. Isaac Beeckman, *Theses de febre tertianâ intermittente* (Caen, 1616), p. 7, one of the propositions in the 'Corollaria': «Ditonus consonantia non consistit in proportione 9 ad 8 duplicata, sed ut 5 ad 4». The same point is found in Beeckman's Journal (1614; Beeckman 1939, pp. 55-56).

[74]. Descartes to Mersenne, January 1630 (Descartes 1897, pp. 110-111; Mersenne 1945, p. 373): «Mais je vous jure que du temps que ce personnage se vante d'avoir écrit de si belles choses sur la musique, il n'en sçavoit que ce qu'il avoit appris dans Faber Stapulensis, et tenir pour un grand secret de sçavoir que la quarte [*sic*; *rectè*: quinte] estoit comme de 2 à 3, et la tierce majeure de 4 à 5, et n'avoit jamais passé plus outre, et trouvoit cela si beau qu'encore qu'il fust tout à fait hors de propos, il l'avoit inséré en des Thèses de Médecine qu'il avoit soutenues peu de temps auparavant». The source for this letter is Claude Clerselier's edition of the *Lettres de Mr Descartes, Tome second* (Paris, 1659), pp. 481-497. Here one reads «quarte», but this is clearly wrong.

[75]. Beeckman 1939, pp. 54-55 (1616).

harmonics in strings[76] and his belief that the higher-frequency sounds travelled faster the low-frequency sounds (commonly held in his time and going back to Aristotle)[77] were disproven by later acoustical insights.

In one respect, however, Descartes's capabilities surpassed those of Beeckman, and that was his capability to organize knowledge into greater coherent schemes. Beeckman never made any attempt to construct a unified theory out of his observations. His hundreds of entries on music and thousands of entries on other topics all remained single, separate and isolated pieces of information and knowledge. Descartes, on the other hand, composed larger texts, in which many individual observations were arranged in coherent discourses, which in turn found their way to publications that would make him famous, as the *Discours de la méthode* (Leiden, 1637), the *Meditationes de prima philosophia* (Paris, 1641) and the *Principia philosophiae* (Amsterdam, 1644). Beeckman had to wait until the rediscovery of his Journal by Cornelis de Waard in 1905 and its subsequent publication by the same scholar in four magnificent volumes from 1939 to 1953 before he could take his rightful place in the history of music and science. And much of what he wrote is still uncharted territory.

Bibliography

August 1965

August, Bertrand. 'Descartes's *Compendium on Music*', in: *Journal of the History of Ideas*, xxvi/1 (1965), pp. 119-132.

Barbieri 2020

Barbieri, Patrizio. *Quarrels on Harmonic Theories in the Venetian Enlightenment*, Lucca, Libreria Musicale Italiana, 2020.

Barker 1989

Greek Musical Writings. 2: Harmonic and Acoustic Theory, translated and edited by Andrew Barker, Cambridge, Cambridge University Press, 1989.

Beeckman 1939

Beeckman, Isaac. *Journal tenu par Isaac Beeckman de 1604 à 1634. Tome premier 1604-1619*, edited by Cornelis de Waard, The Hague, Martinus Nijhoff, 1939.

Beeckman 1942

Id. *Journal tenu par Isaac Beeckman de 1604 à 1634. Tome deuxième 1619-1627*, edited by Cornelis de Waard, The Hague, Martinus Nijhoff, 1939.

[76]. Descartes to Mersenne, 22 July 1633 (Descartes 1897, pp. 267-268; Mersenne 1969, pp. 458-459) and 28 November 1633 (Descartes 1897, p. 272; Mersenne 1969, p. 559).

[77]. Descartes to Mersenne, January 1630 (Descartes 1897, p. 106; Mersenne 1945, p. 370).

Beeckman 1945
Id. *Journal tenu par Isaac Beeckman de 1604 à 1634. Tome troisième 1627-1634 (1635)*, edited by Cornelis de Waard, The Hague, Martinus Nijhoff, 1945.

Beeckman 1953
Id. *Journal tenu par Isaac Beeckman de 1604 à 1634. Tome quatrième: Supplément*, edited by Cornelis de Waard, The Hague, Martinus Nijhoff, 1953.

Boethius 1867
Boethius, Anicius Manlius Torquatus Severinus. *De insititutione arithmetica libri duo, De institutione musica libri quinque*, edited by Gottfried Friedlein, Leipzig, Teubner, 1867.

Cohen 1984
Cohen, H. Floris. *Quantifying Music: The Science of Music at the First Stage of the Scientific Revolution, 1580-1650*, Dordrecht, Reidel, 1984.

Dainville 1954
Dainville, François de. 'L'enseignement des mathématiques dans les Collèges Jésuites de France du xvie au xviiie siècle', in: *Revue d'Histoire des Sciences et de Leurs Applications*, vii/1 (1954), pp. 6-21, 109-123.

Dainville 1978
Id. *L'éducation des jésuites, xvie-xviiie siècles*, Paris, Minuit, 1978, p. 326.

Descartes 1897
Descartes, René. *Correspondance i: avril 1622-février 1638*, edited by Charles Adam and Paul Tannery, Paris, Léopold Cerf, 1897.

Descartes 1901
Id. *Correspondance iv: juillet 1643-avril 1647*, edited by Charles Adam and Paul Tannery, Paris, Léopold Cerf, 1901.

Descartes 1908
Id. *Physico-mathematica. Compendium musicae. Regulae ad directionem ingenii. Recherche de la verité. Supplément à la correspondance*, edited by Charles Adam and Paul Tannery, Paris, Cerf, 1908.

Descartes 1909
Id. *Le monde. Description du corps humain. Passions de l'âme. Anatomica. Varia*, edited by Charles Adam and Paul Tannery, Paris, Cerf, 1909.

Descartes 1961
Descartes, René. *Compendium of Music («Compendium musicae»)*, transated by Walter Robert, Rome, American Institute of Musicology, 1961.

Descartes 1978

Id. *Leitfaden der Musik*, translated by Johannes Brockt, Darmstadt, Wissenschaftliche Buchgesellschaft, 1978.

Descartes 1979

Id. *Compendium musicae*, translated by Patrizia Iandolo, Mestre, Stilo, 1979.

Descartes 1987

Id. *Abrégé de musique «Compendium Musicae»*, edited and translated by Frédéric de Buzon, Paris, Presses Universitaires de France, 1987.

Descartes 1990a

Id. *Abrégé de musique — Éclaircissements physiques sur la musique de Descartes, du R.P. Nicolas Poisson*, translated by Pascal Dumont, Paris, Klincksieck, 1990.

Descartes 1990b

Id. *Breviario di musica*, translated by Luisa Zanoncelli, Florence, Passigli, 1990.

Descartes 1992

Id. *Compendio de música*, translated by Primitiva Flores and Carmen Gallardo, Madrid, Tecnos, 1992.

Descartes 2010

Id. *Samenvatting van de muziekleer; Persoonlijke aantekeningen; Descartes' dromen; Regels om richting te geven aan het verstand*, edited by Erik-Jan Bos and Han van Ruler, Amsterdam, Boom, 2010.

Descartes – Huygens 1926

Correspondence of Descartes and Constantyn Huygens, edited by Leon Roth, Oxford, Clarendon, 1926.

Gatto 2006

Gatto, Romano. 'Christoph Clavius' *Ordo Servandus in Addiscendis Disciplinis Mathematicis* and the Teaching of Mathematics in Jesuit Colleges at the Beginning of the Modern Era', in: *Science & Education*, no. 15 (2006), pp. 235-258.

Gozza 2000

Gozza, Paolo. 'A Renaissance Mathematics. The Music of Descartes', in: *Number to Sound: The Musical Way to the Scientific Revolution*, edited by Paolo Gozza, Dordrecht, Kluwer, 2000, pp. 155-172.

Huygens 2007

Huygens, Constantijn. *Driehonderd brieven over muziek van, aan en rond Constantijn Huygens*, translated and edited by Rudolf Rasch, Hilversum, Verloren, 2007.

Jorgensen 2012

Jorgensen, Larry M. 'Descartes on Music. Between the Ancients and the Aestheticians', in: *British Journal of Aesthetics*, lii/4 (2012), pp. 407-424.

Lukács 1965
Monumenta Paedagogica Societatis Jesu, I (1540-1556), edited by Ladislaus Lukács, Rome, Institutus Historicus Societatis Jesu, 1965.

Mercadier 1901
Mercadier, E. 'Les théories musicales de Descartes', in: *Revue d'Histoire et de Critiques Musicales*, I (1901), pp. 129-137, 189-195, 237-241.

Mersenne 1945
Mersenne, Marin. *Correspondance du P. Marin Mersenne, réligieux minime. 2: 1628-1630*, edited by Mme Paul Tannery, Cornelis de Waard and René Pintard, Paris, Presses Universitaires de France, 1945.

Mersenne 1955
Id. *Correspondance du P. Marin Mersenne, réligieux minime. 4: 1635*, edited by Cornelis de Waard, Paris, Éditions du Centre National de la Recherche Scientifique, 1955.

Mersenne 1967
Id. *Correspondance du P. Marin Mersenne, réligieux minime. 10: Du 6 août 1640 à fin décembre 1641*, edited by Cornelis de Waard, Paris, Éditions du Centre National de la Recherche Scientifique, 1967.

Mersenne 1969
Id. *Correspondance du P. Marin Mersenne, réligieux minime. 3: 1631-1633 (seconde édition)*, edited by Cornelis de Waard, Paris, Éditions du Centre National de la Recherche Scientifique, 1969.

Mersenne 1983
Id. *Correspondance du P. Marin Mersenne, réligieux minime. 15: 1647*, edited by Cornelis de Waard and Armand Beaulieu, Paris, Éditions du Centre National de la Recherche Scientifique, 1983.

Pirro 1907
Pirro, André. *Descartes et la musique*, Paris, Fischbacher, 1907.

Psychoyou 2018
Psychoyou, Théodora. 'Monsieur Descartes by Étienne Loulié. Reception of the *Compendium musicae* in the Second Half of the 17th Century', in: *Musiques en liberté: Entre la court et les provinces au temps des Bourbons*, Paris, École des Chartes, 2018, pp. 335-350.

Rasch 1992
Rasch, Rudolf. 'Six Seventeenth-Century Dutch Scientists and Their Knowledge of Music,' in: *Music and Science in the Age of Galileo*, edited by Victor A. Coelho, Dordrecht, Kluwer, 1992, pp. 185-210.

Rasch 2005
Id. 'Descartes en de Ban-Boësset-controverse', in: *«Née Cartésienne» / «Cartesiaansch gebooren». Descartes en de Utrechtse Academnie 1636-2005*, edited by Willem Koops, Leen Dorsman, and Theo Verbeek, Assen, Van Gorcum, 2005, pp. 178-195.

Rochemonteix 1889
Rochemonteix, Camille de. *Un collège de jésuites aux* xviie *&* xviiie *siècles. Le Collège Henri* iv *de La Flèche*, 4 vols., Le Mans, Leguicheux, 1889.

Sassen 1964
Sassen, Ferdinand. *De reis van Marin Mersenne in de Nederlanden (1630)*, Brussel, Vlaamse Academie voor Wetenschappen, Letteren en Schone Kunsten van België, 1964.

Seidel 1970
Seidel, Wilhelm. 'Descartes' Bemerkungen zur musikalischen Zeit', in: *Archiv für Musikwissenschaft*, xxvii/4 (1970), pp. 287-303.

Van Berkel 1983
Berkel, Klaas van. *Isaac Beeckman (1588-1637) en de mechanisering van het wereldbeeld*, Ph.D. Diss., Utrecht, Utrecht University, 1983.

Van Berkel 1990
Id. 'Isaac Beeckman 1588-1637', in: *Van Stevin tot Lorentz. Portretten van achttien Nederlandse natuurwetenschappers*, edited by A. J. Kox, Amsterdam, Bakker, 1990, pp. 20-33.

Chapter Twelve

The Trumpet as Nature's Voice: Marin Mersenne and the Nature of Music

Leendert van der Miesen
(Bibliotheca Hertziana –
Max Planck Institute for Art History, Rome)

In his *Harmonie universelle* (Paris, 1636), the French friar and polymath Marin Mersenne (1588-1648; Ill. 12.1) devoted several pages to proving that the diatonic scale was the most natural of all scales[1]. He included observations on the singing of shepherds and non-European musicians that seemed to affirm exactly this. He also considered the fact that some animals seemed to make the same intervals as humans. But ultimately such evidence could not be conclusive, Mersenne argued, since one could always argue that music was dependent on custom rather than nature. Eventually, a musical instrument — an inanimate object — was considered the most definitive proof. He wrote: «for although the animals could have learned these intervals, inanimate bodies could not have acquired them, and nonetheless they produce these consonances, after which they form the whole tone and the larger semitone, as I have remarked in the treatise on trumpets and flutes»[2]. It was not the singing of those who are considered part of the natural world — animals or children — but rather an artificial object

[1]. The publication history of Mersenne's *Harmonie universelle* is more complicated that the simple year '1636' suggests. For the sake of simplicity, we have just maintained the single year. See Van der Miesen 2021.

[2]. Marin Mersenne, *Harmonie universelle* (Paris, 1636), 'Livre troisiesme des genres de la musique', p. 149: «car encore que les animaux eussent apris ces intervalles, les corps inanimez ne peuvent les avoir apris, & neanmoins ils font lesdites Consonances, apres lesquelles ils font le ton & le demiton Diatonique, comme j'ay remarqué au traité des Trompettes & des Flustes».

Ill. 12.1: Claude Duflos (1675-1727), *Marin Mersenne* (engraving, 25 × 9 cm). NL-Ar, RP-P-1910-5007.

that here received a central role in displaying the order of nature. Mersenne continued this line of inquiry in the books on instruments in the *Harmonie universelle*. After having spent more than a decade investigating the trumpet, he decided that the range of this instrument is the best proof that nature consists of harmony. He wrote: «so it seems that nature or its movements are nothing but a ravishing harmony»[3]. Mersenne's use of the trumpet is remarkable, as the instrument received little attention from music theorists before him. Moreover, his use of artificial instruments to establish the nature of music complicates the distinction between what can be considered natural and what artificial. This raises the question of how a man-made object was turned into the primary evidence for the nature of music.

Throughout the early modern period, certain musical practices and theories were considered more natural than others. As Suzannah Clarke and Alexander Rehding (2001) have put it: «the question was not so much *whether* music was connected with nature but *how* it was connected with it»[4] (emphasis original). A wide variety of connections between music and the natural world have been explored in recent decades. Daniel Chua (2001) has investigated the grounding of music in nature as a symptom of modernization[5]. Thomas Christensen (1993) and David E. Cohen (2001) have investigated Jean-Philippe Rameau's efforts to naturalize the principles of European harmonic theory[6]. As these studies show, the early modern period saw a wide variety of efforts to solve musical questions by means of legitimation in nature, from the singing of birds and newly discovered animals to encounters with musicians from outside Europe. Similarly, historians of science and early modernists have explored the variety of meanings of nature in the period. Besides philosophers like René Descartes and Francis Bacon, who challenged the distinction between natural and mechanical movement, artists, engineers, chemists, and collectors all increasingly played with the boundaries between art and nature[7]. While categories like nature and artifice continued to dominate scholarly discourse in the early modern period, their relations and interactions were newly configured by automata, *trompe-l'œil* paintings, and alchemical laboratories.

By foregrounding the natural valveless trumpet, Mersenne positioned himself into a long history that asked to what extent musical instruments could function as objects of knowledge. According to Boethius, the mythical Pythagoras did not devote himself to instruments, which he believed to be inconsistent and uncertain. In the words of Boethius, Pythagoras considered «all

3. *Ibidem*, 'Livre cinquiesme des instrumens à vent', p. 251: «qu'il semble que la nature ou ses mouvemens ne soient autre chose qu'une ravissante harmonie».

4. CLARKE – REHDING 2001, p. 1.

5. CHUA 2001.

6. CHRISTENSEN 1993, COHEN 2001, and BLOECHL 2008.

7. DASTON 1998.

instruments as unreliable» and granted them «a minimum of trust»[8]. Although instruments like the monochord were used to demonstrate the tonal system, Boethius's suspicion of instruments percolated into many Renaissance philosophical and theoretical texts. Over the course of the sixteenth century, this slowly changed due to several factors such as a growing body of instrumental repertoire, the status of instrumentalists, and the physical investigation of instruments. As others have already pointed out, aside from telescopes, air-pumps, and microscopes, musical instruments were used as tools to investigate the natural world and were themselves objects of investigation[9]. Mersenne's use of the trumpet can be seen within this broader attention to instruments from scholars and theorists, although this chapter makes clear that his approach cannot solely be captured by categories such as «empirical» or «experimental». Mersenne was careful to select evidence from his engagement with the trumpet that could be utilized for demonstrating the harmonies present in the world.

Mersenne's works on music contain many references to the physics of sound. He was, for example, interested in overtones or the harmonic series. Each tone of a voice, trumpet, or violin consists of a fundamental frequency with higher partials or overtones. For musical notes, the series usually includes the octave, the fifth, the fourth, and the major third. The first tones constitute a diatonic triad, but eventually the series includes more dissonant notes. The seventh and the eleventh harmonic are, for example, quite flat. Mersenne's understanding of the harmonic series was limited. He had often heard these «small sounds» and confirmed his observations with musicians. But exactly which notes were audible in the higher registers was difficult to tell. Mersenne was convinced that they should be consonant since sound was created by God and thus must in principle be harmonious. He also seemed hesitant to connect his observations on overtones to the phenomena of overblowing on wind instruments and the playing of flageolet tones on stringed instruments. It would be the next generation of musical scientists that would connect these different phenomena with one another, building on Mersenne's descriptions[10].

By highlighting Mersenne's interest in trumpet tones, this chapter will shed new light on the relationship between his acoustical observations and musical practice. I will first show how Mersenne positioned musical instruments within his *Harmonie universelle* when referring to the debate between Gioseffo Zarlino and Vincenzo Galilei in the late sixteenth century. I will then explain how «naturalness» was discussed by music theorists in the early seventeenth century and how Mersenne's work related to this debate. This leads to a discussion of the trumpet, before concluding on the role of musical instruments in general as objects of knowledge within Mersenne's epistemology.

8. Boethius, *De institutione musica*, Liber I, Cap. 10 (BOETHIUS 1867, p. 197; BOETHIUS 1989, p. 18): «omnia haec [instrumenta] inconsulta minimaeque eastimans fidei».

9. COHEN 1984, GOUK 1999, GAUVIN 2011, and CYPESS 2016A.

10. DOSTROVSKY 1975.

HARMONIE
VNIVERSELLE,
CONTENANT LA THEORIE
ET LA PRATIQVE
DE LA MVSIQVE,
Où il est traité de la Nature des Sons, & des Mouuemens, des Consonances, des Dissonances, des Genres, des Modes, de la Composition, de la Voix, des Chants, & de toutes sortes d'Instrumens Harmoniques.
Par F. MARIN MERSENNE de l'Ordre des Minimes.
A PARIS,
Chez Sebastien Cramoisy, Imprimeur ordinaire du Roy, ruë S. Iacques, aux Cicognes.
M. DC. XXXVI.
Auec Priuilege du Roy, & Approbation des Docteurs.

Ill. 12.2ab: Marin Mersenne, *Harmonie universelle* (Paris, 1636), Title page and frontispiece. The engraving on the frontispiece is by the French engraver Henry Le Roy. The printer's mark on the title page is by the French engraver Jean Picquet. In most copies the frontispiece precedes the title page. Copy E-Mn.

Voices and Instruments, or Nature and Art

Mersenne's *Harmonie universelle* and *Harmonicorum libri xii* (Paris, 1636) gathered a decade's worth of research, containing many topics that today are no longer considered within the domain of music theory (Ill. 12.2ab; Table 12.1). There are discussions of the physiology of the voice; mechanical problems; vibrations of all kinds of objects; and the use of music for architects, engineers, and preachers. The books on instruments take up almost a third of the *Harmonie universelle*, forming in many ways a culmination of the previous books. Mersenne included four books on stringed instruments (including one on the monochord), and separate books on wind instruments, the organ, and percussion instruments. As he himself noted, his books on instruments in many ways continue the discussions of the previous chapters, but now make them more palpable and empirical[11]. Its Latin counterpart — in Mersenne's own words written for 'foreigners' who could not read French — is much shorter. As he reveals in a letter to the antiquarian and patron Nicolas-Claude Fabri de Peiresc (1580-1637), he wished to make the Latin version as voluminous as the French version but was constrained by the costs of printing[12]. Both books had been in the making for a decade. Mersenne had already announced his research on instruments in the *Traité de l'harmonie universelle* (Paris, 1627), where he wrote that the perfect musician (*parfait Musicien*) also had to consider the animal skins, wood, and metal of instruments, otherwise the study of music would always remain imperfect[13].

TABLE 12.1

Composition of Marin Mersenne's *Harmonie universelle* (Paris, 1636), according to his own copy in F-Pa, after the facsimile edition of this copy published by the Centre National de la Recherche Scientifique (Paris, 1965). Typography has been regularized and partly modernized, because of the large variations of the original typography concerning letter forms (i/j, u/v), capitalization and diacritics. Roman page numbers refer to pages without original page numbers. Prop. = number of propositions. In the column 'Latin' the corresponding books in the *Harmonicorum libri XII* (Paris, 1636, ²1648) are listed.

	PAGES	PROP.	LATIN
Harmonie universelle, Contenant la théorie et la pratique de la musique			
Harmonie universelle	iii		
Première préface générale au lecteur	i-xvi		
Table des propositions des dix-neuf livres de l'Harmonie universelle	xvii-xxxii		
Table des xix. livres de musique	xliii-li		
Première observation	lii		
Seconde observation	liii-liv		
Traitez de la nature des sons, et des mouvemens de toutes sortes de corps			
A [...] Monseigneur Louis de Valois, Conte d'Alais (dedication)	iii-vi		
Préface au lecteur	vii-viii		
Livre premier de la nature & des proprietez du son	1-84	34	I
Livre second du mouvement des corps	85-156	22	II
Livre troisiesme des mouvemens & du son des chords	157-228	23	III
Advertissement au Lecteur	[229]		
[Gilles Personne de Roberval], Traité de méchanique	1-36	3	

[11]. Mersenne, *Harmonie* (1636), 'Préface au Lecteur' before the 'Livre premier des instrumens'.

[12]. Mersenne to Nicolas-Claude Fabri de Peiresc, 2 February 1635 (MERSENNE 1959, p. 47) and 12 October 1635 (*ibidem*, p. 423). The seven books on instruments, for example, were reduced to four.

[13]. Marin Mersenne, *Traité de l'harmonie universelle* (Paris, 1627), pp. 10-11.

	Pages	Prop.	Latin
Traitez de la voix, et des chants			
A Monsieur [Jacques] Hallé, Seigneur de Boucqueval (dedication)	iii-v		
Preface au lecteur	vi-vii		
Fautes surveniües en l'impression	vii-viii		
Livre premier de la voix	1-88	53	
Livre second des chants	89-180	27	VII
Traitez des consonances, des dissonances, des genres, des modes, & de la composition			
A Monsieur Nicolas Claude Fabry, Sieur de Peiresc et de Callas (dedication)	iii-vi		
Préface, & advertissement au lecteur	vii-xi		
Fautes surveniües en l'impression	xii		
Livre premier des consonances	1-112	40	IV
Livre second des dissonances	113-140	14	V
Livre troisiesme des genres de la musique	141-196	19	VI
Livre quatriesme de la composition	197-282	27	VIII
Livre cinquiesme de la composition	283-330	12	
Livre sixiesme de l'art de bien chanter	331-440	34	
Traité des instrumens à chords			
A Monsieur [Henri] de Refuge (dedication)	iii-v		
Préface au lecteur	vi-viii		
Livre premier des instruments	1-44	29	I
Livre second des instruments à chords	45-100	16	
Livre troisiesme des instrumens à chords	101-176	27	
Livre quatriesme des instrumens à chords	177-228	18	
Livre cinquiesme des instrmens à vent	225-308	35	II
A Monsieur [Étienne] Pascal	i-ii		
Préface au lecteur	iii-iv		
Livre sixiesme des orgues	309-412	45	III
Livre septiesme des instrumens de percussion	1-62	31	IV
Éloge de Jacques Mauduit, excellent musician	63-72		
Fautes de l'impression, & quelques advis	73-79		
Livre de l'utilité de l'harmonie, & des autres parties des mathématiques	1-68	18	
Nouvelles observations physiques et mathématiques	1-27	14	
Advertissement	28		

Throughout Mersenne's writings, one finds evidence of his close contact with instrument makers and musicians. In Paris, he sought out the many instrument builders centred around the rue des Arcis and he asked his correspondents repeatedly about makers in their vicinity. In the *Harmonie universelle*, he included the smallest details of how to make a spinet, one of the most elaborate descriptions of making gut strings from the early modern period, and reports on the expert hearing of instrument makers and tuners. He also was a close observer of performance practices, incorporating the playing techniques of several Parisian musicians into his writings. At the same time, he stressed the superior role of theory and his own methods based on reason. Instrument makers, Mersenne argued, relied on their ears, often «without knowing the reason»[14] why they did something in a certain way. In the case of bell founders, he concluded that they did not rely on «a certain and infallible science»[15] like the one he presented in his own work. Within Mersenne's correspondence, there are numerous statements of disappointment or distrust between (amateur) scholars and artisans. His correspondents often disparaged local instrument makers for their secrecy or lack of knowledge. Only a few makers are cited by name and Mersenne seemed hesitant to base himself completely on their testimony.

In Proposition iii of the 'Livre Premier des Instruments' of the *Harmonie universelle*, Mersenne discussed the late sixteenth-century debate between the lutenist and composer Vincenzo Galilei (ca. 1520-1591) and the music theorist and composer Gioseffo Zarlino (1517-1590). The question Mersenne aimed to answer was «whether musical instruments were made in imitation of the voice or if the intervals were governed by those of instruments», concluding from this, «whether art perfects nature, or nature perfects art»[16]. This debate between Zarlino and his former student Galilei centred around the use of the syntonic diatonic (Ptolemaic) tuning — or just intonation in modern terminology — in singing, but quickly turned into a discussion of what can be called natural and what artificial. The debate started when in 1588 Gioseffo Zarlino published his *Sopplimenti musicali*; the following year Galilei published his rebuttal, the *Discorso intorno all'opere di Messer Gioseffo Zarlino da Chioggia* (Venice, 1589)[17]. The debate between the two has been re-told many times and will here be briefly summarized with respect to the categories of nature and art. Their positions can be seen as emblematic for the changing meanings of art and nature in the early modern period. In his writings, Zarlino based himself on Aristotle, who had famously maintained that natural objects have an internal principle, while artificial objects lack such principles of change[18]. The arts could imitate nature

[14]. Mersenne, *Harmonie* (1636), 'Livre premier des instruments', p. 41: «Cette division du manche des instrumens de Musique n'est pas nouvelle, puis que tous les Facteurs, et ceux qui les touchent en usent ordinairement sans en sçavoir la raison».

[15]. *Ibidem*, 'Livre septiesme des instrumens de percussion', p. 11: «car les Fondeurs [...] n'en sçavent pas la fabrique par une science certaine et infaillible que j'explique icy».

[16]. *Ibidem*, 'Livre premier des instruments', p. 7: «iii. Proposition. Determiner si l'on a fait les Instrumens de Musique à l'imitation des voix, ou si l'on a reglé les intervalles des voix par ceux des Instrumens; & consequemment si l'Art peut perfectionner la Nature, ou si la Nature perfectionne l'Art: & s'il faut juger des choses artificielles par les naturelles». Mersenne's *Harmonicorum libri xii* does not contain this discussion; instead, it offers some materials that are not present in *Harmonie universelle*, as, for example, his reference to the classification of instruments in *El melopeo y maestro* (Naples, 1613) by the Italian/Spanish music theorist Pietro (Pedro) Cerone (1566-1625); see Mersenne, *Harmonicorum* (1636), 'De instrumentis harmonicis', p. 2.

[17]. Vendrix 1995, Korrick 2009, and Fix 2019.

[18]. Aristotle, *Physics*, Book ii, Chap. 1 (Aristotle 2008, p. 33).

or improve it, but always with reference to nature. It was this question of the dependence of art on nature's principles that was increasingly interrogated.

Throughout his writings, Zarlino emphasized the primacy of nature. Music, he declared, depended «first on nature, then on art»[19]. Without nature, there would exist no consonances and no harmony. Nature appeared for Zarlino as a philosophical idea and norm, to which all things should aspire. And it was the human voice that demonstrated this norm. In contrast to instruments like the lute and viol, the voice was given to humans by nature and should act as the guiding principle in music. The voice gave forth the perfect forms of the intervals and consonances, such as 2:1 for the octave, 3:2 for the fifth, 4:3 for fourth, 5:4 major third, and 6:5 for minor third. As such, it held a special epistemological power for Zarlino. In contrast, musical instruments were considered artificial objects which could only be corrected by following the principles of nature:

> And to apply what we have discussed to what follows, we should know that inasmuch as artificial instruments in imitation of those that nature uses, every time their artisans or fabricators wish to correct or improve something that they see lacking in them, they seek to correct it with no other means than the example and model made by nature herself; and when it is necessary for them to want to render some reason for their work, they never serve themselves if not by those principles they have derived from the things they wish to imitate[20].

Instruments, Zarlino argued, were always made in imitation of nature and could only be corrected by means of imitating its principles. Just as painting would always try to approach nature, artificial musical instruments would always be imperfect compared to music based on nature. It becomes clear here that Zarlino based himself again on Aristotelian thought, as is evident from his references to Aristotle's *Posterior Analytics*, which does not allow one to apply conclusions across fields, for example from arithmetic to geometry[21]. Zarlino interpreted this Aristotelian maxim to mean that one is not able to draw conclusions about nature from art

---

19. Gioseffo Zarlino, *Sopplimenti musicali* (Venice, 1588), Libro I, Cap. 4, p. 18: «che la Musica dipende prima dalla Natura che dall'Arte».

20. *Ibidem*, Libro I, Cap. 6, p. 23: «Et per applicare quello, c'habbiamo discorso à quello che segue, dobbiamo sapere; che essendo gli Istrumenti Arteficiali fatti ad imitatione di quelli, che usa la Natura; tutte le fiate che i loro Artefici & Fabricatori uogliono correggere ò migliorare alcuna cosa, laquale uedono mancare in essi, cercano di correggerla non con altro mezo, che con l'essemplare & modello fatto da essa Natura; & quando li fà dibisogno di uoler rendere alcuna ragione dell'opere loro, non si seruono mai se non di quei Principij, c'hanno cauato dalle cose che uogliono imitare». Quoted in Vincenzo Galilei, *Discorso intorno all'opere di Messer Gioseffo Zarlino da Chioggia* (Venice, 1589), pp. 72-73. Translation from Goldberg 2011, p. 332.

21. Aristotle, *Posterior Analytics*, Book I, Chap. 7, 75a38 (Barker 1989, p. 71).

and vice versa: «one cannot conclude properly from the things of art about those of nature»[22]. And the voice, Zarlino found, naturally tended toward Ptolemy's syntonic diatonic scale. These consonances are not «products of art nor inventions of man» but lie enclosed in nature itself[23].

Vincenzo Galilei had studied with Zarlino in Venice but would increasingly question his former teacher's adherence to just intonation. His *Discorso* is a point-by-point rebuttal of Zarlino's *Sopplimenti*. In the *Discorso*, Galilei especially criticized Zarlino's use of the terms 'artifice' and 'nature'. Galilei was keen to point out that a tempered scale was no less natural than the syntonic diatonic scale or even just intonation. He showed the impracticality and impossibility of using the system as it was calculated on the monochord. Musicians mixed their tuning systems, and no musical practice could be fully encapsulated in such a musical system. Galilei cited numerous examples in which musicians departed from the rules that Zarlino set. Lutenists, for example, often followed the diatonic tuning of Aristoxenus. Players of wind instruments like flutes or cornettos adjusted on the spot, the same as singers did.

As Daniel Walker (1978) has pointed out, the biggest disagreement was not the fact that tuning would always remain imperfect; rather, the concern was what was considered natural and what artificial[24]. Whereas Zarlino would always see the human voice as an ontologically different sound object than an instrument, Galilei rejected such classifications. The syntonic diatonic scale, hailed by Zarlino as the natural norm for all music, was according to Galilei «a wholly artificial thing, so made by artifice of Ptolemy»[25]. If the human voice possessed the perfection that Zarlino ascribed to it, Galilei argued that this perfection was acquired through practice and was not naturally given. Moreover, Galilei saw no difference between a sound from an instrument and a sound from a voice since both were composed of the material of sound. He wrote: «sounds and voice, as the material of these intervals, are natural, as much in instruments as in voices»[26]. The sound of a flute and the sound of a human voice were both sound. A man-made object like an instrument did not necessarily have to imitate nature, but rather had to fulfil the function for which the artisan designed it. Ultimately, he decided that «one can conclude properly from the things of art about those of nature»[27].

---

22. Zarlino, *Sopplimenti* (1588), Libro i, Cap. 6, p. 23: «& che non si può concluder bene dalle cose dell'Arte in quelle della Natura». Translation from Goldberg 2011, p. 331.

23. *Ibidem*, Libro i, Cap. 1, p. 8: «non sono cosa dell'Arte, ne invention dell'Huomo». Translation from Palisca 1985, p. 272.

24. Walker 1978, pp. 14-26; Fix 2019, p. 552.

25. Galilei, *Discorso* (1589), p. 31: «il sintono è un solo tutto artifiziale, fatto cosi dall'artifizio di Tolomeo». Translation from Goldberg 2011, p. 286.

26. Galilei, *Discorso* (1589), p. 94: «il suono, & la voce, come materia di essi intervalli é naturale, tanto in questi quanto in quelli». Translation from Goldberg 2011, p. 351.

27. Galilei, *Discorso* (1589), p. 72: «Et che si può concluder bene dalle cose dell'arte in quelle della natura». Translation from Goldberg 2011, p. 331.

The Trumpet as Nature's Voice: Marin Mersenne and the Nature of Music

When Mersenne wrote the books on musical instruments for the *Harmonie universelle*, Zarlino's and Galilei's disagreement was very much on his mind. Although he would rely on Zarlino for his writing on counterpoint and composition, Mersenne nevertheless sided here with Galilei. In his dedication, he echoed Galilei by saying that the properties of air had to be considered both «in the mechanics of art and nature»[28]. Like Galilei, Mersenne emphasized that each object had a different goal. A flute had the goal of making sounds, not imitating nature. And like Galilei, he doubted that the voice possessed the perfection Zarlino ascribed to it, writing: «There is no doubt that the degrees and intervals of the voice take after art». If the intervals of the voice were natural, all children would naturally intone the exact intervals without having received previous education. Mersenne found this was «against experience»[29]. Moreover, a variety of fields had already demonstrated the perfection of art over nature, such as painting, medicine, agriculture, and fashion. A surgeon who set a dislocated bone did not learn this method from nature; a painter did not solely imitate nature but rather everything he could imagine[30]. Like Galilei, Mersenne saw musical instruments as especially suitable objects from which to learn about the nature of music. Instruments were even more revealing than human singers since no element of custom or training was involved. As such, they became central objects in his investigations of music, capable of revealing music's nature and the universal harmony at large.

The Musical Scale of Nature

The debate surrounding nature and artifice in music did not only centre around the use of just intonation. In Mersenne's work, the problem regarding naturalness in music seemed more pressing in his discussion of scales. Most texts from Antiquity and the Middle Ages distinguished at least three genera: the diatonic, the chromatic, and the enharmonic. Although there were several attempts to revive the enharmonic genus in the sixteenth and seventeenth centuries — most famous those by Nicola Vicentino (1511-1576) and Giovanni Battista Doni (1595-1647) — the diatonic genus was widely considered the most natural. For example, Zarlino argued that the voice naturally tended towards the diatonic genus, calling the chromatic and

[28]. Mersenne, *Harmonie* (1636), 'A Monsier [Henri] de Refuge', f. ã-ii-recto: «dans les Mechaniques tant de l'art que de la nature».

[29]. *Ibidem*, 'Livre premier des instruments', p. 8: «Or il n'y a nul doute que les degrez, & les intervalles des voix tiennent de l'Art, quoy que les paroles en tiennent davantage, car s'ils ne dependoient de l'Art, ceux qui enseignent à chanter aux enfans, n'auroient point de difficulté à les faire entonner, & chanter juste; & tous les hommes feroient les justes intervalles sans les avoir appris: ce qui est contre l'experience, qui monstre que les apprentifs ont de la peine à faire les demi-tons, & les autres intervalles tant consonans que dissonans, jusques à ce qu'ils les ayent accoustumez par l'Art».

[30]. *Ibidem*.

enharmonic genera artificial ornaments[31]. This question was also raised in France, where fewer attempts to revive the enharmonic genus existed than in Italy. For example, Salomon de Caus (1576-1626), in his *Institution harmonique* (Frankfurt, 1615), contrasted music to sciences like geometry, arithmetic, and perspective, all of which, he said, were invented to fulfil human needs. In contrast to these sciences, De Caus writes, «the science of music is not a human invention, as everyone knows, because the consonances of music are natural and not of our invention»[32]. Not all the consonances were natural, only those which were part of the diatonic scale. It was clear to De Caus that the diatonic genus was the most natural, based on the fact that the chromatic and enharmonic genera could not be sung properly with the voice. De Caus writes the following on the enharmonic genus:

> It is not possible for the human voice to know how to intone them, and even if it could be done with the help of instruments, it would not be possible to find the consonances to make any pleasant harmony with those intervals, thus we will also hold this musical genus useless[33].

According to De Caus, the voice was not able to make the small inflections necessary for the chromatic and enharmonic genera and therefore these divisions had to be regarded as unnatural. Other music theorists spread similar ideas. The composer and music theorist Antoine Parran, for example, held that the pure chromatic genus was not singable, and the ear could not judge the distances of the enharmonic genus[34]. Echoing the widespread idea that the diatonic scale was natural, the Jesuit Athanasius Kircher, for example, argued that all cultures preferred the diatonic scale in their songs[35].

Like De Caus, Parran, and Kircher, Mersenne aimed at defending the rationality and universality of the diatonic system. In this, he was especially influenced by the rise of scepticism in France. He included several sceptical statements in his works, which argued that music depended on custom and taste, not reason. In *La verité des sciences* (Paris, 1625) and *Questions harmoniques* (Paris, 1634), he adopted the sceptical position on music that musical judgement is subjective; each listener has a different preference. There is no reason why an octave, fifth,

---

31. Zarlino, *Sopplimenti* (1588), Libro III, Cap. 1, p. 84.

32. Salomon de Caus, *Institution harmonique* (Frankfurt, 1615), 'A [...] Anne, Royne de la Grande Bretaigne': «mais ceste science de Musique ne vient d'aucune invention humaine, car chacun sçait, que les consonnantes de ladite Musique, sont naturelles, et non de nostre invention».

33. *Ibidem*, f. 22v: «qu'il n'est possible à la vois humaine les sçavoir entonner, et encores que lon le pourroit faire à l'aide des instrumens, il ne se pourroit pas trouver des consonnantes contre lesdites intervalles pour rendre aucune harmonie agreable, ainsi nous tiendrons aussi ce genre de Musique inutile».

34. Antoine Parran, *Traité de la musique théorique et pratique* (Paris, 1639), p. 37.

35. Athanasius Kircher, *Musurgia universalis* (Rome, 1650), Tomus primus, Liber VII, p. 568: «dixi diatonicum prae coeteris esse naturale, quia video hoc omnes totius orbis populos naturaliter suis in cantibus proferre».

or fourth would be more consonant than the other intervals, the sceptic holds in *La verité des sciences.* And many believed that music was not a science at all, according to Mersenne[36]. In order to ground the consonances as rational and universal, Mersenne looked for arguments to demonstrate that they were natural, considering the same forms of evidence as De Caus, Parran, and Kircher had investigated.

In the 'Livre premier des consonances' in the *Harmonie universelle*, Mersenne turned to the same question Zarlino and De Caus had raised: which of the three ancient genera is most natural? He wrote that he had found that the human larynx seemed just as well adjusted to form the chromatic and enharmonic degrees as the tongue or the reed of a flute[37]. The preference for the diatonic genus at his time could very well come from custom rather than nature: «it appears that our belief that the diatonic degrees are more natural and easier than the enharmonic and the other arises only from the custom that preoccupies almost all of us»[38]. It was most likely custom that the diatonic scale was more widespread. Arguments that Zarlino had used — for example, that peasants or shepherds sang in the diatonic scale — were not enough evidence for Mersenne. He had heard many countrymen sing quartertones and God had given humans a flexible voice[39]. And singers could have always picked up their melodies somewhere. Although the enharmonic genus was not widespread in France, if practitioners put more effort into educating children in this manner, there would be no reason why it could not become more widely used. In contrast to other music theorists, Mersenne seemed to consider the role of custom and learning more than was usual in his time. In his writings on language, he similarly emphasized the role of custom, rejecting the idea of a natural language[40]. At the same time, for example in his *Harmonicorum libri XII*, he did suggest that nature corrected art, citing Isaac Beeckman's observations on the singing of psalms by «commoners»[41].

Mersenne also posed the question whether one could find out which scale was natural by investigating cultures that were considered more primitive by the French population at the time. This idea had already been explored by Salomon de Caus. According to De Caus, the intervals commonly used in Europe were given by nature, so there would be no reason to think that other cultures would use different intervals[42]. In his *Questions harmoniques* and *Harmonie universelle*,

36. Marin Mersenne, *La verité des sciences* (Paris, 1625), p. 32, *Questions harmoniques* (Paris, 1634), p. 80.

37. Mersenne, *Harmonie* (1636), Livre troisiesme des genres de la musique, pp. 147-150.

38. *Ibidem*, p. 147: «il semble que ce que l'on juge que les degrez Diatoniques sont plus naturels & plus aisez que les Enharmoniques, & les autres, ne vient que de la coustume qui nous preoccupe quasi tous».

39. *Ibidem*, 'Livre premier de la voix', p. 16: «car il n'y a point d'intervalles ausquels la voix humaine ne puisse s accommoder [...]. Mais puis que Dieu nous a donné la voix si flexible, qu'elle peut passer par tous ces degrez».

40. Koch 2006.

41. Mersenne, *Harmonicorum* (1636), 'Liber sextus de speciebus consonantiarum deque generibus & modis', p. 108.

42. De Caus, *Institution* (1615), f. 23rv: «Car veu que nous avons tout ce que la nature peut donner, ce seroit erreur de penser que leurs proportions fussent autres que les nostres».

Mersenne cited reports from French colonizers on the North American coast that confirmed the idea that other cultural traditions also used the diatonic scale. In his *Harmonie universelle*, he also printed the songs published by the French pastor Jean de Lery (1536-1613) in the account of his travels in Brazil[43]. But at the same time, such evidence could not be decisive, since there was always the chance that the consonances were learned. In a similar way, Mersenne rejected the evidence that birds or other animals sang in a diatonic scale. Ultimately it was a difficulty he could not resolve: only when one encountered «someone who sings without having heard singing» the question what is natural could be answered[44].

It is there that Mersenne turned to trumpets. Whereas humans and animals could have learned the consonances of the fifth or fourth, an «inanimate object» like the trumpet could not have done so[45]. And nonetheless, by changes in embouchure, the trumpet jumped to the octave, the fifth, the fourth, the major third, the minor third, the whole tone, and the semitone. Following Galilei's lead, Mersenne focused on instruments as special objects of investigation. But whereas Galilei had emphasized the difficulty of designating any consonance as natural, Mersenne's investigations into the trumpet led him to a meditation on the Divine, showing that nature was truly composed of harmony.

The Case of the Trumpet

Players of the natural trumpet (Ill. 12.3) can play the notes of the harmonic series by means of embouchure. Some of these harmonics are quite flat and need to be adjusted. Players often skip the flatter notes, as, for example, the seventh and fourteenth harmonic, which is a low B-flat for a trumpet in C; they go directly to the eight harmonic, a C for a trumpet in C. By means of altering air pressure and lip embouchure, trumpeters can also play outside the series, making smaller intervals and altering the jumps. As such, entire major and minor diatonic scales can be played in the higher registers, «tuning» the impure notes of the harmonic series. Although some of the trumpet notes confirm the belief in an underlying diatonic system of music, some difficulties remain in explaining the less harmonic notes.

43. Mersenne, *Harmonie* (1636), 'Livre troisiesme des genres de la musique', pp. 148-149. He refers to Jean de Lery, *Histoire d'un voyage fait en la terre de Brazil* (Paris, 1578). Lery did not include these melodies in the first edition but added them only in the third edition (Geneva, 1585).

44. Mersenne, *Harmonie* (1636), 'Livre troisiesme des genres de la musique', p. 148: «si l'on ne rencontre premierement quelqu'un qui chante sans jamais avoir ouy chanter». This question continued to be raised in the eighteenth century; see Bloechl 2008, pp. 177-215.

45. Mersenne, *Harmonie* (1636), 'Livre troisiesme des genres de la musique', p. 149: «car encore que les animaux eussent apris ces intervalles, les corps inanimez ne peuvent les avoir apris, & neanmoins ils font lesdites Consonances, apres lesquelles ils font le ton & le demiton Diatonique, comme j'ay remarqué au traité des Trompettes & des Flustes».

Ill. 12.3: After Jacob de Gheyn (ii), *Three Mounted Trumpeters* (engraving, 16.5 × 20.5 cm). NL-Ar, RP-P-OB-10.424.

Within the history of acoustics, wind instruments like flutes or trumpets were notoriously difficult to explain. In contrast to a stringed instrument or monochord, their proportions were not immediately visible. Whereas the shortening of a string made the immediate effect on its pitch understandable, wind instruments were an entirely different category of sounding bodies. The same body of air in a flute could produce different pitches by overblowing and by closing or opening the air holes. The idea of an air column was not yet understood in the early seventeenth century, nor were the notes of the trumpet or flageolet tones. In his *Harmonics*, Ptolemy advised the musical scholar not to rely on wind instruments in the construction of the harmonic canon, since he considered them imprecise and irregular[46]. When the proportions of wind instruments

46. Ptolemy, *Harmonics*, Book i, Chap. 8 (Barker 1989, p. 291).

were discussed, the numbers of the monochord were often transmitted to different sounding bodies, for example in the famous image included in Franchinus Gaffurius's *Theorica music*e (Milan, 1492; ILL. 0.5). The question if the same proportions of strings also governed pipes remained unresolved for much of the early modern period[47].

No early modern scholar devoted so much space and effort to explain the trumpet as Mersenne. The idea that the trumpet — «one of the most ancient instruments» — could demonstrate the universal distribution of intervals occupied Mersenne for a long time[48]. He wrote about the phenomenon as early as in his *Quaestiones celeberrimae in Genesim* (Paris, 1623). There he recounted how the French organist and composer Jean Titelouze observed someone playing through the different sounds of the trumpet: first the octave and the fifth, finally the whole tone and the semitone[49]. In his unpublished *Livre de la nature des sons* (Book on the Nature of Sounds), Mersenne continued this idea, writing that «the fifth is given to us by nature and is the first interval that the trumpet makes»[50]. After the fifth, the instrument would jump to the fourth, the major third, the minor third, the whole tone and the semitone, only in that order[51]. This was the proof that «these intervals were natural and proper for music, since nature gave them to us, and it was necessary to reject the other artificial ones»[52]. The artificial consonances would be the sixth, the seventh, and anything smaller than a semitone. Similarly, in his *Les préludes de l'harmonie universelle* (Paris, 1634), Mersenne noted that the best reason to use solely the diatonic scale was that it was taken from nature. The trumpet did not give the smaller semitone or the enharmonic diesis, whereas it did give the whole tone and the larger semitone[53].

That the trumpet occupied such a central role in Mersenne's considerations is remarkable. Historically, the instrument received little attention from music theorists. In sixteenth-century musical writings, such as Lodovico Zacconi's *Prattica di musica* (Venice, 1596), it is described as a signal instrument with little musical capabilities by itself. Zacconi excluded the trumpet from his overview of instruments because it was a field or a battle instrument, one that could not play

---

47. On the research of Vincenzo Galilei and Mersenne into organ pipes, see KONOVAL 2018.

48. Mersenne, *Harmonie* (1636), 'Livre cinquiesme des instrumens à vent', p. 247.

49. Marin Mersenne, *Questiones celeberrimae in Genesim* (Paris, 1623), p. 1699.

50. F-Pa, MS 2884, dated around 1627. TAVEAU 2018, p. 405: «La quinte nous est enseignée par la nature, et est le premier intervalle que fait la trompette». I would like to thank Laurence Taveau for making his unpublished edition of this manuscript available to me.

51. TAVEAU 2018, pp. 405-406: «sans qu'il soit possible de la faire passer à ces sons sans garder l'ordre que nous venons de décrire».

52. *Ibidem*, p. 407: «ces seuls intervalles étoient naturels, et propres pour la musique, puisque la nature nous les donne, et qu'il failloit rejetter les autres artificiels».

53. Marin Mersenne, *Les préludes de l'harmonie universelle* (Paris, 1634), p. 166.

Ill. 12.4ab: Claudio Monteverdi, *L'orfeo* (Venice, 1609), title page and page iv, with the *Toccata* for five trumpets. Copy I-MOe.

all the notes of the diatonic scale[54]. This reflected common musical practice, where trumpets indeed were most often used as a signal instrument for important events. It is notable that while Mersenne would argue that the diatonic scale was natural because the trumpet was able to play it, Zacconi did not attribute the diatonic scale to the trumpet at all. This demonstrates the change of position that the instrument underwent around 1600. Most books of early trumpet calls require no more than the first six or eight harmonics. Over the course of the late sixteenth and early seventeenth centuries, however, the range of the trumpet was expanded. Perhaps the most famous example is the opening toccata of Claudio Monteverdi's *L'Orfeo* (first performed in Mantua, 1607; Ill. 12.4), which requires harmonics up to the twelfth. A few decades later, players such as Girolamo Fantini were able to extend the range to non-harmonic notes and unlock the higher register of the harmonic series. Fantini's own instruction book, the *Modo per imparare a sonare di tromba* (Frankfurt, 1638), utilizes the harmonic series as far as the eighteenth harmonic[55]. By making use of the full range of the trumpet and a wide variety of non-

[54]. Lodovico Zacconi, *Prattica di musica* (Venice, 1596), Libro iv, Cap. 40, p. 213: «& nota che sotto gli'Istrumenti di fiato quantunque vi sieno ancora le Trombe, nondimeno perche le Trombe, sono Istrumenti da campi ò da bataglie, & non havendo tutte le voci Diatoniche, per questo si escludano». See Singer 1968, p. 59.

[55]. Wallace – McGratton 2011, p. 98.

harmonic tones, Fantini and other trumpeters at the time transformed the signal instrument into a solo instrument.

Although Mersenne had already announced in his early writings that the trumpet demonstrated how the diatonic intervals were given by nature, he continued to search for the reason why. Between his first writings on the topic and the publication of the *Harmonie universelle*, he extensively investigated the phenomenon of overblowing. In Paris, he observed numerous musicians and experimented himself with different pipes and flutes. The trumpet was also the topic of numerous exchanges in his correspondence. In a list of questions that Mersenne circulated among his correspondents, he asked why it was not possible to intone *ut*, *re*, *mi*, *fa*, and *sol* in the lower range of the trumpet[56]. He also asked Nicolas-Claude Fabri de Peiresc to conduct the experiment himself, if he were to encounter «an excellent trumpet player who knows a bit of music»[57]. By circulating such questions, Mersenne received numerous reports on the skills of local players. From Rome, for example, Mersenne received the message that some trumpeters could play entire scales[58]. And the physician Christophe Villiers from Sens asked a trumpeter to help him answering Mersenne's question, but the investigation was halted because the trumpeter could not read music notation[59].

News of the star trumpeter Fantini also reached Paris. In his *Harmonicorum libri XII*, Mersenne cited a report from the French physician Pierre Michon Bourdelot from Rome about Fantini. According to Bourdelot, Fantini could perform all notes on his trumpet, and play them in tune (*concinnè*) along with notes that Girolamo Frescobaldi played on the organ[60]. Through such accounts and his observations of local trumpeters, Mersenne was well-aware that expert players could manipulate the tones, departing from the harmonic series. For example, he wrote in the *Harmonie universelle*: «I have met someone who descends by a major third lower than the second sound [= second harmonic], instead of dropping an entire octave»[61]. And in his *Harmonicorum libri XII*, Mersenne similarly wrote that expert trumpeters could adjust their

[56]. Théodore Deschamps to Pierre Trichet, January 1635 (Mersenne 1959), p. 571. In this letter, Deschamps provided answers to a series of musical questions posed by Mersenne to the chemist Jean Rey.

[57]. Mersenne to Fabri de Peiresc, 2 February 1635 (Mersenne 1959, p. 48): «Si jamais il passe quelque excellent trompete qui sçache un peu de musique, si vous pouvés en faire faire l'experience, vous m'obligeres».

[58]. Mersenne to Fabri de Peiresc, 2 February 1635 (*ibidem*, p. 47).

[59]. Christophe Villiers to Mersenne, mid-November 1633 (Mersenne 1969, p. 548).

[60]. Marin Mersenne, *Harmonicorum libri XII* (Paris, 1636), 'Liber secundus de instrumentis harmonicis', p. 109: «per gradus ascendant: Quae suspicio litteris domini Bourdelotij Medici doctissimi ad me Roma missis vehementer augetur, quibus asserit se ab Hieronymo Fantino Tubicine totius Italiae excellentissimo audiuisse quòd tonos omnes sua tubâ caneret, eosque sonis Organi Cardinalis Burguessij iunxisse, quo Hieronymus Fresco Baldi Ducis Hetruriae, & Ecclesiae Romanae D. Petri Organiste concinnè ludebat». On this letter, see Conforzi 1990. Peter Downey has pointed out several mistakes in Conforzi's interpretation regarding Fantini; see Downey 1994.

[61]. Mersenne, *Harmonie* (1636), 'Livre cinquiesme des instrumes à vent', p. 256: «j'aye rencontré quelqu'un qui descend d'une Tierce majeure plus bas que le second son, au lieu de descendre d'une Octave entiere».

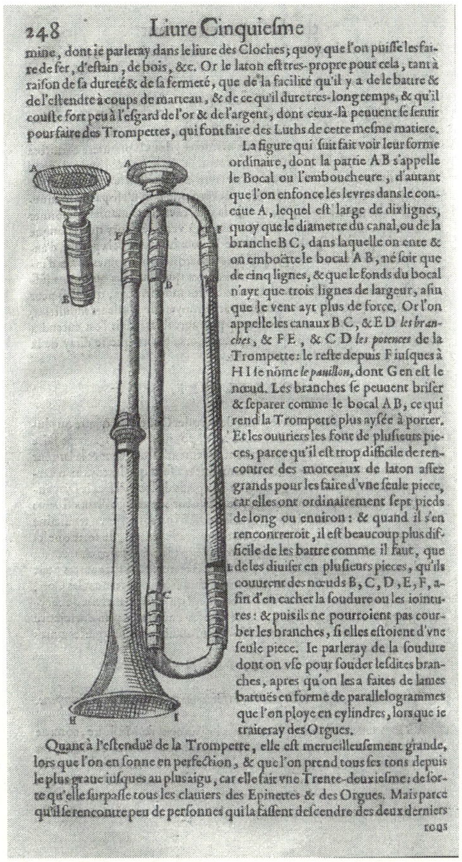

248 Liure Cinquieſme

mine, dont ie parleray dans le liure des Cloches; quoy que l'on puiſſe les faire de fer, d'eſtain, de bois, &c. Or le laton eſt tres-propre pour cela, tant à raiſon de ſa dureté & de ſa fermeté, que de la facilité qu'il y a de le battre & de l'eſtendre à coups de marteau, & de ce qu'il dure tres-long temps, & qu'il couſte fort peu à l'eſgard de l'or & de l'argent, dont ceux-là peuuent ſe ſeruir pour faire des Trompettes, qui font faire des Luths de cette meſme matiere.

La figure qui ſuit fait voir leur forme ordinaire, dont la partie A B s'appelle le Bocal ou l'emboucheure, d'autant que l'on enfonce les levres dans le concaue A, lequel eſt large de dix lignes, quoy que le diametre du canal, ou de la branche B C, dans laquelle on entre & on emboëtte le bocal A B, ne ſoit que de cinq lignes, & que le fonds du bocal n'ayt que trois lignes de largeur, afin que le vent ayt plus de force. Or l'on appelle les canaux B C, & E D *les branches*, & F E, & C D *les potences* de la Trompette: le reſte depuis F iuſques à H I ſe nôme *le pauillon*, dont G en eſt le nœud. Les branches ſe peuuent briſer & ſeparer comme le bocal A B, ce qui rend la Trompette plus ayſée à porter. Et les ouuriers les font de pluſieurs pieces, parce qu'il eſt trop difficile de rencontrer des morceaux de laton aſſez grands pour les faire d'vne ſeule piece, car elles ont ordinairement ſept pieds de long ou enuiron: & quand il s'en rencontreroit, il eſt beaucoup plus difficile de les battre comme il faut, que de les diuiſer en pluſieurs pieces, qu'ils couurent des nœuds B, C, D, E, F, afin d'en cacher la ſoudure ou les iointures: & puis ils ne pourroient pas courber les branches, ſi elles eſtoient d'vne ſeule piece. Ie parleray de la ſoudure dont on vſe pour ſouder leſdites branches, apres qu'on les a faites de lames battuës en forme de parallelogrammes que l'on ploye en cylindres, lors que ie traiteray des Orgues.

Quant à l'eſtenduë de la Trompette, elle eſt merueilleuſement grande, lors que l'on en ſonne en perfection, & que l'on prend tous ſes tons depuis le plus graue iuſques au plus aigu, car elle fait vne Trente-deuxieſme: de ſorte qu'elle ſurpaſſe tous les claniers des Epinettes & des Orgues. Mais parce qu'il ſe rencontre peu de perſonnes qui la faſſent deſcendre des deux derniers

tons

ILL. 12.5: Marin Mersenne, *Harmonie universelle* (Paris, 1636), 'Livre cinquiesme des instrumens à vent', p. 248. Copy E-Mn.

breath in such a way that it allowed them to emit the tones between those of the standard series[62]. But he did not pay much attention to them, writing: «But, apart from the fact that this is not common, and that these sounds are made with great difficulty, they are worth nothing and therefore they do not require any attention»[63]. Rather, he was interested in what tones the instrument could naturally bring forth.

Mersenne's depiction of the range of the trumpet in the *Harmonie universelle* does not include the non-harmonic notes that expert players like Fantini were able to play (ILL. 12.5). Rather, he emphasized that the trumpet naturally played the diatonic scale, also excluding the dissonant seventh harmonic. He seems not to have observed that trumpet players exclude this

62. Mersenne, *Harmonicorum libri XII* (1636), 'Liber secundus de instrumentis harmonicis', p. 109.

63. Mersenne, *Harmonie* (1636), 'Livre cinquiesme des instrumes à vent', p. 256: «mais outre que cela n'est pas ordinaire, & que ces tons se font avec une grande difficulté, ils ne valent rien, c'est pourquoy il n'en faut faire nul estat».

des Instrumens à vent. 249

tons, qui sont marquez souz le C *sol vt fa* dans la figure de l'autre Trompette en taille douce que ie donneray apres, il suffit de luy donner l'estenduë d'vne Vingt-neufiesme, c'est à dire des quatre Octaues, qui sont l'estenduë des claviers: & parce que ie veux examiner pourquoy la Trompette fait les interualles des consonances deuant ceux des dissonances, & les plus grandes consonances deuant les moindres, il faut icy considerer les raisons desdits interualles, afin de rechercher pourquoy elle les choisit tellement que l'on n'en peut faire d'autres en leur place.

Or ie commence ces interualles par son ton plus graue, qui est en C *sol vt fa*, dans lequel ie les termine aussi, apres qu'il a esté trois fois repeté: mais il faut remarquer que les premiers nombres à main gauche ne vont que iusques à la Vingt-quatriesme, d'autant que l'on ne peut continuer les raisons des autres interualles sans vser de fraction, lors que l'on retient les termes radicaux: les seconds nombres qui sont à main droite, commencent par le 9, parce qu'il faut multiplier tous les precedens par 9, afin de continuer tous les interualles de l'estenduë d'vne Vingt-neufiesme sans aucune fraction.

Estenduë de la Trompette.

16	C sol vt fa	144
	demiton maieur	
15	♮ mi	135
	ton maieur	
	A mi la re	120
	ton mineur	
12	G re sol vt	108
	ton maieur	
	F vt fa	96
	demiton maieur	
10	E mi la	90
	ton mineur	
9	D la re sol	81
	ton maieur	
8	C sol vt	72
	Quarte	
6	G re sol vt	54
	Tierce mineure	
5	E mi la	45
	Tierce maieure	
4	C sol vt	36
	Quarte	
3	G re sol vt	27
	Quinte	
2	C sol vt fa	18
	Octaue	
1	C sol vt fa	9

Or il est si aysé d'en comprendre l'estenduë par cette table, qu'il n'est pas besoin de l'expliquer puis qu'outre les nombres qui sont d'vn costé & d'autre, i'ay marqué chaque interualle par son propre nom: de sorte qu'il ne reste plus qu'à expliquer pourquoy la Trompette passe tout d'vn coup à l'Octaue, & puis à la Quinte sans pouuoir passer par les degrez du milieu: ce qui semble estre contraire à la maxime, qui asseure que l'on ne peut passer d'vne extremité à l'autre sans passer par le milieu: quoy que l'on puisse respondre que cet instrument, & quant & quant la voix, & le vent ont vn mesme priuilege que l'esprit, qui passe souuent d'vne raison à l'autre sans considerer celles qui sont entre-deux: mais cette difficulté merite vne Proposition particuliere.

PROPOSITION XII.

Expliquer pourquoy la Trompette ne peut faire les degrez en bas, comme en haut: & pourquoy elle fait l'Octaue dans son premier interualle, la Quinte dans le second, & ainsi des autres.

IL faut icy supposer l'experience qui est vniforme par tout le monde, à sçauoir que l'on ne peut faire *vt*, *re*, *mi*, *fa*, &c. depuis le premier ton de la Trompette, car ceux qui en sonnent font tousiours *vt*, *sol*, *fa*, c'est à dire la Quinte & puis la Quarte en haut: ce qui arriue semblablement aux Cors & aux Trompes. Il faut aussi remarquer que le ton, que l'on appelle ordinairement le premier, ou le plus bas de la Trompette, n'est pas celuy dont

Y

des Instrumens à vent. 267

qu'il arriue la mesme chose à la Musique qu'à chaque autre science, qui est cultiuée par vn esprit excellent & laborieux, lequel trouue toutes les autres sciences dans celle qu'il embrasse, car lors qu'vn homme d'esprit s'est proposé d'arriuer à la perfection de quelque instrument, & qu'il le pratique toute sa vie, il y transporte toutes les gentillesses des autres, & le rend aussi vniuersel qu'il se le peut imaginer, comme Iean Price a fait voir auec sa Fluste à trois trous, & le sieur Bocan auec son Violon.

PROPOSITION XX.

Expliquer toutes les circonstances de la Trompette, son estenduë en toutes sortes de manieres, & ses chants militaires par les notes de la Musique.

CETTE Proposition seruira pour expliquer tout ce qui n'est pas contenu dans l'autre figure, & ce que cette plâche represente beaucoup mieux auec tous ses ornemens. Or le corps de la Trompette est marqué par ABC

Z iiij

Ill. 12.6ab: Marin Mersenne, *Harmonie universelle* (Paris, 1636), Livre cinquiesme des instruments à vent, pp. 249 and 267. Coypy E-Mn.

note by choice. The fact that the eleventh and thirteenth harmonics were adjusted to conform to the just intonation scale seems to have eluded him. Rather, Mersenne explained the trumpet tones by imagining that a single tone is caused by a single vibration of air. To this, another vibration of air is added, doubling the motion, and sounding an octave higher. To this, another vibration of air was added, and the sound rose a fifth above the first octave. He explained the first series of trumpet tones by this principle (2:1, 3:2, 4:3, 5:4, 6:5; Ill. 12.6a). There was, however, a problem: instead of a 7:6 (a flat minor third), the trumpet sounded here a fourth higher (the 'seventh harmonic' in Ill. 12.6a). The proportion 7:6 was excluded, rejected by nature, in Mersenne's words, because it was considered a dissonance:

> It seems that the sesquisexta [7:6] should follow the sesquiquinta [6:5], since it is the lesser interval or the lesser ratio that follows the minor third. [...]. Because it is neither a consonance nor the difference of consonances, nature, which is harmonic, rejects it and prefers to interrupt the succession of its intervals and of its songs rather than to pass through an interval that is worth nothing except to offend the ear and the mind[64].

Nature, not wanting to «offend the ear and the mind» skipped the seventh partial, which formed a low minor seventh with the fundamental tone. The number 7 was «useless to harmony» in Mersenne's words[65], and instead of this impure number, the trumpet jumped to the eighth harmonic. Similarly, the eleventh harmonic was adjusted (so that it made a semitone with the previous note, 'F vt fa 96' and 'E mi la 90' in Ill. 12.6a). Following his rule, one would expect here a 11:10 ratio, not a 16:15 (=96:90). But here the semitone was the result of the difference in the previous series, between the fourth and the major third[66]. In his table, Mersenne gave the numbers 72, 81, 90, 96, 108, 120, 135, 144 for the upper range. Together these numbers form the ratios 8:9, 9:10, 15:16, 8:9, 9:10; 8:9, 15:16, resulting in a scale in just intonation. Combining different arguments, Mersenne went a long way to explain how the trumpet could produce such a scale.

According to Mersenne, the notes that expert trumpeters like Fantini were able to play did not require the attention of the musical scientist, since they were not natural to the instrument and could therefore be considered artificial. Skilled performers could manipulate the pitches of the instrument, just as «the industry of man can trouble the movements of nature»[67]. Nature appeared as a stable entity, emitting diatonic tones, showing that the diatonic genus was the most natural genus. He wrote: «So it seems that nature or its movements are nothing but a ravishing harmony, which invites us to consider the first source, from which it takes its birth, that is to say to contemplate God, who is the Author, and to love Him above all things»[68]. The

64. *Ibidem*, p. 251: «La Sesquisexte devroit ce semble suivre la Sesquiquinte, puis qu'elle est le moindre intervalle, ou la moindre raison qui suit la Tierce mineure [...]. Mais parce qu'elle n'est ny consonance, ny difference des consonances, la nature qui est harmonique, la rejette & ayme mieux rompre la suite de ses intervalles & de ses chansons, que de passer par un intervalle qui ne vaut rien, que pour blesser l'oreille & l'esprit».

65. *Ibidem*, p. 252: «inutile à l'harmonie».

66. *Ibidem*, p. 253: «Ce qu'elle fait afin de donner la troisiesme difference des consonances, car le demiton majeur est la difference de la Tierce majeure, & de la Quarte, comme le ton majeur est la difference de la Quarte & de la Quinte, & le ton mineur celle de la Quinte & de la Sexte majeure».

67. *Ibidem*, p. 256: «l'industrie des hommes peut troubler les mouvemens de la nature».

68. *Ibidem*, pp. 250-251: «Par où l'on void que le progrez de la nature est amy de l'harmonie, qu'elle gouverne ou dont elle depend: de sorte qu'il semble que la nature ou ses mouvemens ne soient autre chose qu'une ravissante harmonie, qui nous invite à consider la premiere source, dont elle prend sa naissance, c'est à dire à contempler Dieu, qui en est l'Autheur, & à l'aymer sur toutes choses».

trumpet showed the order of the natural intervals, starting with the octave and moving to the smaller intervals, rejecting artificial and dissonant notes[69].

After the publication of the *Harmonie universelle*, the trumpet continued to be discussed in Mersenne's correspondence[70]. For example, Mersenne received an explanation of the trumpet which suggested that trumpeters avoided the notes that were not part of the diatonic scale[71]. *Musicians* — rather than nature — rejected these notes. Mersenne shared this explanation with René Descartes, who found sound propagation in trumpets obvious: sound is dependent on vibrations of air and their different proportions make the different intervals, the consonances before the dissonances: «when one blows harder, by means of which it makes higher sounds, but which are all in tune with the first, and therefore also with each other»[72]. But this explanation could not explain the jumps. Why the divisions of 6:7 or 7:8 were skipped, ultimately remained unknown.

INSTRUMENTS AND THE DISCIPLINING OF EXPERIENCE

Mersenne would always remain convinced that the creation in principle was harmonious. If nature seemed chaotic or random, it meant that humans were just not able to perceive the harmony around them. In his *Les préludes*, he cited Ptolemy (*Harmonics*, Book I, Chap. 2) that it was the task of the scholar to demonstrate that «the works of nature are well-ordered, and that there is nothing confused, or done by chance»[73]. Mersenne felt obliged to show that order reigned, not chaos. With his conception of a harmonious universe, he was a follower of

[69]. *Ibidem*, p. 251: «D'où il est aysé de conclure que l'ordre des Consonances est natural, & que la manière dont nous contons en commençant par l'unité jusques au nombre de six, & au delà, est fondée dans la nature». («From which it is easy to conclude that the order of the consonances is natural, and that the way in which we count, beginning with unity to the number six, and further, is founded in nature».)

[70]. Théodore Deschamps expressed dissatisfaction with Mersenne's explanation of the trumpet tones. See Deschamps to Mersenne, 29 August 1640 (MERSENNE 1967, p. 72) and 26 March 1642 (MERSENNE 1970, pp. 84-85).

[71]. Gaudais (first name unknown) to Mersenne, December 1638? (MERSENNE 1964, p. 252): «Toutesfois, je croys que les joueurs les fuient en cest endroit comme n'estant pas intervalles diatoniques».

[72]. Descartes to Mersenne, 9 February 1639 (DESCARTES 1898, p. 500; MERSENNE 1964, p. 296): «lorsqu'on souffle plus fort; au moyen de quoy elle fait des sons plus aigus, mais qui sont tous accordans avec le 1er, et par consequent aussy entre eux».

[73]. Mersenne, *Les préludes* (1634), p. 163: «Car il appartient aux hommes sçauans qui employent leur vie, & leur estude à la contemplation, de monstrer que les œuvres de la nature sont bien ordonnées, & qu'il n'y a rien qui soit confus, ou qui se fasse par hazard, particulierement dans ce qui concerne la veuë & l'oüie, qui approchent plus de la raison, que les autres sens, et qui nous servent pour apprendre les sciences, & pour loüer, contempler & admirer les œuvres de Dieu, & l'excellence, & la grandeur de l'ouvrier».

the Church Father St. Augustine. Mersenne and Augustine shared the belief that music and sound offered privileged ways of obtaining access to this harmony. In his *De ordine*, for example, Augustine started reflecting on the meaning of order after hearing trickling water[74]. Perceiving sounds, music, or a spoken verse was for Augustine the clearest example of how sensory perception could lead to the perception of ordered numbers. Mersenne similarly saw music as the order that existed between the different parts of the world. As he wrote at one point, if humans could understand the songs of birds, the sounds of animals, and the noise of thunder or wind, then «we would find an admirable harmony»[75]. Nature was in principle harmonious, but humans could not always perceive this harmony.

This harmonious vision of nature coincided with Mersenne's interest in mechanics and artificial instruments. Robert Lenoble, in his study on Mersenne (1943), has emphasized his contribution to what he describes as mechanistic philosophy. According to Lenoble, nature «was reduced to a play of mechanical forces»[76], codifying mechanistic philosophy before the work of Descartes. According to Peter Dear (2000), Mersenne adhered less to a particular philosophical position but rather gathered arguments from different traditions, combining empirical, mathematical, and mechanical methods to argue for the harmonious creation of the world and the existence of certainties. Mersenne was less concerned with «natures» in the Aristotelian sense. A physical science dealing with the true causes of things as envisioned by Aristotle was untenable, Mersenne argued, although he never would openly fully dismiss Aristotle. Mersenne thought that human knowledge could never fully grasp the essence of things; only a science of appearances was possible[77]. He sought this in a variety of mathematical or physico-mathematical sciences such as optics, mechanics, and music. Mersenne's interest in the trumpet demonstrates how he tried to offer quantitative explanations of physical and mechanical movements.

Mersenne's interest in the trumpet as a method of demonstrating the naturalness of the first six integer numbers allows us to re-evaluate existing claims regarding his acoustical interests. In his study of Mersenne, Claude Palisca (1995) already pointed to the central significance of the debate between Zarlino and Galilei. As Palisca noted, Mersenne rhetorically followed Galilei in the latter's emphasis to put musical knowledge to the test. Both men found that the mythical anecdote of Pythagoras and the smithy could not have been correct. The large role of experimentation within musical science is described as a move away from numerology. Mersenne's and Zarlino's differences are explained by «accurate observation and experimental evidence» on the one hand and «myth, metaphysics, or legend» on the other[78]. As this chapter

---

[74]. Augustine, *De ordine*, Liber I, Cap. 6 (Augustine 2007, pp. 8-9).

[75]. Mersenne, *Traité* (1627), p. 63: «nous y trouverions une admirable harmonie».

[76]. Lenoble 1943, p. 381: «La nature réduite à un jeu de forces mécaniques».

[77]. Dear 2000, pp. 161-162.

[78]. Palisca 1995, p. 61.

attempts to demonstrate, their goals were closer to each other than one might anticipate: both theorists mobilized different forms of evidence to assert the universality of a certain set of proportions. Although Mersenne disagreed with Zarlino at times, he similarly emphasized the role of numbers and argued that just intonation was founded in nature. Whereas Mersenne's use of instruments can be seen within a broader turn towards empiricism, his use of experimental evidence was selective.

At the same time, Mersenne's use of the trumpet to demonstrate the nature of music would have been unacceptable for Zarlino: to use an artificial object to make claims about the natural world was mixing analytical categories. That does not mean Aristotelian thought had disappeared: in his discussion of the trumpet, Mersenne still cited the Aristotelian maxim that nature took the shortest routes and did nothing in vain[79]. This shows the complexity of meanings and associations attached to nature and artifice in the early modern period. Although the Aristotelian distinction between objects of nature and objects of art was increasingly challenged, naturalness and artificiality remained central categories for scholars to approach the world around them. As Lorraine Daston (1998) wrote: «During the sixteenth and seventeenth centuries, the art/nature opposition was attacked on several fronts — aesthetic, philosophical, and artisanal — but the intellectual and practical reflex of classifying objects as one or the other persisted»[80]. Numerous scholars in Mersenne's vicinity worked hard to re-define the relationship between artificial objects and nature at large. Claude Picot, for example, paraphrased René Descartes when he wrote that «all things that are artificial are like those natural»[81], comparing the movement of a clock to a tree producing fruits. The science of mechanics belonged to the realm of physics. Whereas in the Aristotelian tradition, the internal motion of a tree or plant could not be compared to a mechanical device, both were increasingly considered as a type of motion that could be quantified. Natural philosophers like Robert Boyle and Christiaan Huygens viewed the natural world as an artifact, a clockwork or automaton that was ingeniously put together by the creator.

Musical instruments were very much a part of these shifting configurations in the early modern period. Instruments were increasingly seen as objects to deconstruct nature, blurring the boundaries between nature and artifice, as Rebecca Cypess (2016A and 2016B) has emphasized in her study on musical instruments as experimental objects. Someone like Giovanni Battista Della Porta (1535-1615), Cypess has argued, showed «the beginnings of a rigorous, empirical

[79]. Mersenne, *Harmonie* (1636), 'Livre cinquiesme de Instrumes à vent', p. 250.

[80]. DASTON 1998, p. 156.

[81]. Abbé Claude Picot, *Les principes de la philosophie* (Paris, 1647), Quatrième partie, [Article] CCIII, p. 480: «Et il est certain que toutes les regles des Mechaniques appartiennent à la Physique, en sorte que toutes les chose qui sont artificielles sont avec cele naturelles». This last phrase is not found in Descartes' original: *Principia philosophiae* (Amsterdam, 1644), Pars quarta, [Articulum] CCIII, p. 307: «Et sanè nullae sunt in Mechanicâ rationes, quae non etiam ad Physicam, cujus pars vel species est, pertineant».

method to the study of the natural world», centred around instruments[82]. Although Mersenne followed his Italian predecessors like Della Porta, for him mechanics and the study of artificial objects were always mobilized to re-affirm the belief in a harmonious world put together by a Divine creator. Mersenne's trumpet moved between the world of artificial movements and natural order: a man-made object that could demonstrate the true harmony of nature, and at the same time this harmony could be troubled through human intervention, especially by virtuoso players like Fantini. Instruments like the trumpet were investigated because they contained clues on the harmonies present in the natural world[83].

CONCLUSION

Mersenne's depiction of the trumpet with the French *fleur de lis* on the banner below it (ILL. 12.6B) made the military and ceremonial nature of the early modern trumpet crystal clear. Although the instrument was increasingly used within ensembles and for solo sonatas, it was still primarily a royal and military signal instrument. As Kate van Orden (2005) wrote: «For listeners, the associations between the trumpet's brassy tones, mounted knights, and the king were unequivocal»[84]. For Mersenne, we can add to these associations the image of an all-encompassing universal harmony and the belief that nature in principle brought forth only consonances. No instrument demonstrated the presence of harmony within nature so exceptionally well as the trumpet. Strings and organ pipes could be adjusted and tuned, but the trumpet proved that the diatonic intervals were simply hidden in animate objects, waiting to be discovered.

The example of the trumpet also demonstrates how Mersenne utilized empirical evidence for proving that nature was fundamentally harmonious. The ability of the trumpet to play the intervals of the diatonic scale was seen as proof of the diatonic scale, ignoring the seventh harmonic. Numerological convictions and experimental methods operated side by side in Mersenne's work. The ability of trumpeters to play outside the diatonic scale was considered an example of art destroying the harmony of nature. In many ways, Mersenne utilized the

82. CYPESS 2016B, p. 160; CYPESS 2016A.

83. This side of Mersenne has already been stressed by Peter Dear in his study on Mersenne. Dear demonstrated that Mersenne altered the evidence of his experiment on string tensions to come to the proportion 4:1 instead of the less perfect 4.5:1. This demonstrates that there was still a prevalence of whole numbers in early modern experiments and that experimental data were adjusted to fit these numbers. Brandon Konoval has continued this line of inquiry in his study of Vincenzo Galilei and Mersenne's research on organ pipes. As Konoval convincingly shows, empiricism and experimentation were not ruptures from an older Pythagorean/Platonic worldview, but a continuation. See DEAR 1988, pp. 159-160, and KONOVAL 2018.

84. VAN ORDEN 2005, p. 24.

trumpet not as an instrument for discovery, but to establish something of which he was already convinced. Investigating which tones were sounded, he came to conclude that the trumpet was the best proof for the fact that the diatonic scale was given to humans by nature.

BIBLIOGRAPHY

ARISTOTLE 2008

ARISTOTLE. *Physics*, translated by Robin Waterfield, edited by David Bostock, Oxford, Oxford University Press, 2008.

AUGUSTINE 2007

AUGUSTINE, Aurelius. *On Order/De Ordine*, translated and edited by Silvano Borruso, South Bend (IN), St. Augustine's Press, 2007.

BARKER 1989

Greek Musical Writings. 2: Harmonic and Acoustic Theory, translated and edited by Andrew Barker, Cambridge, Cambridge University Press, 1989.

BLOECHL 2008

BLOECHL, Olivia. *Native American Song at the Frontiers of Early Modern Music*, Cambridge, Cambridge University Press, 2008.

BOETHIUS 1867

BOETHIUS, Anicius Manlius Severinus. *De insititutione arithmetica libri duo, De institutione musica libri quinque*, edited by Gottfried Friedlein, Leipzig, Teubner, 1867.

BOETHIUS 1989

ID. *Fundamentals of Music*, translated by Calvin M. Bower, edited by Claude V. Palisca, New Haven (CT), Yale University Press, 1989.

CHRISTENSEN 1993

CHRISTENSEN, Thomas. *Rameau and Musical Thought in the Enlightenment*, Cambridge, Cambridge University Press, 1993.

CHUA 2001

CHUA, Daniel. 'Vincenzo Galilei, Modernity and the Division of Nature', in: CLARKE – REHDING 2001, pp. 17-29.

CLARKE – REHDING 2001

Music Theory and Natural Order from the Renaissance to the Early Twentieth Century, edited by Suzannah Clark and Alexander Rehding, Cambridge, Cambridge University Press, 2001.

Cohen 2001
Cohen, David E. 'The «Gift of Nature»: Musical 'Instinct' and Musical Cognition in Rameau', in: Clarke – Rehding 2001, pp. 68-92.

Conforzi 1990
Conforzi, Igino. 'Girolamo Fantini «monarca della tromba». Nuove acquisizioni biografiche', in: *Recercare*, ii (1990), pp. 225-241.

Cypess 2016a
Cypess, Rebecca. *Curious and Modern Inventions: Instrumental Music as Discovery in Galileo's Italy*, Chicago (IL), University of Chicago Press, 2016.

Cypess 2016b
Ead. 'Giovanni Battista Della Porta's Experiments With Musical Instruments', in: *Journal of Musicological Research*, xxxv/3 (2016), pp. 159-175.

Daston 1998
Daston, Lorraine. 'The Nature of Nature in Early Modern Europe', in: *Configurations*, vii/2 (1998), pp. 149-172.

Dear 1988
Dear, Peter. *Mersenne and the Learning of the Schools*, Ithaca (NY), Cornell University Press, 1988.

Dear 2000
Id. 'Method and the Study of Nature', in: *The Cambridge History of Seventeenth-Century Philosophy. 1*, edited by Daniel Garber and Michael Ayers, Cambridge, Cambridge University Press, 2000, pp. 147-177.

Descartes 1898
Descartes, René. *Correspondance. 2: mars 1638 - décembre 1639*, edited by Charles Adam and Paul Tannery, Paris, Léopold Cerf, 1898.

Dostrovsky 1975
Dostrovsky, Sigalia. 'Early Vibration Theory. Physics and Music in the Seventeenth Century', in: *Archive for the History of the Exact Sciences*, xiv/3 (1975), pp. 169-218.

Downey 1994
Downey, Peter. 'Fantini and Mersenne. Some Additions to Recent Controversies', in: *Historic Brass Society Journal*, vi (1994), pp. 355-362.

Fix 2019
Fix, Adam. 'Esperienza, Teacher of All Things. Vincenzo Galilei's Music as Artisanal Epistemology', in: *Nuncius: annali di storia della scienza*, xxxiv/3 (2019), pp. 535-574.

Gauvin 2011

Gauvin, Jean-François. 'Instruments of Knowledge,' in: *The Oxford Handbook of Philosophy in Early Modern Europe*, edited by Desmond M. Clarke and Catherine Wilson, Oxford, Oxford University Press, 2011, pp. 315-337.

Goldberg 2011

Goldberg, Randall. *Where Nature and Art Adjoin: Investigations into the Zarlino-Galilei Dispute, Including an Annotated Translation of Vincenzo Galilei's 'Discorso intorno all'opera Di Messer Gioseffo Zarlino'*, Ph.D. Diss., Bloomington (IN), Indiana University, 2011.

Gouk 1999

Gouk, Penelope. *Music, Science and Natural Magic in Seventeenth-Century England*, New Haven (CT), Yale University Press, 1999.

Koch 2006

Koch, Erec. 'Voice, Aurality, and the Natural Language of Passion in Mersenne's *Harmonie universelle*', in: *Seventeenth-Century French Studies*, xxviii/1 (2006), pp. 77-89.

Konoval 2018

Konoval, Brandon. 'Pythagorean Pipe Dreams? Vincenzo Galilei, Marin Mersenne, and the Pneumatic Mysteries of the Pipe Organ', in: *Perspectives on Science*, xxvi/1 (2018), pp. 1-51.

Korrick 2009

Korrick, Leslie. 'Vincenzo Galilei's Revision of Renaissance Tuning. Trading on Art and Nature', in: *The Sounds and Sights of Performance in Early Music: Essays in Honour of Timothy J. McGee*, edited by Maureen Epp and Brian E. Power, Abingdon-New York, Routledge, 2009, pp. 193-214.

Lenoble 1943

Lenoble, Robert. *Mersenne ou la naissance du mécanisme*, Paris, J. Vrin, 1943.

Mersenne 1959

Mersenne, Marin. *Correspondance du P. Marin Mersenne, réligieux minime. 5: 1635*, edited by Cornelis de Waard, Paris, Éditions du Centre National de la Recherche Scientifique, 1959.

Mersenne 1964

Id. *Correspondance du P. Marin Mersenne, réligieux minime. 8: août 1638-décembre 1639*, edited by Cornelis de Waard, Paris, Éditions du Centre National de la Recherche Scientifique, 1964.

Mersenne 1967

Id. *Correspondance du P. Marin Mersenne, réligieux minime. 10: Du 6 août 1640 à fin décembre 1641*, edited by Cornelis de Waard, Paris, Éditions du Centre National de la Recherche Scientifique, 1967.

Mersenne 1969

Id. *Correspondance du P. Marin Mersenne, réligieux minime. 3: 1631-1633 (seconde édition)*, edited by Cornelis de Waard, Paris, Éditions du Centre National de la Recherche Scientifique, 1969.

Mersenne 1970
Id. *Correspondance du P. Marin Mersenne, réligieux minime. 9: 1642*, edited by Cornelis de Waard, Paris, Éditions du Centre National de la Recherche Scientifique, 1970.

Palisca 1985
Palisca, Claude V. *Humanism in Italian Renaissance Musical Thought*, New Haven (CT), Yale University Press, 1985.

Palisca 1995
Id. 'Mersenne pro Galilei contra Zarlino', in: *Essays in Honour of David Evatt Tunley*, edited by Frank Callaway, Nedlands (Western Australia), Callaway International Resource Center for Music Education, 1995, pp. 61-72.

Singer 1968
Singer, Gerhard. *Lodovico Zacconi's Treatment of the Suitability and Classification of all Musical Instruments in the «Prattica di musica» of 1592*, Ph.D. Diss., Los Angeles (CA), University of Southern California, 1968.

Taveau 2018
Taveau, Laurence. *Le manuscrit 2884 du Père Mersenne à la bibliothèque de l'Arsenal: Étude et édition critique*, Ph.D. Diss., Paris, Sorbonne Université, 2018.

Van der Miesen 2021
Van der Miesen, Leendert. '«Unbelievably Hard Work»: Marin Mersenne's *Harmonie universelle* at the Printers', in: *Early Printed Music and Material Culture in Central and Western Europe*, edited by Andrea Lindmayr-Brandl and Grantley McDonald, Abingdon-New York, Routledge, 2021, pp. 231-244.

Van Orden 2005
Van Orden, Kate. *Music, Discipline, and Arms in Early Modern France*, Chicago, The University of Chicago Press, 2005.

Vendrix 1995
Vendrix, Philip. 'La dichotomie vocal/instrumental dans les écrits théoriques aux confins des xvie et xviie siècles', in: *Le concert des voix et des instruments à la Renaissance*; edited by Jean-Michel Vaccaro, Paris, CNRS Éditions, 1995, pp. 71-81.

Walker 1978
Walker, Daniel Pickering. *Studies in Musical Science in the Late Renaissance*, London, The Warburg Institute, 1978.

Wallace – McGrattan 2011
Wallace, Johan – McGrattan, Alexander. *The Trumpet*, New Haven (CT), Yale University Press, 2011.

Chapter Thirteen

Joachim Jungius and the Question of Music as an Experimental 'Mixed Science'

Roberta Vidic
(Hochschule für Musik und Theater, Hamburg)

> *And is it not for almost two centuries that the most distinguished modern philosophers*[1] *have been searching for the true principles of musical harmony? Were these principles ever established, or do they remain controversial more than ever*[2]*?*

For centuries, Aristotle's *Posterior Analytics* served as a model for extremely divergent approaches to scientific methodology, ranging from pure rationalism to an extreme form of radical empiricism. The *Posterior Analytics* is the fourth of the six texts in Aristotle's *Organon* and deals with demonstration, definition, and scientific knowledge. The *Organon* is a collection of works on logic and has been known, commented, and studied since Antiquity, depending on the local availability of versions and translations. From the sixteenth century onward, interest in Aristotle's logic was closely related to developments in scientific methodology that finally led to our concept of science. These methodological innovations affected the development of a science of music as well.

[1]. Wallace 1992a, p. 26: «(Natural) philosophers» are the Renaissance equivalent of scientists in our own day.

[2]. Giuseppe Tartini, *De' principj dell'armonia musicale contenuta nel diatonico genere* (Padua, 1767), p. iv: «E poi non sono forse due secoli da che si cercano i veri principj della musical armonia dai più insigni Filosofi moderni? Si sono anche accertati questi principj, o sono attualmente in disputa più che mai?».

Given the vast literature on this topic, the aim of this article is merely to shed light on some specific issues concerning the interaction between the Veneto culture and Northern Germany, with a special focus on the German scholar Joachim Jungius (Jung; 1587-1657). Jungius completed his studies in Padua and was acquainted, as we will see, with the work of the two Galileis, Vincenzo (1520s-1591) and Galileo (1564-1642). Additionally, he has been called in our times a 'radical empiricist'[3], whereas the expression 'radical empiricism' should be considered within the framework of the dichotomy in epistemology between rationalism and empiricism[4]. In a nutshell, the main distinction in epistemology concerns the role of sensory experience and the subsequent definition of knowledge: for rationalists there are ways such as intuition and deduction to gain knowledge independently of sensory experience. In this case, we speak of knowledge *a priori*. Empiricists claim, on the contrary, that our knowledge is totally dependent upon sensory experience. This leads to the definition of knowledge *a posteriori*[5] — but it is only half the story.

A key role in defining an empiricist component in the rise of modern science is certainly played by the change of the semantic field around the Latin *experientia* compared to the modern concept of experiment. As Mordechai Feingold (2016) has recently observed, historians of science have outlined the importance of medieval and Renaissance thought for the emergence of the methodological and experimentalist tradition behind Francis Bacon, Galileo Galilei, and René Descartes. But, as Feingold continued, «[t]his awareness of the significance of experiments [...] largely failed to transcend a philosophically-grounded, and anachronistic, bifurcation of 'empiricism' — the nineteenth-century appellation of 'experimental philosophy' — and the post-Kantian conception of 'rationalism'»[6]. On the other side, the historiography of music has often focused on the dichotomy between speculative and practice-oriented music theory, where practice coincided with composition and performance. The present discussion of issues such as musical temperament departs from methodological questions, in order to open different perspectives on music as an experimental «mixed science».

Medieval and Early Modern Theories of Demonstration

Aristotle's *Organon* became available for Western scholars in its complete form by the middle of the twelfth century. At that time, scholasticism prepared the ground for a systematic

---

3. Meinel 2013, p. 45.

4. An interpretation as a 'pure experience' in terms of William James, *Essays in Radical Empiricism* (1912), would be anachronistic and is generally excluded from the present discussion.

5. For a general definition of the dispute between rationalists and empiricists, see the *Stanford Encyclopedia of Philosophy Archive* (<https://plato.stanford.edu>, accessed May 2022), article 'Rationalism vs. Empiricism' by Peter Markie.

6. Feingold 2016, p. 2.

science, about one century before the foundation of teh first universities. In a context of increasing complexity in economy and society, the scholastic method of reasoning faced the task of finding a common denominator among the disparate authorities inherited from the past. The reorganization of traditional knowledge and the foundation of the sciences were achieved through a logical technique that aimed at reconciling apparently incompatible authorities with each other. Students exercised this technique in debates, where they were asked to elaborate argumentation schemes that could resist refutation[7].

Medieval theories of demonstration were generally based on Aristotle's syllogistic logic and were handed down either in form of commentaries and discussions of the *Posterior Analytics* or (sometimes) as independent opuscula on specific problems. A syllogism is a logical argument that applies deductive reasoning. It consists of three sentences, departing from two (true) premises or assumptions (axioms) and leading to a logical conclusion. As the theory of demonstration is the part of logic that is specifically concerned with scientific knowledge, a demonstration is a syllogism that produces scientific knowledge: that is, not only knowledge in the sense *that* a fact is true, but also knowledge of the reason *why* it is true. Nevertheless, some sorts of demonstration were allowed by Aristotle that do not totally fulfill the second condition[8]. Medieval authors therefore refer to a demonstration from an established fact as a so-called demonstration «that it is the case» (*demonstratio quia*), in contradistinction to a demonstration from a reasoned fact that is called «why it is the case» (*demonstratio propter quid*). *Quia* demonstrations require principles from another science and are said to be subaltern to that science[9]. In this sense, the science of music or harmonics is proved by and is in this way subaltern to arithmetic, which is a subdiscipline of mathematics. In Aristotelian terms, music is called a «middle science» (*scientia intermedia*) within the mathematical disciplines, since the 'that it is' belongs to arithmetic (another part of mathematics), and the 'why it is' to the «superordinate» genre of mathematics[10].

Greek philosophy encompasses, of course, not only Aristotelianism, but also various currents that were later differently received and combined[11]. A tendency to syncretism is characteristic for Humanism and Early Modern science and was cultivated in Venice and at the

[7]. On scholasticism in general, and the historical background of scholastic reasoning, see Hankins 2007, p. 33.

[8]. Aristotle, *Posterior Analytics*, i, 13, 78a22-79a17 (Aristotle 1994, pp. 19-21).

[9]. For a general survey on the medieval theories of demonstration, see the *Stanford Encyclopedia of Philosophy Archive* (<https://plato.stanford.edu>, accessed May 2022), article 'Medieval Theories of Demonstration' by John Longeway. On the *quia* and *propter quid* demonstration, see also the article 'Giacomo Zabarella' by Heikki Mikkeli in the same online resource.

[10]. Aristotle, *Posterior Analytics*, i, 9, 76a9-12 (Aristotle 1994, p. 14).

[11]. For a definition of 'mathematical (Neo-)Pythagoreanism', see Hicks 2014, pp. 418-419. On the (Neo-) Platonic philosophy of mathematical understanding', see Nolan 2002, pp. 202-203. On the application of a Euclidean model to other disciplines than geometry, see Gozza 2000, p. 87.

university of Padua[12]. A turning point in this development occurs when scholastic reasoning was replaced by the Euclidean axiomatic method, or as Paolo Gozza (2000) has described it, «by the axiomatic order of the deductive method, which passes from the simple to the complex via the preliminary definition of the terms and postulates of reasoning and the subsequent demonstration of the theorems and corollaries»[13]. The axiomatic method had been already extended from Euclidean geometry to other mathematical sciences in Antiquity and had become an integral part of a general reevaluation of ancient mathematics in Humanism. This applies also to Boethius's *De institutione musica*, when he passes from the fundamental propositions in Book I to their mathematical demonstrations in Books II and III[14].

The Venetian music theorist Gioseffo Zarlino (1517-1590) resorted to Boethius in his *Istitutioni harmoniche* (Venice, 1558) and *Dimostrationi harmoniche* (Venice, 1571). For Gozza, these works mark the point of transition between two kinds of reasoning on music. Other than the 'neutral' *Institutioni harmoniche*, where Zarlino took inspiration mainly from Aristotle's *Metaphysics*, the *Dimostrationi harmoniche* were especially addressed to scholars of the arts and sciences, showing various important models from Antiquity to his own time. Yet, Zarlino's attempts at a demonstrative science of music were not limited to generic cross-references. In analogy to the definition of professional roles in Boethius[15], he literally embodied the speculative *musicus* and holder of the *propter quid* as the opposite of his illustrious colleagues Adrian Willaert, Alfonso Dalla Viola, and Claudio Merulo, who in their proximity to musical practice were not able to fully explain the «reason» of musical matters. The title *Dimostrationi harmoniche* already indicated the certainty achieved in music through the application of the axiomatic method to musical reasoning. In this, Zarlino explicitly adhered to the ideas of the mathematicians from the Paduan university who applied the method of Euclidean geometry to all other disciplines. Moreover, the *Dimostrationi harmoniche* show a synthesis between the humanistic dialogue and the Euclidean method. For this reason, Zarlino's work is more likely to have served as a model for Galileo Galilei's scientific dialogues than the *Dialogo della musica antica, et della moderna* (Florence, 1581) of his father Vincenzo Galilei[16].

Vincenzo Galilei studied with Zarlino around 1563 and engaged fifteen years later in a long dispute with his former teacher on questions of pitch and temperament. This coincided with the publication of his *Dialogo della musica antica, et della moderna*. A specific matter of controversy regarded Zarlino's use of the legend of Pythagoras at the blacksmith's shop as the

---

12. Gozza 2000, p. 93.

13. *Ibidem*, p. 87.

14. Boethius 1989, p. xxix.

15. Boethius, *De institutione musica*, Liber I, Cap. 34, 223.27-225.15 (Boethius 1989, pp. 50-51).

16. Gozza 2000, pp. 81-82, 86 and 92-95. Proclus's *Commentarium in primum Euclidis Elementorum librum* (Commentary on the First Book of Euclid's Elements; 5th century AD) belonged together with the *Posterior Analytics* to the canonical texts in the then current debate on mathematics and demonstrative reasoning.

mathematical basis for the theory of intervals. Accordingly, the relative weight of the hammers used by the blacksmiths were supposed to be the reason of the differences between the sounding intervals. Pythagoras should have tried to verify his assumption by comparing the sound of plucked strings[17]. Vincenzo repeated the narrated «experience» and communicated the result in his *Discorso intorno all'opere di Zarlino* (Florence, 1589). It turned out that when weights are attached to strings of equal thickness and length, it is necessary to stretch a string with a weight that corresponded not to the simple ratio, but to its square, to obtain the expected interval: not 2:1 but 4:1 for the octave, not 3:2 but 9:4 for the fifth, and so on[18].

The controversies between Zarlino and Vincenzo Galilei are documented not only in letters and printed treatises, but also in several manuscripts from about 1590 that Vincenzo eventually left unfinished or unpublished. These manuscripts were edited only in 1980 by Frieder Rempp and suffered, in this way, the same fate as the early contemporary writings of Vincenzo's son Galileo[19]. The latter's contribution on musical acoustics appeared in his last book, *Discorsi e dimostrazioni matematiche intorno a due nuove scienze* (Leiden, 1638). The book covers, however, four decades of research on local motion and mechanics, starting in the years 1588-1592[20]. The fact that harmonics stood between mathematics and physics came to the fore especially during the Scientific Revolution.

Galileo Galilei: «Logician or Scientist»[21]?

Galileo Galilei studied at the University of Pisa in the early 1580s and returned there for a teaching position at the end of the decade. At that time the faculty presented a variety of approaches to Aristotelian philosophy. According to William A. Wallace (1992), the local situation and the current hierarchy in the teaching staff can be delineated as follows: The less-distinguished professors of logic are barely known today because they did not publish their lectures. The professors of natural philosophy covered an entire spectrum of views toward Aristotle's *Physics* and further related works[22]. Galileo, who held a chair of mathematics, certainly aspired to be regarded as a scientist and not as a mere logician. He was aware of the traditional

17. For a commented summary of the legend and essential research literature, see Nolan 2002, p. 272.

18. Palisca 1996, pp. 171-172. According to Stillman Drake (1992, pp. 10-11), the experiments conducted by Vincenzo Galileo c. 1590 on the vibrating string probably enabled his son Galileo to learn various aspects of experimental design. For more literature about Vincenzo's influence on Galileo's methodology, see Finocchiaro 2020, pp. 741-742.

19. See the editions in Rempp 1980 and Palisca 1989.

20. Salvia 2017, p. 108.

21. Wallace 1992a, p. 25.

22. *Ibidem*, p. xv.

distinction between *logica docens* and *logica utens* and used it in arguing against the contemporary supporters of the Aristotelian science. The Latin expressions for 'logic as doctrine' and 'logic in use' are quite frequently used in medieval and Renaissance works for discussing the question whether logic is a science or an art. They are associated to three different kinds of scholars and scholarship — taking us back to the initial ranking within the teaching staff at the University of Pisa. The *logica docens* is proper to the logicians, whether engaged in teaching logic or not. The *logica utens* is proper to the scientists and stays for the application of the precepts from the *Posterior Analytics* to investigate a particular subject matter, and to demonstrate it to attain scientific knowledge. If a scientist failed in the application of the canon of the *Posterior Analytics* to a certain subject matter, he seems only to have opinions about them and was considered merely a dialectician. In Aristotelian terms, Galilei had a scientific methodology in mechanics for the local motion, but ultimately lacked one in astronomy for the Copernican issue, as an example. In any case, Galileo's resorting to logic or dialectics should not be confused with his use of the hypothetical-deductive method of modern science[23].

Galileo left several unpublished manuscripts that were written during his teaching in Pisa. One of them includes a commentary on Aristotle's *Posterior Analytics* and derives from lecture notes of the teaching by the Jesuits at the Collegio Romano in Rome[24]. Another one is now commonly known as 'De motu antiquiora' (Older Writings on Motion) and contains Galileo's earliest attempts to a science of local motion[25]. The so-called *Iuvenilia* manuscripts were rediscovered at the National Library of Florence in the 1870s and showed, for Stefano Salvia (2017), «a new [and] unexpected Galileo». They contain, in fact, «not just written notes, but also drawings, calculations, diagrams, tables, often full of revisions — a true 'laboratory on paper' made of changing ideas, problems to be solved and possible solutions, experiments to be thought and maybe performed»[26]. Nevertheless, due to the strict separation of the curricula at a late-sixteenth century university, we should not imagine of public experiments performed during academic lessons, but rather of private and informal discussions with some disciples and colleagues. According to Salvia, Galileo pursued with his «repeated experiments» not a didactic but a dialectic goal, as they were part of an «argumentative strategy to show the contradictions between Aristotle's physics and what could be directly observed»[27].

Galileo apparently felt the need first to destroy the old science by means of logical reasoning, before the building of a new science upon a new scientific method could be

23. For a survey on the terms *logica docens* and *logica utens* in Galileo, see *ibidem*, pp. 21-26.

24. I-Fn, MS Gal. 27. See Wallace 1992a, pp. xi-xii.

25. I-Fn, MS Gal. 71, was compiled after I-Fn, MS Gal. 27 (Wallace 1992b, p. 59), probably between 1589 and 1591: thus, it must be contemporary with the «repeated experiences» described by Galileo's disciple Vincenzo Viviani (Salvia 2017, p. 111) and the late theoretical work of his father Vincenzo.

26. Salvia 2017, p. 108.

27. *Ibidem*, p. 111.

undertaken. Below, we will discover an equally twofold — destructive and constructive — use of logic by the German scholar Joachim Jungius.

MEDIEVAL AND EARLY MODERN THEORIES OF TUNING

Medieval and Early Modern theories of tuning were commonly formulated in terms of relative string lengths on a monochord. Towards the middle of the sixteenth century, there was a growing need to quantify the string lengths related not only to tuning systems, but above all to various temperaments. The breakthrough came, however, only about hundred years later, when logarithms made it possible to simplify the calculation of tempered intervals and solve the problem of incommensurable ratios. Temperament is a procedure, in which the just intervals of a certain tuning system are slightly altered to produce the desired effect for the ears. This implies that a mathematical problem meets a matter of sense perception and judgement of musical intervals. In other words: just intervals are made up of whole ratios, while slightly tempered intervals are often expressed by irrational ratios. Apart from this, some issues regarded exact measurement (or error). Finally, listening and perception can vary between different individuals, and the tension of strings (influencing the pitch) may depend on the ambient temperature.

Authors have been aware of all these issues since Antiquity. At the same time, harmonics had to be founded on certain principles: that is, on a reliable measurement of intervals. In Early Modern times, it was not only a matter to give music a scientific definition and justification in the frame of the mathematical sciences, but also to find an explanation for a temperament that had already been in place for some time. This required — in turn — not only further developments in mathematics, but also a wider acceptance of the results obtained through the new computational methods.

It is worth taking a closer look at Boethius to understand which issues remained topical until the time in question. Boethius wrote on the error of the human sense in Book I, Chapter 9, of *De institutione musica*. In Book V, Chapter 2, he described the function of harmonics:

> Harmonics is the faculty that weighs differences between high and low sounds using the sense of hearing and reason. For sense and reason are, as it were, particular instruments for the faculty of harmonics. The sense perceives a thing as indistinct, yet approximate to that which it is; reason exercises judgment concerning the whole and searches out ultimate differences. So, the sense discovers something confused, yet close to the truth, but it receives the whole through reason. Reason itself comes to know the whole, even though it receives an indistinct and approximate likeness of truth. For sense brings nothing whole to itself but arrives only at an approximation. Reason makes the judgment[28].

[28]. Boethius, *De institutione musica*, Liber V, Cap. 2, 352.4-14: «Armonica est facultas differentias acutorum et gravium sonorum sensu ac ratione perpendens. Sensus enim ac ratio, quasi quaedam facultatis armonicae

Approximation therefore played a key role in the formation of an opinion, but it was still not sufficient to ascertain the truth. The novelty was in this case not represented by the inadequacy of hearing in judging intervals precisely, which Boethius has borrowed for the most part from Book v of Ptolemy's *Harmonics*, but by the notion of cumulative error he added in the same chapter[29]. The cumulative error becomes particularly significant when several intervals are added up to form an octave — as we will see in the last part of this chapter.

More generally, relative string lengths could be calculated by either arithmetic or geometrical methods. Temperaments of the musical octave provided for the division of the syntonic comma 81:80 into *n* parts, so that we obtain 1/4-comma, 1/5-comma, 2/7-comma temperaments, and so on. However, traditional arithmetic methods allowed to divide intervals into equal parts only in exceptional cases. From the middle of the sixteenth century, the formal, exact solution could be obtained — with some restrictions — by geometrical constructions. The methods used until the seventeenth century only allowed the calculation of one or two geometric means between two terms, which made it possible to divide a ratio into 2, 3, 4, 6, 8, 9, 12, etc. equal parts. This was sufficient to obtain equal temperament graphically. But this solution did not allow Zarlino in 1558, for instance, to subdivide the comma into seven equal parts. For his 2/7-comma temperament, Zarlino recommended instead the use of a mesolabium, a measurement instrument known from Antiquity. The introduction of mathematical tools such as root extraction allowed then — from the end of the sixteenth century onwards — to divide intervals of any size. At the time of Galileo, another instrument enjoyed a particular popularity: the sector or proportional compass. This compass can be related to the military compass or even to a reduction compass that it still in use in lute making. Although it offered some advantages over the mesolabium, it seems to have been surpassed by logarithms when these began to be used for musical calculations. The logarithmic transformation of linear pitch values was originally proposed to avoid the overcomplicated root calculations necessary for the different temperaments[30].

According to Rudolf Rasch, early research including logarithms was scarcely publicized and was carried out «seemingly independently by different researchers»[31]. Around the mid-seventeenth century, the writings of the Italian mathematician Lemme Rossi and the German

instrumenta sunt. Sensus namque confusum quiddam ac proxime tale, quale est illud quod sentit, advertit. Ratio vero diiudicat integritatem atque imas persequitur differentias. Itaque sensus invenit quidem confusa ac proxima veritati, accipit vero ratione integritatem. Ratio vero ipsa quidem invenit integritatem, accipit vero sensu confusam ac proximam veri similitudinem. Namque sensus nihil concipit integritatis, sed usque ad proximum venit, ratio vero dijudicat». Translation from Boethius 1989, p. 163.

29. Boethius 1989, p. 164, note 7.

30. On the mathematical methods for the calculation of temperaments, see Rasch 2002, pp. 210-214. On the geometric methods, see Barbieri 1996, pp. 201-202. On regular temperaments, see Lindley 1987, pp. 201-224.

31. Rasch 2002, p. 212.

scholar Joachim Jungius have several similarities (see Table 13.1). Among others, both theorists elevated mean-tone temperament from a mere practice to a theoretical system. For this, especially Jungius introduced a specific terminology for tempered intervals, beginning with the quarter of the syntonic comma: the «quartula»[32]. In general, equal temperament is associated with Aristoxenus and Vincenzo Galilei.

Table 13.1

Systems of tuning and temperament in comparison: Lemme Rossi, *Sistema musico* (Perugia, 1666), p. 83, and Joachim Jungius, 'Harmonica', in *Praecipuae opiniones physicae* (Hamburg, 1679; posthumous), Statements [158-165] and [228].

Tuning System	Lemme Rossi (1666)	Joachim Jungius (1679; posthumous)
Systems with Exact Calculation		
Pythagorean tuning	Diatono ditonico (Pitagorico)	Scala diatonica antica
Just intonation	Diatono sintono (Tolomeo)	Scala diatonica nova
Systems with Geometric Division of Intervals		
Mean-tone temperament	Sistema temperato Sistema participato	Scala diatonica reformata
Equal tempermant	Divisione uguale, sia d'Aristosseno	Semitonia Vincentii Galilaei

When it is not about the exact measurement of intervals but about tolerance values for temperaments, mathematical science of music is «mixed» with the properties of perception. The notion of «mixed (mathematical) sciences» was introduced following Aristotle to indicate branches like music[33], astronomy, and optics which are particularly close to physics. The fifteenth-century German astronomer and mathematician Johannes Müller, better known as Regiomontanus, used the term «middle sciences» (*scientiae intermediae*) of Thomas Aquinas in his *Oratio introductoria in omnes scientias mathematicas* (Introductory Oration on All the Mathematical Sciences; published Nuremberg, 1537), based on an introductory lecture he held at the University of Padua in 1464. Later, in his influential works *The Advancement of Learning* (Oxford, 1605) and *De dignitate et augmentis scientiarum* (About the Dignity and Growth

32. Barbieri 1987, p. 119, note 3; Joachim Jungius, *Praecipuae opiniones physicae* (Hamburg, 1679), 'Harmonica', [Statement] 118.

33. Regarding the previous passage from Boethius's Book v, Calvin Bower (Boethius 1989, p. 163, notes 3-4) has made a further distinction between «music» and «harmonics». Accordingly, *musica* is «one of the four mathematical disciplines preparatory to the study of philosophy», *harmonica*, on the other hand, «not merely a discipline, but a power or capability of the mind». It remains to be seen if this distinction also applies to Early Modern reception.

of Knowledge; London, 1623), the English philosopher Francis Bacon made the distinction between 'pure mathematics' and 'mixed mathematics'[34]. The question if it is possible to establish some correlation between the field of science classification, methodological issues, and further developments of the theories of tuning, will now be investigated into more detail with a focus on Joachim Jungius.

JOACHIM JUNGIUS: A «RADICAL EMPIRICIST»[35]?

Joachim Jungius began his academic career as a mathematician. He interrupted his later studies of medicine at the University of Rostock to complete them in Padua, where he received the doctorate after only a few months (11 August 1618 - 1 January 1619)[36]. Jungius was generally well acquainted with Galileo's work and studied at a faculty that was still under Galileo's influence after the latter's eighteen years of teaching and research there from 1592-1610. Nevertheless, Jungius offers a good example of how complex interrelationships within the scientific community functioned in the early seventeenth century and beyond. According to Christoph Meinel (2013), who defined Jungius «[a]s a radical empiricist who wanted to continue the dissection and fragmentation down to the lowest building blocks of the physical world», Jungius never adopted a method comparable with those of Galileo or Descartes, although he had almost all the major works of these authors in his own library[37].

After his return to Rostock, Jungius became professor of mathematics in 1624. His inaugural lecture contained a detailed taxonomy of the sciences and the arts under the umbrella term 'mathesis'. Therefore, sciences were divided into «pure or abstract sciences» (*scientiae purae sive abstractae*) and «mixed or concrete sciences» (*scientiae mixtae sive concretae*). In the former group Jungius newly introduced a 'protomathesis': that is, a mathematical discipline being still more abstract than arithmetic and geometry. Music (*harmonica*) and astronomy then belonged together with optics, kinematics, and other disciplines to the «mixed sciences»[38]. The addition of further disciplines to the traditional *quadrivium* and the distinction among 'pure' and 'mixed' mathematical sciences were at the time not new at all, as we have seen in relation to Regiomontanus's and Bacon's works.

Meinel's choice of the term «radical empiricist» has followed, however, less Jungius's public lectures on mathematics and more a modern analysis of his teaching and scientific work.

34. On the history of the term «mixed mathematics», see BROWN 1991, especially pp. 81-83.

35. MEINEL 2013, p. 45.

36. DI LISCIA 2002, pp. 217-219.

37. MEINEL 2013, p. 45.

38. HÜBNER 1996, pp. 187-189.

Jungius's first independent and published work, the *Geometria empirica* (Rostock, 1627), proposed an explicitly «empirical» approach to geometry, in which theorems should not be derived by means of definitions, axioms, and proofs, but by *experientia.* Jungius aimed, in this way, to promote the student's independent ability to make a judgement through trial and error and starting from one's own sensory perception. In 1629, Jungius was appointed professor of logic and natural science at the Akademisches Gymnasium in Hamburg and assumed the rectorship of the Latin school, the Johanneum. He was charged by the *Scholarchat* (the school board, consisting of representatives of the council, the merchants, the main pastors and the senior elders) with a reform of the Hamburg school system[39].

The new school regulations of 1634, which were enacted under Jungius's rectorship, showed the typical teaching offer of a seventeenth-century Latin school. In the first class, for example, the younger pupils were taught in Greek and Latin literature and grammar, Greek Bible study, logic, rhetoric, and music[40]. The students at the Gymnasium were taught, besides classical languages, logic, and rhetoric, also mathematics (including harmonics), metaphysics, history, natural philosophy (physics), and practical philosophy (ethics)[41]. Already from 1635 onwards, Jungius was exposed to increasing criticism of clergy members in the *Scholarchat*, who opposed his ideas and renewals. This reached a peak with the publication of his *Logica Hamburgensis* (Hamburg, 1638; Ill. 13.1). The textbook was commissioned by the *Scholarchat* for use at the Akademisches Gymnasium and then sharply criticized by the Wittenberg Aristotelian Johannes Scharf — one should bear in mind that the University of Wittenberg has long been the stronghold of orthodox Lutheranism. In view of the ongoing polemics, Jungius gave up the rectorship and retained from 1640 onwards only his teaching of logic and physics at the Akademisches Gymnasium[42].

Also in 1640, Samuel Hartlib, a German emigré in England, quoted Joachim Hübner's opinion that Jungius was destined to «prove a most rationall & experimentall philosopher»[43]. Hartlib did not apply, on the contrary, the adjective 'experimental' to members of the Oxford Group, which leads us to a further observation. Jungius died on 23 September 1657 in Hamburg. In February 1658, Hartlib asked the natural philosopher, physicist, and chemist Robert Boyle to distribute Martin Vogel's biography of Jungius among the Oxford Group[44]. This group of

---

[39]. The Akademisches Gymnasium was added to the Johanneum in 1613 without the right to confer academic degrees. Although similar institutions existed in various Northern-European cities, their function between school and university remained eventually unclear. On Jungius's biography and context, see Meinel 2013, pp. 31-34.

[40]. Kremer 1991, p. 131.

[41]. Meinel 2013, p. 33: «Mathesis, Logica, Metaphysica, Physica, Philosophia practica, Eloquentia, Historia» and Classical Languages.

[42]. On the controversies that led to Jungius's resignation, see *ibidem*, p. 36.

[43]. Hartlib, quoted in Feingold 2016, p. 4.

[44]. *Ibidem*, p. 9.

Ill. 13.1: Joachim Jungius, *Logica Hamburgensis* (Hamburg, 1681), frontispice, engraving (16 x 9 cm) by Joachim Wichmann, with likenesses of Socrates, Plato, Aristotle, and Joachim Jungius. «Eluctor anfractuum vertigines» = I overcome the dizzy spells of complexities. «Elicuit, Excepit, Aptavit, Explevit» = He selected, accepted, adjusted, completed. Copy D-Hs, P 22: J 20.

scholars, also known as the Invisible College, constituted a precursor group of the famous Royal Society, formally The Royal Society of London for Improving Natural Knowledge. With the Societas Ereunetica sive Zetetica, Jungius had founded himself an early scientific society in Rostock in 1622[45]. The two societies differed enormously from one another in terms of their scientific methods. While Jungius and his Societas Ereunetica attached a great value to logic

[45]. To compare, the Royal Society of London was founded in 1660 after previous informal circles.

and the development of so-called doxoscopic research, members of the Royal Society excluded all disciplines of the *trivium*: logic and therefore also the philosophy of language, philosophy of science and «method» (regressive or non-regressive)[46]. In particular, the question of method will be discussed in the context of the Paduan tradition.

The Role of the Paduan *regressus* for Galileo and Jungius

The scholastic method of reasoning aimed at reconciling apparently incompatible authorities with each other and, as we have seen before, drew heavily on the reception of Aristotle's *Posterior Analytics*. In this, the whole of Aristotle's *Organon* has undergone very different interpretations over a long period of time. With these premises, we can easily imagine how urgent it was to further develop the scientific methodology during this time of the rise of modern sciences. As Daniel Di Liscia (2002) has summarized, only two paths seem to have been generally available at this point: «either one firmly closes one's eyes to what has been handed down so far and accordingly one searches for the new knowledge with the best possible independence from the counter-opinions of the past, or one accepts the challenge to work out a general procedure to deal with the doctrines that are still common and rigid in scientific practice»[47]. Such bifurcation can be observed very well in a comparison of Galileo and Jungius since it has a common ground in the further development of theories of demonstration in the Veneto culture. This involves a reevaluation of the Euclidean axiomatic method and also the introduction of the Paduan *regressus*, with which both scholars were well acquainted.

A *regressus* is a logical argument in which any sentence requires a justification and any justification itself requires support, resulting in a potentially infinite regress. As a method, the *regressus* can be traced back to the Aristotelian tradition of the twelfth-century polymath Averroës, which became popular among Italian Averroists and medical authors. In its simplest form, a first demonstration proceeds from a known effect to the existence of the cause of this effect, and a second from the cause to the effect. The two-step procedure combines in this order both demonstration methods, that is a *demonstratio quia* with a *demonstratio propter quid*. On the other hand, it makes it difficult to prove that the cause demonstrated in the first step is really the cause for the effect under investigation. For this reason, an intermediary step was introduced from the early sixteenth century. The Paduan philosopher and logician Giacomo (Jacopo) Zabarella (1533-1589) called this step a mental examination (*examen mentale*) and explained it in terms of his psychology of knowledge. As Heikki Mikkeli has recently observed, the three-step *regressus demonstrativus* as presented in Zabarella's commentary on the *Posterior Analytics*

46. Di Liscia 2002, pp. 242-243.

47. *Ibidem*, p. 231.

«is based on the cognitive structure of the knowing subject rather than on the ontological structure of the object of knowledge»[48].

Both Galileo and Jungius developed their respective methods either in a certain agreement with or in open opposition to the Paduan philosophers. Galileo explains the *regressus* in his 'Tractatio de demonstratione' (Treatise on Demonstration) in full length[49]. Different from the Paduan Aristotelians, he makes an important distinction between *logica docens* and *logica utens*. In his teaching, the *quia* and *propter quid* demonstrations are separated, like in Zabarella's work, by an intermediate step. When he applied the *regressus* to his scientific investigations, however, this intermediate step included not only time but also work for estimating the need of experimentation or computation[50]. Jungius's challenge of Aristotelianism was strictly connected with a discussion technique that he called doxoscopy (*doxoscopia*). This term refers to a preparatory critical discussion of opposing opinions for the purpose of their complete rejection. The criticism concerns three points in the following order: inaccurate and ambiguous expression, lack of consistency, and the falsity or groundlessness of premises, even if the logical deduction is correct. Lack of consistency is proved by incompatibility (a) with other statements by the same author; (b) with statements by other authors who are assumed to be adherents of the same theory; and (c) with the words of Aristotle himself — in case the theories under discussion are of Aristotelian origin[51].

As was the case for Galileo, Jungius's usage of logic or dialectics should not be confused with a method for knowledge construction. His syndiacritic method (*syndiakritische Methode*) was named after the Greek terms for analysis (*diacrisis*) and synthesis (*syncrisis*). As it did not distinguish between the basic operations of the intellect, a didactic order of the sciences, and the basic operations of nature, this method aimed at a universal science, in which ontology was essentially provided by physics. With the exception of Galileo, mathematics was not intended as a means of quantifying or as a legitimate correlation of research data. For Descartes and Leibniz, it served as a model for proper method and the systematic construction of science. It moved from an undefined pre-scientific everyday experience towards a clearly defined, scientific experience. Yet, its conceptualization relied neither on «experiments», nor in «repeated experiences», such as in Galileo's circle, but on reading books. Apprehending the world empirically meant here to learn its details from the written sources[52]. Overall, empirical geometry, doxoscopy, and the syndiacritic method are the three important premises that must be understood for classifying Jungius's approach to musical matters, in contrast to Vincenzo Galilei.

48. *Stanford Encyclopedia of Philosophy Archive* (<https://plato.stanford.edu>, accessed May 2022), article 'Giacomo Zabarella' by Heikki Mikkeli.

49. I-Fn, MS Gal. 27. See Wallace 1995, p. 76.

50. Wallace 1995, p. 95.

51. Di Liscia 2002, p. 228.

52. On Jungius's syndiacritic method, see Meinel 2013, pp. 37-39.

Joachim Jungius and the 18:17 Rule of Vincenzo Galilei

In his *Dialogo della musica antica, et della moderna* (1581), Vincenzo Galilei gave a simple but very effective rule for placing the frets for to equal temperament on the lute (or the viola da gamba). In summary, the length of a vibrating string corresponding to each fret must be assumed to be 17/18 of the length of the fret preceding it[53]. Because of its simple formulation and application, the rule is very convenient for practitioners. On the other hand, it confronts theorists with a dilemma without a simple solution. In fact, if the 'quantity of sound' is considered, the whole tone cannot be divided arithmetically into two equal parts. Vincenzo's solution was to consider, instead, the 'quality of sound', as Aristoxenus already had done in his «chromatic tonikon» system (Galilei: *Cromatico Toniaco*)[54]. Following this principle, it was easy to divide the whole tone in *n* parts. The problem is that the whole string is not divided into two equal parts after placing twelve frets, for the addition results to fall short a little of the ratio 2:1. A stated aim of this rule is the satisfaction of the sense; it is a means to remedy the inconvenience in practice by approximation:

> I therefore warn the industrious person for removing, with his discretion and diligent attempts, that small inconvenience which exists between the measurer and the measured[55].

The procedure basically depends on the judgement of the ear, while an exact measurement of the intervals plays a secondary role[56]. In defense of Aristoxenus, Vincenzo chose the following words:

> For Aristoxenus knew very well — as I have demonstrated — that he had to distribute the quality of sound into equal parts, not the quantity of the line, string, or space, since he was acting at that time as a musician concerned with the sound[ing] body, and not as a simple mathematician concerned with the continuous quantity [of sound][57].

---

53. Barbieri 1996, p. 219.

54. Vincenzo Galilei, *Dialogo* (1581), p. 49. See Herman 1973, pp. 288-289.

55. Vincenzo Galilei, *Dialogo* (1581), p. 49: «La onde auuertisco l'industrioso agente che con la sua discretione & diligenza cerchi ouuare à quella poca disconuienienza, che è tra il misurante & il misurato». Translation Herman 1973, p. 301.

56. On the 18:17 rule, see Lindley 1987, p. 175, and Barbieri 1996, pp. 219-220.

57. Vincenzo Galilei, *Dialogo* (1581), p. 53: «Imperoche molto bene (secondo che vi ho dimostrato) sapeua Aristosseno, d'hauere à distribuire in parti vguali la qualità del suono, & non la quantità della linea, corda, & spatio: operando allhora come Musico intorno al corpo sonoro, & non come semplice Matematico intorno la

Upon further analysis, the Aristoxenian approach of Vincenzo turns out not only to invert the ideals of Boethius and Zarlino, but more generally to be the opposite of mathematical Pythagoreanism. His concept of approximation is based, in this sense, not on an abstraction to the simplest ratio, but on the principle of similar proportion: a correct ratio is chosen, in this case, for the semitone and then applied twelve times to a string[58]. The divisor necessarily had to be 18, since smaller numbers as 17, 16, and 15 would just give a number of frets smaller than desired, while larger numbers as 19, 20, and 21 would result in a number of frets larger than desired[59]. In practice, Vincenzo's tight octave works better than any further attempt of mathematical correction of the temperament, because it can better compensate for the influence of string tension on the frequency of the sound; this can vary considerably based on the position along the neck and, for fretted instruments, the finger pressure on the strings. For the same reason, Zarlino would deny some years later in the *Sopplimenti musicali* (Venice, 1588) that an «Aristoxenian scale» could be played on a lute, a viola da gamba or a similar instrument[60]. An exact rendition of equal temperament was, however, not what Vincenzo really aimed at, even not in his late manuscript treatises on the enharmonic genus — where he resorted to a psychoacoustical conception of interval qualities. He did not supplement his explanation of the 18:17 rule with detailed calculations, nor did his later acoustic construction aim at the reproduction of fixed pitches in a tuning system[61].

The calculation of semitones according to Vincenzo Galilei's rule is part of the surviving reflections on harmonics by Joachim Jungius. His reference for this calculation was not Galilei's *Dialogo*, but Johannes Kepler's *Harmonices mundi libri v* (Five Books on the Harmonics of the World; Linz, 1619)[62]. Jungius did not publish on musical issues during his lifetime. His 'Harmonica' is a posthumous edition of lecture notes, consisting of a numbered series of individual statements (from 1 to 239) without page numbering. The definitions have, in addition, their own numbering (from (1) to (38)), as do the theorems (from i to xli). Together, the statements form

continua quantità». Translation Herman 1973, pp. 312-313. Herman translated «la qualità del suono» with «the quantity [*sic*] of sound».

58. On the principle of similar proportion, see Herman 1973, p. 299, note 71.

59. Vincenzo Galilei, *Dialogo* (1581), p. 49. See Herman 1973, p. 301.

60. Gioseffo Zarlino, *Sopplimenti musicali* (Venice, 1588), pp. 220 and 225, quoted in Lindley 1987, p. 177.

61. Rempp 1980, pp. 200-202. Zarlino's inclusion of thirds and sixths into the tonal system also resulted in the distinction between a greater and a lesser whole tone. What Vincenzo Galilei described in his late treatises on the enharmonic genus is, on the contrary, the acoustic construction of a tonal system that could not be reproduced exactly in any tuning. Thus, a distinction between a greater and a lesser whole tone became meaningless both in terms of the scale — the way the tonal system was represented — and in musical reality.

62. Johannes Kepler, *Harmonices mundi libri v* (1619), Liber iii, pp. 48-50: «Sectio Monochordi expedita pro chely eiusque reprobatio per Auditus sensum». (Division of the monochord suitable for the lute and its rejection by the sense of hearing.)

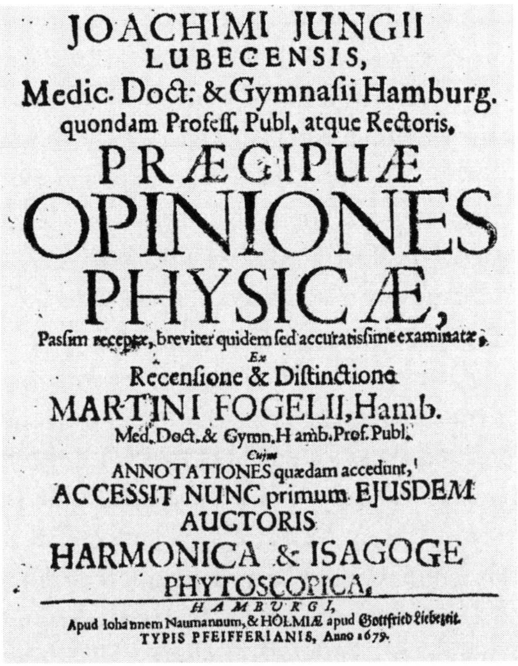

JOACHIMI JUNGII
LUBECENSIS,
Medic. Doct: & Gymnaſii Hamburg.
quondam Profeſſ. Publ. atque Rectoris,
PRÆCIPUÆ
OPINIONES
PHYSICÆ,
Paſſim receptæ, breviter quidem ſed accuratiſſime examinatæ;
Ex
Recenſione & Diſtinctione
MARTINI FOGELII, Hamb.
Med. Doct. & Gymn. Hamb. Prof. Publ.
Cujus
ANNOTATIONES quædam accedunt,
ACCESSIT NUNC primum EJUSDEM
AUCTORIS
HARMONICA & ISAGOGE
PHYTOSCOPICA,
HAMBURGI,
Apud Iohannem Naumannum, & HOLMIÆ apud Gottfried Liebezeit.
TYPIS PFEIFFERIANIS, Anno 1679.

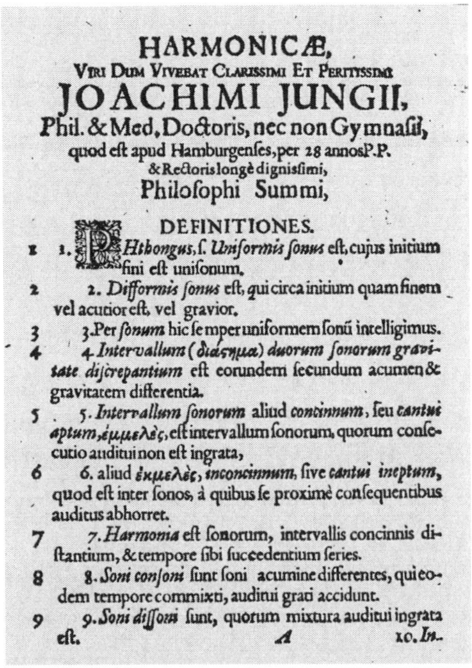

HARMONICÆ,
VIRI DUM VIVEBAT CLARISSIMI ET PERITISSIMI
JOACHIMI JUNGII,
Phil. & Med. Doctoris, nec non Gymnaſii,
quod eſt apud Hamburgenſes, per 28 annos P.P.
& Rectoris longè digniſſimi,
Philoſophi Summi,
DEFINITIONES.

1 1. *Phthongus, ſ. Uniformis ſonus* eſt, cujus initium fini eſt uniſonum.

2 2. *Difformis ſonus* eſt, qui circa initium quam finem vel acutior eſt, vel gravior.

3 3. Per *ſonum* hic ſemper uniformem ſonũ intelligimus.

4 4. *Intervallum* (διάστημα) *duorum ſonorum gravitate diſcrepantium* eſt eorundem ſecundum acumen & gravitatem differentia.

5 5. *Intervallum ſonorum* aliud *concinnum*, ſeu *cantui aptum*, ἐμμελές, eſt intervallum ſonorum, quorum conſecutio auditui non eſt ingrata,

6 6. aliud ἐκμελές, *inconcinnum*, ſive *cantui ineptum*, quod eſt inter ſonos, à quibus ſe proximè conſequentibus auditus abhorret.

7 7. *Harmonia* eſt ſonorum, intervallis concinnis diſtantium, & tempore ſibi ſuccedentium ſeries.

8 8. *Soni conſoni* ſunt ſoni acumine differentes, qui eodem tempore commixti, auditui grati accidunt.

9 9. *Soni diſſoni* ſunt, quorum mixtura auditui ingrata eſt. A 10. *In-*

ILL. 13.2AB: Joachim Jungius, *Praecipuae opiniones physicae* (Hamburg, 1679), general title and first page of the 'Harmonica' (with Statements 1-9). Copy D-Mbs.

demonstrations, with the theorems left unproven[63]. The edition was supervised by Johann Vaget (1633-1691), a former student of Jungius. Based on manuscripts in the handwriting of Christian Buncke (??-1659) that are now part of the Jungius Archive at the Hamburg State Library[64], the 'Harmonica' appeared as an appendix in the postumous *Praecipuae opiniones physicae* (Excellent Physical Ideas; Hamburg, 1679; ILL. 13.2). The short preface to the handwritten lecture notes explains that the demonstrations do not necessarily follow the Aristotelian tripartite syllogism or topics. Instead, definitions are explained through experience («per experientiam») before hypotheses and theorems are added[65].

Upon closer examination of the 'Harmonica', the aforementioned scheme appears not to have been applied generally, but specifically to the three portions of the 'Harmonica' that introduce the main part of the treatise (Statements 1-15, 16-112 and 113-140 respectively). The schemes consist, in their simplest form, of (basic) definitions followed by an 'experience' and a

[63]. See also BÜHLER 2010, p. 151.

[64]. D-Hs, MS NJJ Wo. 29, 'Collegium Apodicticum et Exemplare', which includes parts of a lecture on musical intervals (Section [3]) and 'Definitiones musicae' (Section [3a]) in the handwriting of Christian Buncke, as well as a partial fair copy of anonymous notes on a lecture about harmonics (Section [6]). Buncke's 'Definitiones' show a selection of contents from the lecture about harmonics and represent a draft for the printed 'Harmonica'. For matters of cataloguing, see MEINEL 1984, p. 118.

[65]. D-Hs, Ms. NJJ Wo. 29, Section [3], fol. 15r.

series of theorems. The third and last experience is decisive for the justification of mean-tone temperament, as it establishes the syntonic comma «as a perceptible interval» (132: «Comma est intervallum sensile»). Jungius had previously given the definition of «quartula», the quarter of a comma, as a measuring unit for the mean tone (*tonus medius*) as well as for the tempered intervals in mean-tone temperament (Statements 117-131). After the sensory experience of the comma, he could proceed with the comparison of tempered intervals (*Instrumenta simplicis concinnationis*) (Statements 133-140). The next part of the 'Harmonica' is dedicated to a series of «problems» of tuning and temperament on different musical instruments. Jungius distinguished between instruments with «simple tuning» (*Instrumenta simplicis concinnationis*: instruments with strings or pipes that produce a single pitch) like harpsichord and organ (Statements 152-154), or instruments with «double tuning» (*Instrumenta duplicis concinnationis*; instruments with strings that produce multiple pitches) like lutes and viols (Statements 155-157). The former can be tuned according to Pythagorean tuning (*scala diatonica antiqua*), just intonation (*scala diatonica nova*) or (quarter-comma) mean-tone temperament (*scala diatonica reformata*). As a final problem, the neck of the lute or the viola da gamba should be «divided into semitones, in such a way so that the sense will be satisfied»[66]. The solution begins with a short description of Galilei's 18:17 rule of equal temperament (Statements 210-215), followed by a series of theorems (see Table 13.2).

Table 13.2

Joachim Jungius's theorems 33-38 on lute tuning: original text from 'Harmonica', in *Praecipuae opiniones physicae* (Hamburg, 1679), Statements 216-228, and translation.

THEOREMATA	THEOREMS
XXXIII. [216] Octo Commata tonum Minorem non explent.	33. Eight commas do not fulfill the minor whole tone.
XXXIV. [217] Novem Commata tonum Minorem excedunt, [218] Majorem non complent.	34. Nine commas exceed the smaller whole tone, do not complete the major one.
XXXV. [219] Decem Commata tonum Majorem excedunt.	35. Ten commas exceed the larger whole tone.
XXXVI. [220] Semitonium Naturale maius est, quàm Quinque, [221] & minus, quàm Sex Commata.	36. The natural semitone is larger than five, and smaller than six commas.
XXXVII. [222] Semitonium Reformatum minus est, quam sex Commata, [223] majus est Naturali, [224] quia adjicitur quartula.	37. The reformed semitone is smaller than six commas, larger than the natural semitone, because a quarter-comma is added.
XXXVIII. [225] Octava (διὰ πασῶν) quae ex duodecim semitoniis sesquidecimae octavae proportionis componitur, [226] deficit à Verâ Octavâ dimidio commate & amplius.	38. An octave (diapason), which is composed of twelve semitones of ratio 18:17, falls short the true octave by more than half a comma.

[66]. Joachim Jungius, *Praecipuae opiniones physicae* (Hamburg, 1679), 'Harmonica', Statement 209: «*Collum* Testudinis aut Violae in semitonia sua *distinguere*, ita, ut sensui satisfiat».

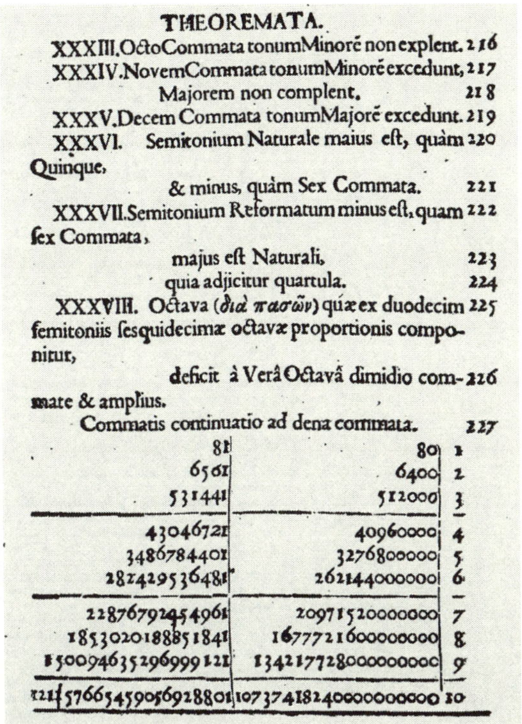

THEOREMATA.

XXXIII. OctoCommata tonumMinorẽ non explent. 216
XXXIV. NovemCommata tonumMinorẽ excedunt, 217
Majorem non complent. 218
XXXV. Decem Commata tonumMajorẽ excedunt. 219
XXXVI. Semitonium Naturale maius eſt, quàm 220 Quinque,
& minus, quàm Sex Commata. 221
XXXVII. Semitonium Reformatum minus eſt, quam 222 ſex Commata,
majus eſt Naturali, 223
quia adjicitur quartula. 224
XXXVIII. Octava (διὰ πασῶν) quæ ex duodecim 225 ſemitoniis ſesquidecimæ octavæ proportionis componitur,
deficit à Verâ Octavâ dimidio com- 226 mate & amplius.
Commatis continuatio ad dena commata. 227

81	80	1
6561	6400	2
531441	512000	3
43046721	40960000	4
3486784401	3276800000	5
282429536481	262144000000	6
22876792454961	20971520000000	7
1853020188851841	1677721600000000	8
150094635296999121	134217728000000000	9
121{5766545905692880 1	10737418240000000000	10

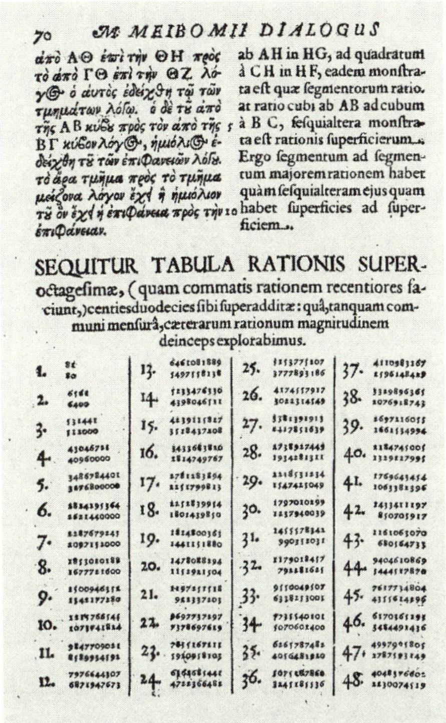

70 M. MEIBOMII DIALOGUS

ἀπὸ ΑΘ ἐπὶ τὴν ΘΗ πρὸς τὸ ἀπὸ ΓΘ ἐπὶ τὴν ΘΖ λόγος ὁ αὐτὸς ἐδείχθη τῷ τῶν τμημάτων λόγῳ. ὁ δὲ τοῦ ἀπὸ τῆς ΑΒ κύβου πρὸς τὸν ἀπὸ τῆς ΒΓ κύβον λόγος, ἡμιόλιος ἐδείχθη τοῦ τῶν ἐπιφανειῶν λόγου. τὸ ἄρα τμῆμα πρὸς τὸ τμῆμα μείζονα λόγον ἔχει ἢ ἡμιόλιον τοῦ ὃν ἔχει ἡ ἐπιφάνεια πρὸς τὴν ἐπιφάνειαν.

ab AH in HG, ad quadratum à CH in HF, eadem monſtrata eſt quæ ſegmentorum ratio. at ratio cubi ab AB ad cubum à BC, ſeſquialtera monſtrata eſt rationis ſuperficierum. Ergo ſegmentum ad ſegmentum majorem rationem habet quàm ſeſquialteram ejus quam habet ſuperficies ad ſuperficiem.

SEQUITUR TABULA RATIONIS SUPER-octageſimæ, (quam commatis rationem recentiores faciunt,) centiesduodecies ſibi ſuperadditæ: quâ, tanquam communi menſurâ, cæterarum rationum magnitudinem deinceps explorabimus.

1.	81 / 80	13.	6461081889 / 5497558138	25.	5153775207 / 3777893186	37.	4110983167 / 2596148429
2.	6561 / 6400	14.	5233476330 / 4398046511	26.	4174557917 / 3022314549	38.	3329896365 / 2076918743
3.	531441 / 512000	15.	4239115827 / 3518437208	27.	3381391913 / 2417851639	39.	2697216055 / 1661534994
4.	43046721 / 40960000	16.	3433683820 / 2814749767	28.	2738927449 / 1934281311	40.	2184745005 / 1329227995
5.	3486784401 / 3276800000	17.	2781283894 / 2251799813	29.	2218531234 / 1547425049	41.	1769643454 / 1063382396
6.	2824295364 / 2621440000	18.	2252839914 / 1801439850	30.	1797010299 / 1237940039	42.	1433411197 / 850705917
7.	2287679245 / 2097152000	19.	1824800363 / 1441151880	31.	1455578342 / 990352031	43.	1161063070 / 680564733
8.	1853020188 / 1677721600	20.	1478088294 / 1152921504	32.	1179018457 / 792281625	44.	9404610869 / 5444517870
9.	1500946352 / 1342177280	21.	1197251518 / 922337203	33.	9550049507 / 6338253001	45.	7617734804 / 4355614296
10.	1215766545 / 1073741824	22.	9697737297 / 7378697629	34.	7735540101 / 5070602400	46.	6170361298 / 3484491436
11.	9847709021 / 8589934592	23.	7855167211 / 5902958103	35.	6265787482 / 4056481920	47.	4997995805 / 2787593149
12.	7976644307 / 6871947673	24.	6362685441 / 4722366482	36.	5075287860 / 3245185536	48.	4048376602 / 2230074519

Ill. 13.3ab: Powers of the ratio of the syntonic comma (81:80) as calculated by Joachim Jungius, in 'Harmonica', Statement 227, and Marcus Meibom, *De proportionibus dialogus* (Copenhagen, 1655), p. 70. Copies D-Mbs and US-AA respectively.

The theorems begin with the description of the larger and smaller whole tones in terms of commas, and then continue with the natural and tempered semitones (Statements 216-224, Theorems 33-37). The next theorem states that an octave consisting of twelve 18:17 semitones is over half a comma smaller than a natural octave (Statement 226, Theorem 38). In other words, the issue here is whether the cumulative error exceeds the half comma, which for Zarlino represented the maximum tolerance for a good temperament[67]. We do not know if this is exactly what Jungius meant with his last theorem, but the statement is in any case true and can be verified using two tables for calculation, which were attached to the fair copy by Buncke and later included in the printed 'Harmonica'. The first table contains the calculation of the first ten powers of the ratio of the comma (Statement 227)[68]. It resembles a table in Marcus Meibom's *De proportionibus dialogus* (Copenhagen, 1655), p. 70, but Meibom truncated the

[67]. Lindley 1987, p. 175.

[68]. Joachim Jungius, *Praecipuae opiniones physicae* (Hamburg, 1679), 'Harmonica', Statement 227: a table headed «Commatis continuatio ad dena commata». Compare with D-Hs, MS NJJ Wo. 29, Section [3], fol. 9ar in the hand of Buncke. The lecture notes in D-Hs, MS NJJ Wo. 29, Section [6], contain similar calculations, only in a different context.

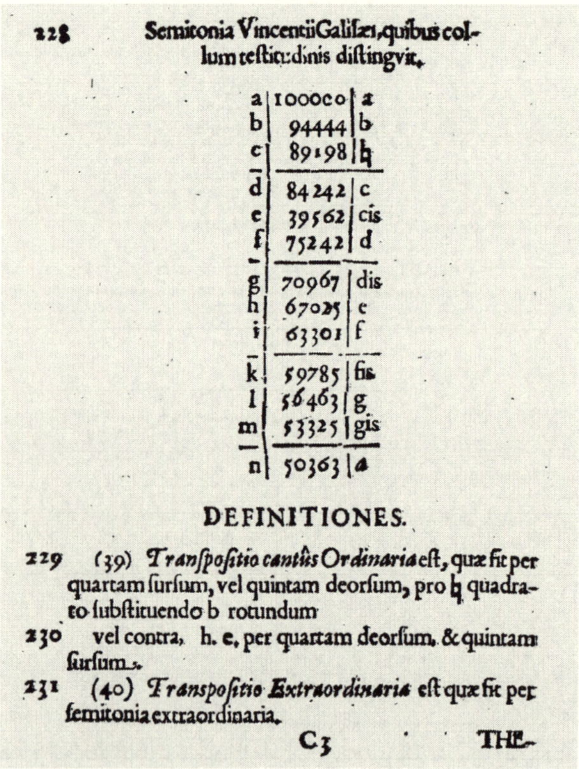

228 Semitonia Vincentii Galilæi, quibus collum testitudinis distinguit.

a	100000	a
b	94444	b
c	89198	♮
d	84242	c
e	39562	cis
f	75242	d
g	70967	dis
h	67025	e
i	63301	f
k	59785	fis
l	56463	g
m	53325	gis
n	50363	a

DEFINITIONES.

229 (39) *Transpositio cantûs Ordinaria* est, quæ fit per quartam sursum, vel quintam deorsum, pro ♮ quadrato substituendo b rotundum

230 vel contra, h. e, per quartam deorsum, & quintam sursum.

231 (40) *Transpositio Extraordinaria* est quæ fit per semitonia extraordinaria.

C3 THE-

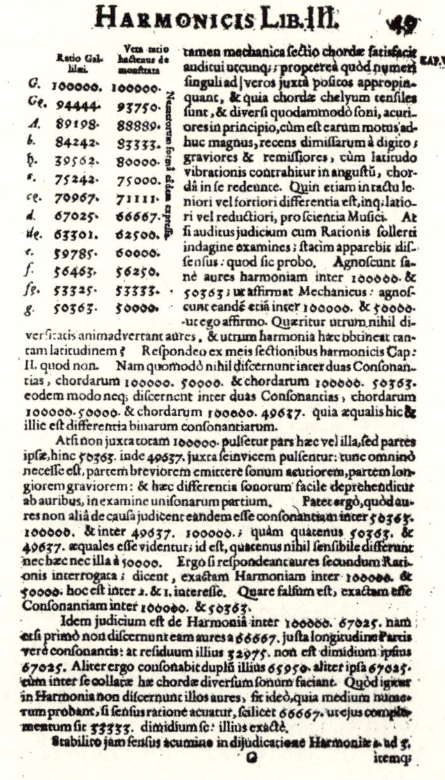

HARMONICIS LIB. III. 49

	Ratio Galilæi.	Vera ratio hactenus demonstrata
G.	100000.	100000.
Ge.	94444.	93750.
A.	89198.	88889.
b.	84242.	83333.
♮.	39562.	80000.
c.	75242.	75000.
ce.	70967.	71111.
d.	67025.	66667.
de.	63301.	62500.
e.	59785.	60000.
f.	56463.	56250.
fe.	53325.	53333.
g.	50363.	50000.

tamen mechanica sectio chordæ satisfacit auditui utcunq; propterea quòd numeri singuli ad veros juxtà positos appropinquant, & quia chordæ chelyum tensiles sunt, & diversi quodammodò soni, acutiores in principio, cùm est earum motus adhuc magnus, recens dimissarum à digito; graviores & remissiores, cùm latitudo vibrationis contrahitur in angustū, chordâ in se redeunte. Quin etiam in tactu leniori vel fortiori differentia est, inq; latiori vel reductiori, pro scientia Musici. At si auditus judicium cum Rationis sollerti indagine examines; statim apparebit dissensus: quod sic probo. Agnoscunt sanè aures harmoniam inter 100000. & 50363; ut affirmat Mechanicus: agnoscunt eandē etiā inter 100000. & 50000. ut ego affirmo. Quæritur utrum nihil diversitatis animadvertant aures, & utrum harmonia hæc obtineat tantam latitudinem? Respondeo ex meis sectionibus harmonicis Cap: II. quod non. Nam quomodò nihil discernunt inter duas Consonantias, chordarum 100000. 50000. & chordarum 100000. 50363. eodem modo neq; discernent inter duas Consonantias, chordarum 100000. 50000. & chordarum 100000. 49637. quia æqualis hic & illic est differentia binarum consonantiarum.

Atsi non juxta totam 100000. pulsetur pars hæc vel illa, sed partes ipsæ, hinc 50363. inde 49637. juxta seinvicem pulsentur: tunc omninò necesse est, partem breviorem emittere sonum acutiorem, partem longiorem graviorem: & hæc differentia sonorum facile deprehenditur ab auribus, in examine unisonarum partium. Patet ergò, quòd aures non aliâ de causa judicent eandem esse consonantiam inter 50363. 100000. & inter 49637. 100000.; quàm quatenus 50363. & 49637. æquales esse videntur; id est, quatenus nihil sensibile differunt nec hæc nec illa à 50000. Ergo si respondeant aures secundum Rationis interrogata; dicent, exactam Harmoniam inter 100000. & 50000. hoc est inter 2. & 1. interesse. Quare falsum est, exactam esse Consonantiam inter 100000. & 50363.

Idem judicium est de Harmoniâ inter 100000. 67025. nam etsi primò non discernunt eam aures a 66667. justa longitudine Partis verè consonantis: at residuum illius 32975. non est dimidium ipsius 67025. Aliter ergo consonabit duplū illius 65950. aliter ipsa 67025. cùm inter se collatæ hæ chordæ diversum sonum faciant. Quòd igitur in Harmonia non discernunt illos aures, fit ideò, quia medium numerum probant, si sensus ratione acuatur, scilicet 66667. ut ejus complementum sit 33333. dimidium sc: illius exactè.

Stabilito jam sensus acumine in dijudicatione Harmoniæ 2. ad 3, itemq;

ILL. 13.4AB: Relative string-length ratios for an octave from 100000 to 50363, calculated with a semitone ratio of 18:17, by Jungius, in 'Harmonica', in *Praecipuae opiniones physicae* (1679), and Johannes Kepler, in *Harmonices mundi libri V*, (Linz, 1619), Liber III, p. 49. The fourth figure in both tables should be 79562 (instead of 39562), the fifth figure 75142 (instead of 75242). Kepler's second table has, for comparison, the string lengths for just intonation. Copies D-Mbs.

figures after the tenth decimal, showing that Jungius cannot simply have borrowed the table from Meibom's work (ILL. 13.3)[69]. (Meibom continued the series until the 112th power and added comparisons between the sizes of ratios and the number of commas that would best approximate those sizes.)

The second table contains the string lengths of a chromatic octave down from 100000 to 50363, according to the 18:17 rule of Vincenzo Galilei (Statement 228)[70]. It is clearly

[69]. Marcus Meibom, *De proportionibus dialogus* (Copenhagen, 1655), pp. 70-71: «Tabula rationis superoctagesimae». Jungius owned a copy of this work. See MEINEL 1992, p. 154, catalogue no. 696. In 1655 Meibom sent a copy to Peter Lambeck (1628-1680), the son of Heino Lambeck (1586-1662), professor of mathematics at the St. Jakobsschule in Hamburg. According to Meibom's accompanying letter of 22 March 1655, he wanted his friend Jungius to make a critical statement on his work. See WESSELY 1980, pp. 46-47.

[70]. Joachim Jungius, *Praecipuae opiniones physicae* (Hambrug, 1679), 'Harmonica', Statement 228: a table headed «Semitonia Vincentii Galilaei, quibus collum testudinis distinguit». Compare with the table in D-Hs,

taken from Kepler's *Harmonices mundi libri v* (Linz, 1619), Book III, Chapter 8, p. 49, which deals with the 18:17 rule of Galilei, because it contains the same typographical errors for the fourth and fifth semitones (ILL. 13.4)[71]. The question arises of why the theorems remained unproven. A plausible explanation is that subsequent editing has transformed the lecture materials into a textbook, in which learners are expected to do the rest. At any rate, the edited text is conceived in terms of Jungius' «empirical geometry», but a further examination of the unedited manuscripts should help clarifying whether Jungius subjected the excerpts to a doxoscopic critique.

CONCLUSION

At this stage of research, any conclusion made from a very small sample taken from a complex and far-reaching history would obviously be premature. Further studies should certainly go deeper into the matters of reception and knowledge transfer between Italian and German music theory. More generally, we have seen the different ways in which harmonics could be defined as a «middle science»: first, between arithmetic and mathematics, and second, as a «mixed science» between mathematics and physics. The traditional distinction between *logica docens* and *logica utens*, as well as the various redefinitions and the use of demonstration, calculation, and experiment, should keep us thinking about the nature and scope of some propositions and methodological approaches. Vincenzo Galilei did not hesitate to make use of the — as Marc Lindley (1987) has put it — «well-known handmaiden of science», data embellishment, when addressing the half-comma tolerance value for musical temperament in his controversy with Zarlino[72]. German scholars like Kepler and Jungius decided to get to the bottom of the issue of Galilei's practical rule for equal temperament. This required, however, a practical approach to mathematics that was strictly dependent on experiences of hearing. Finally, we have seen how Jungius has been regarded in our times as a «radical empiricist», while he was sometimes called during his lifetime an «experimental philosopher».

The use of adequate terminology is a general problem that occurs not only in the historical research on natural sciences, but also and more specifically in the discourse about the science of music. Floris Cohen concludes in his early book *Quantifying Music* (1984) that «[t]he revolution that occurred in the science of music cannot possibly be subsumed under the heading

---

MS NJJ Wo. 29, Section [3], fol. 9a-recto, in the hand of Buncke.

71. Johannes Kepler, *Harmonices mundi libri v* (Linz, 1619), Liber III, p. 49. Jungius owned a copy of this work. See MEINEL 1992, p. 144, catalogue no. 580. The lecture notes in D-Hs, MS NJJ Wo. 29, Section [6], contain calculations until the second fret.

72. LINDLEY 1987, p. 175.

of an increase in empiricism». The main point consisted, indeed, in «a novel way of framing theories, not an increased respect of empirical data»[73]. Cohen's findings can be compared with those in Claude Palisca's well-known article on Vincenzo Galilei (1992)[74] and with Maurice Finocchiaro's fundamental distinction between active experimentation and passive observation (2020). As Finocchiaro resumes, Vincenzo Galilei was regarded by Palisca as an experimental scientist for two main reasons: first, because his experimental investigations «involved tinkering of the type required when tuning a lute or a keyboard instrument». Secondly, because these investigations also had «a theoretical aspect — namely, the search for truth or the acquisition of knowledge». Beyond that, his active experimentation was — as distinguished from the passive observation of Aristotle and his followers — «as much aprioristic as empirical. For it involved the active intervention of the investigator into the natural order and the manipulation of various natural conditions so as to create artificially produced and reproducible conditions»[75]. Reproducibility is, however, not an unproblematic point when experiences of hearing are involved: according to Cohen, «[f]or *unlike all other sciences the science of music deals with a changing subject*». In other words, the perception of a musical interval has changed over the course of history, especially «after seven centuries of harmonic development has passed»[76]. Vincenzo Galilei, Kepler, Jungius, and later the above-mentioned Tartini were themselves part of this long history. In this, they were all differently skilled in scientific research and musical practice at different stages of their respective development. As a result, the search after natural principles for a science of music was almost predestined to remain controversial, even in terms of eighteenth-century harmony and beyond.

Bibliography

Aristotle 1994

Aristotle. *Posterior Analytics*, translated and edited by Jonathan Barnes, Oxford, Clarendon Press, [2]1994.

Barbieri 1987

Barbieri, Patrizio. *Acustica, accordatura e temperamento nell'Illuminismo veneto*, Rome, Torre d'Orfeo, 1987.

Barbieri 1996

Id. 'Il mesolabio e il compasso di proporzione. Le applicazioni musicali di due strumenti matematici (1558-1675)', in: *Musica, scienza e idee nella Serenissima durante il Seicento*, edited by Francesco Passadore and Franco Rossi, Venice, Fondazione Levi, 1996, pp. 201-220.

73. Cohen 1984, p. 253.
74. Palisca 1992.
75. Finocchiaro 2020, p. 743.
76. Cohen 1984, p. 250.

Boethius 1989

Boethius, Anicius Manlius Severinus. *The Fundamentals of Music*, translated by Calvin M. Bower, edited by Claude V. Palisca, New Haven (CT), Yale University Press, 1989.

Brown 1991

Brown, Gary I. 'The Evolution of the Term «Mixed Mathematics»', in: *Journal of the History of Ideas*, lii/1 (1991), pp. 81-102.

Bühler 2010

Bühler, Walter. 'Musikalische Skalen und Intervalle bei Leibniz unter Einbeziehung bisher nicht veröffentlichter Texte (Teil i)', in: *Studia Leibnitiana*, xlii/2 (2010), pp. 129-161.

Cohen 1984

Cohen, H. Floris. *Quantifying Music: The Science of Music at the First Stage of the Scientific Revolution, 1580-1650*, Dordrecht, Springer, 1984.

Di Liscia 2002

Di Liscia, Daniel A. 'Operosum negotium: Jungius' doxoskopische Betrachtung des Aristotelismus von Zabarella', in: *La presenza dell'Aristotelismo padovano nella filosofia della prima modernità*, edited by Gregorio Piaia, Rome, Antenore, 2002, pp. 215-255.

Drake 1992

Drake, Stillman. 'Music and Philosophy in Early Modern Science', in: *Music and Science in the Age of Galileo*, edited by Victor A. Coelho, Dordrecht, Kluwer, 1992, pp. 3-16.

Feingold 2016

Feingold, Mordechai. "Experimental Philosophy'. Invention and Rebirth of a Seventeenth-Century Concept', in: *Early Science and Medicine*, xxi/1 (2016), pp. 1-28.

Finocchiaro 2020

Finocchiaro, Maurice A. 'Vincenzo's Galilei Musicology and Galileo's Science. Methodological Comparison and Contrast', in: *Isis*, cxi/4 (2020), pp. 740-758.

Gozza 2000

Gozza, Paolo. '«Desiderio da Pavia» and Renaissance Musical Theory', in: *Number to Sound: The Musical Way to Scientific Revolution*, edited by Paolo Gozza, Dordrecht, Kluwer, 2000, pp. 79-96.

Hankins 2007

Hankins, James. 'Humanism, Scholasticism, and Renaissance Philosophy', in: *The Cambridge Companion to Renaissance Philosophy*, edited by James Hankins, Cambridge, Cambridge University Press, 2007, pp. 30-48.

Herman 1973

Herman, Robert H. *Dialogo della musica antica et della moderna of Vincenzo Galilei: Translation and Commentary*, Ph.D. Diss., Denton (TX), North Texas University, 1973.

HICKS 2014

HICKS, Andrew. 'Pythagoras and Pythagoreanism in Late Antiquity and the Middle Ages', in: *A History of Pythagoreanism*, edited by Carl Huffman, Cambridge, Cambridge University Press, 2014, pp. 416-434.

HÜBNER 1996

HÜBNER, Gaby. 'Die mathematischen Reden von Joachim Jungius (1587-1657)', in: *Suddhoff's Archiv*, LXXX/2 (1996), pp. 184-197.

KREMER 1991

KREMER, Joachim. 'Schulische Musikpflege im Zeichen pädagogischer Neuorientierung. Zur Musik an der Gelehrtenschule Hamburgs im 18. Jahrhundert', in: *Archiv für Musikwissenschaft*, XLVIII/2 (1991), pp. 126-152.

LINDLEY 1987

LINDLEY, Mark. 'Stimmung und Temperatur', in: *Hören, Messen und Rechnen in der frühen Neuzeit*, edited by Frieder Zaminer, Darmstadt, Wissenschaftliche Buchgesellschaft, 1987, pp. 109-331.

MEINEL 1984

MEINEL, Christoph. *Katalog der Handschriften der Staats- und Universitätsbibliothek Hamburg. Bd. IX. Der handschriftliche Nachlaß von Joachim Jungius*, Stuttgart, Ernst Hauswedell & Co, 1984.

MEINEL 1992

ID. *Die Bibliothek des Joachim Jungius. Ein Beitrag zur Historia litteraria der frühen Neuzeit*, Göttingen, Vandenhoeck & Ruprecht, 1992.

MEINEL 2013

ID. 'Wissenschafts- und Schulreform aus dem Geist der Naturforschung. Joachim Jungius' «Syndiakritische Methode»', in: *Das Akademische Gymnasium: Bildung und Wissenschaft in Hamburg 1613-1883*, edited by Dirk Brietzke, Franklin Kopitzsch and Rainer Nicolaysen, Berlin, Reimer, 2013, pp. 31-46.

NOLAN 2002

NOLAN, Catherine. 'Music Theory and Mathematics', in: *The Cambridge History of Western Music Theory*, edited by Thomas Christensen, Cambridge, Cambridge University Press, 2002, pp. 272-303.

PALISCA 1989

PALISCA, Claude V. *The Florentine Camerata: Documentary Studies and Translations*, New Haven (CT), Yale University Press, 1989.

PALISCA 1992

ID. 'Was Galileo's Father an Experimental Scientist?', in: *Music and Science in the Age of Galileo*, *op. cit.*, pp. 143-151.

PALISCA 1996

ID. 'Legami tra Venezia e Francia nella scienza musicale del Seicento', in: *Musica, scienza e idee nella Serenissima durante il Seicento*, *op. cit.*, pp. 201-220.

Rasch 2002

Rasch, Rudolf. 'Tuning and Temperament', in: *The Cambridge History of Western Music Theory*, *op. cit.*, pp. 193-222.

Rempp 1980

Rempp, Frieder. *Die Kontrapunkttraktate Vincenzo Galileis*, Cologne, Arno Volk, 1980.

Salvia 2017

Salvia, Stefano. 'From Archimedean Hydrostatics to Post-Aristotelian Mechanics. Galileo's Early Manuscripts *De motu antiquiora* (ca. 1590)', in: *Physics in Perspective*, xix (2017), pp. 105-150.

Wallace 1992a

Wallace, William A. *Galileo's Logic of Discovery and Proof*, Dordrecht, Kluwer, 1992.

Wallace 1992b

Id. *Galileo's Logical Treatises: A Translation, with Notes and Commentary, of His Appropriated Latin Questions on Aristotle's Posterior Analytics*, Dordrecht, Kluwer, 1992.

Wallace 1995

Id. 'Circularity and the Paduan Regressus. From Pietro d'Abano to Galileo Galilei', in: *Vivarium*, xxxiii/1 (1995), pp. 76-97.

Wessely 1980

Wessely, Othmar. 'Zwei unveröffentlichte Briefe von Marcus Meibom', in: *Studien zur Musikwissenschaft*, xxxi (1980), pp. 35-48.

CHAPTER FOURTEEN

TRAJECTORIES OF MUSICAL ACOUSTICS IN SEVENTEENTH-CENTURY FRENCH THEORY AND EXPERIMENT: SOUND, NOISE, AND THE AUTHORITIES OF THE PAST

Théodora Psychoyou
(SORBONNE UNIVERSITY, PARIS)

DURING THE SEVENTEENTH CENTURY very intense research on sound was undertaken, which shed new light on music and its theory, and widened the disciplinary scope to new territories. This research took place in a context of a paradigm shift in aesthetics and musical language, but also, more broadly, in knowledge and models of science, as well as in the context of the rise of experimental methods. At the end of the century, these surveys culminated, at the institutional level in France, in the creation of a new discipline: indeed, 'Acoustics' was introduced as a distinct domain in the activities of the French Académie Royale des Sciences at the dawn of the eighteenth century. In this chapter we will discuss some aspects of this long and collective process, which is also that of exploring the nature of sound in its materiality, whether approached through the prism of music (see ILL. 14.1), mechanics, or through natural sciences. The emergence of acoustics as a discipline also required inevitably the redefinition of the boundaries of music as a field of thought and experiment. Studied systematically in relation to music, the notion of sound widened gradually, till the confines of noise; the latter notion, present among Greek authors but excluded from discussion during the Middle Ages, was proposed again, both as a concept and an object of study[1]. Thus, we will discuss some aspects of

[1]. While the Greek *psophos* (ψόφος) accounted for the physical aspect of sound, the medieval *musica speculativa* neglected the matter. The seventeenth-century causal — rational — approach revitalized the topic. I discussed the mechanisms and positions of these transitions in a previous survey (PSYCHOYOU 2006).

Ill. 14.1: Anonymous, *Hearing* (*L'ouie*) (engraving, 17.5 × 19 cm; sold by Henri Bonnart, Paris, probably 1690s). The lines below the picture read: «A tant d'appas vous joignes des accords | Qui font en nous une merveille | Car sans violenter les coprs | Vous enleues nos ames par l'oreille». (To so much charm you join your chords, Which make in us a wonder, For without hitting the bodies, You elevate our souls through the ear.) NL-Ar, RP-P-2008-452.

the status of sound and noise as discussed by seventeenth-century French authors, in particular Marin Mersenne and Claude Perrault, up to Joseph Sauveur in the early eighteenth century.

Music as 'Sound', or Music as 'Pleasant Sound'?

In 1701, Joseph Sauveur (1653-1716), holder of the first chair of Acoustics at the Académie Royale des Sciences, declared that there subsisted a field larger than music, and actually introduced the term of the new discipline: «I thus believed that there was a science superior to Music, which I called Acoustics, whose object is Sound in general, instead of Music, whose object is Sound inasmuch as it is pleasant to the sense of hearing»[2].

«The object of Acoustics is Sound», he said. This statement paraphrased, probably deliberately, through an intertextuality process, the famous opening phrase of René Descartes's *Compendium musicae* (Utrecht, 1650; written in 1618), in an early English translation: «The Object of Musick is a Sound»[3]. Let us note that when Descartes declared that the object of music was sound, we must understand, already, 'sound in its materiality'; but he clearly remained in the united disciplinary logic of the *ars musica*. Therefore, Sauveur's definition stands like a 'correction' of the one by Descartes: in 1701, sound was not the object of Music, but of Acoustics.

As we mentioned, the foundation of this academic chair, at the dawn of the eighteenth century, marked the official — that is, institutional — birth of a discipline and represented the culmination of a century of work on sound, beyond the musical work as an aesthetic object. Before Acoustics was identified and established within its proper disciplinary boundaries finally in 1700, the period that separated the statements of Descartes and Sauveur, had been one of intense development of the knowledge of sound, its nature, and the mechanisms of its generation, propagation and perception.

Indeed, questions about the nature and properties of sound were the object of numerous considerations and experiments in the entire seventeenth century, contributing to an epistemological paradigm shift, providing a new basis for the aesthetics of music derived from perception. There is a natural connection between the paradigm shift from a formal musical cosmology based on analogies to a new science based on experiment, and to the new perception-based aesthetics of music[4]. Descartes's second part of the introductory phrase of his *Compendium musicae* — «The End [of music] [is], to delight, and move various Affections in us» — evokes the mechanics of perception in order to codify — objectively, if possible — the effects of music: the real purpose of this Art. With a great economy of means, in these two sentences, Descartes drew up the theoretical and aesthetical requirements for seventeenth-century music: music in its materiality and in its ability to move the listener from one emotional state to another. The challenge then was to understand how all this worked.

---

2. Joseph Sauveur, 'Système général des intervalles des sons, & son application à tous les systèmes & à tous les instrumens de musique', in: *Histoire de l'Académie Royale des Sciences, Année M.DCCI* (Paris, 1704), Mémoires, pp. 299-366: 299: «J'ai donc crû qu'il y avoit une science superieure à la Musique, que j'ai appellée *Acoustique*, qui a pour objet le Son en general, au lieu que la Musique a pour objet le Son entant qu'il est agreable à l'oüie». Here, the adjective «superieur» should not be understood in a qualitative sense, but in the sense of generic (*versus* specific). All translations are mine unless otherwise stated. On the functioning of the Academy, see COHEN 1981. On Sauveur's work in general, see SAUVEUR 1984.

3. René Descartes, *Excellent Compendium of Musick* (London, 1653), p. 1. The full phrase reads: «A Compendium of Musick. The Object of this Art is a Sound. The End; to delight, and move various Affections in us».

4. On this connection see, in particular, PALISCA 1961.

Let us indicate here that the perspective that guided my investigations was musicological, that of the history of music theory. Reading seventeenth-century writings on music invites us, beyond the literality of the texts, to consider them in their historicity as well, in other words, in their context: the purpose of the enterprise, the intended audience, the discursive strategies, the materiality of the documents, and so on. In other words, it is also about analysing the 'discourse on'; from this perspective, we are fundamentally involved in the formulation and therefore in the construction of a system of representation. Beyond the field of musical expression, composition and performance, we are dealing with another field, that of the utterances that aim to codify the musical act, and to describe it: that it is a normative frame (a treatise or a method), a didactic, a historical or, later, an aesthetic framework, in the sense of the assumed subjectivity.

'Sound' as Music, or 'Sound' as Noise?

Intimately linked to the materiality of sound, the concept of noise, present in ancient texts on music (the *psophos*), disappears from the theoretical canons of the Middle Ages. At the beginning of the seventeenth century, research on noise gradually joined research concerning sound, until then exclusively understood as musical sound.

As René Descartes had done in his *Traité de l'homme* (Treatise on Man), written in 1633 but published only posthumously[5], Marin Mersenne explicitly distinguished the two objects, sound and noise, but did not study the latter in a specific way. The numerous pages of his *Harmonie universelle* (Paris, 1636) devoted to the nature and properties of sound essentially consider sound in the musical comprehension of the term. The study of sound occupies *in extenso* the 'Livre premier de la nature et des proprietez du son' (First Book on the Nature and Properties of Sound — that is to say, here, 'musical sound'; pp. 1-84); it is followed by the two books on «Mouvements» (pp. 85-228), to which is added the 'Traité de Mechanique' (Treatise on Mechanics) by Gilles Personne de Roberval (pp. 1-36)[6]. These texts together make up the 'Traitez de la nature des sons, et des mouvemens de toutes sortes de corps' (Treatises on the Nature of Sounds, and Movements of All Kinds of Bodies), the first major section of *Harmonie universelle* (see Ill. 14.2ab).

5. First in Latin as *De homine* (Leiden, 1662), then in French as *L'homme* (Paris, 1664).

6. On several occasions, Mersenne integrated texts by other authors — generally his friends — in his own publications, as is the case here with this short treatise on mechanics by Roberval. Another example is found in Mersenne's *Questions harmoniques* (Paris, 1634), which includes a long essay by François de La Mothe Le Vayer, 'Discours sceptique sur la musique, au R. P. Mersenne' (pp. 84-165).

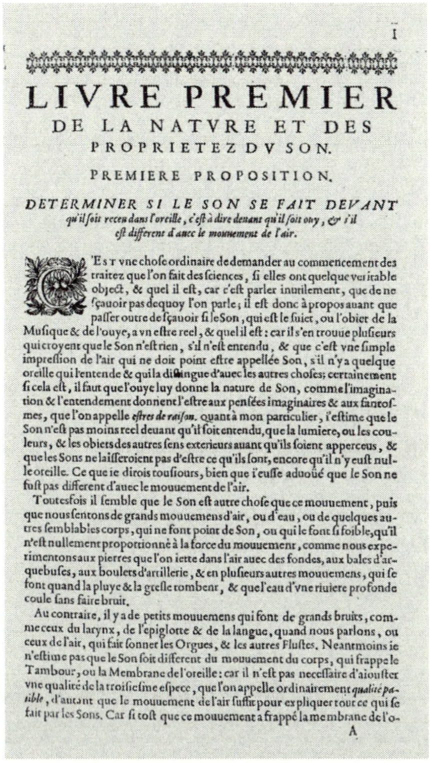

1

LIVRE PREMIER
DE LA NATVRE ET DES PROPRIETEZ DV SON.

PREMIERE PROPOSITION.

DETERMINER SI LE SON SE FAIT DEVANT qu'il ſoit receu dans l'oreille, c'eſt à dire deuant qu'il ſoit ouy, & s'il eſt different d'auec le mouuement de l'air.

C'Est vne choſe ordinaire de demander au commencement des traitez que l'on fait des ſciences, ſi elles ont quelque veritable object, & quel il eſt, car c'eſt parler inutilement, que de ne ſçauoir pas dequoy l'on parle; il eſt donc à propos auant que paſſer outre de ſçauoir ſi le Son, qui eſt le ſuiet, ou l'obiet de la Muſique & de l'ouye, a vn eſtre reel, & quel il eſt : car il s'en trouue pluſieurs qui croyent que le Son n'eſt rien, s'il n'eſt entendu, & que c'eſt vne ſimple impreſſion de l'air qui ne doit point eſtre appellée Son, s'il n'y a quelque oreille qui l'entende & qui la diſtingue d'auec les autres choſes; certainement ſi cela eſt, il faut que l'ouye luy donne la nature de Son, comme l'imagination & l'entendement donnent l'eſtre aux penſées imaginaires & aux fantoſmes, que l'on appelle *eſtres de raiſon*. quant à mon particulier, i'eſtime que le Son n'eſt pas moins reel deuant qu'il ſoit entendu, que la lumiere, ou les couleurs, & les obiets des autres ſens exterieurs auant qu'ils ſoient apperceus, & que les Sons ne laiſſeroient pas d'eſtre ce qu'ils ſont, encore qu'il n'y euſt nulle oreille. Ce que ie dirois touſiours, bien que i'euſſe aduoüé que le Son ne fuſt pas different d'auec le mouuement de l'air.

Toutesfois il ſemble que le Son eſt autre choſe que ce mouuement, puis que nous ſentons de grands mouuemens d'air, ou d'eau, ou de quelques autres ſemblables corps, qui ne font point de Son, ou qui le font ſi foible, qu'il n'eſt nullement proportionné à la force du mouuement, comme nous experimentons aux pierres que l'on iette dans l'air auec des fondes, aux bales d'arquebuſes, aux boulets d'artillerie, & en pluſieurs autres mouuemens, qui ſe font quand la pluye & la greſle tombent, & que l'eau d'vne riuiere profonde coule ſans faire bruit.

Au contraire, il y a de petits mouuemens qui font de grands bruits, comme ceux du larynx, de l'epiglotte & de la langue, quand nous parlons, ou ceux de l'air, qui fait ſonner les Orgues, & les autres Fluſtes. Neantmoins ie n'eſtime pas que le Son ſoit different du mouuement du corps, qui frappe le Tambour, ou la Membrane de l'oreille : car il n'eſt pas neceſſaire d'aiouſter vne qualité de la troiſieſme eſpece, que l'on appelle ordinairement *qualité patible*, d'autant que le mouuement de l'air ſuffit pour expliquer tout ce qui ſe fait par les Sons. Car ſi toſt que ce mouuement a frappé la membrane de l'o-

A

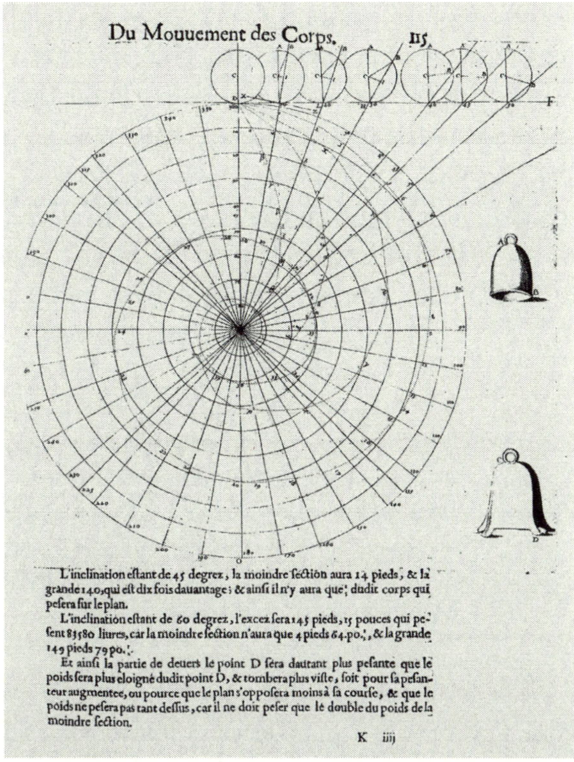

L'inclination eſtant de 45 degrez, la moindre ſection aura 14 pieds, & la grande 140, qui eſt dix fois dauantage; & ainſi il n'y aura que dudit corps qui peſera ſur le plan.

L'inclination eſtant de 60 degrez, l'excez ſera 145 pieds, 15 pouces qui peſent 83580 liures, car la moindre ſection n'aura que 4 pieds 64. po., & la grande 149 pieds 79 po..

Et ainſi la partie de deuers le point D ſera dautant plus peſante que le poids ſera plus eloigné dudit point D, & tombera plus viſte, ſoit pour ſa peſanteur augmentee, ou pource que le plan s'oppoſera moins à ſa courſe, & que le poids ne peſera pas tant deſſus, car il ne doit peſer que le double du poids de la moindre ſection.

K iiij

Ill. 14.2ab: Marin Mersenne, *Harmonie universelle* (Paris, 1636), 'Livre premier de la nature et des propriétez du son', p. 1, and 'Livre second des mouvemens de toutes sortes de corps', p. 115. Copy F-Pn.

Therefore, by extending the notion of sound to the physical object beyond musical meaning, Mersenne integrated mechanics and geometry — the ways in which musical sounds were generated — in his research programme, and not noise — the matter — as did, some years later, Claude Perrault and Joseph Sauveur. The widening of the field of investigation brought Mersenne to mechanism. In the field of sound propagation, for example, the trajectory of the echo was studied geometrically, just like, elsewhere, catoptrics[7]. We may distinguish, albeit schematically, two different approaches as follows:

Sound as a paradigm of the physical mechanisms of generation, propagation and perception	}{	Its properties constitute an application of the laws of mechanics and geometry (Mersenne [and Kircher])
Sound as 'matter' (an object generated, propagated and perceived)	}{	Its properties constitute a special case of *noise* (Perrault, Sauveur, also Francis Bacon and Kircher)

[7]. On this question in general see Lenoble 1971 and Dear 1988.

During the seventeenth century, however, it became necessary to distinguish and qualify the two terms, noise and music, explicitly, and to give them specific meanings. This need emerged primarily in the field of scientific research, more than in music theory: in a physical approach (as the production and propagation of sound) and a physiological (as the perception of sound via the hearing device up to the brain). Claude Perrault (1613-1688) devoted the second volume of his *Essais de physique* (Essays on Physics; Paris, 1680; Ill. 14.3ab) entirely to the study of noise. In the introductory note to this treatise, whose internal title was unique and unfamiliar, 'Du bruit' (On Noise), Perrault explained his terminological choice:

> I am using the word 'noise' and not that of 'sound', because having to speak in general of the quality expressed by these two words, I believed that that of noise, which, if I am not mistaken, is the generic term, should be used rather than that of sound, which I believe to mean a kind of noise: for all sound is noise and not all noise is sound. We do not properly speak of the sound of a cannon, a carriage, or a mill, because these noises are not of the kind designated by the word sound, which signifies a kind of noise whose duration surpasses that of the blow that produced it[8].

This preliminary definition, based on a physical criterion, distinguishes noise from sound as caused by a shock, for noise was a sound whose duration did not exceed «that of the blow that produced it». A few pages later, Perrault characterised more precisely what noise was, according to his approach:

> I call 'noise' the effect of a particular agitation which is produced by the encounter of two bodies, first in the surrounding air, and almost at the same time more distantly, and up to the organ of hearing. The explanation of this definition, which is the whole subject of this treatise, consists of three things, which are to know: what this particular agitation of the air is; how the encounter of two bodies produces it; and how it is made audible to the animal by the organ of hearing[9].

---

[8]. Claude Perrault, 'Du Bruit' = *Essais de physique, Tome II* (Paris, 1680), p. 1: «Je me sers du mot de Bruit et non de celuy de Son, parce qu'ayant à parler en general de la qualité exprimée par ces deux mots, j'ay crû que celuy du Bruit, qui, si je ne me trompe, est le mot du genre devoit estre plutost employé que celuy de Son, que je croy signifier une espece de Bruit: car tout son est bruit & tout bruit n'est pas son. On ne dit pas proprement le son d'un canon, d'un carrosse, ni d'un moulin, parce que ces bruits ne sont point de l'espece désignée par le mot de son, qui signifie une espece de bruit dont la durée surpasse celle du coup qui l'a produit».

[9]. *Ibidem*: «J'appele Bruit l'effet d'une agitation particuliere que la rencontre de deux corps produit premierement dans l'air voisin, & presque en mesme temps dans un plus éloigné, & jusques dans l'organe de l'ouie. L'explication de cette definition, qui est tout le sujet de ce Traité, consiste en trois choses, qui sont de sçavoir quelle est cette Agitation particuliere de l'air; comment la rencontre de deux corps la produit; & comment elle est renduë sensible à l'animal par l'organe de l'ouïe».

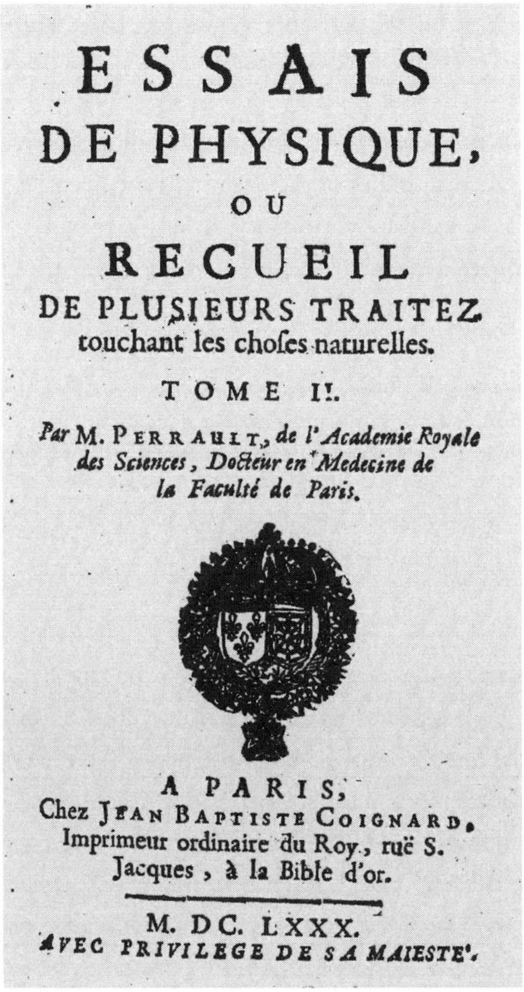

ESSAIS
DE PHYSIQUE,
OU
RECUEIL
DE PLUSIEURS TRAITEZ
touchant les choſes naturelles.
TOME II.
Par M. PERRAULT, *de l'Academie Royale des Sciences, Docteur en Medecine de la Faculté de Paris.*

A PARIS,
Chez JEAN BAPTISTE COIGNARD, Imprimeur ordinaire du Roy, ruë S. Jacques, à la Bible d'or.
M. DC. LXXX.
AVEC PRIVILEGE DE SA MAIESTE'.

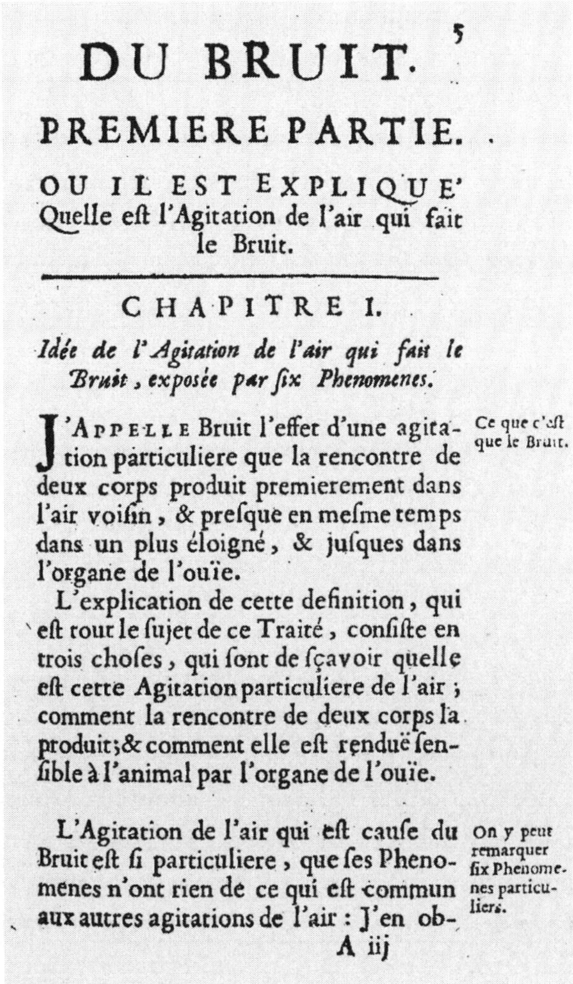

5

DU BRUIT.
PREMIERE PARTIE.
OU IL EST EXPLIQUE'
Quelle eſt l'Agitation de l'air qui fait le Bruit.

CHAPITRE I.
Idée de l'Agitation de l'air qui fait le Bruit, exposée par ſix Phenomenes.

J'APPELLE Bruit l'effet d'une agitation particuliere que la rencontre de deux corps produit premierement dans l'air voiſin, & preſque en meſme temps dans un plus éloigné, & juſques dans l'organe de l'ouïe. Ce que c'eſt que le Bruit.

L'explication de cette definition, qui eſt tout le ſujet de ce Traité, conſiſte en trois choſes, qui ſont de ſçavoir quelle eſt cette Agitation particuliere de l'air; comment la rencontre de deux corps la produit; & comment elle eſt renduë ſenſible à l'animal par l'organe de l'ouïe.

L'Agitation de l'air qui eſt cauſe du Bruit eſt ſi particuliere, que ſes Phenomenes n'ont rien de ce qui eſt commun aux autres agitations de l'air: J'en ob- On y peut remarquer ſix Phenomenes particuliers.

A iij

ILL. 14.3AB: Claude Perrault, *Essais de physique, Tome II* (Paris, 1680), title page and first page of text. Copy F-LYm.

The main object of this treatise was noise, but this conducted Perrault to consider some questions of organology and musical acoustics as well, in particular concerning the shape of musical instruments and their sounds: the sound of an instrument, even musical, could also be, in terms of physics, potentially related to noise. Perrault pursued his studies along three axes. The first one focused on vibratory phenomena linked to noise and its propagation. The second studied objects that caused noise, in other words, how their shape, their material and the way they were agitated conditioned the emitted sound. In these chapters, Perrault studied the properties of certain families of musical instruments: not surprisingly percussion, but also stringed and wind instruments, in particular brass instruments. Finally, Perrault focused on the physiology and the psychology of perception: first on the anatomy of the ear, both in humans and in certain animals — birds and fish —, and then on the relationship between the sense of hearing and the brain, notably via the functions of reasoning and memory, as well as the resulting reactions.

This voluminous essay 'Du bruit', the first of its kind ever published, in 1680, is impressive both for the quality of the demonstrations and for the originality of the ideas and the precision of the discourse. It was unanimously praised by the scientific community. Barely a year later, in 1681, Jean de Hautefeuille, engineer and inventor, confessed that he abandoned his own project for an essay on the same subject, after having read Claude Perrault's work:

> I will only add that I had started working on a small treatise upon sound, in which I explained its nature, its propagation, its reflection, and several things that belong to this subject. But the learned treatise *Du bruit*, which has just been published in the second volume of the *Essais de Physique* by Mr Perrault, of the Royal Academy of Sciences, replete with very curious remarks and experiments, and with a very exact description of the organ of hearing, surpasses anything I could write about it[10].

Following the publication of Perrault's 'Du bruit', several other essays of physical-physiological nature explored the same route, that of explorating the phenomena of sound generation and perception. In the same year, 1681, the *Journal des sçavans* — the French scholarly weekly periodical of the time — announced the publication of a book by the anatomist Joseph-Guichard Du Verney (1648-1730), the object of which would be the study of the auditory apparatus, namely the morphology and anatomy of the ear[11]. Du Verney's *Traité de l'organe de l'ouïe, contenant la structure, les usages & les maladies de toutes les parties de l'oreille* (Treatise on the Organ of Hearing, Containing the Structure, Uses and Diseases of All the Parts of the Ear) was published two years later, in 1683, and considerably complemented the study of anatomy as developed in Perrault's 'Du bruit' (ILL. 14.4AB-14.5).

---

[10]. Jean de Hautefeuille, *L'art de respirer sous l'eau* (Paris, 1681), pp. 21-22: «J'ajoûterai seulement que j'avois commencé à travailler à un petit traité du Son, dans lequel j'expliquois sa nature, sa propagation, sa réflexion, & plusieurs choses qui appartiennent à céte matiere, mais le Savant traité du bruit qui vient de paroître dans le second tome des essais de Phisique de Monsieur Perrault, de l'Académie Royale des Sciences, rempli de remarques & d'experiences tres curieuses, & d'une description tres-exacte de l'organe de l'oüie, surpasse tout ce que j'en pourrois écrire, [...]».

[11]. This announcement, in the form of a letter, appeared in the *Journal des sçavans* of 23 June 1681: 'Extrait d'une lettre escrite à l'auteur du Journal par ... touchant les Observations que M. du Verney a faites sur l'Organe de l'ouye' (Extract from a Letter Sent to the Author [=Editor] of the Journal by ... Concerning the Observations that Mr. Du Verney Made on the Organ of Hearing). The anonymous '...' could be Du Verney himself.

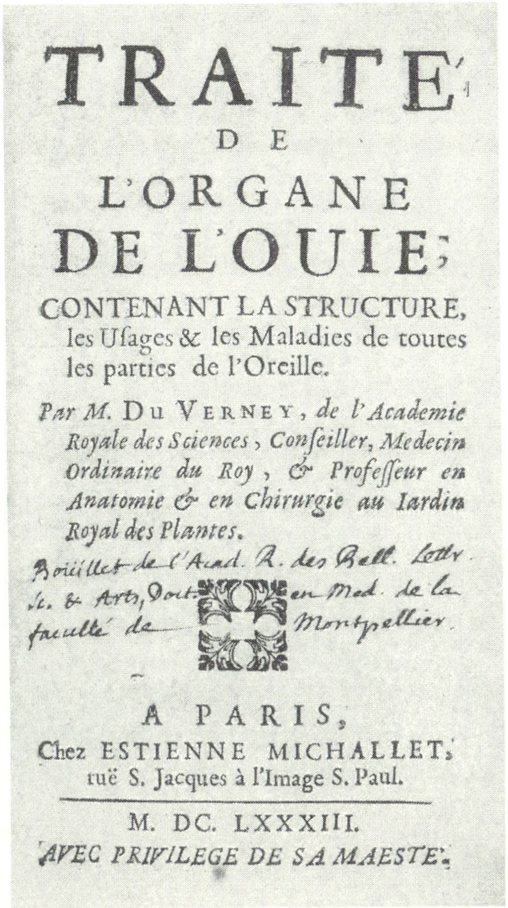

TRAITE DE L'ORGANE DE L'OUIE; CONTENANT LA STRUCTURE, les Uſages & les Maladies de toutes les parties de l'Oreille.

Par M. DU VERNEY, *de l'Academie Royale des Sciences, Conſeiller, Medecin Ordinaire du Roy, & Profeſſeur en Anatomie & en Chirurgie au Iardin Royal des Plantes.*

A PARIS, Chez ESTIENNE MICHALLET, ruë S. Jacques à l'Image S. Paul.

M. DC. LXXXIII.

AVEC PRIVILEGE DE SA MAJESTE.

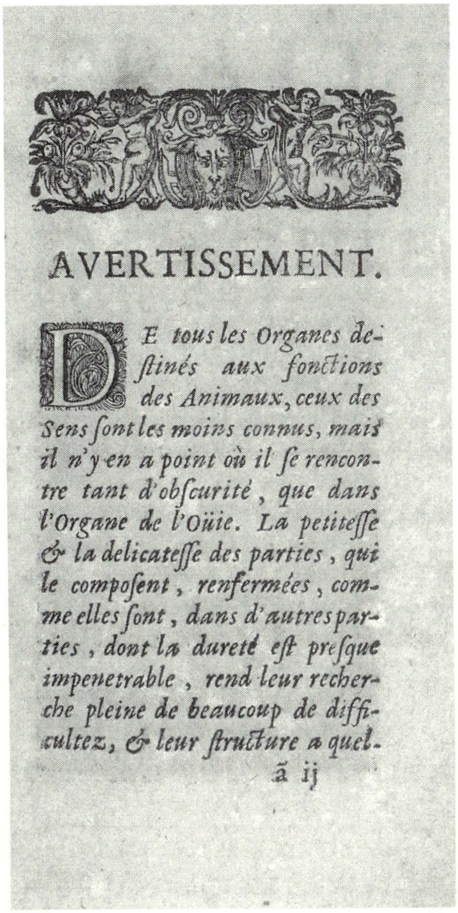

AVERTISSEMENT.

De tous les Organes deſtinés aux fonctions des Animaux, ceux des Sens ſont les moins connus, mais il n'y en a point où il ſe rencontre tant d'obſcurité, que dans l'Organe de l'Oüie. La petiteſſe & la delicateſſe des parties, qui le compoſent, renfermées, comme elles ſont, dans d'autres parties, dont la dureté eſt preſque impenetrable, rend leur recherche pleine de beaucoup de difficultez, & leur ſtructure a quel-

ã ij

ILL. 14.4AB: Joseph-Guichard Du Verney, *Traité de l'organe de l'ouïe* (Paris, 1683), title page and first page of 'Avertissement'. Copy F-Pn.

Mention must be made also of the 'Mémoire sur les Causes de la Voix de l'Homme, & de ses différens Tons' (Memoire on the Causes of the Human Voice, and it Different Tones) that the physician Denis Dodart (1634-1707) read at the Royal Academy of Sciences on 13 November 1700[12]. Contrary to Du Verney, Dodart did not study the physiology of perception, but that of the sound production, such as the vocal cords, the oral cavity and the respiratory system. In a second lecture, given on 14 April 1706, he made a comparison between the singing voice and ordinary speech, and between sound and noise[13].

12. Denis Dodart, 'Mémoire sur les Causes de la Voix de l'Homme, & de ses différens Tons', in: *Histoire de l'Académie Royale des Sciences, Année M.DCC* (Paris, 1703), Mémoires, pp. 244-283.

13. Denis Dodart, 'Suite de la Première Partie du Supplément au Mémoire sur la Voix & sur les Tons', in: *Histoire de l'Académie Royale des Sciences, Année M.DCCVI* (Paris, 1707), 'Mémoires', pp. 388-410. Particularly

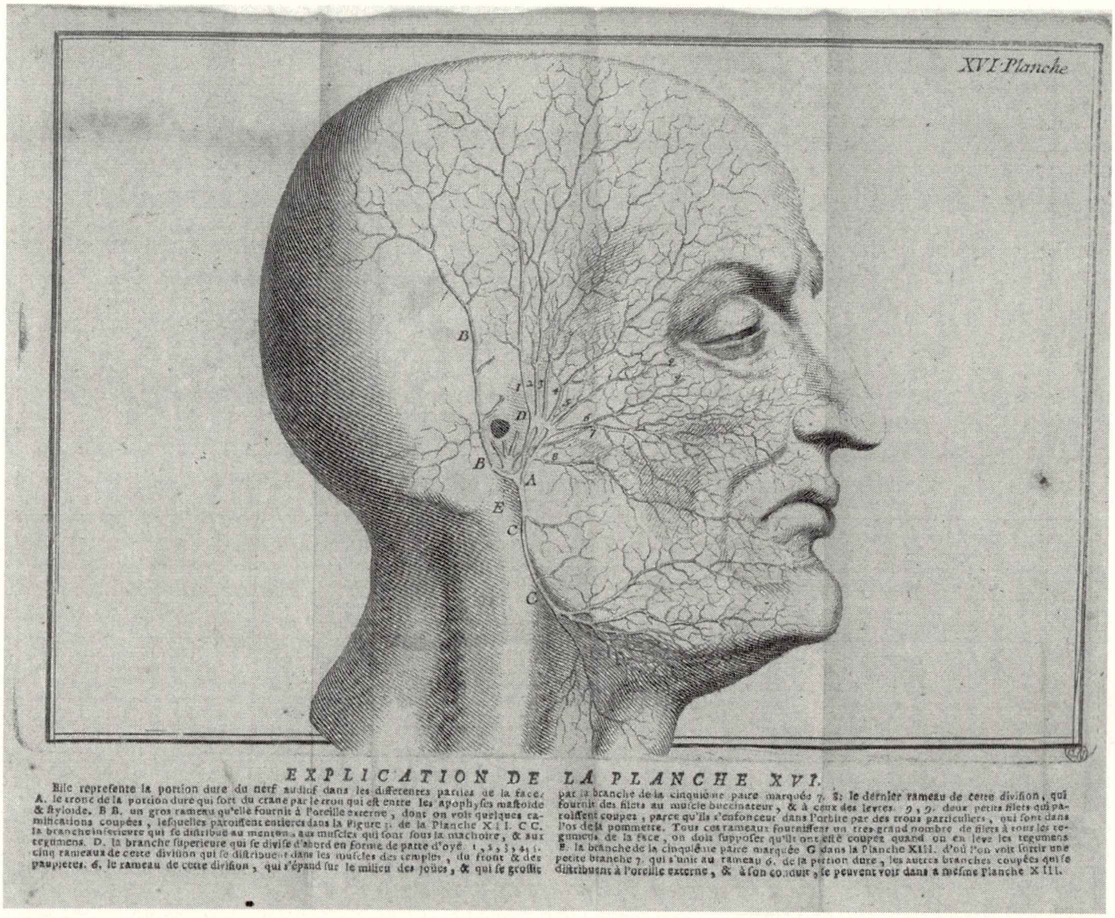

Ill. 14.5: Joseph-Guichard Du Verney, *Traité de l'organe de l'ouïe* (Paris, 1683), Plate xvi. Copy F-Pn.

In the last decades of the seventeenth century, the term 'noise' (*bruit*) seems to have acquired its current meaning, quite distinct and independent from the term 'sound' (*son*). In his *Dictionnaire universel, contenant généralement tous les mots François* (Universal Dictionary, Containing Generally All French Words), published posthumously in 1690, Antoine Furetière (1619-1688) defined noise as follows:

> Noise. Cluster of several confused and violent sounds that offend the ear. Noise is different from sound, as Mr. Perrault Physician remarked very well in the

interesting is the opening section (pp. 388-394), which is headed 'iv. Addition. De la différence des tons de la parole & de la voix du chant par rapport au récitatif, & par occasion, des expressions de la Musique antique & de la Musique moderne' (About the difference in the tones of the speech and of the singing voice compared to the recitative, & occasionally, about the expressions of ancient Music and modern Music).

> Dissertation he wrote on the sense of hearing: because we do not call sound the Noise of a cannon, a carriage, a mill, or an assembled populace[14].

Following this general definition, Furetière added some additional meanings to the term, considered figuratively. Similarly, for François Le Gallois (1613-1693), in his *Lettre à Mademoiselle Regnault de Solier touchant la musique* (Letter to Mlle Regnault de Solier on Music; Paris, 1680), 'noise' was an image referring to, for example, excessive and inappropriate virtuosity, to a type of «playing which makes a lot of noise by the quantity of passages and inappropriate diminutions»[15]. Applied to music, the term therefore is pejorative, a value judgment of taste, not only of the performer, but also of the listener, following the opinion of several commentators, such as Jean-Pierre de Crousaz (1663-1750), in his aesthetics essay *Traité du beau* (Treatise on Beauty; Amsterdam, 1715):

> The People want Vaudevilles and airs to dance to; but as far as far as overtures, chaconnes and other airs of this level are concerned, they perceive nothing but noise[16].

It appears clearly that with this new definition of noise, an object distinct from sound and a fortiori from music, the issue is a redefinition of criticism which allows subjectivity of taste. Music should always be 'beautiful and good', but there is a multitude of ways to achieve this, which the commentator could not precisely define: and in this indecipherable space unfolds the '*je ne sais quoi*' of the sublime, a new level of poetics at the end of the seventeenth century[17].

While, on the side of literature and art criticism, this concept particularly allows to theorize subjectivity — which is an oxymoron that the authors are called upon to resolve in the

---

[14]. Antoine Furetière, *Dictionnaire universel, contenant généralement tous les mots François* (The Hague, 1690), Tome premier, p. 339: «Bruit. Amas de plusieurs sons confus & violents qui offencent l'oreille. Le bruit est different du son, comme a fort bien remarqué M[r]. Perrault Medecin dans la Dissertation qu'il a fait du sens de l'ouye: car on n'appelle point son le bruit d'un canon, d'un carrosse, d'un moulin, ou d'une populace assemblée».

[15]. François Le Gallois, *Lettre* [...] *à Mademoiselle Regnault de Solier touchant la musique* (Paris, 1680), p. 55: «Combien en voit-on qui sont plus touchez d'un jeu qui fait beaucoup de bruit par quantité de passages & de diminutions faites mal à propos que par un jeu bien reglé; [...]».

[16]. Jean-Pierre de Crousaz, *Traité du beau, où l'on montre en quoi consiste ce que l'on nomme ainsi* (Amsterdam, 1715), p. 171: «Le Peuple veut des Vaudevilles & des airs à danser; mais pour ce qui est des ouvertures, des Chacones & d'autres airs de cette force, il n'y aperçoit que du bruit».

[17]. The theory of the sublime, inspired by this Greek treatise attributed to Longinus, *On the Sublime*, occupied an important place in aesthetic theory during the seventeenth and eighteenth centuries. Nicolas Boileau-Despréaux, *Œuvres* [...] *avec Le traité du sublime ou du merveilleux dans le discours, Traduit du Grec de Longin* (Paris, 1674). In a previous study, I have discussed the connection of the concept of the sublime with seventeenth-century French music and music theory (Psychoyou 2015a).

normative framework of a treatise on music — the mechanist philosophers, including Claude Perrault, and before him Descartes, Mersenne and others, investigated the objective reasons for the effects of perception.

Authorities, Modern Thought and the Crisis of the Order of Knowledge

> *The rules were invented for beauty and for grace: but if by observing them [the rules] one does not find them [beauty and grace], one is allowed to invent others in order to achieve this goal. We are no longer in the days of Pythagoras, when it was enough to say the 'The Master said it in this way': If it is permitted to dispute today against Aristotle, I believe that it is no less permitted to argue against those who established the rules of Music, considering that they have much less foundation than the principles of Philosophy*[18].

This apologetic statement by the French composer Jacques de Gouy, is highly emblematic of a status shift concerning norm and the authority of tradition. The discursive strategies of writings on music throughout the seventeenth century produced, among other things, a profound modification in the dialectics of pedagogy and persuasion (now more demonstrative than axiomatic dialectics and those based on ancient authority); this change affected directly the status of references made to the models of the past, and this because of the evolution, or even the crisis, of several parameters, and in particular three of them that we would like to mention here, inherent to seventeenth-century modern science, and interconnected. The first one is that of the authority of the Ancients and predecessors in general, as Gouy expressed in an eloquent way. In this regard, René Descartes's attitude towards the knowledge of the past is highly emblematic and significant. A second parameter concerns the status of modern writings on music. The form of the treatise on music underwent a notable change during the seventeenth century, marked by the abandoning of monumental and encyclopaedic projects, by a disciplinary reorganization and, increasingly, by the specialization and concentration on specific subjects, conveyed by lighter means, and intended for an increasingly large audience of amateurs and not reserved to some scholars. This new form authorized abandoning the

---

[18]. Jacques de Gouy, *Airs à quatre parties* (Paris, 1650), fol. [xii^{v}]: «Les reigles ont esté inventées pour la beauté & pour la grace: mais si en les observant on ne la rencontre pas, il est permis d'en inventer d'autres pour arriver à ce but. Nous ne sommes plus du temps de Pytagoras, où c'estoit assez de dire le Maistre l'a dit ainsi: S'il est permis de disputer aujourd'huy contre Aristote, je croy qu'il n'est pas moins permis de disputer contre ceux qui ont estably les reigles de la Musique, veu qu'elles ont bien moins de fondement, que les principes de la Philosophie».

formal obligation to summon the archaeology of the discipline and a whole obligatory course before reaching the main subject. Besides the structure of the writings, we can also observe transitions on the discursive and 'mental' level, that can be especially observed through the way in which new experimental methods integrated discourses on music and music theory: there was a tension between modern thought, leaning on experimental methods, and the modern way of writing, which took more and more distance from the traditional apologetic forms and, above all, from the status accorded to the authorities of the past. The figure of Mersenne is, thus, emblematic for this paradoxical situation, since he provided modern thought through an ancient form of writing — in several respects —, while Descartes adopted the opposite attitude[19]. Finally, we should keep in mind a third aspect, which is the restructuring of the *quadrivium*, and therefore of the place of *musica* as a disciplinary field within the new orders of knowledge since the beginning of the seventeenth century. Indeed, the new proposals and methodologies in the field of theory of knowledge and philosophy of sciences—those of Francis Bacon, of René Descartes — also redefined the place of music within it, in a long shift towards typologies mobilizing, eventually, from the end of the century and onwards, the idea of fine arts. In any case, the art-science dichotomy concerning music was (and is) permanent but with a different balance according to the time.

Conclusion: The Object and the End, and the Notion of Progress

Physiological research on the mechanisms of perception through the external senses, as well as the attempts to rationalize the passions of the soul, in particular by Descartes and Claude Perrault, showed that impressions by «the senses which very well knew the things that belonged to the union of soul and body», to paraphrase Descartes[20], could become known but not understood.

Firstly, they escaped the process of precise and unequivocal definitions: these «sense impressions» could not be demonstrated like a theorem of physics, they could not be totally and absolutely «reduced to its natural principles», to use Rameau's expression (and aspiration)[21]. Therefore, the power of music was the power that was felt and appreciated by each listener. This perception could not be codified and normalized in a universal form; it was subjective for each individual, it was also *relative* in terms of taste, style or fashion, place, time or nation. Beside

[19]. I have discussed this tension between thought and discursive modalities in a previous study (Psychoyou 2015b).

[20]. René Descartes to Elizabeth of Bohemia, 21 May 1643 (Descartes 1899, pp. 691-692): «et enfin, les choses qui appartiennent à l'union de l'âme et du corps, ne se connoissent qu'obscurément par l'entendement seul, ni même par l'entendement aidé de l'imagination; mais elles se connaissent très clairement par le sens».

[21]. An idea postulated in the title of Rameau's *Traité de l'harmonie réduite à ses principes naturels* (Paris, 1722).

these parameters, the main characteristics of perception that seventeenth-century theorists have opposed to universality were memory[22], imagination and habituation of the senses. In other words, perception, more particularly that of art, of 'genius', could not be definitively quantified, as was, at that same time, the circumference of the earth. The observation of the subjectivity of perception has profoundly conditioned the new scientific discourse: the truth cannot be a question of taste or fashion. Like the truth of science, that of the public also evolves. More than any other discipline, music concerns both at the same time, it is art and science.

Beyond the contingencies of the Quarrel of the Ancients and the Moderns, beyond even the epistemological shift claimed by Descartes[23] and quickly assimilated by the particularly favourable ground of music theory[24], the heritage of the Ancients became a historical object, from which the present fed itself. This is the very principle of progress: the Moderns of the seventeenth century were inevitably the Ancients of the eighteenth and later centuries to come. Their awareness took place in the realm of science; that of music played a considerable, even determining role. At the middle of the seventeenth century, in his preface to the *Traité du vide* (Treatise on Vacuum), Blaise Pascal foreshadowed how much the notion of 'progress' was an object of history:

> This is how geometry, arithmetic, music, physics, medicine, architecture, and all the sciences that are subject to experiment and reasoning, must be augmented to become perfect. The Ancients found them only sketched out by those who came before them; and we will leave them to those who come after us in a more complete state than we have received them[25].

22. And especially the memory that allowed the structuring of the discourse, the memory of the form.

23. René Descartes, *Les passions de l'âme* (Paris, 1649), p. 2: «ce que les Anciens en ont enseigné est si peu de chose, & pour la plus part si peu croyable, que je ne puis avoir aucune espérance d'approcher de la vérité qu'en m'éloignant des chemins qu'ils ont suivis». (What the Ancients have taught about it is so little, and for the most part so unplausible, that I cannot have any hope of approaching the truth, except by straying away from the paths they followed.)

24. The examples are numerous, such as Sébastien de Brossard in the article 'Temperamento' in his *Dictionaire de musique* (Paris, 1703), fol. āāā2-recto, sub verbo 'Temperamento': «Il est bon quelques fois de ne pas suivre servilement les vestiges des Anciens, & qu'il est beau même souvent d'encherir sur leurs inventions & de s'écarter de leurs traces pour suivre des routes nouvelles». (It is sometimes good not to follow slavishly the footsteps of the Ancients, and it is often beautiful even to improve their inventions and to move away from their traces to follow new paths.)

25. From 'Fragment de préface pour le *Traité du vide* [1647]' (Pascal 1954, pp. 529-535): «C'est ainsy que la geometrie, l'arithmetique, la musique, la physique, la medecine, l'architecture, et toutes les sciences qui sont soumises à l'experience et au raisonnement, doivent estre augmentées pour devenir parfaictes. Les anciens les ont trouvées seulement esbauchées par ceux qui les ont precedés; et nous les laisserons à ceux qui viendront apres nous en un estat plus accompli que nous ne les avons receues».

Bibliography

Cohen 1981

Cohen, Albert. *Music in the French Royal Academy of Sciences: A Study in the Evolution of Musical Thought*, Princeton (NJ), Princeton University Press, 1981.

Dear 1988

Dear, Peter. *Mersenne and the Learning of the Schools*, Ithaca (NY), Cornell University Press, 1988.

Descartes 1899

Descartes, René. *Correspondance. 3: Janvier 1640-Juin 1643*, edited by Charles Adam and Paul Tannery, Paris, Léopold Cerf, 1899.

Lenoble 1971

Lenoble, Robert. *Mersenne ou la naissance du mécanisme*, Paris, Vrin, [2]1971.

Palisca 1961

Palisca, Claude V. 'Scientific Empiricism in Musical Thought', in: *Seventeenth Century Science and the Arts*, edited by Hedley Howell Rhys, Princeton (NJ), Princeton University Press, 1961, pp. 91-137 [Reprinted in: Id. *Studies in the History of Italian Music and Music Theory*, Oxford, Clarendon Press, 1994, pp. 200-235].

Pascal 1954

Pascal, Blaise. *Œuvres complètes*, edited by Jacques Chevalier, Paris, Gallimard, 1954.

Psychoyou 2006

Psychoyou, Théodora. 'Du *psophos* au bruit: Sur les origines et les transformations de l'objet au xviie siècle', in: *Musurgia*, xiii/4 (2006), pp. 17-31.

Psychoyou 2015a

Ead. 'Du canon au sublime: Normes et interstices dans les écrits sur la musique en France au xviie siècle', in: *À la croisée des arts. Sublime et musique religieuse en Europe (xviie-xviiie siècle)*, edited by Sophie Hache and Thierry Favier, Paris, Classiques Garnier, 2015, pp. 367-393.

Psychoyou 2015b

Ead. 'Penser en «Moderne» en musique au xviie siècle: théorie, pratique et notion du progrès (Mersenne, Perrault, Loulié)', in: *Écrire et penser en Moderne (1687-1750)*, edited by Claudie Poulouin and Christelle Bahier-Porte, Paris, Honoré Champion, 2015, pp. 129-144.

Sauveur 1984

Sauveur, Joseph. *Collected Writings on Musical Acoustics (Paris 1700-1713)*, edited by Rudolf Rasch, Utrecht, The Diapason Press, 1984.

Chapter Fifteen

The Acoustical Paradox: How Music Was Unmoored and Set Adrift from the Sciences in the Seventeenth Century

Adam Fix[*]
(University of Minnesota, Minneapolis MN)

Regarding your question, whether the reason of beauty can be established...[1]

The science of music has long been recognized as an important facet of seventeenth-century science and natural philosophy. Isaac Beeckman, René Descartes, Christiaan Huygens, and Robert Hooke studied music's mathematical theory, while others like Johannes Kepler and Marin Mersenne contemplated its philosophical and cosmic implications. Key developments included physicalized theories of sound production and reception that rendered sensations of consonance as the mechanical stimuli of sound fell upon the eardrum. These were buttressed by an expansive experimental program in which instruments — from strings and organ pipes to drums, trumpets, and keyboards — helped philosophers link particular sounds to properties of bodies in motion. Taken at face value, these changes were predictable extrapolations of well-documented trends during the period known as the scientific revolution: the mathematization and mechanization of nature, the rise of experimentalism, and their

*. This chapter may also be found in Fix 2019, pp. 93-151.

1. René Descartes to Marin Mersenne, 18 March 1630 (Descartes 1897, p. 133; Mersenne 1945, p. 417): «Pour vostre question, sçavoir si on peut establir la raison du beau...».

synthesis into Newtonian physico-mathematical sciences. Scholars have often assumed that the scientific study of sound followed a parallel trajectory to that of optics in the seventeenth-century physicalization of sensory phenomena. Yet, while one case concerned light and vision, the other involved not just sound and hearing but *music*. Music stood apart from contemporary sciences in one crucial respect: the passions. Beyond explaining sounds and consonances in isolation, philosophers deployed mathematics and experimentalism to understand the nature of music and its myriad effects on the human soul. Their successes and failures in this quest became their lasting legacy.

This chapter investigates a particular epistemological undercurrent of the scientific revolution, one that dragged philosophers away from traditional conceptions of music and its relation to the sciences. I first analyze contemporary speculations regarding sound production and reception while juxtaposing them against older models inherited from the scholastic tradition. I then show how a new theory of consonance drove a wedge between the new natural philosophy and the concerns of musical science. Floris Cohen (1984), Jamie Kassler (2004), and others have noted a certain 'unbridgeable gap' between mathematical theories of consonance and perception that initially hindered philosophical investigations into the causes of human musical experiences[2]. Previous scholars have seen this gap as an a priori aspect of all musical science throughout history. In contrast, I investigate the historically contingent origins of this gap and its ramifications for musical science in the seventeenth century. In an era in which philosophers strove to infuse their speculations with mathematical reasoning backed by concrete experiments, musical science inquired into mysterious intangibles such as individual tastes, pleasures, and passions. I label this incompatibility of physico-mathematical theories of perception with the effects of music on the soul the 'acoustical paradox'. The paradox disrupted the traditional relationship between mathematical theory and sense perception and introduced a host of fundamental challenges to musical science. These challenges played out in dramatic fashion in the writings of three pivotal mechanical philosophers of the era: Descartes's *Compendium musicae* (MS, 1618; Utrecht, 1650), Mersenne's *Harmonie universelle* (Paris, 1636), and Huygens's 'Lettre touchant le cycle harmonique' (Rotterdam, 1691) and his scattered yet masterful writings on music that remained in manuscript. Through a close reading of these works, I argue that the acoustical paradox engendered an epistemological reorganization of musical science, practice, and the ever-tense link between the two.

The paradox was born when the new philosophy of the scientific revolution collided with the new music of the Baroque period. It began with the *mechanization of sound*, as philosophers came to depict sound as pulsations of air and the ear as a mechanism that interpreted aerial pulsations as perceptions. The accompanying theory of consonance (called the 'coincidence theory' today) rendered musical pleasure as resulting from coinciding patterns of stimuli upon

2. Cohen 1984, p. 2; Kassler 2004, p. 3.

the eardrum[3]. This theory aimed to reconcile musical mathematics with aural experiences; its most famous explication occurred in Galileo's *Discorsi e dimostrazioni matematiche intorno a due nuove scienze* (Leiden, 1638). The paradox emerged when philosophers tried to connect this physical model to musical sensations that defied such explanations, residing only within the immaterial soul. Fixed, universal laws were invoked to explain experiences that varied dramatically from place to place, era to era, and even from one observer to another. Some philosophers, like Mersenne, insisted that a reconciliation of musical theory and experience was possible. Others, including Descartes and Huygens, cautiously considered the coincidence theory but engaged with musical experience on its own terms. Having witnessed the rise of polyphony, counterpoint, and myriad other techniques of the Baroque period that stirred the passions of listeners as never before, these philosophers conceded that key aspects of music lay beyond physico-mathematical inquiry. Descartes, Mersenne, and Huygens all questioned the possibility of a musical artisanal epistemology. Their reactions to the acoustical paradox revealed that the three noble goals of the new philosophy — mathematical models, mechanico-corpuscular hypotheses, and empirical observation — could not coexist within a single musical science. Beyond any new model of sound or consonance, this schism between theory and experience was the pivotal paradigm shift in early modern musical science.

This thesis of an acoustical paradox illustrates how the new sciences, in recalibrating the parameters of proper knowledge, excised older forms of learning from philosophy altogether. Recent studies have made similar points in different contexts. Ofer Gal and Raz Chen-Morris (2012) have chronicled 'science's disappearing observer' in Johannes Kepler's optics, a process they dubbed the 'optical paradox'. As they tell it, the physicalized model of the eye (conceived as a device, a natural camera obscura) clouded the long-held conviction that vision directly rendered objects to the soul. This alienated geometric optics from human vision cast doubt about what kinds of knowledge empirical natural philosophy could grasp. Hence, «the naturalization of the eye estranges the observer, and a deeper understanding of optics turns vision into a mystery»[4]. At the heart of both the optical and the acoustical paradox lays the impossibility of mechanizing the soul. However, the science of musical sound encompassed more than mere sensory stimuli. It dealt with experiences that would today be considered aesthetic — morals, passions, and wonderous effects invoked by sound — and that had no analogue in optics. Musical science sought a rational account of these experiences in absolute, mathematical, and later physical terms (that is, not as subjective or culturally contingent). However, under a philosophy that demanded empiricism but sealed off the soul from physical nature, this goal suddenly made no sense. The privileged philosophical status of music as sensible mathematics with direct access

[3]. To the best of my knowledge, the term 'coincidence theory' originated from Cohen (1984, p. 31).

[4]. Gal – Chen-Morris 2012, p. 26. Note that the mechanical philosophy's inability to bridge the physical/perceptual chasm has been known since Eduard Dijksterhuis's work (Dijksterhuis 1950, 1961).

to human nature was nullified. The optical paradox denied men knowledge of the world; the acoustical paradox denied them knowledge of themselves.

Previous interpreters have ignored the aesthetic facets of music and flattened this historical dynamic. Cohen especially has treated the «problem of consonance» as the study of «scientific as distinguished from aesthetic» factors. Cohen's approach imposed an anachronistic division between science and aesthetics upon early modern times. In considering the seventeenth century, this led him to accuse Descartes and Huygens — both of whom expressed reservations about mathematizing musical experience — of misunderstanding the subject matter and goals of musical science. «In Descartes's view, there is no hope for science to solve the age-old problem of consonance, nor is there any point in pursuing it», Cohen (1984) noted with some condescension. Cohen overlooked that it was his own, presentist definition of consonance that he was accusing Descartes of misunderstanding[5]. By contrast, a contextualized history of music embraces the cultural embeddedness of music and asks how the scientific/aesthetic divide Cohen took as self-evident first came about. Many seventeenth-century philosophers absolutely believed that aesthetic musical experience could be analyzed scientifically; those who dissented shouldered the burden of proof. The acoustical paradox was born from this endeavor. It revealed the unique challenges early modern musical scientists faced and the kinds of knowledge they could, or could not, grasp[6].

This chapter explores music's distinct place within the scientific revolution while depicting it as one of the new sciences' first casualties. Music, mathematics, and natural philosophy all changed meanings during the seventeenth century, and the shifting dynamics between them offer key insights into the development of sciences and arts. Musical scientists, in striving to reconcile abstract theory with real experience, were instrumental in bringing together mathematical and experimental scientific traditions. However, it was this very process that stripped music of much of its former scientific credibility. Hence music and natural philosophy — interwoven since the age of Pythagoras — seemed to become unmoored and set adrift over the course of the seventeenth century. Musical science still existed, but its emphasis shifted away from the acquisition of philosophical knowledge and towards the practical needs of composers and musicians. This development set the stage for two major eighteenth-century events in the histories of science and music: the birth of *acoustics* as a science of sound distinct from music, and the reinvention of musical practice as a *fine art*, irreducible to mathematics or philosophy. Indeed, it is perhaps because of music's modern identity as a fine art that the acoustical paradox has thus far gone unnoticed. Few historians have thought to ask how music lost its once-eminent scientific status. By revealing the twin historical trajectories of seventeenth-century science and music, this chapter enriches the context for understanding the scientific revolution and its relation to the Baroque arts.

---

5. Cohen 1984, pp. 169-170.

6. For a more thorough critique of Cohen's book, see Kassler's 1985 review.

Origins of the Paradox: Scholastic Background

In scholastic philosophy, human sensation was a self-evident teleological process. The senses conveyed essential aspects of the perceived object to the soul, free from mediation or distortion. Gal and Chen-Morris (2012) have defined the 'teleological' character of medieval optics as «a direct acquaintance of the visual faculty with visible objects» that guaranteed the veridicality of vision. Under this view, optics and vision were two perspectives on the same process, which «aimed at providing adequate images of visible objects for the intellect» and portrayed God's creation — as it actually existed — directly to the soul[7]. Objects produced immaterial 'species' ('visual rays' in optics) that replicated their particular characteristics. Light conveyed species to the eye but did not itself produce perceptions. There was no possibility of a flawed or inaccurate perception because the physical properties of light and the eye played no causal role in vision. To use modern terminology, the eye, like all sense organs, was transparent: it directly interfaced with the sensible world, without tampering with perceived images, and had no agency of its own[8]. As the most accessible kind of species, visual rays provided the scholastic template for understanding other sensory faculties. Sound and hearing, however, presented altogether different challenges.

The teleology of hearing manifested itself in the Pythagorean conviction that nature favored simple number proportions, and that humans could experience these proportions through sound. Ptolemy (2nd century AD) explained consonance as arising from a correspondence between the proportions of audible music and the harmonies of the soul: «Indeed, our souls are quite plainly affected in sympathy with the actual activities of the melody, recognizing the kinship, as it were, of the ratios belonging to its particular kind of constitution»[9]. Boethius furthered this idea in the early sixth century:

> For when we hear what is properly and harmoniously united in sound in conjunction with that which is harmoniously coupled and joined together within us and are attracted to it, then we recognize that we ourselves are put together in its likeness. For likeness attracts, whereas unlikeness disgusts and repels[10].

---

7. Gal – Chen-Morris 2012, p. 21.

8. For this terminology see Tresch – Dolan 2013.

9. Ptolemy, *Harmonics*, Book iii, Chap. 7, (Düring 1930, pp. 99.25-100.7): «τοιγάρτοι καὶ ταῖς ἐνεργείαις αὐταῖς τῆς μελῳδίας συμπάσχουσιν ἡμῶν ἄντικρυς αἱ ψυχαί, τὴν συγγένειαν ὥσπερ ἐπιγινώσκουσαι τῶν τῆς ἰδίας συστάσεως λόγων». English translation from Barker 1989, p. 379.

10. Boethius, *De institutione musica*, Liber i, Cap. 1 (Boethius 1867, p. 180): «Cum enim eo, quod in nobis est iunctum convenienterque coaptatum, illud excipimus, quod in sonis apte convenienterque coniunctum est, eoque delectamur, nos quoque ipsos eadem similitudine compactos esse cognoscimus. Amica est enim similitudo. dissimilitudo odiosa atque contraria». English translation from Boethius 1989, p. 2.

Neither Ptolemy nor Boethius identified any mediation between the external phenomenon of sound and internal hearing. Just as visual rays emanated from objects and arrived at the soul through the eye, harmonious proportions emanated from strings or other sonorous objects measured by those proportions, were 'united in sound' and conveyed to the ear, and finally transmitted to the soul. If such proportions emulated the harmony of the soul, itself governed by the principles of *musica humana*, musical pleasure would be experienced. Sound (defined as 'a percussion of air') and the ear (which was 'subject to change') played only indirect roles in the soul's perception of the proportions of music. Instead, Boethius appointed reason, «which, holding itself to fixed rules, does not falter by any error», the final arbiter of musical experience[11]. He granted audible species privileged access to the soul and praised the Pythagoreans for understanding «that the whole structure of our soul and body has been joined by means of musical coalescence»[12]. Hearing thus maintained the singular *telos* of joining the incorporeal harmony of the rational soul to the corporeal harmony of the world.

In upholding this teleological understanding, subsequent scholastics were at pains to differentiate sound as perceived by the soul from the physical motions of air. Albertus Magnus (c. 1200-1280) drew on Avicenna's *Liber de anima* (c. 980-1037) and insisted that sound emanated from vibrating bodies and was conveyed by pulsations of air but was *essentially* neither of those things. As Avicenna wrote, «sound is an accident that comes from such motion, resulting from it and occurring with it», and thus possessed a different «specific nature». Avicenna seems, at first, to quibble about semantics: should 'sound' refer to external motions of air or to internal sensations of the soul? These logical maneuvers, however, were necessary to preserve Aristotle's dichotomy between sensible qualities (to which musical sound belonged) and sensible quantities (encompassing motion and extension)[13]. In choosing the latter, Avicenna and Albertus Magnus went beyond Boethius and held that qualitative musical experiences could not, by definition, be identified with quantitative motions. They thus denied the causal link between audible physical phenomena and immaterial aural perceptions. It followed that the material constitutions of different ears did not interfere with hearing: the ear was a passive receiver and transmitter but never an interpreter of sounds. Alongside their shared belief in the kinship between harmonious proportions and the harmony of the soul, this ensured the reliability of hearing, given a rational soul, to discern audible species, with no further physical cause required. As long as sound was a species only incidental to the motions of air, hearing — specifically, the hearing of music — remained a teleological process.

---

11. Boethius, *De institutione musica*, Liber I, Cap. 9 (BOETHIUS 1867, p. 195): «Quae certis regulis sese tenens nunquam ullo errore prolabitur». English translation from BOETHIUS 1989, p. 17.

12. Boethius, *De institutione musica*, Liber I, Cap. 1 (BOETHIUS 1867, p. 186): «[...] quod tota nostrae animae corporisque compago musica coaptatione coniuncta sit». English translation from BOETHIUS 1989, p. 7.

13. For more details see PASNAU 2000; the Avicenna translations are on page 33.

IO· BAPTISTAE
BENEDICTI
Patritij Veneti Philoſophi.
DIVERSARVM SPECVLATIONVM
Mathematicarum, & Phyſicarum
Liber.
Quarum ſeriem ſequens pagina indicabit.

AD SERENISSIMVM CAROLVM EMANVELEM
ALLOBROGVM, ET SVBALPINORVM
DVCEM INVICTISSIMVM.

TAVRINI, Apud Hæredem Nicolai Beuilaquæ, MDLXXXV.
Superioribus permiſſum.

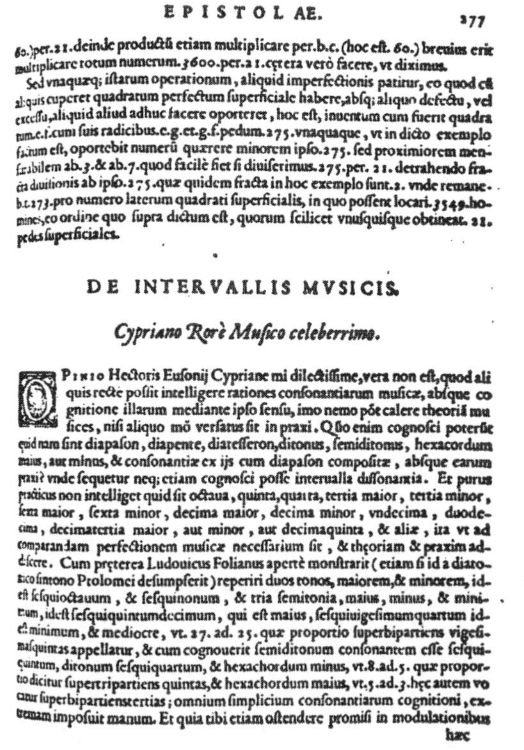

EPISTOLAE. 277

60.)per.21.deinde productũ etiam multiplicare per.b.c.(hoc eſt. 60.) breuius erit multiplicare totum numerum.3600.per.21.cętera verò facere, vt diximus.

Sed vnaquæq; iſtarum operationum, aliquid imperfectionis patitur, eo quod ſi aliquis cuperet quadratum perfectum ſuperficiale habere,abſq; aliquo defectu, vel exceſſu,aliquid aliud adhuc facere oporteret, hoc eſt, inuentum cum fuerit quadratum.e.f.cum ſuis radicibus.e.g.et.g.f.pedum.275.vnaquaque, vt in dicto exemplo factum eſt, oportebit numerũ quærere minorem ipſo.275. ſed proximiorem menſurabilem ab.3.& ab.7.quod facilè fiet ſi diuiſerimus.275.per. 21. detrahendo fracta diuiſionis ab ipſo.275.quæ quidem fracta in hoc exemplo ſunt.2. vnde remanebit.273.pro numero laterum quadrati ſuperficialis, in quo poſſent locari.3549.homines,eo ordine quo ſupra dictum eſt, quorum ſcilicet vnuſquiſque obtineat. 21. pedes ſuperficiales.

DE INTERVALLIS MVSICIS.

Cypriano Rorè Muſico celeberrimo.

OPINIO Hectoris Euſonij Cypriane mi dilectiſſime,vera non eſt,quod aliquis rectè poſſit intelligere rationes conſonantiarum muſicæ, abſque cognitione illarum mediante ipſo ſenſu, imo nemo põt calere theoriã muſices, niſi aliquo mõ verſatus ſit in praxi. Q ſſo enim cognoſci poterſſe quid nam ſint diapaſon, diapente, diateſſeron,ditonus, ſemiditonus, hexacordum maius, aut minus, & conſonantiæ ex ijs cum diapaſon compoſitæ, abſque earum praxi? vnde ſequetur neq; etiam cognoſci poſſe interualla diſſonantia. Et purus practicus non intelliget quid ſit octaua, quinta,quarta, tertia maior, tertia minor, ſexta maior, ſexta minor, decima maior, decima minor, vndecima, duodecima, decimatertia maior, aut minor, aut decimaquinta, & aliæ, ita vt ad comparandam perfectionem muſicæ neceſſarium ſit, & thęoriam & praxim addiſcere. Cum pręterea Ludouicus Folianus apertè monſtrarit (etiam ſi id à diatonico ſintono Ptolomei deſumpſerit) reperiri duos tonos, maiorem,& minorem, id eſt ſeſquioctauum, & ſeſquinonum, & tria ſemitonia, maius, minus, & minimum, id eſt ſeſquiquintumdecimum, qui eſt maius, ſeſquiuigeſimumquartum id eſt minimum, & mediocre, vt. 27. ad. 25. quæ proportio ſuperbipartiens vigeſimaſquintas appellatur, & cum cognouerit ſemiditonum conſonantem eſſe ſeſquiquintum, ditonum ſeſquiquartum, & hexachordum minus, vt.8.ad.5. quæ proportio dicitur ſupertripartiens quintas,& hexachordum maius, vt.5.ad.3.hęc autem vocatur ſuperbipartienstertias; omnium ſimplicium conſonantiarum cognitioni, extremam impoſuit manum. Et quia tibi etiam oſtendere promiſi in modulationibus hæc

ILL. 15.1AB: Giovanni Battista Benedetti, *Diversarum speculationum mathematicarum et physicarum liber* (Turin, 1585), title page and page 277, with the beginning of the first letter to Cipriano de Rore. Copy GB-Lbl.

PHYSICO-MATHEMATICAL THEORIES OF CONSONANCE

The acoustical paradox emerged when species and teleology could no longer be taken for granted. It originated in two theories that laid the foundations of a physico-mathematical science of sound: Giovanni Battista Benedetti's corpuscular theory of consonance and Isaac Beeckman's mechanical model of the ear. Benedetti (1530-1590; ILL. 15.1AB), an Italian engineer who pioneered the mathematical study of motion before Galileo, penned a series of letters starting around 1563 in which he explained consonance as «a certain equalization of percussions, or an equal concurrence of air waves» striking the ear[14]. These, in turn, were

[14]. Giovanni Battista Benedetti to Cipriano de Rore, two undated letters ('De intervallis musicis' and 'De eodem subiecto'; c. 1563?), in Giovanni Battista Benedetti, *Diversarum speculationum mathematicarum et physicarum liber* (Turin, 1585), pp. 277-283: 283: «ex quadam aequatione percussionum, seu aequali concursu vndarum aeris, [...]». See CROMBIE 1994, vol. II, pp. 789-790.

produced by coinciding vibrations of strings or other bodies. Because Benedetti explained consonance without recourse to species, Cohen has deemed this 'coincidence theory' the first physical explanation of consonance[15]. For example, an octave, formed by strings vibrating in a 2:1 frequency ratio, produces a coincidence every other vibration of the higher-frequency body; in contrast, the larger whole tone, with ratio 9:8, produces a coincidence only one out of nine vibrations and sounded dissonant. «The pleasure that consonances afford the hearing is due to the fact that the senses are softened, just as on the other hand the pain that arises from dissonances is due to their harshness», Benedetti concluded, couching his theory in explicitly physical terms with no reference to species or abstract numbers[16]. Benedetti's theory gained popularity because it neatly mapped onto the ordering of consonances implied by Gioseffo Zarlino's *senario*. Benedetti simply supplied a physical cause to a well-known mathematical principle.

A physical theory of consonance required a deeper understanding of the anatomy of the ear. Beeckman (1588-1637) is most famous today for his correspondences with Descartes and for acting as a kind of godfather to Cartesian mechanical philosophy in its early years. In 1619, Beeckman proposed a physicalized theory akin to Benedetti's and analogized the ear to a drum beaten upon by pulses of sound[17]. He noted how «the skin of a drum, when struck by a stick, gives sound because the inner air is squeezed and compressed» and later postulated an analogous «ear-drum» (*tympanum in aure*) that reverberated sound to the brain[18]. Likewise, the physicality of sound was indisputable: «never would I have called sound an immaterial species», Beeckman furiously scrawled in a 1628 journal entry, «for how can something immaterial move something material?»[19]. Contemporary species theory, articulated by Francis Bacon in 1627, posited «spiritual species (both visible and audible)» and highlighted the immateriality of music. «The sense of hearing striketh the spirits more immediately than the other senses, and more incorporeally than the smelling», Bacon noted, adding that (invoking *musica humana*) melodies pleased people «according to the sympathy they have with their spirits»[20]. Beeckman, in contrast, defined sound (*sonus*) not as a perception (caused by species)

[15]. I am indebted to Cohen's (1984, pp. 75-78) analysis of Benedetti's musical science.

[16]. Benedetti to Cipriano de Rore, undated letter 'De eodem subiecto' [= De intervallis musicis] (c. 1563?), in Benedetti, *Diversarum speculationum liber* (1585), p. 283: «Voluptas autem, quam auditui afferunt consonantiae, quia leniuntur sensus, quemadmodum contra, dolor qui à dissonantijs oritur, ab asperitate nascitur, [...]». See Crombie 1994, vol. ii, pp. 789-790.

[17]. Beeckman's rendition of the coincidence theory can be found in his *Journal*, 1615 (Beeckman 1939, p. 52).

[18]. Beeckman, *Journal*, June 1619 (*ibidem*), p. 313: «Sic corium tympani baculo pulsum, edit sonitum quia aer intrinsecus in se pellitur, cogitur, [...]». I am also indebted to Cohen's (1984, p. 140) analysis and translations of Beeckman's journal. Beeckman explicitly compares the ear to a drum in 1631 (Beeckman 1945, p. 187).

[19]. Beeckman, *Journal*, fol. 321r, July 1638 (Beeckman 1945, p. 67): «Nunquam enim ego sonum vocaverim speciem immateriatam. Quî enim immateriatum movere possit materiatum»? English translation from Cohen 1984, p. 134.

[20]. Francis Bacon, *Sylva Sylvarum, or A Natural History, in Ten Centuries* (London, 1627), pp. 36, 32.

but as a material phenomenon caused by the motion of projectile-like «globules» (*globuli*) of air and their impingement against the eardrum[21]. He concluded that the mind (*animus*) perceived not sound itself but the physical beatings on the device of the eardrum, transmitted via vital spirits through auditory nerves to the brain. Beeckman never attempted a physicalized theory of the mind, which he regarded as a perceptual faculty (divorced from any notion of *musica humana*) whose materiality was beyond the scope of his inquiry. He claimed only that, if the beatings of two sounds sufficiently coincided, consonance would — somehow — be perceived by the mind.

Galileo, who was familiar with Benedetti's work, published his take on the coincidence theory in his *Discorsi e dimostrazioni matematiche intorno a due nuove scienze*. Sagredo, the inquisitee layman of the dialogue, observed that the pitch and tension of strings varied in square proportions rather than direct, linear proportion. He proclaimed that the number ratios of the *senario* could not «contain» (*contenere*) musical intervals[22]. Galileo's mouthpiece Salviati answered that it was the «titillation of the eardrum (*timpano*) which in the mind (*anima*) becomes sound»[23]. Salviati then explained his physico-mathematical theory of consonance: «those pairs of sounds will be consonant, and heard with pleasure, which strike the eardrum with good order; this requires first that the impacts made within the same period are commensurable in number, so that the cartilage of the eardrum need not be in a perpetual torment of bending in two different ways to accept and obey ever-discordant beatings»[24]. To illustrate, he offered two experiments: a goblet filled with water, whose ripples doubled as its pitch jumped an octave; and a copper plate, which when scraped by a chisel at two different speeds sounded a fifth while the chisel left, respectively, 30 and 45 streaks on the plate (a 2:3 frequency ratio). Galileo's (questionable) experimental trials mattered less than his line of argumentation from reason, sensation, and instruments to hidden natural principles — in this case, the hidden link between frequency and consonance[25]. According to Galileo, coinciding vibrations tickled the

[21]. Although natural philosophers had not yet concurred on whether sound was a wave or projectile, all accepted that it involved some motion or impulse through physical air. Beeckman's word is 'globulus', that is, little sphere, globule, or corpuscle. See Beeckman to Marin Mersenne, 30 May 1633 (Beeckman 1953, p. 214; Mersenne 1969, p. 403).

[22]. Galileo Galilei, *Discorsi e dimostrazioni matematiche intorno a due nuove scienze* (Leiden, 1638), p. 100. English translation from Galilei 1974, p. 101.

[23]. Galileo, *Discorsi e dimostrazioni* (1638), p. 101: «la titillazione sù 'l timpano del nostro orecchio, la quale nell'anima ci doventa suono». English translation from Galilei 1974, p. 102.

[24]. Galileo, *Discorsi e dimostrazioni* (1638), p. 104: «Consonanti, e con diletto riceuute saranno qualle coppie di suoni, che verranno à percuotere con qualche ordine sopra 'l timpano; il qual'ordine ricerca prima, che le percosse fatte dentro all'istesso tempo siano commensurabili di numero, acciò che la cartilagine del timpano non habbia à star'in un perpetuo tormento d'inflettersi in due diverse maniere per acconsentire, & ubbidire alle sempre discordi battiture». English translation from Galilei 1974, p. 104.

[25]. Apparently, making glasses and chisels behave in this way is much harder than Galileo let on. See Walker 1978, p. 30, and Cohen 1984, pp. 88-90.

ear (without directly affecting the soul), dissonance arose from excessive and painful bending (*inflettersi*) of the eardrum, and number ratios themselves were not intrinsically consonant or dissonant. These ideas echoed Galileo's earlier conviction from 'De motu antiquiora' (MS c. 1590, unpublished during his lifetime), that he did «not believe that there exists anything in external bodies for exciting tastes, smells, and sounds, etc. except size, shape, quantity, and motion» of material particles[26].

Galileo completed the shift of musical ontology away from numerology and into sonorous bodies and human experiences. An interval could not be said to exist without ears to interpret the 'good order' of sounds. Moreover, by pondering the 'kiss', 'bite', and 'fire' conveyed by certain tones, Galileo never let mathematical theory blind him to music's incalculable affective powers that only audible experience could judge[27]. Just as he reduced color to the shapes and speeds of light corpuscles, Galileo reduced musical tones to the pulsations of air striking the ear. Like Beeckman, Galileo did not investigate the perceptual faculty that lay beyond the ear, where aural sensations were converted into pleasures and passions in the soul. Upon receiving Galileo's endorsement both the coincidence theory and the mechanical depiction of the eardrum became commonplace within early modern natural philosophy.

Following Galileo, the problem of consonance shifted from a purely mathematical to a physical and anatomical question. Joseph-Guichard du Verney's *Traité de l'organe de l'ouïe* (Treatise about the Organ of Hearing; Paris, 1683) provided copious illustrations in which the ear was disassembled and analyzed like a device (ILL. 15.2). The English anatomist Thomas Willis (1621-1675) noted how «the impulse of the sound, shaking this membrane like a drum, delivers the impression to the sonorifick particles planted beyond»[28]. Kenelm Digby (1603-1665) analogized the ear to another instrument: «in a lute, as long as a string that hath been strucken shakes sensibly to our eye; so long, and to the same measure, the sound shakes in our ear»[29]. Robert Hooke (1635-1703) had defined sound as «nothing else but a tremulous motion of the drum and organ of the ear, excited by the like motion of the sonorous medium». Marveling at «all the curious mechanisms of that admirable organ», he portrayed the ear as the pinnacle of nature's ingenuity. «How rarely nature has in this play'd the part both of an excellent geometrician & mechanick», he concluded with appreciable awe[30]. In defining the ear as both mechanical and geometrical, Hooke rendered it akin to artificial constructs such as the telescope (and hence the eye) or his own microscopes and pocket watches. This conviction inspired Hooke to design ear trumpets and other hearing aids — auditory equivalents of the

26. GALILEI 1960, p. 48.

27. Galileo, *Discorsi e dimostrazioni* (1638), p. 106. English translation from GALILEI 1974, p. 107.

28. Thomas Willis, *Two Discourses Concerning the Soul of Brutes* (London, 1683). See WARDHAUGH 2008, p. 79.

29. Kenelm Digby, *Of Bodies, and of Mans Soul* (London, 1669). See WARDHAUGH 2008, p. 76.

30. Robert Hooke, 'A Curious Dissertation Concerning the Causes of Power & Effects of Musick' (MS, London, Royal Society, Classified Papers, Vol. II, no. 31). See GOUK 1980, p. 601.

telescope — that might replicate nature's ingenuity. All these natural philosophers explored the role of instruments in the reception and interpretation of sound. In their accounts, the distinction between auditory sensory organs and manmade instruments collapsed, and the soul was written out of the equation[31].

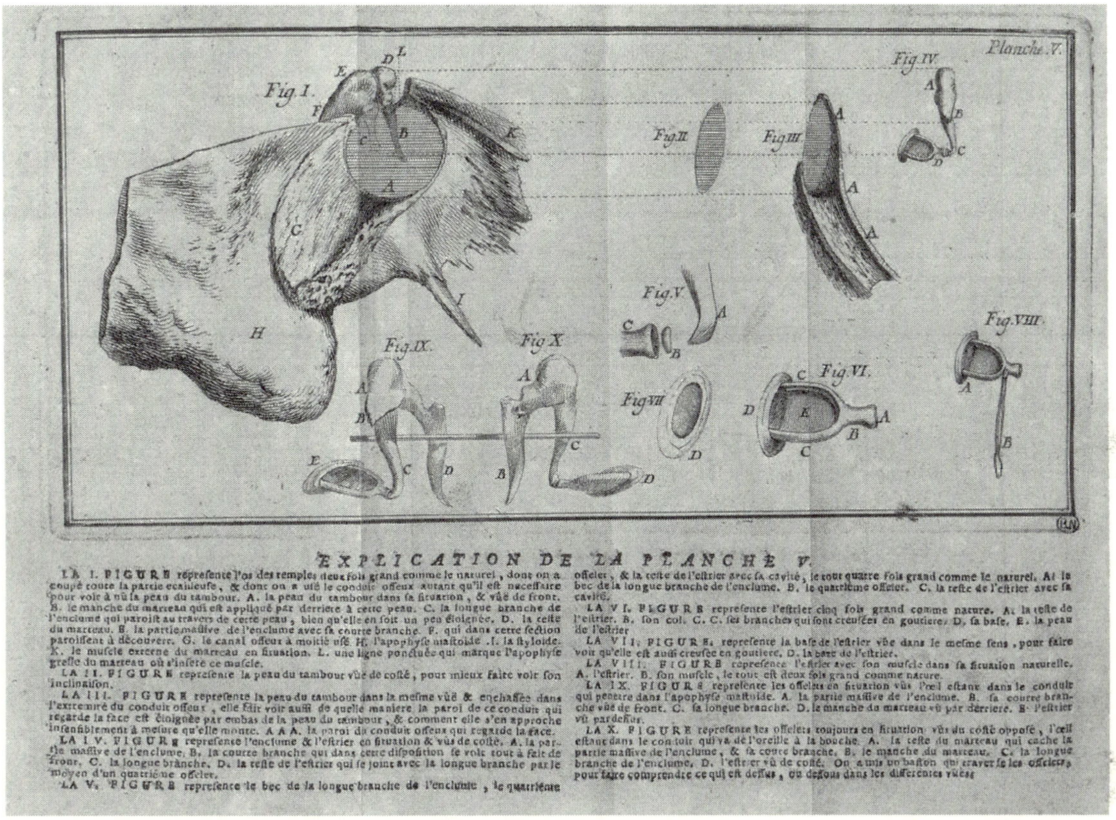

Ill. 15.2: Joseph-Guichard Du Verney, *Traité de l'organe de l'ouïe* (Paris, 1583), Planche v, with a diagram of the ear, showing its mechanical structure. Copy F-Pn.

These physico-mathematical theories redefined the ontology and epistemology of music. Ofer Gal and Raz Chen-Morris (2012) have stressed how Johannes Kepler (1571-1630), in his *Dioptrice* (Dioptrics; Augsburg, 1611), redefined 'species of things' as mere 'light and shadow' cast upon the blank screen of the retina. This abolished the «privileged relation between source and image» inherent in medieval optics, and thus «the intentionality and teleology of vision

31. The most notable seventeenth-century study of the ear on the Continent was Joseph-Guichard du Verney's *Traité de l'organe de l'ouïe* (Paris, 1683), available in an English translation by John Marshal published as *A Treatise of the Organ of Hearing* (London, 1737).

are lost»[32]. Likewise in music, audible species were revealed to be nothing but sound, nothing but percussions on the eardrum. Consonances were transformed from harmonious proportions in abstraction to frequency ratios of concrete vibrating bodies. Hence, an unprecedented chasm opened up between audible phenomena and aural perceptions. Each new discovery Hooke made, evidencing the marvelous mechanics of the ear, eroded any belief that hearing directly transmitted sensations to the soul. The eardrum, no matter how immaculately designed, insulated human listeners from music, filtering their experiences and mediating their interactions. Teleology of design supplanted teleology of sensation: the structure of the ear — *not* species, now rendered meaningless — mediated hearing. That the design was Godly, and therefore perfect, did not alter the reality that the soul had lost its privileged relationship with musical number and proportion. The disturbing question of how the soul converted movements of the eardrum into musical pleasures and passions became a mystery and was pushed beyond the goals and aims of natural philosophy. Thus, much as the optical paradox severed the epistemological ties between optics and vision, the acoustical paradox disrupted the concord between musical science and musical experience. The physicalization of the ear estranged the human listener. Philosophers all agreed that the soul never even directly experienced sound.

KEPLER'S OBJECTION

All, that is, except Kepler. In his *Harmonices mundi libri v* (Five Books on the Harmonics of the World; 1619; ILL. 15.3AB), Kepler not only rejected the physicalized theory of consonance but seemed offended by the idea:

> For what, I ask, is the proportion of the titillation of the hearing, a corporeal thing, to that unbelievable pleasure, which we feel totally within the mind from harmonic consonances? [...] The operations and motions of bodies, which imitate the harmonic proportions, are on the side of the soul and the mind, assigning them a cause for their delight in consonance [...] It is Mind or the human intellect by the judgment or instinct of which the sense of hearing discriminates pleasant, that is consonant proportions from the unpleasant and the dissonant [...] Proportions are entities of reason, perceptible by reason alone, not by sense, and that to distinguish proportions, as form, from that which is proportioned, as matter, is the work of mind[33].

32. GAL – CHEN-MORRIS 2012, pp. 23-24.

33. Johannes Kepler, *Harmonices mundi libri v* (Linz, 1619) Liber III, Cap. 1, pp. 15-16: «Nam quae quaeso proportio titillationis auditus, rei corpereae, ad incredibilem illam voluptatem, quam ex harmonicis Consonantijs intus in animo penitissimè percipimus. [...] Opera Motusque corporum, aemuli proportionum harmonicarum, ab Animae, Mentisque partibus stent, ijs causam assignantes, cur Consonantiae delectent. [...] Mentem esse,

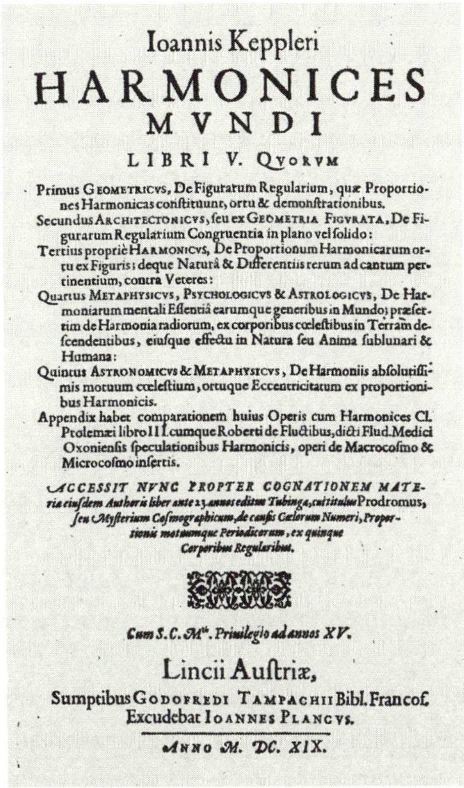

Ioannis Keppleri

HARMONICES MVNDI

LIBRI V. QVORVM

Primus GEOMETRICVS, De Figurarum Regularium, quæ Proportiones Harmonicas constituunt, ortu & demonstrationibus.
Secundus ARCHITECTONICVS, seu ex GEOMETRIA FIGVRATA, De Figurarum Regularium Congruentia in plano vel solido:
Tertius propriè HARMONICVS, De Proportionum Harmonicarum ortu ex Figuris; deque Naturâ & Differentiis rerum ad cantum pertinentium, contra Veteres:
Quartus METAPHYSICVS, PSYCHOLOGICVS & ASTROLOGICVS, De Harmoniarum mentali Essentiâ earumque generibus in Mundo; præsertim de Harmonia radiorum, ex corporibus cœlestibus in Terram descendentibus, eiusque effectu in Natura seu Anima sublunari & Humana:
Quintus ASTRONOMICVS & METAPHYSICVS, De Harmoniis absolutissimis motuum cœlestium, ortuque Eccentricitatum ex proportionibus Harmonicis.
Appendix habet comparationem huius Operis cum Harmonices Cl. Ptolemæi libro III. cumque Roberti de Fluctibus, dicti Flud. Medici Oxoniensis speculationibus Harmonicis, operi de Macrocosmo & Microcosmo insertis.

ACCESSIT NVNC PROPTER COGNATIONEM MATERIÆ eiusdem Authoris liber ante 23 annos editus Tubingæ, cui titulus Prodromus, *seu Mysterium Cosmographicum, de causis Cælorum Numeri, Proportionis motuumque Periodicorum, ex quinque Corporibus Regularibus.*

Cum S. C. Mtis. Privilegio ad annos XV.

Lincii Austriæ,

Sumptibus GODOFREDI TAMPACHII Bibl. Francof.
Excudebat IOANNES PLANCVS.

ANNO M. DC. XIX.

ILL. 15.3AB: Karl Barth (1828-1853), *Johannes Kepler* (steel engraving, 14.5 × 16 cm). NL-Ar, RP-P-1914-1102; Johannes Kepler, *Harmonices mundi libri v* (Linz, 1619), title page. Copy D-Mbs.

While Kepler saw no difficulty in supplanting visible rays with physical light and shadow, the equivalent reduction of harmonious proportions to physical sound proved beyond the pale. There was no depersonalized blank screen, no auditory equivalent of the camera obscura by which to analyze sounds apart from human listeners. What physical device, after all, could embody musical pleasures and emotions? Instead, Kepler explicitly defined sound as species caused by — but distinct from — motions of material things. «Sound is a species given out by a body», he affirmed in *Harmonices mundi libri v*, «in accordance with [the body's] size, and to a certain extent how it is shaped, and how it is established in motion»[34]. He upheld the

Animumque humanum, cujus seu judicio seu instinctu, sensus auditus proportiones suaves, hoc est consonas, ab insuavibus & dissonis discernat; [...] Proportiones esse Rationis Entia, ratione sola, non sensu, perceptibilia, & proportiones ceu formam, à re proportionata, ceu materiâ, distinguere, Mentis opus esse». English translation from KEPLER 1997, pp. 149-150. Emphasis mine.

34. Kepler, *Harmonices mundi* (1619), Liber IV, Cap. 3, p. 127: «[...] nam sonus est species emissa è corpore, & ut quantum illud, & ut figuratum quadamtenus, & in motu constitutum est.». English translation from KEPLER

traditional, teleological view of hearing by portraying musical intervals as immaterial 'entities of reason', which only the 'human intellect' could judge. To Kepler, as to the scholastics, audible music was the sensation of proportions, conveyed directly to the soul.

Kepler based his theory of consonance on an unyielding faith in the essential geometry of nature. Locating the ontology of music in abstract mathematical archetypes rather than physico-mathematics, Kepler derived the consonances geometrically by comparing inscribed polygons and the arc lengths of their respective circles[35]. For instance, the pentagon, dividing a circle into five parts, produces arc lengths four-fifths and three-fifths of the circumference, and thus generates the major third (5:4) and major sixth (5:3). Kepler's famous pronouncement that «geometry [...] is coeternal with God» assured his readers that rational souls could interpret these proportions because God had crafted soul and universe alike according to primordial geometric archetypes[36]. This theory was mathematical but *not* physical: the physics of sound played no causal role in the sensation of consonance. As such, Cohen has labeled this Kepler's 'archetypical', or 'intellectual' harmony', which, as an activity of the soul, existed outside the sense of hearing proper[37]. This passage affirms Kepler's attitude towards the ontology of music: certain sounds (and other phenomena such as the orbits of the planets) were intrinsically harmonious; observers came to know this harmony through a mental, not sensible or bodily, process. By appealing to the metaphysical primacy of geometry, Kepler maintained the teleological connection between reason and musical sensation.

Kepler's objection exemplified that the obstacles presented by hearing and vision within the seventeenth-century reconceptualization of the senses were not at all symmetrical. Abolishing visual rays, and with them the veridicality of human vision, was for Kepler trivial compared to relinquishing musical proportions as the bond between God's universe and the human soul. Referencing earlier work in his *Dioptrice*, he allowed that other faculties could be construed as «titillations» upon sense organs:

1997, p. 317. I have altered the translation of Aiton *et alii*. Kepler wrote 'sonus est species' but the translation replaces 'species' with 'emanation'. Aiton *et alii* explain in their choice in the «Translators' notes» section as accentuating the Neoplatonic notion of 'species' as 'a power propagated by a body'. I have chosen to preserve Kepler's term, which invokes the scholastic understanding of sense perception as immaterial species that mimic properties of objects. See KEPLER 1997, p. xli. David LINDBERG (1986, p. 35), in considering Kepler's optics, has argued that Kepler would never «employ the archaic terminology of species while divesting it of its traditional content». In the case of sound and music, at least, I agree with Lindberg.

[35]. Kepler defined 'constructible' in the Euclidean sense: only polygons that could be drawn with compass and straightedge were truly knowable and valid bases for consonance. See Kepler, *Harmonices mundi* (1619), Liber III, Cap. 1, p. 10 (KEPLER 1997, p. 143: «The Causes of Consonances»).

[36]. Kepler, *Harmonices mundi* (1619), Liber III, Cap. 1, Axioma VI, p. 13 (KEPLER 1997, p. 146). For Kepler's notion of geometrical archetypes, see WALKER 1967.

[37]. COHEN 1984, p. 25.

> Surely if any pleasure does come from the titillation, the chief participant in that pleasure is the organ which undergoes the titillation? For it seemed to me that every sense should be designed in this way, in the *Dioptrics*, because the particular sensation is complete, generating pleasure or pain, when the emanation (*species*) of the organ which is ordained from that sensation, as it is affected by the external circumstance, comes within the tribunal of the common sense, by the passage of the spirits[38].

Not so for hearing. «Yet in the hearing of consonant notes or sounds», Kepler interrupted himself, «what parts of the pleasure, I ask, are attached to the ears?»[39]. Aware of the explanatory power of the coincidence theory, Kepler feared the Pandora's box it would open by sequestering the soul behind the organ of the ear. He sought to beat back the acoustical paradox via a teleological account of hearing that preserved music's status as direct, sensible mathematics — an intermediary between corporeal and incorporeal realms[40]. To Kepler, the 'unbelievable pleasure' of music transcended other senses; it required transcendental treatment.

As with much of *Harmonices mundi libri v*, Kepler pushed this line of research as far as it could go. Later philosophers, denouncing species and situating consonance within a mechanico-corpuscular worldview, had no such recourse. Musical science in the following decades was characterized by efforts to reconcile newly physicalized theories with the vast range of musical experience in the soul. Descartes, whose philosophical career began with his oft-overlooked treatise on music, became the first to tackle the acoustical paradox head on.

Reactions to the Acoustical Paradox: Descartes's *Compendium Musicae*

Descartes wrote his *Compendium musicae* in 1618 — one year before Kepler published his *Harmonices mundi libri v* — as a New Year's gift to his friend Beeckman. He closed the treatise by begging Beeckman to keep it «forever hidden in the privacy of your desk or your library», fearing that others «might not, as I trust you will, turn from these fragments and look with good will at those writings in which I can say that at least some characteristics of

---

38. Kepler, *Harmonices mundi* (1619), Liber iii, Cap. 1, p. 15: «Nonne si qua voluptas est ex titllatione, de ea voluptate primas tenet membrum, quod titillationem sustinet? Sic enim definiendus mihi visus est sensu omnis, in Dioptricis, quod ea demum sit absoluta sensio, voluptatem aut dolorem pariens, cum species Membri sensioni destinati, ut id est affectum ab externa re, venit introrsum ad sensus communis tribunal, commeatu spirituum». English translation from Kepler 1997, p. 149.

39. Kepler, *Harmonices mundi* (1619), Liber iii, Cap. 1, p. 15: «Iam verò in auditione consonantium vocum vel sonorum, quae quaeso partes voluptatis in auribus haerent?». English translation from Kepler 1997, p. 149.

40. In various alchemical and magical traditions, the same place was given to a 'spirit' of sorts, that partook of the material and immaterial realms. By this analogy, music was for Kepler 'spiritual'.

my talent find accurate expression»[41]. Descartes did not wish posterity to judge him based on this work alone, fears many historians have validated by dismissing it as youthful enthusiasm rather than serious natural philosophy. Recently, however, Kate van Orden (2013) has shed new light on the *Compendium* through the lens of cultural history. She emphasizes music's place in seventeenth-century French society as a cultural institution and bodily experience: it charmed noblemen and commoners alike, invigorated listeners into song and dance, and a steady drumbeat could (in an almost mechanical fashion) spur even the most ragtag band of soldiers to march in lockstep. As Van Orden put it, Descartes practiced «a dualism of rational inquiry and bodily perception»[42]. This was no estranged dualism, wherein reason and perception presided over mutually exclusive realms, but a complementary dynamic that pointed the way towards a new conception of mathematics and its relation to natural philosophy. In the *Compendium*, Descartes deployed music as an instantly familiar, paradigmatic case to test this mathematical study of sensation. In contrast to Kepler's cosmic vision, Descartes built his musical science from the ground up, from common human experiences. He offered music as a new kind of sensible mathematics: number as it was manifested in corporeal motions and experienced as pleasures and passions in the soul.

The *Compendium* opens with a simple statement that was far from self-evident in Descartes's time: «the object of music is sound; its aim is to please and to arouse various passions in us»[43]. With these words, Descartes disavowed the Pythagoreanism emphasis on abstract numerology over audible experience. Adding that «the quality of tone itself (from what body and by what means it emanates in the most pleasing manner) is in the domain of the physicist», Descartes renounced the scholastic distinction between perceived sound and motions of air[44]. Like Beeckman, he defined music as sound and sound as motion. Descartes then inserted one of the more esoteric claims of his philosophical career:

> The human voice seems most pleasing to us because it is most directly attuned to our souls. By the same token, the voice of a close friend is more agreeable

---

[41]. René Descartes, *Compendium musicae* (MS, 1618; Descartes 1908, p. 89) «Hac tamen, si placet, conditione, ut perpetuo in scriniorum vel Musæi tui vmbraculis delitescens, aliorum iudicia non perferat. Qui, sicut te facturum mihi polliceor, ab hujus truncis partibus benevolos oculos non diverterent ad illas, in quibus nonnulla certe ingenij mei lineamenta ad vivum expressa non inficior». English translation from Descartes 1961, p. 53.

[42]. Van Orden 2013, p. 19.

[43]. Descartes, *Compendium* (1618; Descartes 1908, p. 89): «Finis, vt delectet, variosque in nobis moveat affectus». English translation from Descartes 1961, p. 11. Walter Robert translates the Latin 'obiectum' as 'basis' but I have changed the word to 'object'. I have also exchanged Robert's word 'emotions' for 'passions'. This is to better correspond with Descartes's Latin term 'affectus', as well as with the contemporary philosophical discussions of 'les passions'.

[44]. Descartes, *Compendium* (1618; Descartes 1908, p. 89): «Nam de ipsius soni qualitate, ex quo corpore & quo pacto gratior exeat, agant Physici». English translation from Descartes 1961, p. 11.

> than the voice of an enemy because of sympathy or antipathy of feelings — just as it is said that a sheepskin stretched over a drum will not give forth any sound when struck if a wolf's hide on another drum is sounding at the same time[45].

This passage, so incongruous with the rest of Cartesian philosophy, has perhaps discouraged scholars from taking Descartes's first written work too seriously. Here Descartes gave a gentle nod to the wondrous powers and effects that, at the dawn of the Baroque period, were still commonly attributed to music. However, by referencing sympathy and antipathy (typically associated with magical traditions) in a treatise on sensible mathematics, Descartes did more than tip his hat to an outmoded school of thought. He implied that such occult forces might, through mathematics, be made intelligible. If, «as it is said», sheep's skin fell silent to the sound of wolf's hide, this fact had to originate from the physics and mathematics of sound. Thus, in two paragraphs, Descartes spurned both scholastic theorists such as Zarlino, who were bound by the quadrivial *musica* discipline, and Neoplatonic philosophy, which obsessed over cosmic harmony but hardly spared a thought for music as humans experienced it. Descartes's musical science concerned only sound: its physics, mathematical properties, and effects on listeners.

The *Compendium* exhibited an innovative fusion of physical and mathematical reasoning that hewed closer to Descartes's mature philosophy than historians of science have traditionally assumed. Not yet aware of the coincidence theory, Descartes derived the consonances much as Zarlino did but replaced numerical with geometrical ratios. Unlike Kepler's abstract circumscribed polygons, Descartes assembled these ratios from lines representing real musical strings; «pitch is related to pitch like string to string», he observed[46]. The octave, for instance, was not composed of the abstract numerals 2 and 1 but of lines in a 2:1 length proportion. Strings, in turn, were nothing other than lines made material. By freely shifting terminology between «lines» and «strings», Descartes let the once intractable divide between geometry and corporeal objects fade into the background. In doing so, he geometrized *and* physicalized consonance in one fell swoop. Years later, Descartes wrote Mersenne that he had quit «abstract geometry [...] in order to cultivate another sort of geometry, which takes as its questions the explanation of the phenomena of nature». He famously added that «my entire physics is nothing other than geometry»[47]. The *Compendium* set the stage for this transition. In it,

---

[45]. Descartes, *Compendium* (1618; Descartes 1908, p. 92): «Id tantùm videtur vocem humanam nobis gratissimam reddere, quia omnium maxime conformis est nostris spiritibus. Ita forte etiam amicissimi gratior est quàm inimici ex sympathiâ & dispathiâ affectuum, eâdem ratione quâ aiunt ovis pellem tensam in tympano obmutescere, si feriatur, lupinâ in alio tympano resonante». English translation from Descartes 1961, p. 11.

[46]. Descartes, *Compendium* (1618; Descartes 1908, p. 97): «Sonus se habet ad sonum, ut nervus ad nervum». English translation from Descartes 1961, p. 16.

[47]. Descartes to Mersenne, 27 July 1638 (Descartes 1898, p. 268; Mersenne 1962, p. 417): «Mais ie n'ay resolu de quitter que la Geometrie abstracte, c'est a dire la recherche des questions qui ne seruent qu'a exercer

Descartes suggested that the physical and mathematical causes of consonance were, in truth, one and the same. It was this unbroken causal chain — from physicalized mathematics, to sensation, to bodily motion and affections — that fascinated Descartes and inspired him to contemplate music. Thus, while not discussing the cause of sound or consonance in any mechanico-corpuscular sense, this codependence between physical objects, geometry, and sensible phenomena marked the *Compendium* as a bona fide, if embryonic, work of physico-mathematics.

The *Compendium* also foreshadowed the obstacles Descartes's new philosophy would face. This was most evident in his scant discussions of the passions. Descartes did not question the veridicality of the senses, treating sense organs as unproblematic interfaces between reality and the soul. Sensibility and individual musical tastes were ignored. Moreover, in contemplating «the various powers which the consonances possess of evoking passions», Descartes was maddeningly evasive. «These powers are so varied and based on such imponderable circumstances», he wrote, «that a whole book would not suffice for the task»[48]. In his section on rhythm, he compared lively and languid drumbeats before concluding that «a more thorough investigation of this question depends on a detailed study of the movements of the soul, of which [I say] no more»[49]. Even the modes — the traditional wellsprings of music's mythical powers and effects — were brushed aside:

> Many conclusions might be drawn concerning the nature of the modes, but this would lead too far afield. I should be forced to deal in detail with the various passions which music can arouse; it would be necessary to show which steps, consonances, meters, etc., are instrumental in arousing these passions. But this would exceed the scope of this small volume[50].

l'esprit; & e ce affin d'auoir d'autant plaus de loysir de cultiuer vne autre sorte de Geometrie, qui se propose pour questions l'explication des phainomenes de la nature. [...] toute ma Physique n'est autre choseque Geometrie».

[48]. Descartes, *Compendium* (1618; Descartes 1908, p. 111): «Nunc sequeretur, vt de varijs consonantiarum virtutibus ad movendos affectus loqueremur; sed huius rei disquisitio exactior potest elici à iam dictis, & compendij limites excedit. Illæ enim tam variæ sunt, et tam levibus circumstantijs fultæ, ut integrum volumen ad id perficiendum non sufficeret». English translation from Descartes 1961, p. 27.

[49]. Descartes, *Compendium* (1618; Descartes 1908, p. 95): «Sed huius rei magis exacta disquisitio pendet ab exquisitâ cognitione motuum animi, de quibus nihil plura». English translation from Descartes 1961, p. 15. Here the translators have rendered 'animus' as 'soul', by which Descartes usually meant the physical part of the brain responsible for thought and perception. Descartes said this after discussing the emotions aroused by various meters, such as how slow music engendered «quieter feelings such as languor, sadness, fear, pride, etc.» (*lentiores* [...] *motus,* [...] *quales sunt languor, tristitia, metus, superbia, et cætera*) These experiences would be familiar to the community of *gentilhommes* with whom Descartes preferred to socialize.

[50]. Descartes, *Compendium* (1618; Descartes 1908, p. 140): «Ex quibus & similibus varia de horum [= modorum] naturâ possent deduci, sed longum foret. Et iam quidem sequeretur, vt de singulis animi motibus, qui

Descartes's long and well-documented struggle to fit the soul within his new philosophy began here. Unlike Kepler, he made no last-ditch appeal to geometrical metaphysics or *musica humana*. His physico-mathematization of music forced him to shun the soul, to disregard the passions that were, by his own declaration, a core aim of music. No sooner did Descartes broach the topic of the passions than he brought his *Compendium* to an abrupt end, hastily apologizing to Beeckman for whatever errors «this immature offspring of my mind» (*ingenii mei partus*) might have produced[51]. Though he explored the passions later in his career, Descartes never fulfilled his promise of «a detailed study» of music and the soul. Later he circumvented the issue by redefining consonance and, by extension, musical science itself.

Descartes's Later Musical Writings

Descartes did this by breaking the concept of consonance into distinct mathematical and sensorial components. In a correspondence with Mersenne spanning from 1629 to 1634 — while Mersenne was hard at work on his *Harmonie universelle* — Descartes questioned which aspects of musical sound were open to scientific investigation. He had by then adopted the coincidence theory, defining sound as «nothing other than a certain trembling of the air, which comes to titillate our ears» and consonance as «two sounds [that] strike the air at the same time»[52]. Yet, Descartes cautioned against an overreliance on theory: «when we judge [music] by reason, this reason must always consider the capacity of the ear»[53]. He advised Mersenne that «there are two things to distinguish» regarding consonances: «that which renders them simpler and more concordant (*accordantes*) and that which renders them more agreeable (*agreables*) to the ear»[54]. Concordant intervals appealed to reason. They possessed simple ratios and produced elegant patterns of coinciding pulsation on the eardrum; the smaller the

à musicâ possunt excitari, separatim agerem, ostenderemque per quos gradus, consonantias, tempora, & similia, debeant illi excitari; Sed excederem compendij institutum». English translation from Descartes 1961, p. 52.

51. Descartes, *Compendium* (1618; Descartes 1908, p. 140). English translation from Descartes 1961, pp. 52-53.

52. Descartes to Mersenne, early 1631 (Descartes 1897, pp. 223-224; Mersenne 1969, p. 25): «Et pour entendre cecy bien clairement, il fault supposer que le son n'est autre chose qu'un certain tremblement d'aer qui vient chatoiuller nos oreilles [...] Enfin il fault supposer que lorsque deus sons frappent l'aer en mesme tems, ilz sont d'autant plus accordans que leurs tremblemens se recommencent plus souvent l'un avec l'autre, & qu'ilz causent moins d'inesgalité en tout le cors de l'aer».

53. Descartes to Mersenne, 18 December 1629 (Descartes 1897, p. 88; Mersenne 1945, p. 338): «Lorsque nous en jugeons par raison, cete raison doit tousjours supposer la capacité de l'oreille».

54. Descartes to Mersenne, early 1631 (Descartes 1897, p. 223; Mersenne 1969, p. 24): «Touchant la douceur des consonances, il y a deus choses a distinguer: a sçavoir, ce qui les rend plus simples & accordantes, & ce qui les rend plus agreables a l'oreille».

numbers in the ratio, the more concordant the interval. Agreeable intervals, on the other hand, moved the passions. Steeped in tastes and impervious to calculation, they formed a nebulous and indeterminate collection that the ear alone could judge. Historically, consonance had been treated as a single attribute of sound possessing a dual mathematical and sensorial existence. Following the physicalization of sound and mechanization of the ear, consonance referred to *two* attributes, stranded on opposite banks of a physical/perceptual chasm, with at most a tenuous correlation between them. The acoustical paradox was on full display.

By no longer requiring that mathematics and experiences of music match, Descartes shrouded music in the veil of mystery that we today call subjectivity. He continued his letter:

> We can say absolutely, which consonances are the simplest and most concordant, since this depends only on whether their sounds unite better with each other and most closely approach the nature of the unison. Thus, we can say absolutely that the fourth is more concordant than the major third, even though ordinarily it is not as agreeable, much as senna is much sweeter than olives but no more agreeable to our taste (*goust*).

As for agreeability, «this depends on the places where they are employed; and there are places where the diminished fifth and other dissonances are more agreeable than the consonances, so one cannot hope to determine absolutely that one consonance is more agreeable than another»[55]. Agreeability depended on «the capacity of the listener, which varies as taste of individual people». Descartes framed monophony and polyphony as a matter of preference: «some will prefer hearing a single voice [that is, monophony], while others prefer an ensemble [that is, polyphony]; much like some prefer that which is sweet, while others prefer something slightly sour or bitter»[56]. Here, Descartes likened the sense of hearing to taste and smell rather than its usual epistemological analogue of vision: it was a corporeal process, ruled by individual preferences rather than reason. Unlike the scholastics and Kepler, he did not have the luxury of depicting the ear as a transparent, teleological interface between sound and soul; the soul remained just as inaccessible as it had been in the *Compendium*. Thus, consonance could be known with certainty only by expunging all traces of human sensibility and reducing it to the mathematical concept of concordance. To absolutely order consonances based on agreeability would be as ludicrous as determining the absolute best-tasting food or most beautiful color.

---

[55]. Descartes to Mersenne, early 1631 (DESCARTES 1897, p. 223; MERSENNE 1969, p. 24): «Or, pour ce qui les rend plus agréables, cela dépend des lieus ou elles sont employées; & il se trouve des lieus ou mesme les fausses quintes & autres dissonances sont plus agréables que les consonances, de forte qu'on ne sçavroit déterminer absolument qu'une consonance soit plus agréable que l'autre».

[56]. Descartes to Mersenne, January 1630 (DESCARTES 1897, p. 108; MERSENNE 1945, p. 371): «Mais pour déterminer ce qui est plus agréable, il faut supposer la capacité de l'auditeur, laquelle change comme le goust, selon les personnes; ainsi les uns aimeront mieux entendre une seule voix, les autres un concert, &c.; de mesme que l'un aime mieux ce qui est doux, & l'autre ce qui est un peu aigre ou amer, &c.».

Practical musical developments buttressed Descartes's contention that mathematizing agreeability would be a fool's errand. Whereas the *Compendium* was rooted in Renaissance theory and practice, Descartes wrote these letters when the Baroque period was well underway. Summarizing the Baroque era of music history is nearly impossible. Sandwiched between Zarlino and Rameau's music theories, the Baroque was a stylistic no-man's-land devoid of standardized theories of composition. Baroque music is better understood by what it aimed to do — stimulate the passions — than by any specific method or style. Descartes took every opportunity to defend this passionate music against the overreach of theory. The ancients, he wrote, composed music «more powerful than our own, not because they were more knowledgeable, but because they were less», adding that they «accomplished more by the sole force of their imaginations than can be done by those who have corrupted this force through theoretical knowledge»[57]. Descartes expressed similar views in a 1640 exchange between himself, Mersenne, and Constantijn Huygens, which critiqued the work of Dutch priest and amateur composer Joan Albert Ban (1597-1644) in the so-called Ban-Boësset controversy. Ban was a true believer in musical science. In 1636, he put forward a system of composition that purportedly isolated the specific emotional effects of all musical intervals. Ban's music — which he, heretically, reckoned *superior* to that of Antiquity — bucked the trends of polyphony and counterpoint gaining prevalence at the time. Descartes saw Ban's entire project as misguided, telling him that «something else is sought for in polyphony», namely agreeability, that his airs could not furnish. Further, by seeking effects solely in isolated intervals, Ban had curtailed music's expressive range. «Since music ought to imitate everything that happens in ordinary life», Descartes added, «and often in fights and tumults several people say different things at the same time, why should we not allow music to imitate even this confusion?»[58]. While Descartes's concordance/agreeability dichotomy arose through philosophical concerns, it suited a Baroque age that treasured the passions as invaluable pieces of musical experience. «If my countrymen are inferior to you in musical science», he told Ban, «at least, they better grasp the frivolities of love»[59].

---

[57]. Descartes to Mersenne, 18 December 1630 (Descartes 1897, pp. 101-102; Mersenne 1945, p. 351): «Pour la Musique des anciens, je croy qu'elle a eu quelque chose de plus puissant que la nostre, non pas pource qu'ilz estoint plus sçavans, mais pource qu'ilz l'estoint moins [...] faisoint plus par le seule force de l'imagination que ne peuvent faire ceus qui ont corrompu cete force par la connoissance de la theorie».

[58]. Descartes to Ban, December 1640 (Descartes – Huygens 1926, pp. 296-297; Descartes 1976, pp. 832-833 (reprinted from previous reference)): «Quod cum non fiat, indicium est aliud quæri ex concentu quam facilitatem perceptionis verborum. [...] Præterea, cum eorum omnium quæ in communi vitâ accidunt musica esse debeat imitatrix, sæpe autem in rixis et tumultibus plures eodem tempore diversa loquantur, quare non etiam confusionis istius imitationem ei concedemus?». Most of the translation originates from Walker 1978, pp. 81-110.

[59]. Descartes to Ban, December 1640 (Descartes – Huygens 1926, p. 294; Descartes 1976, p. 830 (reprinted from previous reference)): «Sed patere Gallos meos, si tibi inferiores sint in musicæ scientiâ, saltem in amatoriis nugis intelligendis esse magis exercitatos».

Here Cohen's (1984) *a priori* definition of musical science «as distinguished from aesthetic principles» does negative historical work. It obscures how original and controversial Descartes's treatment of consonance actually was[60]. Early modern scientific accounts of sound showed considerable interest in qualities that would be deemed aesthetic today. Ban, who understood the passions through a set of fixed mathematical rules, was Zarlino's spiritual successor in this endeavor. We must also recall that Descartes separated concordance and agreeability in reaction to Mersenne's views on music. Mersenne, in *La vérité des sciences* (The Truth of the Sciences) of 1625, implored philosophers to seek «the most beautiful melody (*chant*) of all those that can be created». He postulated an «algebra of music», equivalent to the «cossic art» recently developed in mathematics, that might recover the lost melodies of Antiquity just as mathematicians had begun to recover ancient analytical techniques[61]. In the *Harmonie universelle*, Mersenne endeavored to practice what he preached. He compiled staggering combinatorial tables of intervals with the aim of «composing the best melody of all those that can be imagined»[62]. Mersenne apparently confided this dream in Descartes. «You ask me whether the essence (*raison*) of beauty can be established», Descartes wrote Mersenne. «This is the same as your previous inquiry as to why one sound is more agreeable than another». Descartes offered little encouragement, concluding that, «because judgements of men differ so much, it cannot be said that *beauty* and agreeability have any definite measure»[63]. As for the affective powers of music: «I do not know of any quality of the consonances that speaks to the passions»[64]. Descartes's point was that music pleased the ear and moved the passions,

---

[60]. COHEN 1984, p. xiv.

[61]. Marin Mersenne, *La vérité des sciences* (Paris, 1625), p. 557: «Il faut sçavoir combien il peut y avoir de chants, ou d'airs sur un sujét donné pour pouvoir juger quel est le plus beau chant de tous ceus qui se peuvent faire dans le nombre des sons qu'on qu'on ne veut pas outre-passer». Mersenne referenced the «Algebre de la Musique» on p. 561. In seventeenth-century studies, algebra was sometimes referred to as the 'cossic art'. Note that, as with ancient melodies, Mersenne's knowledge of ancient mathematical analysis was based more in myth than historical reality. The best study of this text is DEAR 1984.

[62]. In Proposition VII in his *Harmonie universelle* (Paris, 1636), Livre second des chants, p. 103, Mersenne sought to «determine if it were possible to compose the best song of all those one could imagine, and, when composed, it could be sung with all the perfection possible» («determiner s'il est possible de composer le meilleur chant de tous ceux qui se peuvent imaginer, & si estant composé il se peut chanter avec toute la perfection possible»).

[63]. Descartes to Mersenne, 18 March 1630 (DESCARTES 1897, p. 133; MERSENNE 1945, p. 417): «Pour vostre question, sçavoir si on peut establir la raison du *beau*, c'est tout de mesme que ce que vous demandiez auparavant, pourquoy un son est plus agreable que l'autre, sinon que le mot de *beau* semble plus particulierement se rapporter au sens de la veuë. Mais generalement ny le beau, ny l'agreable, ne signifie rien qu'un rapport de nostre jugement à l'objet; & pource que les jugemens des hommes sont si differens, on ne peut dire que le beau, ny l'agreable, ayent aucune mesure determinée». Descartes and Mersenne's term 'raison' could be interpreted as 'cause', 'reason', or 'ratio'.

[64]. Descartes to Mersenne, 4 March 1630 (DESCARTES 1897, p. 126; MERSENNE 1945, p. 408): «Je ne connois point de qualitez aux consonances qui répondent aux passions».

but atomized consonance divorced from musical context did not. Aspects of music that are considered subjective, aesthetic, or culturally contingent today were not necessarily seen as such in the seventeenth century. Descartes's separation of concordance and agreeability was radical indeed and anticipated a reimagining of musical science and its connection to human experiences.

Far from misunderstanding musical science, Descartes actually *legitimized* agreeability as a genuine element of musical experience. To Zarlino, Kepler, and anyone else who viewed sound as numerical species, there could be no disparity between agreeability and concordance. The intellect — not ears or tastes — judged musical sounds. Agreeable sensations that deviated from mathematical theory were dismissed as illusory, the results of unrefined reasoning or clouded judgement. As the next section will explore, Mersenne made similar attacks on agreeability, asserting that singing and ensemble music (*concerts*) «are not the principal end of music» at all[65]. But Descartes, ever the *gentilhomme*, could never treat music this way. Its powers to delight and move listeners were what drew him to contemplate music in the first place. He did not dream of mathematizing the smorgasbord of Baroque styles, no more than he could join Ban in denying certain musical effects because they lacked theoretical bases and varied between listeners. His solution was to grant the 'ear' — the physical organ, but also in the figurative sense of an individual's auditory instincts and preferences — agency to judge sound. These auditory judgements were unconstrained by physico-mathematical laws. Artists, unfettered by philosophical concerns, were free to follow their ears towards the composition of agreeable and passionate music. The next century would supply a label for this explicitly unscientific vision of music: fine art. The acoustical paradox, by severing the teleological link between mathematics and experience and affording the ear a unique role in the interpretation of music, made this possible[66].

MERSENNE'S *HARMONIE UNIVERSELLE*

Mersenne could not abide by Descartes's fractured definition of consonance. Though his replies to Descartes lost, in the *Harmonie universelle* Mersenne strove to reconcile concordance and agreeability. He began by compiling a table juxtaposing «simple» or «sweet» (*concordant*)

---

[65]. Mersenne, *Harmonie* (1636), 'Livre premier des consonances', p. 88: «C'est pourquoy la science de la Musique ne doit pas estre negligée, quoy que les chants & les concerts fussent entierement abolis & defendu; car ils ne sont pas la principale fin de la Musique, comme croyent les Practiciens, qui mesprisent, ou ignorent la raison».

[66]. Later in his career, Descartes maintained that the soul was, ultimately, knowable; his final treatise, *Les passions de l'âme* (The Passions of the Soul; Paris, 1649), framed the passions as invaluable windows into a soul that could neither be objectified nor rationalized. However, music did not factor into this inquiry, a striking oversight that revealed how much the acoustical paradox had fundamentally altered the dynamic between sound and soul. See DESCARTES 1989.

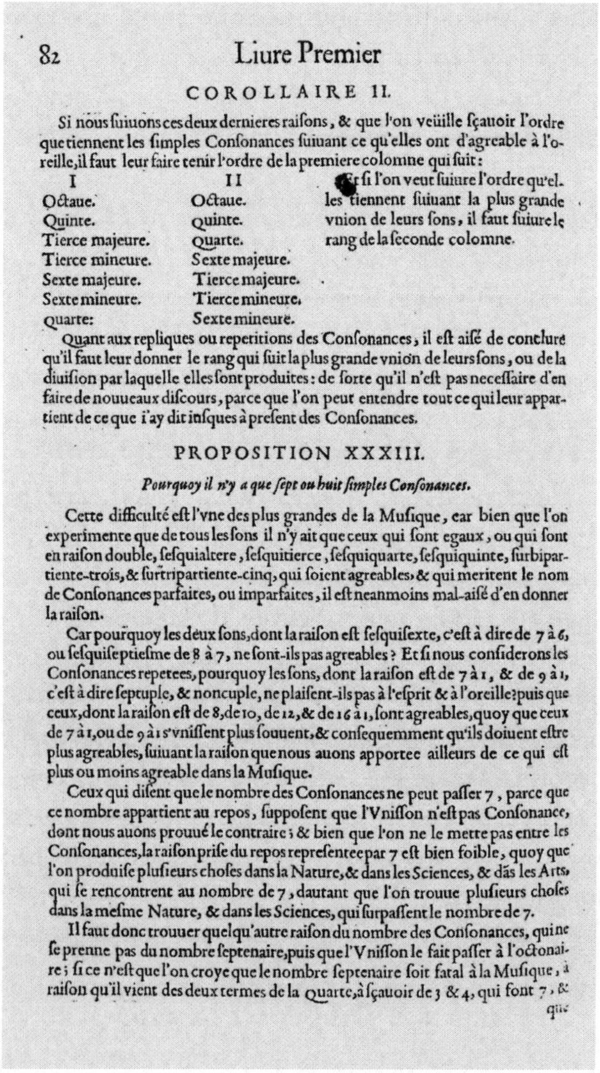

82 Liure Premier

COROLLAIRE II.

Si nous ſuiuons ces deux dernieres raiſons, & que l'on veüille ſçauoir l'ordre que tiennent les ſimples Conſonances ſuiuant ce qu'elles ont d'agreable à l'oreille, il faut leur faire tenir l'ordre de la premiere colomne qui ſuit:

I	II
Octaue.	Octaue.
Quinte.	quinte.
Tierce majeure.	Quarte.
Tierce mineure.	Sexte majeure.
Sexte majeure.	Tierce majeure.
Sexte mineure.	Tierce mineure.
quarte:	Sexte mineure.

Et ſi l'on veut ſuiure l'ordre qu'elles tiennent ſuiuant la plus grande vnion de leurs ſons, il faut ſuiure le rang de la ſeconde colomne.

Quant aux repliques ou repetitions des Conſonances, il eſt aiſé de conclure qu'il faut leur donner le rang qui ſuit la plus grande vnion de leurs ſons, ou de la diuiſion par laquelle elles ſont produites: de ſorte qu'il n'eſt pas neceſſaire d'en faire de nouueaux diſcours, parce que l'on peut entendre tout ce qui leur appartient de ce que i'ay dit iuſques à preſent des Conſonances.

PROPOSITION XXXIII.

Pourquoy il n'y a que ſept ou huit ſimples Conſonances.

Cette difficulté eſt l'vne des plus grandes de la Muſique, car bien que l'on experimente que de tous les ſons il n'y ait que ceux qui ſont egaux, ou qui ſont en raiſon double, ſeſquialtere, ſeſquitierce, ſeſquiquarte, ſeſquiquinte, ſurbipartiente-trois, & ſurtripartiente-cinq, qui ſoient agreables, & qui meritent le nom de Conſonances parfaites, ou imparfaites, il eſt neanmoins mal-aiſé d'en donner la raiſon.

Car pourquoy les deux ſons, dont la raiſon eſt ſeſquiſexte, c'eſt à dire de 7 à 6, ou ſeſquiſeptieſme de 8 à 7, ne ſont-ils pas agreables? Et ſi nous conſiderons les Conſonances repetees, pourquoy les ſons, dont la raiſon eſt de 7 à 1, & de 9 à 1, c'eſt à dire ſeptuple, & noncuple, ne plaiſent-ils pas à l'eſprit & à l'oreille? puis que ceux, dont la raiſon eſt de 8, de 10, de 12, & de 16 à 1, ſont agreables, quoy que ceux de 7 à 1, ou de 9 à 1 s'vniſſent plus ſouuent, & conſequemment qu'ils doiuent eſtre plus agreables, ſuiuant la raiſon que nous auons apportee ailleurs de ce qui eſt plus ou moins agreable dans la Muſique.

Ceux qui diſent que le nombre des Conſonances ne peut paſſer 7, parce que ce nombre appartient au repos, ſuppoſent que l'Vniſſon n'eſt pas Conſonance, dont nous auons prouué le contraire; & bien que l'on ne le mette pas entre les Conſonances, la raiſon priſe du repos repreſentee par 7 eſt bien foible, quoy que l'on produiſe pluſieurs choſes dans la Nature, & dans les Sciences, & dãs les Arts, qui ſe rencontrent au nombre de 7, dautant que l'on trouue pluſieurs choſes dans la meſme Nature, & dans les Sciences, qui ſurpaſſent le nombre de 7.

Il faut donc trouuer quelqu'autre raiſon du nombre des Conſonances, qui ne ſe prenne pas du nombre ſeptenaire, puis que l'Vniſſon le fait paſſer à l'octonaire; ſi ce n'eſt que l'on croye que le nombre ſeptenaire ſoit fatal à la Muſique, à raiſon qu'il vient des deux termes de la Quarte, à ſçauoir de 3 & 4, qui font 7, &

que

ILL. 15.4: Marin Mersenne, *Harmonie universelle* (Paris, 1636), 'Livre premier des consonances', p. 82, with tables comparing agreeability (column I) with concordance (column II). Copy E-Mn.

consonances with those «agreeable to the ear» (ILL. 15.4)[67]. While simple intervals obeyed the ordering dictated by the coincidence theory, the table of agreeability was drawn from his personal observations. He formulated many theories — some ingenious, others *ad hoc* —to account for discrepancies between the two. This was, as he told Descartes, because he sought the «reason» (*raison*) of beauty: a theory of sound production and reception that explained music rationally and in natural terms, without recourse to subjective tastes, unquantifiable

[67]. Mersenne, *Harmonie* (1636), 'Livre premier des consonances', p. 82.

passions, or artistic conventions. Whereas Descartes supposed that the ancients had mastered music by being *less* attuned to theory, Mersenne saw musical science as the key to recovering the mythical powers and effects of ancient songs. That the consonances existed in nature was evident to Mersenne in *La verité des sciences*. He demanded that all «harmonic philosophers» proceed from this unassailable premise[68]. In the *Harmonie universelle*, he pronounced his desire to transcend individual or cultural biases and discover the singular, universal music given in nature. Mersenne thus encountered the same acoustical paradox as Descartes. Unlike his friend, Mersenne did not give up without a fight.

Mersenne contemplated how concordance and agreeability might be unified. Comparing the unison and octave, he made clear that unison «is sweeter, since it unites its sounds more often and more easily [...] the beatings of air unite at each stroke, whereas the beatings of the octave unite only at every second stroke; and one will always find in the operations of all the senses that what unites the most easily is the sweetest»[69]. Mersenne then acknowledged the role of taste: «it does not follow that it is the most agreeable; although sugar and honey are very sweet, they are not agreeable to those who prefer bitter and sour things». However, he immediately confirmed the agreeability of the unison on different grounds: «the unison is more agreeable than the octave because it tickles the ear more, and is comprehended more easily by the imagination, which is the principal seat of pleasure»[70]. He was not deaf to the affective variety and power of polyphony: «if one reflects on the concords (*accords*) which charm the spirit in ensembles (*concerts*), one must admit that one sometimes encounters passages that ravish the listener, which is never done so powerfully by the unison»[71]. But then again:

> As for those who have risen above everything created, and who have experienced a thousand times the distaste received from all the truths of mathematics and physics once they have been discovered, from which is found almost no satisfaction save the toil endured in searching for them, they receive no satisfaction

[68]. *Ibidem*, 'Préface générale au lecteur'.

[69]. *Ibidem*, 'Livre premier des consonances', p. 11: «Je dis premierement qu'il n'y a nul doute qu'il ne soit plus doux, puis qu'il unit ses sons plus souvent & plus aisement, car l'Unison estant d'un à un, tous les battemens de l'air s'unissent à chaque coup, au lieu que les battemens de l'Octave ne s'unissent que de deux en deux coups; & l'on trouvera toujours dans les operations de tous les sens, que ce qui s'unit le plus aisément est le plus doux».

[70]. *Ibidem*: «Mais il ne s'ensuit pas qu'il soit le plus agreable: car encore que le sucre & le miel soit tres-doux, il n'est pourtant pas agreable à ceux qui aiment mieux les choses aigres & ameres: c'est pourquoy il faut voir se l'Unisson est plus agreable que l'Octave [...] Je dis donc secondement qu'il semble que l'Unisson est plus agreable que l'Octave, parce qu'il chatoüille davantage l'oreille, & qu'il se comprend plus facilement par l'imagination, laquelle este principal siege du plaisir».

[71]. *Ibidem*, p. 13: «Si l'on fait reflexion sur les accords qui charment l'esprit dans les Concerts, on sera contraint d'avoür qu'il se rencontre quelquefois de certains endroits qui ravissent l'auditeur; ce qui ne se fait jamais si puissamment par l'Unisson».

> from ensembles (*concerts*), and prefer to hear singing in unison to that in many parts, for the unison represents to them the state (*séjour*) of the blessed, and the perfect union of the three divine persons who are the unison of perfect equality[72].

These ideal and quasi-transcendent listeners would hear no difference between concordance and agreeability. They preferred not variety but unison and relished the simple beauty of monophony over the complex expressions and passions of polyphony. Real human listeners, who «judge things not by how they are simplest or most excellent, but by how they appeal to their appetites and fancies», misjudged music. Their differing tastes testified to the «fragility and inconstancy of men» and «the inability of ordinary human senses to ascertain certain knowledge»[73]. Mersenne lamented that «the spirit of most men is so confined in the body and bound by phantasms that it cannot rise above the senses»[74]. His quest for a unified theory of consonance forced him to disregard music as real people experienced it, including himself.

How could Mersenne do such a thing? How could he begin his *magnum opus* on musical science by undermining the credibility of the entire enterprise? By accusing anyone who enjoyed polyphony of enslavement to the appetites of the senses? As Joscelyn Godwin (1992) has observed, this passage revealed the guilt of a pious natural philosopher torn between the love of music and the love of God[75]. Mersenne's philosophy entailed a kind of symmetry between music, arithmetic, geometry, and theology. «Consonances depend on unison as lines on the point, numbers on the unit, and creatures on God», he wrote[76]. He was no Neoplatonist, dismissing Kepler's geometric theory of consonance and harmony of the spheres as nonsense. Still, Mersenne interspersed his writings with musings on the intrinsic harmony of nature that resonated with Kepler's faith that geometry (and therefore music) was «coeternal with God». In what could have been a reply to Descartes, Mersenne proclaimed:

---

72. *Ibidem*, p. 14: «Quant à ceux qui sont montez au dessus de tout ce qui est creté, & qui ont mille fois experimenté le degoust que l'on a de toutes les veritez des Mathematiques, & de la Physique, lors qu'elles ont esté trouvees, & dont on ne reçoit quasi nul contentement que dans le labeur que l'on souffre en les cherchant, ils ne reçoivent nul contentement des Concerts, & aiment mieux oüir chanter à l'Unisson qu'à plusieurs parties, dautant que l'Unisson leur represente le sejour des Bien-hereux, & la parfaite union des trois personnes divines qui sont à l'Unisson d'une parfaite egalité».

73. *Ibidem*: «J'estime donc que l'Unisson est plus agreable que les Consonances, & qu'il faut porter compassion à la fragilité & inconstance des hommes qui n'ont pas ce sentiment, & qui sont plus grand estat de la diversité & de l'inegalité, que de l'unité & de l'égalité, dautant qu'ils ne jugent pas des choses parce qu'elles ont de plus simple & de plus excellent, mais par ce qui revient le mieux à leur appetit & à leur fantaisie».

74. *Ibidem*, p. 15: «L'esprit de la plus-part des hommes est tellement renfermé dans le corps, & borné par les phantosmes, qu'il ne peut se porter par dessus les sens».

75. Godwin 1992, p. 251.

76. Mersenne, *Harmonie* (1636), 'Livre premier des consonances', p. 15: «Les Consonances dependent de l'Unisson, comme les lignes du point, les nombres de l'unité, & les creatures de Dieu».

> Nature or its movements are nothing but a ravishing harmony, which invites us to consider the first source from which it is born, this is to say, a contemplation of God, who is the author, and to love him above all else. From this it is easy to conclude that the order of the consonances is natural, and that the way in which we count, starting from the unit up until the number six, and beyond, is founded in nature[77].

Contra Descartes, Mersenne saw the mathematical and sensible dimensions of consonance as coextensive. Where Descartes treated agreeability as an artistic concern rather than an object of scientific study, Mersenne held that an absolute ordering of consonances existed and those who disagreed were simply wrong. Polyphony, much as Mersenne personally enjoyed it, did not fit with his vision, for there the whims of taste and artistic license reigned supreme. This ideal listener, perhaps, would possess a transparent ear and could apprehend the mathematics of consonance in its pure, unadulterated form — «look on beauty bare», as Edna St. Vincent Millay (1892-1950) wrote of Euclid in her poem 'Euclid alone has looked on Beauty bare' (1922)[78]. This was Mersenne's solution to the acoustical paradox. The ideal listener, having «risen above everything corporeal», could bypass the physical/perceptual divide, unify concordance and agreeability, and establish a science of music[79].

What pertinence ideal listeners had to real musical experience and practice was another matter. The 'problem of 7' provided an illustrative example of how Mersenne's musical science flew in the face of contemporary Baroque practice. Even more so than the fourth, intervals with 7 (7:5, 7:1, and so on) exhibited great disparity between concordance and agreeability; despite their simple ratios, musicians had historically rejected them as dissonant. Mersenne speculated that such intervals «may become agreeable if one accustoms oneself to hearing and enduring them, and if one use them properly in melodies and ensembles for moving the passions and other effects absent from ordinary music»[80]. Witnessing the freer, more expressive

---

77. *Ibidem*, 'Livre cinquiesme des instruments à vent', p. 251: «Il semble que la nature ou ses mouvemens ne soient autre chose qu'une ravissante harmonie, qui nous invite à considerer la premiere source, dont elle prend sa naissance, c'est à dire à contempler Dieu, qui en est l'Autheur, & à l'aymer surtoutes choses. D'où il est aysé de conclure que l'ordre des Consonances est naturel, & que la maniere dont nous contons en commençant par l'unité jusques au nombre de six, & au delà, est fondée dans la nature».

78. Millay 2020, p. 155.

79. Mersenne (*Harmonie*, 1636, 'Livre premier des consonances', p. 14) wrote that «ceux qui s'eslevent par dessus tout ce qui est corporel» would find the unison most agreeable.

80. *Ibidem* p. 89: «Puisque le long exercice a coustume de rendre doux & facile ce qui sembloit auparavant rude & fascheux, je ne doute nullement que les intervalles dissonans, dont j'ai parlé dans cette proposition, à sçavoir les raisons de 7 à 6, & de 8 à 7, qui divisent la Quarte, ne puissent devenir agreables, si l'on s'accoustume à les oüir & à les endurer, & que l'on en use comme il faut dans les recits & dans les concerts, afin d'emouvoir les passions, & pour plusieurs effets, dont la Musique ordinaire est privée».

use of dissonances characteristic of seventeenth-century composition, Mersenne admitted that practice might evolve to incorporate intervals with 7 as agreeable. Practice proved that the ear — which Mersenne, like Descartes, understood to include habits, tastes, and other mental faculties — could learn to accept new sounds as agreeable[81].

The philosophical status of such intervals was more problematic. In the study of overtones, Mersenne was first to discern additional *petits sons* (as he called them) beyond the octave, each corresponding to harmonic ratios: the twelfth (3:1), fifteenth (4:1) seventeenth (5:1), and nineteenth (6:1). Despite the obvious pattern, Mersenne denied that the next overtone was the seventh harmonic (7:1). «Because it is neither consonance nor a difference of consonances», Mersenne argued, «nature, which is harmonious, rejects it and prefers to break the sequence of its intervals and melodies rather than passing through an interval that is worthless except to harm the ear and mind»[82]. Mersenne's full explanation veered into metaphysics. He claimed that «nature seeks the octave» and would only admit dissonances after the twenty-second (8:1, or the triple octave) in the overtone sequence «because [nature] considers only the solids, without turning to the supersolids and other numbers or quantities called *powers* in algebra, which are in the imagination of men rather than the reality of things» (in plain language, 8 is the first perfect cube after 1 ($8 = 2^3$) and thus, in a universe of three dimensions, seemed a natural choice for partitioning consonant from dissonant overtones)[83]. While practitioners could grow accustomed to intervals with 7, harmonic philosophers had to conclude that the 7:1 interval did not exist as an overtone and was not a natural consonance.

How should we interpret Mersenne's observations? The seventh harmonic *does* occur as an overtone. Could Mersenne — renowned across Europe for his keen ear — have missed it? More likely, he *refused* to hear it. Mersenne approached musical science not as a human listener but as an ideal listener. Ideal listeners, who heard no difference between concordance and agreeability, would concur with nature that intervals with 7 were dissonant and had no place in music. Mersenne could not believe that a harmonious nature would produce sounds that so

---

[81]. Habit, along with skill and handiwork, was then seen as a bodily, not mental, capacity — a matter of physical training rather than rational thought. The physical theory of sensation implied that all aspects of the body, not just the mechanics of sense organs, could play a role in human experience, and hence Mersenne treated the acquiring of new habits similarly to the development of the ear itself. For more, see GAUVIN 2013.

[82]. Mersenne, *Harmonie* (1636), Livre cinquième des instruments à vent, p. 251: «Mais parce qu'elle n'est ni consonance, ni difference des consonances, la nature qui est harmonique, la reiette & ayme mieux rompre la suite des ses intervalles & de ses chansons, que de passer par un intervalle qui ne vaut rien, que pour blesser l'oreille & l'esprit».

[83]. *Ibidem*, p. 253: «L'on peut donc conclure par l'experience, que la nature cherche l'Octave, & qu'elle ne veut pas donner les differences des consonances jusques à ce qu'elle ayt fait la Vingt-deuxiesme, en arrivant à l'octonaire, qui luy sert de borne, parce qu'elle considere seulement les solides, sans se sourcier des sursolides, & des autres nombres ou quantitez que l'on appelle *dignitez* dans l'Algebre, & qui sont plustost dans l'imagination des hommes».

offended his ear, but his solution entailed (1) an *ad hoc*, metaphysical left turn convincing only to those who already believed it, (2) a disregard of recent developments in composition, and (3) a pointblank denial of sensory evidence. As in his polemic against polyphony, Mersenne disregarded real musical experience in favor of imagined ideals.

Mersenne's part in the Ban-Boësset controversy accentuated the disparity between ideal listeners and contemporary practice. Like Descartes, Mersenne was unimpressed by Ban's theory-laden compositions. Whereas Descartes criticized Ban's science for stifling the passions, Mersenne felt that Ban dwelt too much on expression. «Music and consequently airs», Mersenne wrote to Constantijn Huygens, «are made particularly and *principally to charm the mind and ear* and to let us live our lives with some sweetness amidst all the bitterness we encounter». Mersenne preferred simple songs, monodies, and airs such as Constantijn himself composed. These styles «are not made to excite anger and many other passions, but to give enjoyment to the minds of the listeners, and sometimes to bring them to devotion, as occurs during a church service»[84]. It was presumably this kind of music that Mersenne's ideal listener would enjoy. Earlier Ban had written that «whoever does not understand from proportion the unique power of intervals, as I have described, is mad» and «a deaf musician is he who does not clearly perceive them [the proportions] with the ears when they are performed»[85]. But Ban's defense — that his critics misunderstood his theory — missed Mersenne's point. Ban's crime was not an overreliance on science, as Descartes insinuated, but his insistence that music should move the passions. Mersenne judged music from the perspective of an ideal listener. Only compositions that merged concordance and agreeability and «charm the mind and ear» equally were considered musical. Passions, which varied among listeners and had no basis in physical sound, were distractions from a true understanding and enjoyment of music.

This recurring clash between real and idealized experiences shook at the foundation of Mersenne's musical science. His yearning to understand beauty in precise mathematical terms had no place in a world in which music was sound and sound was physical stimuli upon subjective, fallible eardrums. His renunciation of music's passionate affects flew in the face of contemporary practical trends. In this way, Mersenne's polemic on the ideal listener reads

[84]. Mersenne to Huygens, 14 November 1640 (Mersenne 1967, p. 237; Huygens 2007, p. 421): «Il faut premierement supposer que la musique, et par consequent les airs, sont faitez particulierement et principalement pour charmer l'esprit et l'oreille et pour nous faire passer la vie avec un peu de douceur parmy les amertumes qui s'y recontrent [...] Semblablement les airs ne se sont pas pour exciter la cholere, et plusieurs autres passions, mais pour resjouyr l'esprit des auditeurs, et quelquefois pour les porters à devotion comme il arrive aux recits que l'on faict dans les Eglises durant le service divin». Walker (1978, pp. 106-107) has analyzed these quotes and the surrounding controversy. See also Chapter 9 in this volume.

[85]. Joan Albert Ban to Mersenne, 15 May 1638 (Mersenne 1962, p. 206): «Quisquis intervallorum propriam potestatem a me descriptam non intelligit ex proportione, ille delirat; qui non distincte eas auribus percipit dum pronunciatur ad verba, ille surdus est musicus». Translation from Walker 1978, p. 82.

like an apology, meant to assure readers that one could by *transcending* ordinary experience, circumvent the physical/perceptual chasm of the acoustical paradox, and achieve a complete musical science. This was the true meaning behind this well-known corollary from the *Harmonie universelle*:

> Sound can shed more light on Philosophy than any other quality. This is why the science of music must not be neglected, even if singing and ensembles (*concerts*) were entirely abolished and forbidden, for they are not the principal end of music as practitioners, who despise or ignore reason, believe. In fact, if the knowledge of sounds and their proportions can grant us access to the proportions of objects of the eye, of smell, and of taste, there is not a gentleman who does not prefer this knowledge over all the songs and ensembles that could be made following the rules of art[86].

Seeking to save music's scientific status, Mersenne relegated not just polyphony but all effects of musical experience to artists who «despise or ignore reason». The ideal listener would prefer philosophical knowledge over anything «that could be made following the rules of art». Thus, whereas Descartes saved agreeability by carving out epistemological space beyond the reach of science, Mersenne delegitimized any feelings or experiences that deviated from his imagined ideals. After all, Mersenne's whole point was that no such ideal listener could ever exist. Real tastes, passions, and effects of sound were pushed out of Mersenne's science.

CHRISTIAAN HUYGENS'S MUSICAL SCIENCE

It fell to Constantijn's son Christiaan Huygens (1629-1695; ILL. 15.5) to suggest a way out of the acoustical paradox. Both Descartes and Mersenne had nothing but praise for the young Huygens. After the Ban-Boësset controversy, Descartes advised Constantijn that his son «may yet have the occasion to criticize both Ban and myself, and prove that we have understood nothing; for, because our reasons are neither mathematical nor physical but only moral, as I have said, it is easy to produce others that are contrary to them»[87]. Mersenne added

[86]. Mersenne, *Harmonie* (1636), 'Livre premier des consonances', p. 88: «Les sons peuvent apporter plus de lumiere à la Philosophie que nulle autre qualité; c'est pourquoy la science de la Musique ne doit pas estre negligee, quoy que les chants & les concerts fussent entierement abolis & defendus; car ils ne sont pas la principale fin de la Musique, comme croyent les Praticiens, qui mesprisent, ou ignorent la raison. En effet si la cónoissance des sons & de leur proportion nous peut donner l'entree aux proportions des objets de l'oeil, de l'odorat & du goust, il n'y a point d'honneste homme qui ne prefere cette connoissance à tous les chants, & à tous les concerts qui peuvent estre faits suivant les regles de l'art».

[87]. Descartes to Constantijn Huygens, 30 November 1646 (DESCARTES – HUYGENS 1926, p. 248; DESCARTES 1976, p. 788, reprinted from previous reference): «Et si M. vostre second fils se veut exercer en cete

Ill. 15.5: Gerard Edelinck, after Jacques-Antoine Friquet de Vauroze, *Christiaan Huygens* (engraving, 28.5 × 22.5 cm; c. 1688). NL-Ar, RP-P-BI-7524.

that Christiaan's budding mathematical talents «would one day surpass Archimedes»[88]. The capacities Descartes and Mersenne chose to celebrate in Christiaan Huygens spoke to their divergent attitudes towards music and the contributions each hoped he would make. Descartes, who saw music as a matter of subjective aesthetic experiences, lauded Huygens's keen ear and sound judgment. Mersenne, who treated music as mathematical, saw in Huygens a promising candidate for the 'new Archimedes' prophesized in *La verité des sciences* — someone who could «guide mathematics to its former perfection», lost since Antiquity[89]. Both were partially correct.

matiere, il en peut avoir occasion en nous reprenant Bannius et moy et mostrant que nous n'y avons rien entendu ny l'un ny l'autre. Car nos raisons n'estant ny mathematiques ny physiques mais seulement morales, comme i'ay dit là, il est aysé d'en trouver d'autres qui leur soient contraires». See also Tonietti 2014, vol. ii, p. 211.

[88]. Mersenne to Constantijn Huygens, 3 January 1647 (Mersenne 1983, p. 17): «Je ne croy pas s'il continue, qu'il ne surpasse quelque jour Archimede».

[89]. Mersenne, *La vérité des sciences* (1625), p. 750: «Plaise à Dieu de nous faire renaistre en ce siecle quelques nouveaux Archimedes, qui conduisent les Mathematiques jusques à leur derniere perfection».

Christiaan Huygens's musical science is best understood as a unity of opposites. He had no patience for Mersenne's ideal listener, but neither did he accept Descartes's view of agreeability as purely subjective. He never doubted that musical proportions were absolute and given in nature. But he also nurtured an Aristoxenian belief that the nature of music resided in sound rather than abstract number. Huygens described just intervals as 'pure' and 'perfect' but devoted his musical studies towards developing a 31-tone tempered scale aimed at circumventing those intervals. He reckoned this temperament the 'true' tuning[90]. While maintaining that «men all over the world sing the same consonant intervals», Huygens admitted that «not all tastes agree absolutely with this judgment» and that mathematics could not resolve disparities of individual preferences[91]. Thus, like Descartes and Mersenne before him, Huygens confronted the acoustical paradox and was forced to question the basic validity of musical science. While never doubting the mechanico-corpuscular models of sensation, Huygens knew that the resulting estrangement between the sense organs and the soul precluded a complete theory of perception. His solution to the paradox was to invert the epistemological hierarchy between musical science and practice, such that musical science relied on practice — on real melodies that pleased real, human listeners — for its meaning. Huygens offered a new kind of musical science that eschewed philosophical speculation in favor of usefulness towards practice. He sought not a *raison* of beauty but an understanding of sounds that listeners considered beautiful.

What fascinated Huygens about musical sound was not so much its physico-mathematical properties as how it operated on a more minute and subtle level than natural philosophy seemed equipped to discover. In his *Traité de la lumière* (Treatise on Light; Leiden, 1690), Huygens defined the «true [Cartesian] philosophy» as the one that «understands the causes of all natural effects on mechanical grounds». He added: «this must be done, in my opinion, or else we abandon all hope of comprehending anything in physics»[92]. Sound was no exception. Though not discussing sound in the *Traité*, Huygens's private notes affirmed the mechanico-corpuscular model of sound and arranged the consonances based on Mersenne's table of concordance. He worked to resolve discrepancies between concordance and agreeability, attributing, for example, the agreeability of the major third to the total pulsations generated by the third's fundamental tone and its overtones. The following passage was typical:

[90]. Christiaan Huygens, notes on Francisco Salinas, *De musica libri septem* (Salamanca, 1577), (NL-Lu, Cod. Hug. 27; Huygens 1940, p. 113).

[91]. Christiaan Huygens, a treatise on consonance (NL-Lu, Cod. Hug. 27; Huygens 1940, p. 36): «On peut examiner la preference des autres consonances suivant ces mesmes regles et il est utile de connoistre ces degrez de bontè quoy qu'il soit vray que tous les gouts ne s'accordent pas tout a fait en ce jugement».

[92]. Christiaan Huygens, *Traité de la lumière* (Leiden, 1691), p. 2 (Huygens 1937, p. 461): «Dans la vraye Philosophie dans laquelle on conçoit la cause de tous les effets naturels par des raisons de mechanique. Ce qu'il faut faire à mon avis, ou bien renoncer à toute esperance de jamais rien comprendre dans la Physique».

> When we examine the tremblings of strings (I believe [Galileo] Galilei was the first) we find that their returns are more frequent according to whether they are shorter and that the numbers of vibrations are precisely in the inverse ratio of their lengths. I always hear strings similarly when they are equally tense, such as when we hang from them equal weights. Thus, that which is half as long as another makes 2 tremblings while the other makes only one. That which is as 2 to 3 makes 3 tremblings while the other makes 2. And so on for the other consonances. For it is those with lengths as number to number that come to beat the air together in regular intervals by mixing together two beats in the same manner for each reprise; it is this orderly percussion of the air that, acting on our ears, produces the pleasure of the consonances[93].

Huygens concluded with this intriguing claim:

> It is certainly not the case, however, that the mind can discern these pulsations or contemplate their commensurability, for no reasoning intervenes, but rather it finds them agreeable just as the sense of smell perceives the sweetness of perfumes. And I believe that it would be futile to further try to seek the cause of these pleasures[94].

This was the clearest recognition from any philosopher of the estrangement between sound and soul brought about by the acoustical paradox. While Huygens had no problem explaining light or sound in mechanistic terms, music was another matter. By declaring that «no reasoning intervenes» in experiences of agreeability, Huygens repudiated Mersenne's ideal listener and any possibility of discovering the *raison* of beauty. He instead took after Descartes by treating hearing as a corporeal process governed by taste rather than reason — more akin

[93]. Christiaan Huygens, 'Rapports des longueurs des cordes' (MS, NL-Lu, Cod. Hug. 27; HUYGENS 1937, p. 364): «Quand on examine les tremblements des chordes ce que je pense que Galilee a fait le premier, on trouve que leurs retours sont plus frequents, selon qu'elles sont plus courtes et que les nombres des vibrations sont precisement dans la raison converse de leurs longueurs. J'entens tous jours parler de chordes pareilles en tout et egalement tendues comme quand on y pend des poids egaux. Ainsi celle qui est la moitiè aussi longue qu'une autre fait 2 tremblements pendant que l'autre n'en fait qu'un. Celle qui est comme 2 à 3 fait 3 tremblements pendant le temps que l'autre en fait 2. et ainsi des autres consonances. Et parce que cela fait que celles qui ont leur longueur comme nombre a nombre vienent a frapper l'air ensemble par intervalles reglez en y meslant entre deux des battemens confus mais de mesme facon a chaque reprise; c'est cette percussion ordonnée de l'air, qui, agissant dans nostre oreille produit le plaisir des consonances».

[94]. Christiaan Huygens, 'Rapports des longueurs des cordes' (MS, NL-Lu, Cod. Hug. 27; HUYGENS 1937, p. 364): «[...] non pas toutefois que l'esprit puisse aucunement compter ni discerner ces battemens ni contempler leur commensurabilité, car il n'y intervient aucun raisonnement, mais il les trouve agreables de mesme que l'odorast goute la douceur des parfums, et que les autres sens se plaisent à d'autres objects qui les peuvent toucher. Et je croy que ce seroit en vain de vouloir chercher la cause de ces plaisirs plus loin».

to smell than to vision. This was not a statement on the fallibility of human hearing. Huygens described the ear as a godly mechanism, a «marvel» whose «subtlety in the fine discernment of different sounds» proved that experiences of music were «not the production of chance or the accidental stirring of materials, but rather of a supreme and incomprehensible intelligence and power»[95]. In short, physico-mathematical science lacked the tools to explore the causes of agreeability. While not hopelessly subjective as Descartes had believed, musical experience was divinely ordained by an incomprehensible power. Its cause lay beyond the reach of the «true philosophy».

Huygens thoughts on composition reinforced his stance that music should not be reduced to science. In a manuscript written from 1671 to 1674, Huygens performed a detailed analysis of ancient and modern music theorists. Like Mersenne, he bemoaned the Baroque trend of technical, expressive compositions. He begged composers to «not embrace those melodies that are artificial and difficult to invent, but rather those that affect the ear most». In assessing the contrapuntal styles that had gained prevalence in the past century, he asked: «what do these accurately observed imitations, such as fugues, matter to me, if without them the melody becomes more free and gracious?». Here Huygens slipped in a double meaning. A fugue was, by definition, imitative music: it involved a lead voice accompanied by secondary voices imitating the main theme with various transpositions or inversions. However, Huygens also disparaged fugues as artificial imitations of natural music, the epitome of the Baroque era's trend towards elaborate counterpoint at the expense of traditional melodic motion. «Artists say they are delighted by [fugues]», he added, «not so much affected by the sweet harmony as by the consideration of an artificial composition, and thus they misunderstand music, the aim of which is to delight the ear through sound we perceive, not the contemplation of art». Huygens preferred simple songs, monodies, and airs, such as were composed by his father. This music would gratify but not strain listeners, appeal to laymen, artists, and philosophers alike, and bypass the intellect to please the ear. Whereas dissonant and contrapuntal fugues were artificial, simple airs were 'natural'. Just as one could never «contemplate the commensurability» of sounds, so too did the «contemplation of art» hinder natural, pleasing melodies»[96].

---

[95]. Huygens compared vision and hearing in a manuscript text (Huygens 1916, p. 799): «Ce ne sont pas des productions du hazard ou de la matiere fortuitement remuée, mais d'une intelligence et puissance supreme et incomprehensible».

[96]. Christiaan Huygens, notes on Athanasius Kircher, *Musurgia universalis* (Rome, 1650), Liber vii (MS, NL-Lu, Cod. Hug. 27; Huygens 1940, p. 125): «Vellem etiam ut non quae artificiosissima sunt et inventu difficilia sectarentur melopoei nostri sed quae aures maximè afficerent. Quid enim mihi cum imitationibus accurate servatis, quas fugas vocant, quid cum duplicibus, si absque his ut liberior ita et gratior efficitur melopoeia. Artifices istis se delectari dicunt, non tam dulcedine concentus affectos quam consideratione artificiosae compositionis; qui non recte hoc pacto melopoeam aestimant, cujus finis est delectare sono quem auribus percipimus, non contemplatione artis».

Indeed, Huygens depicted music as an almost anti-intellectual process. He continued his notes: «art that does not please us or move our nature should, as Cicero says, be deemed a failure. For this reason, musicians must not consider their rules to be axioms of geometry but should, rather, allow many exceptions». Those who «cling tenaciously to rules», he added, «have no right to judge harmony». Here Huygens disavowed Mersenne's belief in an algebra of music that could generated the most beautiful melody mathematically possible. He beseeched musicians to obey one core tenet: «to bring pleasure to the ears before embarking on any examination of artifice»[97]. Huygens saw aural experience as the primary object of music and felt that abstract, algorithmic approaches to composition distorted music's natural affectations. He could not assume that musical science and practice could neatly converge into a coherent artisanal epistemology. The acoustical paradox had undermined this fundamental premise of traditional musical science. Forced to choose between science and pleasing sound, Huygens took the opposite path from Mersenne. He subordinated science to practice and defined 'natural' music neither as abstract ratios, nor patterns of pulsation upon eardrums, nor judgments of an ideal listener. Natural music, to Huygens, referred to pleasing and affective melodies as experienced by real human listeners.

Huygens's only published musical treatise, the 'Lettre touchant le cycle harmonique' (Letter Concerning the Harmonic Cycle; Rotterdam, 1691)[98], exhibits this reconfigured dynamic between science and practice. The «harmonic cycle», as Huygens called his 31-tone temperament, was his most enduring contribution to music theory. Its purpose was to approximate a different tempered system: meantone temperament, which tempered fifths and fourths while preserving pure major thirds. While maintaining that justly tuned consonances were 'perfect' and 'pure', Huygens declared meantone temperament to be «more abundant in concords (*accords*) and more befitting the nature of melody» than any system of pure intervals devised by the ancients or Zarlino. Such tunings might have sufficed for medieval monophony, but modern polyphony demanded temperament. These ancient systems would be made obsolete by Huygens's harmonic cycle[99]. Meantone was the 'true' tuning because it

---

[97]. Christaan Huygens, notes on Athanasius Kircher, *Musurgia universalis* (Rome, 1650), Liber vii (MS NL-Lu, Cod. Hug. 27; Huygens 1940, pp. 125-126): «Ars autem nisi naturam moveat ac delectet nihil sane egisse videatur, ut Cicero inquit [...] Quamobrem nec praeceptis ita considere debent ac si geometriae axiomata essent, sed multas exceptiones dari existiment semperque fixum illud teneant, propositum delectationem adferre auribus prius quam artis examen instituatur». Both Cohen (1984, p. 225) and Tonietti (2014, vol. ii, p. 204) have analyzed this passage.

[98]. Christiaan Huygens, 'Lettre touchant le cycle harmonique', in: *Histoire des Ouvrages des Sçavans*, September-October 1691, pp. 78-88. Latin translation as 'Novus cyclus harmonicus', in Christiaan Huygens, *Opera varia: Volumen secundum* (Leiden, 1724), pp. 747-745. See Huygens 1986.

[99]. Christiaan Huygens, 'Lettre touchant le cycle harmonique' (1691), p. 79 (Huygens 1905, p. 169): «C'est aussi ce Temperament qui a fait negliger avec raison tous les divers systemes et divisions du monochorde des anciens,

captured the subtle flavouring of intervals with 7, or that «something more plaintive and tender [...] and also something sad» of certain keys and cadences[100]. However, because its tempering generated unequal distances between notes, meantone tuning was incapable of modulation, impractical for many instruments and problematic for keyboards from some point in the seventeenth century onward. Huygens devised his harmonic cycle to emulate the audible effects of meantone while rendering it (with the aid of a hypothetical movable keyboard of his own design) feasible on the harpsichord. This system, Huygens claimed, originated from «the first principle of this science [of music], which is to determine the tones that are observed in singing and in the construction of instruments». It thus captured the «reason of harmony» (*raison harmonique*)[101]. Both *accord* and *raison* were smoothly redefined from their usages in the Descartes/Mersenne correspondence. *Accord* meant 'tuning' rather than the narrow mathematical concept of concordance. *Raison* designated not a physico-mathematical cause of beauty but that beauty itself — the ineffable essence of music all musicians strove to capture. Mathematics could facilitate this pursuit, as Huygens stressed when he deployed logarithms in calculating his harmonic cycle. However, it remained for the musician to employ this temperament and produce the pleasing melodies that were the *raison* of music. Huygens's musical science was thus anchored in musical practice.

Huygens's views on the nature of music were on display in his final, most philosophical work. The *Cosmotheoros, sive De terris cœlestibus earumque ornatu conjecturae* (Spectator of the World, or Conjectures on Heavenly Worlds and Their Attire) was published in 1698 (The Hague; ILL. 15.6A) and immediately translated into English and published as *The Celestial Worlds Discover'd, or, Conjectures Concerning the Inhabitants, Plants and Productions of the Worlds in the Planets* (London, 1698; ILL. 15.6B). Here Huygens forwent his usual abstruse musical-mathematical calculations and embarked on a speculative voyage of the universe, seeking «probable and ingenious conjectures» on the possibility of life on other planets[102]. As Frédérique Aït-Touati (2011) has noted, speculation should not be conflated with fiction, for Huygens claimed to derive his theories through reason and observation rather than

la plupart absurdes et inpracticables; et qui rend nostre systeme plus abondant en accords et plus selon la nature du chant, que n'estoient ceux la».

[100]. Christiaan Huygens, notes on the modes (MS NL-Lu, Cod. Hug. 27; HUYGENS 1940, p. 74): «quelque chose de plus tendre et plus plaintif». He referred to the «vray temperament» in notes on Zarlino's *Dimostrationi harmoniche* (Venice, 1571) in the same manuscript (HUYGENS 1940, p. 115).

[101]. Christiaan Huygens, 'Lettre touchant le cycle harmonique' (1691), pp. 78-79 (HUYGENS 1905, p. 169): «Je vous envoie comme j'avois promis, ma remarque en matiere de Musique. Elle regarde le premier fondement de cette science, scavoir la determination des Tons que l'on observe dans le Chant et dans le fabrique des Instruments qui servent a l'harmonie». «Raison Harmonique» is on p. 82 (1691) and p. 171 (HUYGENS 1905).

[102]. Christiaan Huygens, *Cosmotheoros* (The Hague, 1698), p. [2]: «verisimiles & artificiosas conjecturas». English translation from Christiaan Huygens, *The Celestial Worlds Discover'd* (London, 1698), p. iv.

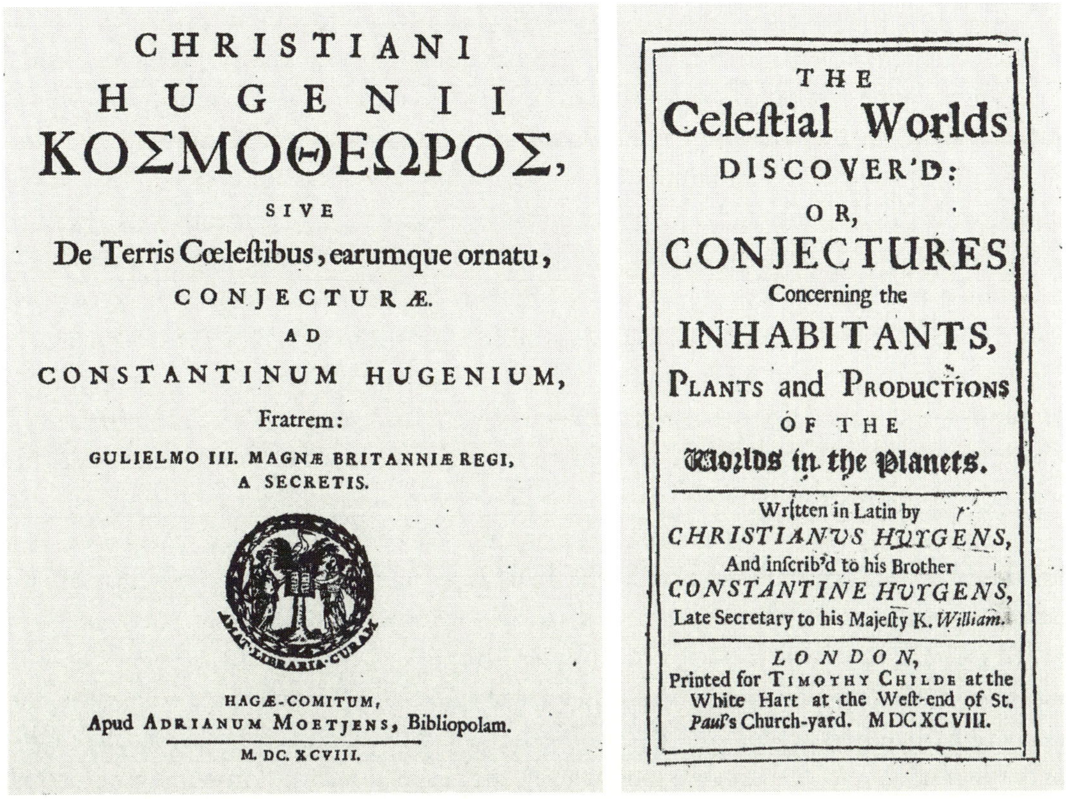

CHRISTIANI
HUGENII
ΚΟΣΜΟΘΕΩΡΟΣ,
SIVE
De Terris Cœleſtibus, earumque ornatu,
CONJECTURÆ.
AD
CONSTANTINUM HUGENIUM,
Fratrem:
GULIELMO III. MAGNÆ BRITANNIÆ REGI,
A SECRETIS.
HAGÆ-COMITUM,
Apud Adrianum Moetjens, Bibliopolam.
M. DC. XCVIII.

THE
Celeſtial Worlds
DISCOVER'D:
OR,
CONJECTURES
Concerning the
INHABITANTS,
Plants and Productions
OF THE
Worlds in the Planets.
Written in Latin by
CHRISTIANUS HUYGENS,
And inſcrib'd to his Brother
CONSTANTINE HUYGENS,
Late Secretary to his Majeſty K. William.
LONDON,
Printed for Timothy Childe at the
White Hart at the Weſt-end of St.
Paul's Church-yard. MDCXCVIII.

Ill. 15.6ab: Christiaan Huygens, *Κοσμοθεωρός* (Cosmotheōros; The Hague, 1698) and *The Celestial Worlds Discover'd* (London, 1698), title pages. Copies NL-Au and GB-Lbl, respectively.

imagination[103]. Mathematics stood tall as Huygens's bastion of certainty. He described geometry as a science of «singular worth and dignity», which ascertained «a full comprehension and infallible certainty of truth»[104]. So too with music:

> It is the same with music as with geometry, it is everywhere immutably the same, and always will be so. For all harmony consists in concord, and concord is all the world over fixed according to the same invariable measure and proportion [...] Very credible authors report that there is a sort of bird in America that can plainly sing in order six musical notes: whence it follows that the laws of music are unchangeably fixed by nature, and therefore the same reason holds valid for [the planetarians'] music, as we have now proposed for their geometry[105].

103. Aït-Touati 2011, pp. 95-129.

104. Christiaan Huygens, *Cosmotheoros* (The Hague, 1698), pp. 71-72: «[...] in qua singularis quidam intelligentiae est usus, ac certa indubitataque veri comprehensio, [...]». English translation from Christiaan Huygens, *The Celestial Worlds Discover'd* (London, 1698), p. 84.

105. Christiaan Huygens, *Cosmotheoros* (The Hague, 1698), pp. 73-74: «Caeterum illud quod uniusmodi & aeternam in Geometrica scientia inesse animadvertimus, similiter quoque in Harmonicis inveniri certum

The same reasoning that moved Huygens to marvel at the 'incomprehensible intelligence' behind the design of the ear convinced him that such design could not be the privilege of humans alone. The example of the bird proved that the capacity to take pleasure in music (even without a rational soul) existed in nature. By the same logic, Huygens speculated that the planetarians would naturally advance from monophony to polyphony, sense the limits of just intonation in expressing that polyphony, and develop temperament in a manner analogous to his invention of the harmonic cycle. Temperament, and musical practice generally, existed in nature — a novel philosophical argument. Thus, when Huygens declared that music was «everywhere immutably the same», he not only meant its perfect proportions but also its effects on listeners. A philosophical account of music could neither disregard agreeability, as Descartes had done, nor idealize it, à la Mersenne. On the contrary, Huygens made agreeability the epistemological focal point of music, both scientific and artistic. Any sounds that failed to please real listeners could not properly be called 'music'.

Huygens brought seventeenth-century musical science to its apex, albeit not in the way Descartes or Mersenne predicted. His analysis of sound and tuning systems surpassed that of his predecessors, while his practical writings evidenced an aesthetic discernment that resisted reduction to mathematics or philosophy. The physico-mathematical theories of sound, coupled with the subjective and irrational nature of agreeability, meant that no absolute causal explanation of musical experience was possible. In a mechanical universe, wherein all sensation was bodily, and the rational soul was locked away, the ear possessed the ultimate agency to judge music. Huygens concluded that a new natural philosophical account of music was needed, one in which the practice of pleasing and moving listeners through melodies was every bit as natural as the perfect consonant ratios that historically formed the ontological basis of music. Musical science, in turn, had an obligation to facilitate — but never subvert — this aim of pleasing the ear; the dissonant imitations of fugues exemplified how displeasing and unnatural music could become when mathematical rules encroached upon aesthetic domains in which they had no place. In short, Huygens demarcated the limits of physico-mathematics as a tool of analysis in natural philosophy, while also granting mathematics more power by defining its new domain. His response to the acoustical paradox was to carve out distinct but equal space for mathematics and aesthetics in music. The physical/perceptual chasm remained unbridged, with Huygens

---

est; cum consonantiae omnes constanti mensura ac proportione constituantur; omnis vero phtongorum ordo, omnisque cantus delectatio, etiam vocis singulae, in consonantiis fundata sit. [...] Imo animal quoddam in terris Americae reperiri fide digni auctores narrant, quod sex musicos tonos deinceps voce exprimat: Ut appareat ipsam naturam immutabili ratione eos praescribere. Quandoquidem igitur quae huc spectant, certa quoque & unica, & necessaria ratione sese habent, verisimile est, non minus quam Geometriae, etiam Musicae oblectationem ad plures quam ad nos pertinere». English translation from Christiaan Huygens, *The Celestial Worlds Discover'd* (London, 1698), p. 86. I have modernized the 1698 English text, which is a very free translation, almost a paraphrase, of the Latin original.

accentuating the crucial, and increasingly separate, roles of musical science and practice. This separation marked a key turning point in the histories of music and science.

Conclusion

This thesis of an acoustical paradox entails the irreducibility of music to physics and mathematics. As in optics, the physicalization of sound and hearing nullified scholastic notions of species and broke the teleological causal link between sensible phenomena and the human observer. Unlike in optics, the acoustical paradox disrupted traditional assumptions regarding passions, the soul, and the relationship between science and practice. Descartes was so annoyed by his and Mersenne's fruitless efforts to rationalize beauty that he broke consonance into concordance and agreeability, declaring the latter a matter of individual taste and impervious to philosophy. Mersenne, determined to piece consonance back together again, imagined an ideal listener who could reconcile concordance and agreeability by only transcending corporeal limitations and experiences. Finally, when Huygens concluded that «no reasoning intervenes» in the process of hearing, it was clear that Kepler's fears from his *Harmonices mundi libri V* had come to pass. Realizing the impossibility of explaining agreeability within the 'true philosophy', Huygens resisted treating music as sensible mathematics. He devised a temperament that was natural and 'true' because it forwent perfect, harmonious intervals in order to please and move real listeners. Descartes and Huygens (more so than Mersenne) envisioned a new science of sound in service to music, which they understood in terms of its effects on listeners rather than universal mathematical principles. They did this not to bury music but to save it. By unshackling agreeability from mathematical natural philosophy, Descartes and Huygens freed composers of the Baroque era to explore music's passionate and expressive effects on real human listeners.

This chapter illuminates how seventeenth-century arts and sciences began to carve out their respective epistemic domains. Music, broadly construed, drifted away from the mathematical sciences and towards the practiced arts. Musical science persisted, but its focus shifted from philosophical speculation and towards a mixed mathematical study of sounds, tuning systems, and vibrational mechanics for the benefit of practicing musicians. Certain kinds of sensations — those that could be quantified and mechanized — were singled out as scientifically meaningful, while agreeability, tastes, and passions were banished to the realm of art and aesthetics. Beauty became the prerogative of artists, whose ears and tastes could not be reduced to mathematical laws. Suffice it to say that musical science at the end of the century was a pale shadow of traditional conceptions of *musica scientia*. This new science, which studied sound without passing judgment on musical experience, would soon be christened *acoustics*. Musical practice, which in subsequent decades reached heights of emotional expressivity once unthinkable, would become a *fine art*. To be sure, attempts would be made in subsequent decades to resolve the acoustical paradox, explain aesthetics in rational, scientific terms, and pull music and science back together. However, the scientific validity of music was now on the defensive, and music never fully regained its once-eminent scientific status.

BIBLIOGRAPHY

AÏT-TOUATI 2011

AÏT-TOUATI, Frédérique. *Fictions of the Cosmos: Science and Literature in the Seventeenth Century*, translated by Susan Emanuel, Chicago, University of Chicago Press, 2011.

BARKER 1989

Greek Musical Writings. 2: Harmonic and Acoustic Theory, translated and edited by Andrew Barker, Cambridge, Cambridge University Press, 1989.

BEECKMAN 1939

BEECKMAN, Isaac. *Journal tenu par Isaac Beeckman de 1604 à 1634, Tome premier 1604-1619*, edited by Cornelis de Waard, The Hague, Martinus Nijhoff, 1939.

BEECKMAN 1945

ID. *Journal tenu par Isaac Beeckman de 1604 à 1634, Tome troisième 1627-1634 (1635)*, edited by Cornelis de Waard, The Hague, Martinus Nijhoff, 1945.

BEECKMAN 1953

ID. *Journal tenu par Isaac Beeckman de 1604 à 1634, Tome quatrième: Supplément*, edited by Cornelis de Waard, The Hague, Martinus Nijhoff, 1953.

BOETHIUS 1867

BOETHIUS, Anicius Manlius Severinus. *De insititutione arithmetica libri duo, De institutione musica libri quinque*, edited by Gottfried Friedlein, Leipzig, Teubner, 1867.

BOETHIUS 1989

ID. *Fundamentals of Music*, translated by Calvin M. Bower, edited by Claude V. Palisca, New Haven (CT), Yale University Press, 1989.

COHEN 1984

COHEN, H. Floris. *Quantifying Music: The Science of Music at the First Stage of the Scientific Revolution, 1580-1650*, Dordrecht, Reidel, 1984.

CROMBIE 1994

CROMBIE, Alistair Cameron. *Styles of Scientific Thinking in the European Tradition: The History of Argument and Explanation Especially in the Mathematical and Biomedical Sciences and Arts*, London, Duckworth, 1994.

Dear 1984

Dear, Peter. 'Marin Mersenne and the Probabilistic Roots of «Mitigated Scepticism»', in: *Journal of the History of Philosophy*, xxii/2 (1984), pp. 173-205.

Descartes 1897

Descartes, René. *Correspondance. 1: avril 1622-février 1638*, edited by Charles Adam and Paul Tannery, Paris, Léopold Cerf, 1897.

Descartes 1898

Id. *Correspondance. 2: mars 1638-décembre 1639*, edited by Charles Adam and Paul Tannery, Paris, Léopold Cerf, 1898.

Descartes 1908

Id. *Physico-mathematica. Compendium musicae. Regulae ad directionem ingenii. Recherche de la verité. Supllément à la correspondance*, edited by Charles Adam and Paul Tannery, Paris, Cerf, 1908.

Descartes 1961

Id. *Compendium of Music («Compendium Musicae»)*, edited by Charles Kent, translated by Walter Robert, Rome, American Institute of Musicology, 1961.

Descartes 1976

Id. *Correspondance. 4: juillet 1643-avril 1647*, edited by Charles Adam and Paul Tannery, Paris, Vrin, 1976 (Œuvres de Descartes, 4).

Descartes 1989

Id. *The Passions of the Soul: Les passions de l'âme*, translated by Stephen H. Voss, Indianapolis (IN), Hackett Publishing Company, 1989.

Descartes – Huygens 1926

Descartes, René – Huygens, Constantijn. *Correspondence of Descartes and Constantyn Huygens 1635-1647*, edited by Leon Roth, Oxford, Clarendon Press, 1926.

Dijksterhuis 1950

Dijksterhuis, Eduard Jan. *De mechanisering van het wereldbeeld*, Amsterdam, Meulenhoff, 1950.

Dijksterhuis 1961

Id. *The Mechanization of the World Picture*, Oxford, Oxford University Press, 1961.

Düring 1930

Düring, Ingemar. *Die Harmonielehre des Klaudios Ptolemaios*, Göteborg, Elander, 1930.

Fix 2019

Fix, Adam. *Sensible Mathematics: The Science of Music in the Age of the Baroque*, Ph.D. Diss., Minneapolis (MN), University of Minnesota, 2019.

GAL – CHEN-MORRIS 2012

GAL, Ofer – CHEN-MORRIS, Raz. *Baroque Science*, Chicago, University of Chicago Press, 2012.

GALILEI 1960

GALILEI, Galileo. *On Motion and On Mechanics*, edited by Marshall Clagett, translated by I. E. Drabkin and Stillman Drake, Madison (WI), University of Wisconsin Press, 1960.

GALILEI 1974

ID. *Two New Sciences, Including Centers of Gravity and Force of Percussion*, translated by Stillman Drake, Madison (WI), University of Wisconsin Press, 1974.

GAUVIN 2013

GAUVIN, Jean-François. 'Instruments of Knowledge', in: *The Oxford Handbook of Philosophy in Early Modern Europe*, edited by Desmond M. Clarke and Catherine Wilson, Oxford, Oxford University Press, 2013, pp. 315-337.

GODWIN 1992

GODWIN, Joscelyn. *The Harmony of the Spheres: The Pythagorean Tradition in Music*, Rochester (VT), Inner Traditions-Bear & Co., 1992.

GOUK 1980

GOUK, Penelope. 'The Role of Acoustics and Music Theory in the Scientific Work of Robert Hooke', in: *Annals of Science*, XXXVII/5 (1980), pp. 573-605.

HUYGENS 1905

HUYGENS, Christiaan. *Correspondance 1691-1695*, edited by Johannes Bosscha Jr., The Hague, Martinus Nijhoff, 1905 (Christiaan Huygens, Œuvres complètes, 10).

HUYGENS 1916

ID. *Dioptrique 1653; 1666; 1685-1692. Fascicule I 1653; 1666*, edited by D. J. Korteweg, The Hague, Martinus Nijhoff, 1916 (Christiaan Huygens, Œuvres complètes, 13).

HUYGENS 1937

ID. *Méchanique théorique et physique de 1666 à 1695*, edited by J. A. Volgraff, The Hague, Martinus Nijhoff, 1937 (Christiaan Huygens, Œuvres complètes, 19).

HUYGENS 1940

ID. *Musique et Mathématique, Musique, Mathématique de 1666 à 1695*, edited by J. A. Volgraff, The Hague, Martinus Nijhoff, 1940 (Christiaan Huygens, Œuvres complètes, 20).

HUYGENS 1986

ID. *Le cyle harmonique (Rotterdam 1691), Novus cyclus harmonicus (Leiden 1724)*, translated and edited by Rudolf Rasch, Utrecht, Diapason Press, 1986.

Huygens 2007

Id. *Driehonderd brieven over muziek van, aan en rond Constantijn Huygens*, translated and edited by Rudolf Rasch, Hilversum, Verloren, 2007.

Kassler 1985

Kassler, Jamie C. Review of H. Floris Cohen, *Quantifying Music. The Science of Music at the First Stage of the Scientific Revolution, 1580-1650*, in: *Journal of Music Theory*, xxix/2 (1985), pp. 328-335.

Kassler 2004

Kassler, Jamie C. *The Beginnings of the Modern Philosophy of Music in England: Francis North's A Philosophical Essay of Musick (1677) with Comments of Isaac Newton, Roger North and in the Philosophical Transactions*, Burlington (VT), Ashgate, 2004.

Kepler 1997

Kepler, Johannes. *The Harmony of the World*, translated by Eric J. Aiton, Alistair Matheson Duncan and Judith Veronica Field, Philadelphia (PA), American Philosophical Society, 1997.

Lindberg 1986

Lindberg, David C. 'The Genesis of Kepler's Theory of Light. Light Metaphysics from Plotinus to Kepler', in: *Osiris* ii (1986), pp. 4-42.

Mersenne 1945

Mersenne, Marin. *Correspondance du P. Marin Mersenne, réligieux minime. 2: 1628-1630*, edited by Paul Tannery, Cornelis de Waard and René Pintard, Paris, Presses Universitaires de France, 1945.

Mersenne 1962

Id. *Correspondance du P. Marin Mersenne, réligieux minime. 7: janvier-juillet 1638*, edited by Cornelis de Waard, Paris, Éditions du Centre National de la Recherche Scientifique, 1962.

Mersenne 1967

Id. *Correspondance du P. Marin Mersenne, réligieux minime. 10: Du 6 août 1640 à fin décembre 1641*, edited by Cornelis de Waard, Paris, Éditions du Centre National de la Recherche Scientifique, 1967.

Mersenne 1969

Id. *Correspondance du P. Marin Mersenne, réligieux minime. 3: 1631-1633 (Seconde édition)*, edited by Cornelis de Waard, Paris, Éditions du Centre National de la Recherche Scientifique, 1969.

Mersenne 1983

Id. *Correspondance du P. Marin Mersenne, réligieux minime. 15: 1647*, edited by Cornelis de Waard and Armand Beaulieu, Paris, Éditions du Centre National de la Recherche Scientifique, 1983.

Millay 2020

Millay, Edna St. Vincent. *The Selected Poetry of Edna St. Vincent Millay*, edited by Nancy Milford, New York, Modern Library, 2020.

PASNAU 2000

PASNAU, Robert. 'Sensible Qualities. The Case of Sound', in: *Journal of the History of Philosophy*, XXXVIII/1 (2000), pp. 27-40.

TONIETTI 2014

TONIETTI, Tito M. *And Yet It Is Heard: Musical, Multilingual and Multicultural History of the Mathematical Sciences*, Berlin, Springer, 2014.

TRESCH – DOLAN 2013

TRESCH, John – DOLAN, Emily I. 'Toward a New Organology. Instruments of Music and Science', in: *Osiris*, XXVIII/1 (2013), pp. 278-298.

VAN ORDEN 2013

VAN ORDEN, Kate. 'Descartes on Musical Training and the Body', in: *Music, Sensation, and Sensuality*, edited by Linda Phyllis Austern, Abingdon-New York, Routledge, 2013, pp. 17-38.

WALKER 1967

WALKER, Daniel Pickering. 'Kepler's Celestial Music', in: *Journal of the Warburg and Courtauld Institutes*, XXX (1967), pp. 228-250.

WALKER 1978

ID. *Studies in Musical Science in the Late Renaissance*, London, Warburg Institute, 1978.

WARDHAUGH 2008

WARDHAUGH, Benjamin. *Music, Experiment and Mathematics in England, 1653-1705*, Farnham, Ashgate, 2008.

Abstracts

David E. Cohen, *«Latet discordantia quartae»: An Early Natural-Scientific Explanation of Upper-Voice Fourths by Franchinus Gaffurius*

The music theorist Franchinus Gaffurius, in his *Practica musice* (1496), addresses an interesting question: Given that the perfect fourth is regarded as a dissonance when it occurs above the lowest sounding voice (and hence always when it occurs in two parts), why is it treated as though it were a consonance — relieved of the requirements of preparation and resolution—when it lies between the upper voices, for example, in those frequently occurring sonorities which figured bass would identify as 6/3 chords? For these were treated as concordant despite their inclusion of such fourths. Gaffurius's answer represents an unusual intrusion of *musica theorica* — and specifically, the natural-scientific (acoustical) side thereof — into a purely *musica practica* context. Based on one of the pseudo-Aristotelian *Problems*, that answer is, briefly, that since lower pitch is produced by 'slower motions', it makes a stronger impression on the ear than higher pitch, while owing to the greater speed of the pitches composing it, 'the discord [of the fourth] is hidden'. The principle underlying this explanation starkly contradicts the conflation of higher pitch with greater auditory salience that is a frequent feature of early acoustic theories — an account that Gaffurius himself had endorsed in his *Theorica musice* just four years earlier (1492). Yet in the *Practica*, Gaffurius cites that earlier discussion in *Theorica* as *supporting* his later, apparently diametrically opposed view. In this chapter Gaffurio's two accounts of pitch are examined in the light of their ancient sources, considering why he might have seen them as concordant rather than conflicting. I conclude by reflecting briefly on the fact that it was an independent, empirical observation about dissonance that led him to raise this question, and with the observation that by, by doing so, he anticipated important music-theoretical developments of much later times, including our own.

Paolo Alberto Rismondo, *Teoria musicale e «patronage» nobiliare a Venezia nel Cinquecento | Music Theory and Noble Patronage in Sixteenth-Century Venice*

At the beginning of the sixteenth century the Venetian nobility readily supported music theory, as evidenced by the dedications of the treatises by Pietro Aaron and Giovanni del Lago. Francesco Zorzi, a Franciscan friar but also a Venetian noble, profoundly influenced the broad cultural and political panorama of the *Serenissima* through his writings. His interpretation of music theory (both in his theological works and in his proposal for the project of the church of San Francesco della Vigna) favoured neo-Pythagorean and neo-Platonic ideas, in strong opposition to the Aristoxenian model. His views fitted very well with the tastes and needs of the Venetian nobility, interested in maintaining music theory worthy of their patrician status, and far above the common concerns of practical music. The circle of music theorists in the Accademia Venetiana basically continued in the same vein. One of their most important and noteworthy purposes was to publish ancient Greek music treatises (from manuscripts held in the Marciana library), some more recent ones (most notably that of Jacques Lefèvre d'Étaples), and others still unpublished, including contemporary treatises still not traced (most likely by Zarlino and Vicentino). The linking of many of the Accademia's members with Spain and the Empire (for example, that of Nicola Vicentino with Cardinal Antoine Perrenot de Granvelle, a Spanish and imperial minister and courtier), which the Venetian political agenda regarded more and more as highly suspicious, was one of the most important causes of its closing. In the second half of the sixteenth century, the broad cultural and political panorama changed; moreover, the increasing controversies that affected music theory led the Venetian nobility, after the early decades of the seventeenth century, to shift its patronage from music theory to *opera in musica*.

Alexander Jakobidze-Gitman, *Nascent Mechanism in Philip Melanchthon's References to Josquin des Prez*

In music history texts, Philip Melanchthon (1497-1560) is usually mentioned as one of those who paved the way for the application of rhetorical theory to music and pointed out how music supported poetry through its impact on the affections. In his three comprehensive textbooks on dialectic, morals, and physics, Melanchthon refers to Josquin des Prez. Strikingly, in none of the citations does Melanchthon say anything about Josquin's handling of the text or influencing emotions. Instead, he utilizes the composer to make different comparisons and to draw analogies. To understand what Melanchthon is thereby aiming at, I suggest contextualizing his Josquin citations within the nascent scientific revolution. It is particularly relevant that Melanchthon was one of the very first to have applied the machine metaphor both to the Universe and to the human body. For Melanchthon, Josquin embodied (1) natural power that is inherent to things, like the force of a magnet to attract iron; (2) divine impetus that makes him akin to ancient heroes such as Achilles, Caesar, or Scipio; and (3) natural motion, which is opposed to the artificial motion of an automaton. In Melanchthon's view, Josquin was vastly superior to common musicians who had acquired habits by training (just as one learns to make clocks). I argue that here Melanchthon revealed the common mechanistic nature of automata and the possession of habits. This insight was a prophetic one, prefiguring Hegel's idea that «a mechanical mode of representation, a mechanical memory, a habit, a mechanical mode of acting, mean that the pervasive presence that is proper to spirit is lacking in what spirit grasps or does» (*The Science of Logic*). Hegel formulated this idea when the mechanistic worldview was already in decline; Melanchthon groped towards the same idea when mechanism was in its infancy.

Carlo Bosi, *«Intra quaternarii limites musicam intelligunt omnem consonantiam»: La musica nel pensiero di Giordano Bruno* | *«Intra quaternarii limites musicam intelligunt omnem consonantiam»: Music in the Thought of Giordano Bruno*

In the ontology of Giordano Bruno (1548-1600), nature — and consequently man — is in a permanent state of transformation. It is only through knowledge and internalization of the mechanisms relating to this uninterrupted metamorphosis that the magician-philosopher can acquire the ability to successfully intervene in natural processes and reorganize or 'manipulate' them according to his will. Music plays a special role in this philosophical system; in it, the art of sound is conceived both as a 'physical' phenomenon and as a symbolic concept. This is despite the fact that Bruno's use of visual rather than aural metaphors has often been emphasized, which is fully understandable, given the primacy of the sense of sight in Bruno's philosophical and scientific conceptualisation. However, at crucial points in his work, he uses musical metaphors to better explain and substantiate the message or notion being examined. This chapter considers a number of passages from Bruno's work where the use of musical terms and concepts, and in any case those relating to the phenomenon of 'sound', serve as a vehicle for the illustration and clarification of important points in the «nolana filosofia».

Gioia Filocamo, *Dissezioni anatomiche con musica a Padova nella prima età moderna* | *Anatomical Dissections with Music in Padua in the Early Modern Period*

Dissection of corpses for scientific purposes was practiced since ancient times, on bodies of people rejected from society. After centuries without a proper documentation, the anatomical practice had a true rebirth in the late Middle Ages, realized privately. Its formalization through public demonstrations at universities could also be achieved thanks to the explicit endorsement of Pope Sixtus IV, who in April 1482 authorized autopsies at the University of Tübingen. The first permanent structure specifically dedicated to anatomical dissections was the anatomical theatre of Padua (1594), encouraged by the legendary anatomist Girolamo Fabrici d'Acquapendente (*c.*1533-1619), who instituted a natural philosophical program of anatomical inquiry and held the chair in anatomy and surgery from 1565 to 1613. Part of the Venetian Republic, Padua welcomed both its *studium* and the traditional Venetian resistance to papal interference. In fact, many protestant foreign students studied in Padua, where they could take a degree despite their faith. Some of them belonged to the *natio Germanica artistarum*, which recorded many facts related to university life. These proceedings recorded that in at least two occasions (in 1597 and 1600) the German students of Padua requested music during the anatomical demonstrations, in order to raise their spirits. Music was in fact present in some anatomical theatres of the modern age, but in connection to celebratory and festive situations (inaugural lectures, carnival parties, etc.), which also included an autopsy to demonstrate the competitive excellence of the teachers. The case of Padua was probably different. In fact, according to the revolutionary vision proposed by Andreas Vesalius (1514-1564) — a professor of anatomy in Padua from 1537 to 1542 — , anatomical dissection should investigate every dead body as a curious detective, not only to confirm the structural details described in the standard Galenic textbooks. This paper explores the complex meaning of the *need* for music expressed by the German students in the context of anatomy lectures, and the possible *function* of music in this peculiar environment.

DANIEL MARTÍN SÁEZ, *Galileo and Opera: Music, Science, Religion and Politics in the Seventeenth Century*

In 1610, Galileo published his *Sidereus nuncius* in the same court where, two decades earlier, Emilio de' Cavalieri and Laura Guidiccioni had performed the first fully sung pastoral plays. In 1613, Galileo had already described the mountains of the Moon, the satellites of Jupiter, the phases of Venus, certain anomalies in Saturn and sunspots, among other things. The new genre of opera received Galileo better than any other art. In 1613, the new discoveries found its first theatrical representation in a Florentine *barriera*, which was transformed into an opera the following year in Rome. The *stelle medicee*, the great emblem of Heliocentrism and the power of the Grand Dukes of Tuscany, appeared on the stage acting and singing. From that moment on, this emblem was used in the editions of several Florentine printed librettos, even after the condemnation of Copernicanism of 1616. This included some religious operas made during the regency of Christina of Lorraine and Maria Magdalena of Austria. The dispute with the Jesuits and the rise of the Barberini explains other operas in the 1620s. Even after the trial of 1633, new operas related to Galileo were sponsored by the papacy. They were written by Cardinal Giulio Rospigliosi and offered an image of Galileo more ambivalent than is sometimes supposed. In total, ten operas can help us to illuminate the most important moments in Galileo's work: the theological debates of 1613-1616, the condemnation of Copernicanism in 1616, the disputes with the Jesuits about the comets, the rise of Urban VIII in the 1620s, the conviction of Galileo in 1633, and the publication of his *Discorsi* in 1638, all of which were marked by the delicate context of the Thirty Years' War and the tension between Protestants and Catholics over the interpretation of the Bible.

JASON STOESSEL – DENIS COLLINS, *«La grandezza del numero sonoro»: Canonic Techniques, Combinatorics, and Early Scientific Thought in Seventeenth-Century Rome*

The cultural significance of the resurgent interest in canonic techniques amongst early seventeenth-century composers, especially those based in Rome, has generally been overlooked by music historians. Situated within the tradition of highly inventive *artificioso* compositions, the vast corpus of canonic ingenuity may also be understood as a witness to new systems for understanding the natural world. The writings on music by Marin Mersenne, and their extensions by Athanasius Kircher, gave expression to the intersections of two profoundly significant areas of early scientific thought, mechanism and combinatorics. These developments advanced tools for the rational induction and mathematical verification of the sensible world. For Kircher, canon became a vehicle for demonstrating the validity of contrapuntal principles that asserted the 'natural' laws of musical consonance. He drew upon the polymorphic compositions of his Roman contemporary, Pier Francesco Valentini (1586-1654) to highlight how these laws could be extended and verified. The outputs of Valentini and Paolo Agostini (c. 1583-1629), another prominent exponent of canon, provide a series of case studies through which it is possible to examine how canonic techniques were a site for experimentation and demonstration of the combinatoric potential of melodic materials. A great wealth of musical invention underlies most especially the processes of polymorphous composition and the simultaneous application of multiple canonic techniques within a single work. The deep engagement by Roman composers with combinatoric processes and their expression through collections of canons and other kinds of *artificioso* composition points to the emergence of a tradition of contrapuntal compendia that became established amongst later generations of composers and writers about music.

Carlos C. Iafelice, *Giovanni Battista Doni on the Tonal Space of Chromatic Music: The Case of Gesualdo's «Tu m'uccidi, o crudele»*

In Rome in the 1630s, the Florentine humanist Giovanni Battista Doni (1595-1647) began to formulate a renewed interpretation of the Greek genera and the modal system, which led him to a particular understanding of tonal space in contemporary music. Possessing a vast erudition, Doni's detailed epistemological approach aimed for what he considered an 'appropriate reconstruction' of the *effetti meravigliosi* of ancient Greek music, which, in his opinion, was abundantly misinterpreted by preceding generations of both composers and theorists. One of the key points of this reconstruction was the presumption that 'chromatic' music was actually a phenomenological mixture of tonoi, not genera. A predilection for the Principe di Venosa among the contemporary composers can be recognized in the *Compendio del trattato de' generi e de' modi della musica* (1635), Doni's first printed work on music, with an *aggiunta* that contains practical examples of his doctrine: an original Duo and an excerpt of Gesualdo's madrigal, *Tu m'uccidi, o crudele*, and a sonnet, *Passa la vita all'abbassar d'un ciglio*, set to music by Pietro Eredia. For several reasons, Gesualdo's madrigal provides an opportunity to determine how Doni's applied his new interpretations of tonal space to a pre-existing composition, that is, one composed without the influence of his doctrine. The aim of this chapter is to obtain conclusions through an application of Doni's doctrine to Gesualdo's *Tu m'uccidi, o crudele*, including a systematic confrontation of the madrigal's score notation (*intavolatura ordinaria*) and with its notation in viol tablature. This will allow us, due to the peculiar nature of this notation, to understand how Doni conceived of the madrigal's modal organization.

Rudolf Rasch, *«Me veux-tu voir mourir»: Joan Albert Ban versus Antoine Boësset (1640-1641)*

The controversy of 'Ban versus Boësset' concerns how to compose vocal airs. The Dutch priest and amateur musician Joan Albert Ban had (1597/8-1644) had developed a theory for the composition of vocal music in which every semantic element of the text had to be expressed in the music. Mersenne, who had heard about this theory through Constantijn Huygens, wanted to put Ban's theory to test, and sent to Ban the words of an air, *Me veux-tu voir mourir*, already set to music before, by the French composer Antoine Boësset. Ban composed music to this text so that it could be compared in France with that of Boësset. Mersenne and several more Frenchmen declared Boësset's composition vastly superior to Ban's. This was hard to accept by Ban, especially since the French did not provide a motivation for their preference. The result was an 'epistolary war' between Ban and his French commentators. This chapter investigated the various arguments that were put forward by the combatants, part of which are 'topical', that is, they have to do with the subject of the controversy, others are 'non-topical' or 'circumstantial', that is, they are rather personal, emotional, or chauvinistic in character. The topical arguments include the interpretation of the text, the choice of mode or key, the role of accents and emphasis in the text, and the role of the representation of semantic elements of the text in the music. The circumstantial arguments include Ban's feeling of standing alone against a collective of French adversaries, his feelings of Dutch inferiority aginst French superiority, argmunets *ad hominem*, and Ban's feeling of being trapped, since Boësset's composition already pre-dated the comparitive test. The analyses presented in this chapter show that in the study of controversies one must never forget that circumstantial arguments can play roles at least as important that the topical arguments.

GALLIANO CILIBERTI, *La Missa «La luna piena» (1657-1658) di Giuseppe Corsi e le teorie galileiane | The Missa «La luna piena» (1657-1658) by Giuseppe Corsi and Galielo's Theories*

From 1657 to 1658, Giuseppe Corsi (1631/32 – *post* 1691) was *maestro di cappella* in the Basilica of Santa Maria Maggiore in Rome. The archive of this institution holds one of his Masses, entitled *La luna piena*, for eight voices divided over two choirs. The Mass is inspired by the image of the Assumption frescoed between 1610 and 1612 by Ludovico Cardi detto il Cigoli (1559-1613) in the Capella Borghese in the Basilica of Santa Maria Maggiore. The image of the Madonna is placed over a wrinkled and spherical full moon dotted with craters. The painter (who was a very close friend of Galileo and maintained an intensive exchange of letters with him) was inspired by Galileo's innovative theories on moon craters as exhibited in the latter's *Sidereus nuncius* (1610). In Corsi's Missa *La luna piena*, we find a close connection with the fresco and inevitably with the thought of Galileo: the use of white and coloured notation and thematic and rhythmic elements often recall the quantity and position of stars and Jupiter and its satellites as described in Galileo's work.

RUDOLF RASCH, *René Descartes and Isaac Beeckman: The Philosopher and the Schoolmaster*

In 1618, while in Breda, in the Netherlands, René Descartes wrote his *Compendium musicae* for his recent friend Isaac Beeckman (1588-1637), later schoolmaster in Utrecht, Rotterdam and Dordrecht. In so doing Descartes created a bond between them that could never be unbound. Twelve years later Descartes complained in letters to Mersenne that Beeckman was apparently claiming the contents of the *Compendium musicae* for himself, as if Descartes just had written down Beeckman's views. Beeckman has left an extensive journal of his observations and ideas about a great many topics, from 1604 to 1634. In it, music takes an important place. A comparison of Beeckman's notes on music from the years 1614-1618 with Descartes's *Compendium musicae* indeed reveals Descartes's tribute to Beeckman regarding certain aspects of the *Compendium musicae*. It also reveals a fundamentally different approach to music by both men: for Descartes, the philosophical insights gained from studying music were the focus; for Beeckman, understanding musical phenomena. This shows that the study of music by non-musicians, let us say the scientific study of music, can have different faces, depending on the goal that the scholar in question is aiming at.

LEENDERT VAN DER MIESEN, *The Trumpet as Nature's Voice: Marin Mersenne and the Nature of Music*

This chapter explores how the French polymath and music theorist Marin Mersenne (1588-1648) turned to musical instruments to investigate the nature of music. He did so in order to establish the 'naturalness' of the consonances and the diatonic scale, closely related to his response to a growing sceptic movement in France. Mersenne investigated techniques of overblowing on wind instruments and observed contemporary musicians. The fact that the valveless trumpet brings forth the intervals of the octave, fifth, fourth, major and minor third, and so on, was for Mersenne important proof that the consonances were given to humans by nature and that the natural world was composed of harmony. In contrast to his contemporaries, he found the nature of music in instruments, not in animals or the human voice. Mersenne's investigation of the trumpet puts him at odds with his intellectual predecessors, such as Gioseffo Zarlino and Vincenzo Galilei. Whereas Zarlino continued to uphold the voice as the most natural instrument, Vincenzo Galilei devoted little attention to the phenomenon of overblowing. This article argues that Mersenne's attention toward instruments signals the increasing role of instruments as objects of knowledge making, while emphasizing that his experiments were used to affirm his belief in universal harmony.

ROBERTA VIDIC, *Joachim Jungius and the Question of Music as an Experimental 'Mixed Science'*

For centuries, Aristotle's *Posterior Analytics* served as a model for extremely divergent approaches to scientific methodology, which range from pure rationalism to radical empiricism. The *Posterior Analytics* is a text from Aristotle's *Organon* that deals with demonstration, definition, and scientific knowledge. In the sense of 'radicalization', the question is to what extent important 'empirical turns' in Italian and German music theory from about mid-sixteenth to mid-eighteenth century can be traced back to the two Galileis, Vincenzo (1520s-1591) and Galileo (1564-1642). Given the vast literature on this topic, the aim of this chapter is to shed some light on specific issues regarding the Veneto culture and Northern Germany, with a special focus on the German scholar Joachim Jungius (1587-1657). Jungius was well acquainted with Galileo's work and studied also in Padua at a faculty, which was still under a strong influence of Galileo. Nevertheless, Jungius' *Harmonica* offer a good example of how complex interrelationships within the scientific community functioned in the seventeenth century. A key role in defining an 'empiricist' component in the rise of modern science is certainly played by the changing semantic field from the Latin *experientia* to the modern concept of experiment. Musical historiography has often focused on the dichotomy of a speculative or practice-oriented theory of music, where practice coincides with composition and performance. The historical discussion of issues such as Vincenzo Galilei's 17:18 rule for equal temperament opens up different perspectives on music as an experimental 'mixed science'. The proposed term 'radicalization' could help, in this frame, to delineate a specific field of research at the interface between music, logic, physics, and mathematics that is strictly related to the big methodical question at the rise of modern science.

THÉODORA PSYCHOYOU, *Trajectories of Musical Acoustics in Seventeenth-Century French Theory and Experiment: Sound, Noise, and the Authorities of the Past*

This chapter discusses some programmatic aspects concerning the relationship between music and science, more precisely that of the emergence of acoustics as a distinct discipline. By tracing the investigations of theorists and philosophers throughout the seventeenth century in France, from Mersenne and Descartes to Joseph Sauveur, we will follow the trajectory of the status of sound, from the paradigm of mechanics to the confines of noise. Sound and noise thus support, simultaneously, two types of definition, one seeking objectivity, by the tools of physics and experiment, and one being an aesthetic, via the filters of perception. The latter, the main tool of the experimental sciences, is synonymous with subjectivity, susceptible to the individuality of the senses, and this generated major epistemological problems for seventeenth-century authors. Their investigations finally induced a crisis in the use of the legacy of the past, in favour of the tools and methods of the Modern Era.

ADAM FIX, *The Acoustical Paradox: How Music Was Unmoored and Set Adrift from Science in the Seventeenth Century*

From Antiquity up to the Scientific Revolution of the seventeenth century, music was ever-present in mathematical and philosophical thought. Known variously as 'speculative music' or 'harmonics', the science of music in the pre-modern world encompassed far more than the audible art form recognized today. However, in a series of letters beginning in 1629, René Descartes and Marin Mersenne questioned whether fundamental principles of music could be known with scientific certainty. While Descartes dismissed music as a matter of taste, emotion, and thus human invention, Mersenne clung to his Pythagorean faith that music existed in nature and was therefore discoverable through natural philosophy. This debate continued in Mersenne's *Harmonie universelle* (1636), in which he struggled to explain musical pleasure through physico-mathematical theories of sound, and Christiaan Huygens's *Cosmotheoros* (1695), which reinterpreted the very notion of natural music. History, however, would declare Descartes the winner, for, by the end of the century, speculative music had largely vanished from the scientific discourse. In an era in which natural philosophers strove to infuse their speculations with mathematical reasoning backed by concrete experiments, music dealt with such nebulous intangibles as emotions, passions, and perceptions. Paradoxically, the more philosophers learned about the mathematics and mechanics of sound, the less they understood about pleasure, consonance, and harmony. This incompatibility between music and science led to the birth of a new science of sound — known as 'acoustics' — and the relegation of music to the domain of fine arts, where it has remained ever since.

Biographies

Carlo Bosi earned a D.Phil. at the University of Oxford (2004) with a dissertation on modality in the songs of Du Fay and Binchois. In 2004-2005 DAAD scholar at the Friedrich-Schiller-Universität Jena-Weimar and in 2006-2007 scholarship of the Norges Forskningsråd at the NTNU Trondheim with a project on chant in medieval Scandinavia. Research Fellow at City University, London between 2007 and 2009. From 2010 until 2017 postdoc and a senior postdoc on two FWF (Austrian Science Fund) projects on the borrowing and citation of monophonic tunes in polyphonic textures (chansonmelodies.sbg.ac.at). Since 2019 senior postdoc in a research project exploring the relationship of libertine literature and libretti in early Venetian opera. Another active research interest is music and Renaissance magic, especially in the work of Giordano Bruno.

Galliano Ciliberti is Professor of Music History at the 'Nino Rota' Conservatory of Monopoli. Graduated in Literature from the University of Perugia, he obtained a Doctorate in Musicology from the University of Liège and a post-doctoral degree from the École Pratique des Hautes Études in Paris. He has received several research contracts from the C.N.R. and was a research fellow at the Department of Musicological and Philological Paleographic Sciences of the University of Pavia, Cremona branch. Author of many essays and numerous books, he has participated in various conferences in Italy and abroad. Winner of the Bertini Calosso Award (1998-2000 edition), he deals with the musical relations between Rome and Paris in the seventeenth century and with Roman sacred music in the seventeenth century.

David E. Cohen researches the history of music theory from Greek antiquity through the nineteenth century. He is currently a Senior Research Scientist with the research group, «Histories of Music, Mind, and Body» at the Max Planck Institute for Empirical Aesthetics in Frankfurt. He has held professorships at Columbia, Harvard, and Tufts Universities, and has also taught at Yale and McGill Universities, as well as at the Hochschule für Musik und Theater Hamburg. His article, '«The Imperfect Seeks its Perfection»: Harmonic Progression, Directed Motion, and Aristotelian Physics' received the 2001 Best Publication award of the Society for Music Theory. Currently, his principal writing project is a book, *Afterlives of Pythagoreanism:* Musica theorica *and Its Legacy.*

Denis Collins is Associate Professor of Musicology and Deputy Head of School at the University of Queensland, Brisbane, Australia. His research interests lie in the history of counterpoint, most especially canon, in late medieval and early modern music. Support for his research on the history of canonic techniques has been provided by two Australian Research Council Discovery Projects, both held jointly with Jason Stoessel. Recent and forthcoming publications include articles in *Acta Musicologica*, *Music Theory Online*, and *Journal of the Alamire Foundation* as well as several chapters in edited volumes. He was co-editor of *J.S. Bach in Australia: Studies in Reception and Performance* (Lyrebird Press, 2018). He is the current Editor of the journal *Musicology Australia.*

GIOIA FILOCAMO received a Diploma in Piano, a Laurea in Drama, Art, and Music Studies, a Ph.D. in the Philology of Music, and a Ph.D. in Modern History. She held post-doc research fellowships and scholarships in Bologna (University), Chicago (Newberry Library), Wolfenbüttel (Herzog August Bibliothek), Cambridge (St John's College), Uppsala (Swedish Collegium for Advanced Study), Gotha (Forschungsbibliothek). She has extensively published on various aspects of musical life in Italy from the fifteenth to the seventeenth century, mainly exploring courts, confraternities, nunneries. Her interest focuses especially on how music interacted with social life. Her more relevant publications include a complete critical edition of an anthology of late fifteenth- and early sixteenth-century music, *Florence, BNC, Panciatichi MS 27: Text and Context* (Brepols, 2010), and *Uno gentile et subtile ingenio*, a co-edited Festschrift in honour of Bonnie Blackburn (Brepols, 2009). In 2001 she organized the first *Medieval and Renaissance Music Conference* ever held outside the United Kingdom, at Spoleto, starting a new vibrant international tradition. She teaches at the Istituto superiore di Studi musicali di Terni.

ADAM FIX received his Ph.D. in History of Science, Technology, and Medicine from the University of Minnesota in 2019. His dissertation and publications have investigated the early modern science of music, focusing on the tension between physico-mathematical theories of sound and human experiences of music. Since leaving academia, he has been working as a technical and content writer at 3M.

CARLOS C. IAFELICE holds a Master's Degree in Musicology from São Paulo State University (UNESP) where he is currently a Ph.D. candidate in Musicology with the support of a research fellowship from FAPESP (São Paulo Research Foundation). During the 2018-2019 academic year, he was granted an internship sponsored by FAPESP at the Department of Musicology of the University of Pavia, Cremona branch. His research is focused on compositional and theoretical aspects of sixteenth-century music grounded especially on Italian treatises. Iafelice is also a member of a research group «Teorias da Música» (UNESP).

ALEXANDER JAKOBIDZE-GITMAN is a faculty member at the University of Witten/Herdecke in Germany, teaching there on aesthetics and phenomenology of music. He holds his Ph.D. in film studies and published the monograph *Rising Phantasms: The Stalinist Era in Post-Soviet Film* (by the Russian publisher New Literary Observer, 2015). His research focuses on the connection between musical thought and the history of science, in particular with mechanistic philosophy. Currently he is working on the monograph on the late treatises by Jean-Philippe Rameau. Having earned degrees at the Royal Academy of Music (London) and the Moscow Tchaikovsky Conservatoire as a pianist, Jakobidze-Gitman also presents public lecture-recitals.

DANIEL MARTÍN SÁEZ holds a Ph.D. in Philosophy and a degree in Musicology. He currently teaches at the University of Granada. After completing his thesis on the birth of opera in 2018, he has published several works on philosophy and music, including articles about Farinelli, Gabriello Chiabrera, the metaphor of *theatrum mundi*, the idea of baroque in music, opera historiography, the poetics of the 17th and 18th centuries, and the philosophy of music from the 19th century to the present, among other topics. He has also taught on the Double Degree in Music and Philosophy at the Autonomous University of Madrid, on the Degree in Musicology at the Granada University and on the Master in Poetics of History at the Complutense University of Madrid. He has participated in numerous international conferences on philosophy, history of science, music, theater and literature.

Biographies

Leendert van der Miesen is a Postdoctoral researcher in the research group «Visualizing Science in Media Revolutions» at the Bibliotheca Hertziana – Max Planck Institute for Art History in Rome. He studied Musicology and Art Studies at the University of Amsterdam and received his Ph.D. in musicology from the Humboldt University of Berlin. His dissertation dealt with the French scholar and music theorist Marin Mersenne and his current project examines the roles of images in early modern acoustics.

Théodora Psychoyou is *maître de conferences* (Associate Professor) at the Sorbonne University in Paris, and member of the Institut de recherche en musicology (IReMus). Formed in Athens, Tours and Paris, she holds a Ph.D. in Music and Musicology at the University of Tours (2003); she has been a fellow at Villa Medici–French Academy in Rome (2005-2007), member of the RISM group (early music manuscripts) at the Bibliothèque nationale de France in Paris (1997-2005) and an associate researcher at the Centre de musique baroque de Versailles. Her research focuses on history of music theory, and sacred music in the seventeenth century. Her current projects concern the quarrel between Ancients and Moderns in music, catholic music practices in the Greek isles, and the relationship between music and science in the early modern era in France.

Rudolf Rasch studied musicology at the University of Amsterdam with Karel Philippus Bernet Kempers and Joseph Smits van Waesberghe. He wrote a dissertation on seventeenth-century polyphonic carols in the Spanish Netherlands (Utrecht, 1985) and was affiliated with the Institute of Musicology, later Department of Media and Culture Studies, of Utrecht University from 1977 to 2010. He is still affiliated to Utrecht University as a guest researcher. His main interests are the musical history of the Netherlands, tuning and temperament, and the works of composers such as Corelli, Vivaldi, Geminiani and Boccherini. He has published a number of articles, books and editions in these fields, including *Music Publishing in Europe 1600-1900* (2005), *Driehonderd brieven over muziek* (letters about music written by and to Constantijn Huygens, 2007), *Understanding Boccherini's Manuscripts* (2014), *Music and Power in the Baroque Era* (2018), *Muziek in de Republiek: Muziek en Maatschappij in de Zeven Verenigde Nederlanden* 1579-1795 (2018) and several volumes in the complete critical editions of the works of Luigi Boccherini and Francesco Geminiani. Since 2015 he is general editor of *Francesco Geminiani Opera Omnia.*

Paolo A. Rismondo obtained his academic degrees from Ferrara and Padua Conservatories and from Bologna University. His research focuses on Venetian music betweeen late sixteenth and seventeenth centuries, and on the biographies of composers of that time (Massimiliano Neri, Francesco Cavalli, Giovanni Rovetta and the Rovetta family of musicians, Biagio Marini, Angelo Notari and others). About these topics he wrote several articles, dictionary entries (*Dizionario Biografico degli Italiani*), books and modern editions (published by A-R Editions).

Jason Stoessel is Senior Lecturer in Musicology and Digital Humanities at the University of New England, Australia. He is Chief Investigator with Denis Collins on an Australian Research Council Discovery Project (2018-2021) that examines the art and science of canons in the music of early 17th-century Rome. His recent research on the music of the Middle Ages and Early Modern Period has appeared in journal articles published in *Intellectual History Review*, *Music Analysis*, *Musica Disciplina* and *Music Perception,* and in chapters in *Music, Myth, and Story in Medieval and Early Modern Europe* (Boydell, 2019) and *Uncovering Music of Early European Women* (1150-1750) (Routledge, 2020).

Biographies

Roberta Vidic is Lecturer for music theory and a Ph.D. candidate at the Hochschule für Musik und Theater Hamburg (HfMT). Her dissertation topic is Palestrina and the origin of harmonic theories in early 18th-century Italy. Degrees in harp and music theory from Udine, Munich and Hamburg, further studies in historical pedal harps and improvisation at the Schola Cantorum Basiliensis. Research interests include the history of music theory from 16th-18th century, musical form in fantasy genres and musical terminology. Regular participation in international conferences and publications. Double winner of the Scientific Competition of the German Society for Music Theory (GMTH), her doctoral research is supported by Fondazione di Venezia and Fondazione Cini in Venice, and the German Historical Institute (DHI) in Rome.

Index of Names

N

O

P